Lecture Notes in Computer Science 9377

Commenced Publication in 1973
Founding and Former Series Editors:
Gerhard Goos, Juris Hartmanis, and Jan van Leeuwen

Xiaolin Hu · Yousheng Xia
Yunong Zhang · Dongbin Zhao (Eds.)

Advances in Neural Networks – ISNN 2015

12th International Symposium
on Neural Networks, ISNN 2015
Jeju, South Korea, October 15–18, 2015
Proceedings

Springer

Editors

Xiaolin Hu
Department of Computer Science
   and Technology
Tsinghua University
Beijing
China

Yousheng Xia
Fuzhou University
Fuzhou
China

Yunong Zhang
School of Information Science and Technology
Sun Yat-sen University
Guangzhou
China

Dongbin Zhao
Chinese Academy of Sciences
Institute of Automation
Beijing
China

ISSN 0302-9743   ISSN 1611-3349   (electronic)
Lecture Notes in Computer Science
ISBN 978-3-319-25392-3   ISBN 978-3-319-25393-0   (eBook)
DOI 10.1007/978-3-319-25393-0

Library of Congress Control Number: 2015950913

LNCS Sublibrary: SL1 – Theoretical Computer Science and General Issues

Springer Cham Heidelberg New York Dordrecht London

Printed on acid-free paper

Springer International Publishing AG Switzerland is part of Springer Science+Business Media
(www.springer.com)

Preface

This volume of *Lecture Notes in Computer Science* constitutes the proceedings of the 12th International Symposium of Neural Networks (ISNN 2015) held during October 15–18, 2015, in Jeju, Korea. During the last 11 years, this annual symposium has become a well-recognized conference in the community of neural networks. This year the symposium was held for the first time outside China, in Jeju, a very beautiful city in Korea. As usual, it achieved great success.

During the last "winter" of neural network research, ISNN was among only a few conferences focusing on theories and applications of neural networks. We never gave up our research because we believe that neural networks are very useful. Now we are witnessing the revival of neural networks. In this context, ISNN 2015 aimed at providing an open academic forum for researchers, engineers, and students to celebrate the spring of neural networks, to discuss the emerging areas and challenges, and to exchange their fantastic ideas. It encouraged open discussion, disagreement, criticism, and debate, and we think this is the right way to push the field forward.

This year, we received 97 submissions from about 188 authors in 19 countries and regions (Australia, Austria, China, Finland, France, Germany, Hong Kong, India, Iran, Macao, Malaysia, New Zealand, Pakistan, Qatar, Sri Lanka, Thailand, The Republic of Korea, Tunisia, USA). Based on the rigorous peer-reviews by the Program Committee members and reviewers, 55 high-quality papers were selected for publication in the LNCS proceedings. These papers cover many topics of neural network-related research including intelligent control, neurodynamic analysis, memristive neurodynamics, computer vision, signal processing, machine learning, optimization etc.

Many organizations and volunteers made great contributions toward the success of this symposium. We would like to express our sincere gratitude to The Chinese University of Hong Kong, Pusan National University, Korean Institute of Intelligent Systems, International Neural Network Society, IEEE Computational Intelligence Society, and Asia Pacific Neural Network Assembly for their technical co-sponsorship. We would also like to sincerely thank all the committee members for all their great efforts and time in organizing the symposium. Special thanks go to the Program Committee members and reviewers whose insightful reviews and timely feedback ensured the high quality of the accepted papers and the smooth flow of the symposium. We would also like to thank Springer for their cooperation in publishing the proceedings in the prestigious *Lecture Notes in Computer Science* series. Finally, we would like to thank all the speakers, authors, and participants for their support.

August 2015

Xiaolin Hu
Yousheng Xia
Yunong Zhang
Dongbin Zhao

Organization

General Chairs

Jun Wang   The Chinese University of Hong Kong,
Hong Kong, SAR China

Steering Chair

Derong Liu   Chinese Academy of Sciences, Beijing, China

Organizing Committee Chairs

Chenan Guo   Dalian University of Technology, Dalian, China
Sungshin Kim   Pusan National University, Busan, Korea
Zhigang Zeng   Huazhong University of Science and
Technology, Wuhan, China

Program Chairs

Xiaolin Hu   Tsinghua University, Beijing, China
Yousheng Xia   Fuzhou University, Fuzhou, China
Yunong Zhang   Sun Yet-sen University, Guangzhou, China
Dongbin Zhao   Chinese Academy of Sciences, Beijing, China

Special Sessions Chairs

Kwang Baek Kim   Silla University, Busan, Korea
Chuandong Li   Southwest University, Chongqing, China
Tieshan Li   Dalian Maritime University, Dalian, China

Publicity Chairs

Yuanqing Li   South China University of Technology,
Guangzhou, China
Yi Shen   Huazhong University of Science and
Technology, Wuhan, China
Zhang Yi   Sichuan University, Chengdu, China

Publications Chairs

Jianchao Fan   National Marine Environmental Monitoring
                                    Center, Dalian, China
Jin Hu   Chongqing Jiaotong University, Chongqing,
                                    China
Zheng Yan   Huawei Shannon Laboratory, Beijing, China

Registration Chairs

Shenshen Gu   Shanghai University, Shanghai, China
Qingshan Liu   Huazhong University of Science and
                                    Technology, Wuhan, China

Program Committee

Salim Bouzerdoum   University of Wollongong, Australia
Binghuang Cai   University of Pittsburgh, USA
Jonathan Chan   King Mongkut's University of Technology
                                    Thonburi, Thailand
Ke Chen   Tampere University of Technology, Finland
Long Cheng   Chinese Academy of Science, China
Zhengbo Cheng   Zhejiang University of Technology, China
Jose Alfredo Ferreira Costa   UFRN – Universidade Federal do Rio Grande
                                    do Norte, Brazil
Ruxandra Liana Costea   Polytechnic University of Bucharest, Romania
Mingcong Deng   Tokyo University of Agriculture
                                    and Technology, Japan
Haibin Duan   Beijing University of Aeronautics
                                    and Astronautics, China
Jianchao Fan   National Marine Environmental
                                    Monitoring Center, Dalian, China
Wai-Keung Fung   Robert Gordon University, UK
Chengan Guo   Dalian University of Technology, China
Ping Guo   Beijing Normal University, China
Haibo He   University of Rhode Island, USA
Zhongsheng Hou   Beijing Jiaotong University, China
Jin Hu   Chongqing Jiaotong University, Chongqing,
                                    China
Jinglu Hu   Waseda University, Japan
Minlie Huang   Tsinghua University, China
Danchi Jiang   University of Tasmania, Australia
Haijun Jiang   Xinjiang University, China
Min Jiang   Xiamen University, China

Contents

Intelligent Control

Neurodynamics Analsysis

Memristor

Computer Vision

Signal Processing

Machine Learning

Optimization

Novel Approaches and Applications

Intelligent Control

A Novel T-S Fuzzy Model Based Adaptive Synchronization Control Scheme for Nonlinear Large-Scale Systems with Uncertainties and Time-Delay

He Jiang and Dongsheng Yang

College of Information Science and Engineering, Northeastern University, Shenyang, China, 110819

Abstract. In this paper, a novel T-S fuzzy model based adaptive synchronization scheme for nonlinear large-scale systems with uncertainties and time-delay is proposed. Based on the universal approximation property of T-S fuzzy model, a nonlinear large-scale system is established and fuzzy adaptive controllers are designed under Parallel Distributed Compensation (PDC) for overcoming the unknown uncertainties in systems and the time-delay in communication. Furthermore, under some certain condition, this synchronization scheme can be transformed into pinning synchronization control, which will indeed save much resource. Finally, a numerical simulation example is taken to show the effectiveness of the proposed adaptive synchronization scheme.

Keywords: nonlinear large-scale systems, uncertainties, time-delay, T-S fuzzy model, adaptive synchronization, pinning control.

1 Introduction

Synchronization of complex systems is a kind of typical basic motions and collective behaviors in nature, which has attracted much attention of researchers from different disciplines, such as mathematics, engineering science and so on [1,2,3,4]. Synchronization problems for complex systems with diverse types of time-delays were discussed by [5,6]. [7,8] proposed two different pinning control approaches for inner synchronization problems, and [9] designed a novel pinning synchronization scheme for outer synchronization problems by using scalar signals.

In recent years, researchers started to use fuzzy models to study the synchronization of complex systems. Mukhija et al.[10] studied a class of fuzzy complex systems with time-delay, and proposed a new synchronization criteria. Mahdavi et al.[11] designed an adaptive pulse controller for synchronization of fuzzy complex systems, and proposed a new method to choose suitable nodes to be controlled. However, both two researches need to calculate a huge amount of linear matrix inequalities (LMIs), which will increase the complexity of obtaining the solution of control laws.

© Springer International Publishing Switzerland 2015
X. Hu et al. (Eds.): ISNN 2015, LNCS 9377, pp. 3–10, 2015.
DOI: 10.1007/978-3-319-25393-0_1

In this paper, a novel adaptive synchronization scheme for fuzzy large-scale systems with time-delay and uncertainties will be proposed and the decentralize adaptive controllers herein will be designed under Parallel Distributed Compensation (PDC). The proposed adaptive method successfully circumvents the LMI-based approaches in the previous works [10,11] and decreases the amount of computation. Furthermore, the nonlinear coupled large-scale systems via T-S fuzzy model with time-delay and uncertainties are taken into the consideration in contrast to the works of [8] and [9] with linearly diffusive couplings, which will make great sense in the practical industrial applications.

2 Background and Preliminaries

In this section, the model of nonlinear large-scale systems with uncertainties and time-delay will be presented based on T-S fuzzy model, and several lemmas which are necessary for derivation and convergence proof will be given.

2.1 Nonlinear Large-Scale Systems with Uncertainties and Time-Delay via T-S Fuzzy Model

Consider a nonlinear large-scale system consisting of J T-S fuzzy subsystems, which can be acquired by the extended approach of [12], as below

$$\begin{cases} If \xi_{i1} \text{ is } M_{i1}^l \text{ and } \cdots \text{ and } \xi_{ig_i} \text{ is } M_{ig_i}^l, \\ Then \ \dot{x}_i(t) = (A_i^l + \Delta A_i^l)x_i(t) + \sum_{j=1, j\neq i}^{J} (C_{ij}^l + \Delta C_{ij}^l)x_j(t-d) + u_i(t), \quad (1) \\ i = 1, 2, \cdots, J, \ \ l = 1, 2, \cdots, r_i. \end{cases}$$

where $x_i(t) \in R^n$ is the state of ith subsystem; $\xi_i(t) = [\xi_{i1}(t), \xi_{i2}(t), \cdots, \xi_{ig_i}(t)]^T$ is the ith subsystem antecedent variables; $M_{iq}^l(q = 1, 2, \cdots, g_i)$ is the fuzzy set; A_i^l is the system matrix with its uncertainty matrix ΔA_i^l and C_{ij}^l is the coupling matrix between the ith subsystem and the jth subsystem with its uncertainty matrix ΔC_{ij}^l under the lth fuzzy rule, respectively; d is the constant parameter of time-delay in the communication; $u_i(t)$ is the control input of ith subsystem.

After applying product inference engine, singleton fuzzification and center average defuzzification to (1), it can be rewritten as

$$\dot{x}_i(t) = \sum_{l=1}^{r_i} h_i^l(\xi_i(t)) \left[(A_i^l + \Delta A_i^l)x_i(t) + \sum_{j=1, j\neq i}^{J} (C_{ij}^l + \Delta C_{ij}^l)x_j(t-d) + u_i(t) \right]$$

$$(2)$$

where

$$\beta_i^l(\xi_i(t)) = \prod_{q=1}^{g_i} M_{iq}^l(\xi_{iq}(t)), h_i^l(\xi_i(t)) = \frac{\beta_i^l(\xi_i(t))}{\sum_{l=1}^{r_i} \beta_i^l(\xi_i(t))}, \sum_{l=1}^{r_i} h_i^l(\xi_i(t)) = 1,$$

$h_i^l(\xi_i(t))$ is the membership function of the ith subsystem under the lth fuzzy rule.

2.2 Mathematical Preliminaries

Lemma 1[13]: For any matrix $Q \in R^{n \times n}$, any constant $\varepsilon > 0$, any positive definite matrix $P \in R^{n \times n}$, and for all $\zeta \in R^m, \eta \in R^n$, one can obtain

$$2\zeta^T Q \eta \leq \varepsilon \zeta^T Q P^{-1} Q^T \zeta + \varepsilon^{-1} \eta^T P \eta.$$

Lemma 2[14]: For the given suitable dimension matrices N, F and G, if F satisfies $F^T F \leq I$, where I represents suitable dimension identity matrix, then for any constant $\varepsilon > 0$, one can obtain

$$NFG + G^T F^T N^T \leq \varepsilon NN^T + \varepsilon^{-1} G^T G.$$

Lemma 3[15]: For any vector $x \in R^n$, if $Q \in R^{n \times n}$ is a symmetric matrix and $P \in R^{n \times n}$ is a positive definite matrix, then one can obtain

$$\lambda_{\min}(P^{-1}Q) x^T P x \leq x^T Q x \leq \lambda_{\max}(P^{-1}Q) x^T P x.$$

where $\lambda_{min}(P^{-1}Q)$ and $\lambda_{max}(P^{-1}Q)$ denote the minimum and maximum eigenvalues of the matrix $P^{-1}Q$ respectively.

3 Fuzzy Adaptive Synchronization for a Nonlinear Large-Scale System with Uncertainties and Time-Delay

In Section 2.1, the fuzzy model of nonlinear large-scale system has been established. Here a novel adaptive control protocol will be utilized to achieve the synchronization of the fuzzy system (2).

Let $S(t) = (s^T(t), s^T(t), \cdots, s^T(t))^T \in R^{n \times J}$ and $S(t-d) = (s^T(t-d), s^T(t-d), \cdots, s^T(t-d))^T \in R^{n \times J}$ be synchronous solution of the controlled system (2), and note that $s(t) \in R^n$ and $s(t-d) \in R^n$ can be the equilibrium point of every subsystem[8]. It is obvious that $s(t)$ and $s(t-d)$ are both n-dimension zero vectors, for there exist a common equilibrium point at origin. Hence, the error vectors can be defined as

$$\begin{cases} e_i(t) = x_i(t) - s(t) \\ e_i(t-d) = x_i(t-d) - s(t-d), i = 1, 2, \cdots, J \end{cases} \tag{3}$$

According to (2) and (3), the error systems can be described by

$$\dot{e}_i(t) = \sum_{l=1}^{r_i} h_i^l(\xi_i(t)) \left[(A_i^l + \Delta A_i^l) e_i(t) + \sum_{j=1, j \neq i}^{J} (C_{ij}^l + \Delta C_{ij}^l) e_j(t-d) + u_i(t) \right] \tag{4}$$

where $\triangle A_i^l = N_i^l F_i^l G_i^l$ and $\triangle C_{ij}^l = N_{ij}^l F_{ij}^l G_{ij}^l$ denote uncertainty parameter matrices, where N_i^l, G_i^l, N_{ij}^l and G_{ij}^l are known matrices, F_i^l and F_{ij}^l are unknown matrices, which satisfy $F_i^{l^T} F_i^l \leq I, F_{ij}^{l^T} F_{ij}^l \leq I$ $(i = 1, 2, \cdots, J, j = 1, 2, \cdots, J$

$i \neq j, l = 1, 2, \cdots, r_i)$; $u_i(t) \in R^n$ $(i = 1, 2, \cdots, J)$ is the adaptive control input, which will be designed later.

To realize the synchronization of the nonlinear large-scale system (2), the controller u_i should guide the error vectors (3) to converge to zero as t goes to infinity, i.e., $\lim_{t \to \infty} \|e_i(t)\| = 0$, $i = 1, 2, \cdots, J$.

By using PDC technique, the adaptive synchronization controllers can be designed as follows:

$$\begin{cases} u_i(t) = - \sum_{m=1}^{r_i} h_i^m(\xi_i(t)) p_i^m e_i(t) \\ \dot{p}_i^m = k_i^m h_i^m(\xi_i(t)) \|e_i(t)\|_2^2, 1 \le i \le J, m = 1, 2, \cdots, r_i. \end{cases} \tag{5}$$

where k_i^m are any positive constants. Thus, the error systems (4) can be rewritten as follows:

$$\dot{e}_i(t) = \sum_{l=1}^{r_i} \sum_{m=1}^{r_i} h_i^l(\xi_{iq}(t)) h_i^m(\xi_{iq}(t)) \left[(A_i^l + \Delta A_i^l) e_i(t) + \sum_{j=1, j \neq i}^{J} (C_{ij}^l + \Delta C_{ij}^l) \times \right.$$
$$\left. e_j(t-d) - p_i^m e_i(t) \right] \tag{6}$$

Definition 1: Let $\tilde{\Delta}_1 = \varepsilon_i^l N_i^l (N_i^l)^T + (\varepsilon_i^l)^{-1} (G_i^l)^T G_i^l$, $\tilde{\Delta}_2 = \sum_{j=1, j \neq i}^{J} \left[N_{ij}^l \right.$

$\times (N_{ij}^l)^T + C_{ij}^l (C_{ij}^l)^T \right]$, $\tilde{\Delta}_3 = \sum_{j=1, j \neq i}^{J} (G_{ij}^l)^T (G_{ij}^l) + (J-1) I_i$, where ε_i^l is any positive constant; S_i is a symmetric positive definite matrix satisfying $S_i > \tilde{\Delta}_3$. Let λ_{imax}^l be the maximum eigenvalues of the matrix $\frac{A_i^l + (A_i^l)^T + \tilde{\Delta}_1 + \tilde{\Delta}_2 + S_i}{2}$. Let k be any positive constant satisfying $k > \lambda_{imax}^l$.

Theorem 1: If there exists a symmetric positive definite matrix S_i satisfying $S_i > \tilde{\Delta}_3$, then the error systems (4) can be asymptotically stable under the controllers (5), i.e., this nonlinear large-scale system with uncertainties and time-delay (2) can be asymptotically synchronized by the proposed fuzzy adaptive controllers (5).

Proof. Construct a Lyapunov candidate as follows:

$$V(t) = V_1(t) + V_2(t) + V_3(t) \tag{7}$$

where

$$V_1(t) = \frac{1}{2} \sum_{i=1}^{J} e_i^T(t) e_i(t), V_2(t) = \frac{1}{2} \sum_{i=1}^{J} \int_{t-d}^{t} e_i^T(\tau) S_i e_i(\tau) d\tau,$$

$$V_3(t) = \frac{1}{2} \sum_{i=1}^{J} \sum_{m=1}^{r_i} \frac{(p_i^m - k)^2}{k_i^m}.$$

By using Lemma 1, one can obtain

$$
\dot{V}(t) \leq \sum_{i=1}^{J}\sum_{l=1}^{r_i}\sum_{m=1}^{r_i} h_i^l(\xi_i(t))h_i^m(\xi_i(t))\times
$$

$$
\left\{ e_i^T(t)\left[\frac{(A_i^l+\Delta A_i^l)^T+(A_i^l+\Delta A_i^l)}{2} \right.\right.
$$

$$
\left. + \frac{\sum_{j=1,j\neq i}^{J}\left[(N_{ij}^l)(N_{ij}^l)^T+(C_{ij}^l)(C_{ij}^l)^T\right]+S_i}{2} \right] e_i(t)-ke_i^T(t)e_i(t)
$$

$$
\left. +e_i^T(t-d)\left[\frac{\sum_{j=1,j\neq i}^{J}(G_{ij}^l)^T(G_{ij}^l)+(J-1)I_i-S_i}{2} \right] e_i(t-d) \right\} \qquad (8)
$$

By using Lemma 2 to simplify uncertainty terms in (9), it can be expressed as below

$$
\dot{V}(t) \leq \sum_{i=1}^{J}\sum_{l=1}^{r_i}\sum_{m=1}^{r_i} h_i^l(\xi_i(t))h_i^m(\xi_i(t))\times
$$

$$
\left\{ e_i^T(t)\left[\frac{(A_i^l)^T+A_i^l+\varepsilon_i^l N_i^l(N_i^l)^T+(\varepsilon_i^l)^{-1}(G_i^l)^T G_i^l}{2} \right.\right.
$$

$$
\left. + \frac{\sum_{j=1,j\neq i}^{J}\left[(N_{ij}^l)(N_{ij}^l)^T+(C_{ij}^l)(C_{ij}^l)^T\right]+S_i}{2} \right] e_i(t)-ke_i^T(t)e_i(t)
$$

$$
\left. +e_i^T(t-d)\left[\frac{\sum_{j=1,j\neq i}^{J}(G_{ij}^l)^T(G_{ij}^l)+(J-1)I_i-S_i}{2} \right] e_i(t-d) \right\} \qquad (9)
$$

By using Lemma 3, one finally has

$$
\dot{V}(t) \leq \sum_{i=1}^{J}\sum_{l=1}^{r_i}\sum_{m=1}^{r_i} h_i^l(\xi_i(t))h_i^m(\xi_{iq}(t))\times
$$

$$
\begin{bmatrix} e_i(t) \\ e_i(t-d) \end{bmatrix}^T \begin{bmatrix} (\lambda_{i\,max}^l-k) & 0 \\ 0 & \frac{\sum_{j=1,j\neq i}^{J}(G_{ij}^l)^T(G_{ij}^l)+(J-1)I_i-S_i}{2} \end{bmatrix} \begin{bmatrix} e_i(t) \\ e_i(t-d) \end{bmatrix} < 0
$$

$$
\qquad (10)
$$

According to (10), it follows that $e_i(t) \to 0$ as $t \to 0,(i = 1, 2, \cdots, J)$. That is, the nonlinear large-scale system with uncertainties and time-delay (2) can be asymptotically synchronized under the adaptive controllers (5). The proof is completed.

Remark 1: For a subsystem i, if there exists $\lambda^l_{imax} < 0$,$(l = 1, 2, \cdots, r_i)$, then this subsystem does not need control input u_i, i.e., $u_i = 0$. That is, for J subsystems in the large-scale system, maybe only $N(N < J)$ controllers are needed. Thus, this adaptive synchronization scheme can be transformed to a pinning adaptive synchronization scheme under some certain conditions. This will save much resource, and be very economical for the real engineering applications.

4   Numerical Simulation

In this section, a numerical simulation example will be taken to verify the effectiveness of the proposed adaptive synchronization scheme. Consider a nonlinear coupling system consisting of three subsystems as follows:

$$A_1^1 = \begin{bmatrix} 7/3 & -5/3 \\ -1 & -1 \end{bmatrix}, C_{12}^1 = \begin{bmatrix} -2/3 & 1/3 \\ 1 & 0 \end{bmatrix}, C_{13}^1 = \begin{bmatrix} 4/3 & -2/3 \\ -1 & 0 \end{bmatrix},$$
$$A_2^1 = \begin{bmatrix} -7/6 & -1/6 \\ -5/3 & 10/3 \end{bmatrix}, C_{21}^1 = \begin{bmatrix} 5/6 & -1/6 \\ -2/3 & 4/3 \end{bmatrix}, C_{23}^1 = \begin{bmatrix} 5/6 & -19/6 \\ 4/3 & -2/3 \end{bmatrix},$$
$$A_3^1 = \begin{bmatrix} 8/3 & 2/3 \\ -2/3 & 1/3 \end{bmatrix}, C_{31}^1 = \begin{bmatrix} -4/3 & -1/3 \\ -2/3 & 10/3 \end{bmatrix}, C_{32}^1 = \begin{bmatrix} -1/3 & 2/3 \\ -5/3 & -2/3 \end{bmatrix}.$$

$$A_1^2 = \begin{bmatrix} 23/5 & -9/5 \\ -2 & -1 \end{bmatrix}, C_{12}^2 = \begin{bmatrix} -2/5 & 1/5 \\ 2 & 0 \end{bmatrix}, C_{13}^2 = \begin{bmatrix} 8/5 & -4/5 \\ -1 & 0 \end{bmatrix},$$
$$A_2^2 = \begin{bmatrix} -39/5 & 7/5 \\ -9/5 & 22/5 \end{bmatrix}, C_{21}^2 = \begin{bmatrix} 6/5 & 7/5 \\ -4/5 & 12/5 \end{bmatrix}, C_{23}^2 = \begin{bmatrix} 6/5 & -28/5 \\ 16/5 & -8/5 \end{bmatrix},$$
$$A_3^2 = \begin{bmatrix} 71/15 & 7/15 \\ -32/15 & -41/15 \end{bmatrix}, C_{31}^2 = \begin{bmatrix} -19/15 & -8/15 \\ -32/15 & -176/15 \end{bmatrix},$$
$$C_{32}^2 = \begin{bmatrix} -4/15 & 7/15 \\ -47/15 & -64/15 \end{bmatrix}.$$

The uncertainty term parameters are chosen as below:

$$N_i^1 = \begin{bmatrix} 0.11 & 0.32 \\ 0.21 & 0.12 \end{bmatrix}, N_i^2 = \begin{bmatrix} 0.13 & 0.36 \\ 0.24 & 0.08 \end{bmatrix}, N_{ij}^1 = \begin{bmatrix} 0.27 & 0.05 \\ 0.13 & 0.28 \end{bmatrix},$$
$$N_{ij}^2 = \begin{bmatrix} 0.28 & 0.35 \\ 0.23 & 0.68 \end{bmatrix}, G_i^1 = \begin{bmatrix} 1.1 & 1.3 \\ 2.1 & 1.2 \end{bmatrix}, G_i^2 = \begin{bmatrix} 1.4 & 1.6 \\ 2.7 & 1.1 \end{bmatrix},$$
$$G_{ij}^1 = \begin{bmatrix} 3.1 & 2.4 \\ 1.1 & 1.7 \end{bmatrix}, G_{ij}^2 = \begin{bmatrix} 3.6 & 2.1 \\ 1.3 & 1.4 \end{bmatrix}, i = 1, 2, 3, j = 1, 2, 3, i \neq j.$$

Let $x_1(0) = \begin{bmatrix} 1 & 0 \end{bmatrix}^T$, $x_2(0) = \begin{bmatrix} 0 & 1 \end{bmatrix}^T$, $x_3(0) = \begin{bmatrix} 1 & 1 \end{bmatrix}^T$ be the initial values and $d = 0.5$ be the time-delay constant.

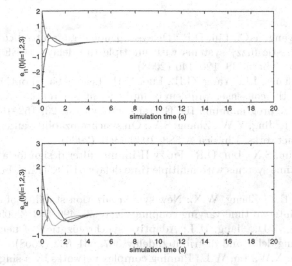

Fig.1. The curves of synchronization errors.

The membership functions are chosen as

$$h_i^1(x_{i1}(t)) = \frac{1 + \cos(x_{i1}(t))}{2}, \quad h_i^2(x_{i1}(t)) = \frac{1 - \cos(x_{i1}(t))}{2}, \quad i = 1, 2, 3.$$

The simulation results are shown in Fig.1. It is obvious that the synchronization errors rapidly converge to zero, and finally the whole nonlinear large-scale system achieves synchronization.

5 Conclusion

Based on T-S fuzzy model, a novel adaptive synchronization control scheme for nonlinear large-scale systems with uncertainties and time-delay has been presented in this paper. Unlike other fuzzy synchronization methods[10,11], the proposed scheme circumvents to calculate huge amounts of LMIs and decreases the complexity of getting the solution. Furthermore, under some certain conditions, this scheme can be transformed to the pinning adaptive synchronization control, which will save much resource. Compared with the researches on the general complex networks in [8,9], this paper takes the nonlinear large-scale systems with uncertainties and time-delay into consideration, which will be very useful and practical for the real engineering applications.

Acknowledgment. This work was supported by the National Natural Science Foundation of China (61273029), the Program for New Century Excellent Talents in University, China(NCET-12-0106) and the Basic Scientific Research Funding of Northeastern University(N130504004).

References

1. Zhang, H.G., Wang, Y.C., Liu, D.R.: Delay-dependent guaranteed cost control for uncertain Stochastic fuzzy systems with multiple tire delays. IEEE Trans. Syst. Man Cybern. B Cybern. 38, 126–140 (2008)
2. Zhang, H.G., Zhang, J.L., Yang, G.H., Luo, Y.H.: Leader-based optimal coordination control for the consensus problem of multi-agent differential games via fuzzy adaptive dynamic programming. IEEE Trans. Fuzzy Syst. 23, 152–163 (2015)
3. Hao, B.B., Yu, H., Jing, Y.W., Zhang, S.Y.: On synchronizability and heterogeneity in unweighted networks. Physica A 388, 1939–1945 (2009)
4. Zhang, H.G., Lun, S.X., Liu, D.R.: Fuzzy H-infinity filter design for a class of non-linear discrete-time systems with multiple time delays. IEEE Trans. Fuzzy Syst. 15, 453–469 (2007)
5. Fei, Z.Y., Gao, H.J., Zheng, W.X.: New synchronization stability of complex networks with an interval time-varying coupling delay 56, 499–503 (2009)
6. Wang, K., Teng, Z.D., Jiang, H.J.: Adaptive synchronization of neural networks with time-varying delay and distributed delay 387, 631–642 (2008)
7. Chen, T.P., Liu, X.W., Lu, W.L.: Pinning complex networks by a single controller. IEEE Trans. Circuits Systems I 54, 1317–1326 (2007)
8. Zhou, J., Lu, J.A., Lu, J.H.: Pinning adaptive synchronization of a general complex dynamical network. Automatica 44, 996–1003 (2006)
9. Fan, C.X., Jiang, G.P., Jiang, F.H.: Synchronization between two complex dynamical networks using scalar signals under pinning control. IEEE Trans. Circuits Systems I 57, 2991–2998 (2010)
10. Mukhija, P., Kar, I.N., Bhatt, K.P.: New synchronization criteria for fuzzy complex dynamical network with time-varying delay. In: 2013 IEEE International Conference on Fuzzy Systems, Hyderabad, pp. 1–7 (2013)
11. Mahdavi, N., Menhaj, M.B., Kurths, J., Lu, J.: Fuzzy complex dynamical networks and its synchronization. IEEE Trans. Cybernetics 43, 648–659 (2013)
12. Marcelo, C.M., Stanislaw, H.: Stabilizing controller design for uncertain nonlinear systems using fuzzy models. IEEE Trans. Fuzzy Syst. 7, 133–142 (1999)
13. Guan, X.P., Chen, C.L.: Delay-dependent guaranteed cost control for T-S fuzzy systems with time delays. IEEE Trans. Fuzzy Syst. 12, 236–249 (2004)
14. Yang, D.D., Zhang, H.G.: Robust H_∞ networked control for uncertain fuzzy systems with time-delay. Acta Automatica Sinica 33, 726–730 (2007)
15. Huang, L.: Linear Algebra in Systems and Control Theory. Science Press, Beijing (1984)

Finite-Time Control for Markov Jump Systems with Partly Known Transition Probabilities and Time-Varying Polytopic Uncertainties

Chen Zheng[1], Xiaozheng Fan[1], Manfeng Hu[1,2],
Yongqing Yang[1,2], and Yinghua Jin[1,2]

[1] School of Science, Jiangnan University, Wuxi 214122, China
[2] Key Laboratory of Advanced Process Control for Light Industry (Ministry of Education), Jiangnan University, Wuxi 214122, China
ChenZ1208@126.com, fanxiaozhengjndx@sina.cn,
humanfeng@jiangnan.edu.cn, yongqingyang@163.com,
jyhmath@jiangnan.edu.cn

Abstract. In this paper, the finite-time control problem for Markov systems with partly known transition probabilities and polytopic uncertainties is investigated. The main result provided is a sufficient conditions for finite-time stabilization via state feedback controller, and a simpler case without controller is also considered, based on switched quadratic Lyapunov function approach. All conditions are shown in the form of LMIs. An illustrative example is presented to demonstrate the result.

Keywords: finite-time stabilization, Markov systems, polytopic uncertainties, partly known transition probabilities, linear matrix inequalities.

1 Introduction

Markov systems are modelled by a set of systems with the transitions between models governed by a Markov chain which takes values in a finite system set. The systems' evolutions in systems are determined by the transition probabilities. As pointed out in [1], the complete knowledge of the transition probabilities may be hard to measure. Therefore, it is of great importance to consider partly known transition probabilities. Some results are researched for Markov jump systems with partly known transition probabilities [2,3,4,5,6,7].

Finite-time stability means once we fix a time interval, its states does not exceed a certain bound over the time interval. In some cases, large values of the state are not acceptable [8]. The concept of finite-time stability has been revisited in the light of linear matrix inequalities and Lyapunov function theory [9,10,11,12,13,14,15]. To the best of our knowledge, the finite-time control for Markov jump systems with partly known transition probabilities has not been fully investigated yet.

Motivated by the above discussions, in this paper, the problem that the finite-time control for Markov jump systems with partly known transition probabilities

© Springer International Publishing Switzerland 2015
X. Hu et al. (Eds.): ISNN 2015, LNCS 9377, pp. 11–18, 2015.
DOI: 10.1007/978-3-319-25393-0_2

and time-varying polytopic uncertainties is investigated. Based on the switched quadratic Lyapunov function approach, a state feedback controller is designed. Further more, a corollary is given based on the main result, which is the sufficient condition for the simpler case of finite-time stability.

The superscript 'T' stands for matrix transposition, $E(\cdot)$ stands for the mathematical expectation. In symmetric block matrices or long matrix expressions, we use $*$ as an ellipsis for the terms that are introduced by symmetry. A matrix $P > 0 \ (\geq 0)$ means P is a symmetric positive (semi-positive) definite matrix.

2   Problem Statement and Preliminaries

Consider the following discrete-time Markov jump linear system:

$$x(k+1) = A(r(k), \lambda)x(k) + B(r(k), \lambda)u(k) \qquad (1)$$

where $x(k) \in R^n$ is the state vector, $u(k) \in R^m$ is the control input vector. $\{r(k), k \geq 0\}$ is a discrete-time Markov chain, which takes values in a finite set $\chi = \{1, 2, \ldots, N\}$ with a transition probabilities matrix $\Lambda = \{\pi_{ij}\}$, for $r(k) = i, r(k+1) = j$, one has $Pr(r(k+1) = j | r(k) = i) = \pi_{ij}$, where $\pi_{ij} \geq 0, \forall i, j \in \chi$, and $\sum_{j=1}^{N} \pi_{ij} = 1$. $N > 1$ is the number of subsystems, and we use $(A_i(\lambda), B_i(\lambda))$ denotes the ith system when $r(k) = i$. The transition probabilities of the jumping process $\{r(k), k \geq 0\}$ are assumed to be partially known. $\forall i \in \chi$, we denote

$$\chi_K^i = \{j : \pi_{ij} \ is \ known\} \ \chi_{UK}^i = \{j : \pi_{ij} \ is \ unknown\} \ . \qquad (2)$$

The matrices of each subsystem have polytopic uncertain parameters. It is assumed that ,at each instant of time k, $(A_i(\lambda), B_i(\lambda)) \in R_i$, where R_i is a given convex bounded polyhedral domain described by

$$R_i = \{ (A_i(\lambda), B_i(\lambda)) = \sum_{m=1}^{s} \lambda_m (A_{i,m}, B_{i,m}) \ ; \ \sum_{m=1}^{s} \lambda_m = 1, \lambda_m \geq 0\} \quad i \in \chi \ , \qquad (3)$$

where $(A_{i,m}, B_{i,m})$ denotes the mth vertex in the ith mode, s means the total number of vertices.

In this paper we derive a state feedback controller of the form $u(k) = K_i(\lambda)x(k)$, such that the Markov jump linear systems (1) is finite-time stabilizable. In particular we have the following definition and lemma.

Definition 1 (Finite-Time Stability [15]). *The discrete-time linear system (1) (setting $u(k) = 0$) is said to be finite-time stable with respect to $(\delta_x, \epsilon, R, N_0)$, where R is a positive-definite matrix, $0 < \delta_x < \epsilon$, if*

$$E(x^T(0)Rx(0)) \leq \delta_x^2 \Rightarrow E(x^T(k)Rx(k)) < \epsilon^2 \quad \forall k \in N_0 \ . \qquad (4)$$

Remark 1. Systems that are Lyapunov asymptotically stable may not be finite-time stable. And the system (1) is said to be finite-time stabilizable if there exists a state feedback controller in the form of $u(k) = K_i(\lambda)x(k)$.

Lemma 1 (Schur Complement). *The linear matrix inequality*

$$S = \begin{bmatrix} S_{11} & S_{12} \\ S_{12}^T & S_{22} \end{bmatrix} < 0,$$

where $S_{11} = S_{11}^T$ and $S_{22} = S_{22}^T$ are equivalent to

$$S_{11} < 0, \quad S_{22} - S_{12}^T S_{11}^{-1} S_{12} < 0 \ . \tag{5}$$

3 Main Result

We consider the state feedback controller with the following structure:

$$u(k) = K_i(\lambda)x(k) \tag{6}$$

Theorem 1. *The Markov jump linear systems (1) is finite-time stabilizable with respect to $(\delta_x, \epsilon, R, N_0)$ if there exist a scalar $\gamma \geq 1$, positive scalars ϕ_1, ϕ_2, matrices $S_{i,m} > 0$, matrices $U_{i,m}$ $\forall i \in \chi$, $1 \leq m < n \leq s$ and $\forall (i,j) \in (\chi \times \chi)$ matrices*

$$\Omega_{m,n}^{i,j} = \begin{bmatrix} X_{m,n}^{i,j} & Y_{m,n}^{i,j} \\ W_{m,n}^{i,j} & Z_{m,n}^{i,j} \end{bmatrix} j \in \chi_{UK}^i, \quad \Xi_{m,n}^{i,j} = \begin{bmatrix} D_{m,n}^{i,j} & E_{m,n}^{i,j} \\ F_{m,n}^{i,j} & G_{m,n}^{i,j} \end{bmatrix} j \in \chi_K^i,$$

satisfying $\forall (i,j) \in \chi \times \chi$, $(1 \leq m < n \leq s)$

$$\begin{bmatrix} -S_{j,m} - S_{j,n} - X_{m,n}^{i,j} - (X_{m,n}^{i,j})^T & S_{UK} \\ * & \nu \end{bmatrix} \leq 0, \ j \in \chi_{UK}^i, \tag{7}$$

$$\begin{bmatrix} -S_{K,m}^i - S_{K,n}^i - D_{m,n}^{i,j} - (D_{m,n}^{i,j})^T & S_K \\ * & \kappa \end{bmatrix} \leq 0, \ j \in \chi_K^i \tag{8}$$

$$\Omega^{i,j} = \begin{bmatrix} \Omega_1^{i,j} & \Omega_{1,2}^{i,j} & \cdots & \Omega_{1,s}^{i,j} \\ * & \Omega_2^{i,j} & \cdots & \Omega_{2,s}^{i,j} \\ \vdots & \vdots & \ddots & \vdots \\ * & * & \cdots & \Omega_s^{i,j} \end{bmatrix} < 0 \quad j \in \chi_{UK}^i, \tag{9}$$

$$\Xi^{i,j} = \begin{bmatrix} \Xi_1^{i,j} & \Xi_{1,2}^{i,j} & \cdots & \Xi_{1,s}^{i,j} \\ * & \Xi_2^{i,j} & \cdots & \Xi_{2,s}^{i,j} \\ \vdots & \vdots & \ddots & \vdots \\ * & * & \cdots & \Xi_s^{i,j} \end{bmatrix} < 0 \quad j \in \chi_K^i, \tag{10}$$

$$\phi_1 R < S_{i,m} < \phi_2 R \ , \tag{11}$$

$$\phi_2 \delta_x^2 < \epsilon^2 \phi_1 / \gamma^{N_0} \ , \tag{12}$$

where

$$\nu = -\gamma S_{i,m} - \gamma S_{i,n} - Z_{m,n}^{i,j} - (Z_{m,n}^{i,j})^T,$$

$$\kappa = (-\gamma \sum_{j \in \chi_K^i} \pi_{ij})(S_{i,m} + S_{i,n}) - G_{m,n}^{i,j} - (G_{m,n}^{i,j})^T, \ \ S_{K,n}^i = (\sum_{j \in \chi_K^i} \pi_{ij})S_{j,n},$$

$$\Omega_m^{i,j} = \begin{bmatrix} -S_{j,m} & A_{i,m}S_{i,m} + B_{i,m}U_{i,m} \\ * & -\gamma S_{i,m} \end{bmatrix} \quad j \in \chi_{UK}^i,$$

$$\Xi_m^{i,j} = \begin{bmatrix} -S_{K,m}^i & (\sum_{j \in \chi_K^i} \pi_{ij})(A_{i,m}S_{i,m} + B_{i,m}U_{i,m}) \\ * & -(\sum_{j \in \chi_K^i} \pi_{ij})\gamma S_{i,m} \end{bmatrix} \quad j \in \chi_K^i,$$

$$S_K = (\sum_{j \in \chi_K^i} \pi_{ij})(A_{i,n}S_{i,m} + A_{i,m}S_{i,n} + B_{i,n}U_{i,m} + B_{i,m}U_{i,n})$$

$$-E_{m,n}^{i,j} - (F_{m,n}^{i,j})^T, \ j \in \chi_K^i,$$

$$S_{UK} = A_{i,n}S_{i,m} + A_{i,m}S_{i,n} + B_{i,n}U_{i,m} + B_{i,m}U_{i,n} - Y_{m,n}^{i,j} - (W_{m,n}^{i,j})^T, \ j \in \chi_{UK}^i.$$

Furthermore, the state feedback controller can be represented as

$$K_i(\lambda) = U_i(\lambda)S_i(\lambda)^{-1}.$$

Proof. We choose the Lyapunov function $V(x(k)) = x(k)^T P_i(\lambda)x(k)$, and assume that $E(x^T(0)Rx(0)) \leq \delta_x^2$. Then we have

$$E(V(x(k+1))) - \gamma E(V(x(k))) = E\{x(k)^T[\tilde{A}_i(\lambda)^T P_j(\lambda)\tilde{A}_i(\lambda) - \gamma P_i(\lambda)]x(k)\} ,$$

where $\tilde{A}_i(\lambda) = A_i(\lambda) + B_i(\lambda)K_i(\lambda)$, and the case $i = j$ denotes that the switched system is described by the ith system $\tilde{A}_i(\lambda)$, and the case $i \neq j$ denotes that the system is being at the switching times from i to j. If

$$\tilde{A}_i(\lambda)^T P_j(\lambda)\tilde{A}_i(\lambda) - \gamma P_i(\lambda) < 0 \quad \forall(i,j) \in (\chi \times \chi) , \tag{13}$$

then we can obtain $E(V(x(k))) < \gamma^k E(V(x(0)))$.

Letting $S_{i,m} = P_{i,m}^{-1}$, $\tilde{S}_{i,m} = R_m^{-1/2}S_{i,m}R_m^{-1/2}$, $\lambda_{sup} = sup\{\lambda_{max}(P_{i,m})\}$ and $\lambda_{inf} = inf\{\lambda_{min}(P_{i,m})\}$,we get

$$\gamma^k E(V(x(0))) \leq \gamma^k (1/\lambda_{inf})E(x(0)^T Rx(0)) \leq \gamma^{N_0}(1/\lambda_{inf})\delta_x^2 ,$$

$$E(V(x(k))) = E(x(k)^T \Sigma_{m=1}^s \lambda_m P_{i,m} x(k)) \geq (1/\lambda_{sup})E(x(k)^T Rx(k)) , \tag{14}$$

where $R = \sum_{m=1}^s \lambda_m R_m$, $\lambda_{max}(P_{i,m})$ and $\lambda_{min}(P_{i,m})$ mean the maximum and minimum eigenvalues. From (11), we can determine $\phi_1 \leq \lambda_{inf}$, $\phi_2 \geq \lambda_{sup}$.

According to the above relations, we obtain

$$E(x(k)^T Rx(k)) < \phi_2/\phi_1 \delta_x^2 \gamma^{N_0} . \tag{15}$$

From (12), the finite-time stability of system (1) is guaranteed. So what we next do is to prove (13) holds.

By Lemma 1, multiplying both sides by diag $(P_j^{-1}(\lambda), P_i^{-1}(\lambda))$, changing the matrix variables with

$$S_i(\lambda) = P_i^{-1}(\lambda), \quad U_i(\lambda) = K_i(\lambda)P_i^{-1}(\lambda) , \tag{16}$$

and according to (3), multiplying by $\lambda_m, \lambda_n \geq 0$ summing up to s, we have

$$\Psi = (\sum_{j \in \chi_{UK}^i} \pi_{ij} + \sum_{j \in \chi_K^i} \pi_{ij})\Psi = (\sum_{j \in \chi_{UK}^i} \pi_{ij})(\sum_{m=1}^{s} \lambda_m^2 \begin{bmatrix} -S_{j,m} & A_{i,m}S_{i,m} + B_{i,m}U_{i,m} \\ * & -\gamma S_{i,m} \end{bmatrix}$$

$$+ \sum_{m=1}^{s-1} \sum_{n=m+1}^{s} \lambda_m \lambda_n \left\{ \begin{bmatrix} -S_{j,m} & A_{i,m}S_{i,n} + B_{i,m}U_{i,n} \\ * & -\gamma S_{i,n} \end{bmatrix} + \begin{bmatrix} -S_{j,n} & A_{i,n}S_{i,m} + B_{i,n}U_{i,m} \\ * & -\gamma S_{i,m} \end{bmatrix} \right\})$$

$$+ \sum_{m=1}^{s} \lambda_m^2 \begin{bmatrix} -S_{K,m}^i & (\sum_{j \in \chi_K^i} \pi_{ij})(A_{i,m}S_{i,m} + B_{i,m}U_{i,m}) \\ * & -\gamma(\sum_{j \in \chi_K^i} \pi_{ij})S_{i,m} \end{bmatrix}$$

$$+ \sum_{m=1}^{s-1} \sum_{n=m+1}^{s} \lambda_m \lambda_n \{ \begin{bmatrix} -S_{K,m}^i & (\sum_{j \in \chi_K^i} \pi_{ij})(A_{i,m}S_{i,n} + B_{i,m}U_{i,n}) \\ * & -\gamma(\sum_{j \in \chi_K^i} \pi_{ij})S_{i,n} \end{bmatrix}$$

$$+ \begin{bmatrix} -S_{K,n}^i & (\sum_{j \in \chi_K^i} \pi_{ij})(A_{i,n}S_{i,m} + B_{i,n}U_{i,m}) \\ * & -\gamma(\sum_{j \in \chi_K^i} \pi_{ij})S_{i,m} \end{bmatrix} \} . \tag{17}$$

From (7),(8), we get

$$\Psi \leq \eta^T \Omega^{i,j} \eta + \eta^T \Xi^{i,j} \eta ,$$

where $\eta = [\lambda_1 I \ \lambda_2 I \ \cdots \ \lambda_s I]^T$. Therefore, $\Psi < 0$.

Conditions similar to those of Theorem 1 can be obtained for the case without controller.

Corollary 1. *The system $x(k+1) = A(r(k), \lambda)x(k)$ is finite-time stable with respect to $(\delta_x, \epsilon, R, N_0)$ if there exist a scalar $\gamma \geq 1$, matrices $P_{i,m} > 0 \ \forall i \in \chi$, $1 \leq m < n \leq s$ and $\forall(i,j) \in (\chi \times \chi)$ matrices*

$$\Omega_{m,n}^{i,j} = \begin{bmatrix} X_{m,n}^{i,j} & Y_{m,n}^{i,j} \\ W_{m,n}^{i,j} & Z_{m,n}^{i,j} \end{bmatrix} \quad j \in \chi_{UK}^i, \quad \Xi_{m,n}^{i,j} = \begin{bmatrix} D_{m,n}^{i,j} & E_{m,n}^{i,j} \\ F_{m,n}^{i,j} & G_{m,n}^{i,j} \end{bmatrix} \quad j \in \chi_K^i,$$

satisfying $\forall(i,j) \in \chi \times \chi$, $(1 \leq m < n \leq s)$

$$\begin{bmatrix} -P_{j,m} - P_{j,n} - X_{m,n}^{i,j} - (X_{m,n}^{i,j})^T & \overline{S}_{UK} \\ * & \overline{\nu} \end{bmatrix} \leq 0, \ j \in \chi_{UK}^i, \tag{18}$$

$$\begin{bmatrix} -P_{K,m}^i - P_{K,n}^i - D_{m,n}^{i,j} - (D_{m,n}^{i,j})^T & \overline{S}_K \\ * & \overline{\kappa} \end{bmatrix} \leq 0, \ j \in \chi_K^i, \tag{19}$$

$$\phi_1 R < P_{i,m} < \phi_2 R , \tag{20}$$

$$\Omega^{i,j} = \begin{bmatrix} \Omega_1^{i,j} & \Omega_{1,2}^{i,j} & \cdots & \Omega_{1,s}^{i,j} \\ * & \Omega_2^{i,j} & \cdots & \Omega_{2,s}^{i,j} \\ \vdots & \vdots & \ddots & \vdots \\ * & * & \cdots & \Omega_s^{i,j} \end{bmatrix} < 0 \quad j \in \chi_{UK}^i, \tag{21}$$

$$\Xi^{i,j} = \begin{bmatrix} \Xi_1^{i,j} & \Xi_{1,2}^{i,j} & \cdots & \Xi_{1,s}^{i,j} \\ * & \Xi_2^{i,j} & \cdots & \Xi_{2,s}^{i,j} \\ \vdots & \vdots & \ddots & \vdots \\ * & * & \cdots & \Xi_s^{i,j} \end{bmatrix} < 0 \quad j \in \chi_K^i, \tag{22}$$

where

$$\overline{S}_K = P_{K,m}^i A_{i,n} + P_{K,n}^i A_{i,m} - E_{m,n}^{i,j} - (F_{m,n}^{i,j})^T,$$

$$\overline{S}_{UK} = P_{j,m} A_{i,n} + P_{j,n} A_{i,m} - Y_{m,n}^{i,j} - (W_{m,n}^{i,j})^T,$$

$$\overline{v} = -\gamma P_{i,m} - \gamma P_{i,n} - Z_{m,n}^{i,j} - (Z_{m,n}^{i,j})^T,$$

$$\Omega_m^{i,j} = \begin{bmatrix} -P_{j,n} & P_{j,n} A_{i,m} \\ * & -\gamma P_{i,m} \end{bmatrix} \quad j \in \chi_{UK}^i,$$

$$\Xi_m^{i,j} = \begin{bmatrix} -P_{K,n}^i & P_{K,n}^i A_{i,m} \\ * & -(\sum_{j \in \chi_K^i} \pi_{ij})\gamma P_{i,m} \end{bmatrix} \quad j \in \chi_K^i,$$

$$P_{K,n}^i = \sum_{j \in \chi_K^i} \pi_{ij} P_{j,n}, \quad \overline{\kappa} = (-\gamma \sum_{j \in \chi_K^i} \pi_{i,j})(P_{i,m} + P_{i,n}) - G_{m,n}^{i,j} - (G_{m,n}^{i,j})^T.$$

4   Illustrative Example

Consider the system (1) , there are two vertices in each subsystem:

$$A_{11} = \begin{bmatrix} 1.413 & -0.652 \\ 0.280 & -0.605 \end{bmatrix}, A_{12} = \begin{bmatrix} -0.475 & 0.013 \\ 0.871 & 0.187 \end{bmatrix},$$

$$B_{11} = \begin{bmatrix} 0.243 & -0.351 \\ 1.286 & 0.92 \end{bmatrix}, B_{12} = \begin{bmatrix} -1.618 & 0.172 \\ -0.406 & -2.418 \end{bmatrix},$$

$$A_{21} = \begin{bmatrix} -1.350 & -0.814 \\ 1.524 & -1.217 \end{bmatrix}, A_{22} = \begin{bmatrix} 0.016 & -1.383 \\ 0.020 & -0.474 \end{bmatrix},$$

$$B_{21} = \begin{bmatrix} 0.165 & -1.642 \\ -0.111 & 1.364 \end{bmatrix}, B_{22} = \begin{bmatrix} -1.585 & -0.447 \\ -2.630 & 0.724 \end{bmatrix},$$

$$A_{31} = \begin{bmatrix} 0.121 & 0.736 \\ -1.496 & -0.73 \end{bmatrix}, A_{32} = \begin{bmatrix} 1.179 & 0.176 \\ 0.399 & -0.196 \end{bmatrix},$$

$$B_{31} = \begin{bmatrix} 0.572 & 0.663 \\ -0.428 & -0.985 \end{bmatrix}, B_{32} = \begin{bmatrix} -1.305 & -0.236 \\ -0.461 & 0.910 \end{bmatrix}.$$

The partly known transition probability matrix is given as follow:

$$\begin{bmatrix} 0.6 & ? & ? \\ ? & ? & 0.5 \\ ? & 0.2 & ? \end{bmatrix}.$$

According to Theorem 1, we assume $R = I$, $N_0 = 31$, $\phi_2 = 1$, $\delta_x^2 = 0.05$, $\epsilon = 8$, $\gamma = 1.04$. It can be seen from the figures that the system (1) with the controller $u(k) = K_i(\lambda)x(k)$ meets the specified requirement, where $K_i(\lambda) = U_i(\lambda)S_i(\lambda)^{-1}$.

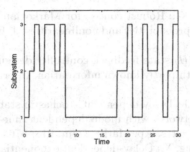

Fig. 1. The switching signal.

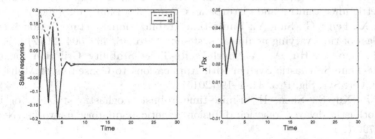

Fig. 2. The state response of system.

5   Conclusion

In this paper, the problem of finite-time stabilization for Markov jump linear systems with partly known transition probabilities and time-varying polytopic uncertainties has been studied. By using the switched quadratic Lyapunov function, all the conditions are established in the form of LMIs. At last, the main result has been demonstrated through an illustrative example.

References

1. Zhang, L., Boukas, E.K., Lam, J.: Analysis and synthesis of Markov jump linear systems with time-varying delays and partially known transition probabilities. IEEE Trans. Auto. Con. 53, 2458–2464 (2008)
2. Zhang, L., Boukas, E.K., Baron, L.: Fault detection for discrete-time Markov jump linear systems with partially known transition probabilities. Int. J. Con. 83, 1564–1572 (2010)
3. Braga, M.F., Morais, C.F., Oliveira, R.C.L.F.: Robust stability and stabilization of discrete-time Markov jump linear systems with partly unknown transition probability matrix. In: American Control Conference (ACC), pp. 6784–6789. IEEE Press, Washington (2013)
4. Tian, E., Yue, D., Wei, G.: Robust control for Markovian jump systems with partially known transition probabilities and nonlinearities. J. Fran. Ins. 350, 2069–2083 (2013)
5. Shen, M., Yang, G.H.: H_2 state feedback controller design for continuous Markov jump linear systems with partly known information. Inter. J. Sys. Sci. 43, 786–796 (2012)
6. Tian, J., Li, Y., Zhao, J.: Delay-dependent stochastic stability criteria for Markovian jumping neural networks with mode-dependent time-varying delays and partially known transition rates. Ap. Mathe. Com. 218, 5769–5781 (2012)
7. Rao, R., Zhong, S., Wang, X.: Delay-dependent exponential stability for Markovian jumping stochastic Cohen-Grossberg neural networks with p-Laplace diffusion and partially known transition rates via a differential inequality. Ad. Differ. Equa. 2013, 1–14 (2013)
8. Amato, F., Ariola, M., Dorato, P.: Finite-time control of linear systems subject to parametric uncertainties and disturbances. Auto. 37, 1459–1463 (2001)
9. Zhang, X., Feng, G., Sun, Y.: Finite-time stabilization by state feedback control for a class of time-varying nonlinear systems. Auto. 48, 499–504 (2012)
10. Hu, M., Cao, J., Hu, A.: A Novel Finite-Time Stability Criterion for Linear Discrete-Time Stochastic System with Applications to Consensus of Multi-Agent System. Cir. Sys. Sig. Pro. 34, 1–19 (2014)
11. Zhou, J., Xu, S., Shen, H.: Finite-time robust stochastic stability of uncertain stochastic delayed reaction-diffusion genetic regulatory networks. Neu. 74, 2790–2796 (2011)
12. Zuo, Z., Li, H., Wang, Y.: Finite-time stochastic stabilization for uncertain Markov jump systems subject to input constraint. Trans. Ins. Mea. Con. 36, 283–288 (2014)
13. Yin, Y., Liu, F., Shi, P.: Finite-time gain-scheduled control on stochastic bioreactor systems with partially known transition jump rates. Cir. Sys. Sig. Pro. 30, 609–627 (2011)
14. Bhat, S.P., Bernstein, D.S.: Continuous finite-time stabilization of the translational and rotational double integrators. IEEE Trans. Auto. Con. 43, 678–682 (1998)
15. Luan, X., Liu, F., Shi, P.: Neural-network-based finite-time H_∞ control for extended Markov jump nonlinear systems. Inter. J. Adap. Con. Sig. Pro. 24, 554–567 (2010)

Hybrid Function Projective Synchronization of Unknown Cohen-Grossberg Neural Networks with Time Delays and Noise Perturbation

Min Han and Yamei Zhang

Faculty of Electronic Information and Electrical Engineering,
Dalian University of Technology, Dalian 116023, China
minhan@dlut.edu.cn

Abstract. In this paper, the hybrid function projective synchronization of unknown Cohen-Grossberg neural networks with time delays and noise perturbation is investigated. A hybrid control scheme combining open-loop control and adaptive feedback control is designed to guarantee that the drive and response networks can be synchronized up to a scaling function matrix with parameter identification by utilizing the LaSalle-type invariance principle for stochastic differential equations. Finally, the corresponding numerical simulations are carried out to demonstrate the validity of the presented synchronization method.

Keywords: synchronization, Cohen-Grossberg neural network, delays, noise perturbation.

1 Introduction

In the past few decades, various types of artificial neural networks have been widely applied in many areas such as image processing, pattern recognition, optimization problems and so on [1-3]. Among these, the Cohen-Grossberg neural network model[4], has been recognized to be one of the most popular and typical neural network models, and some other models such as Hopfield neural networks, cellular neural networks, and recurrent neural networks are special cases of this model.

Chaos synchronization has been well studied because of its potential applications [5-6]. Moreover, it has been shown that neural networks can exhibit complicated behaviors with strange chaotic attractors. Therefore the synchronization of chaotic neural networks[7,8], especially the Cohen-Grosssberg neural networks, has received considerable attention and has been extensively investigated. By using adaptive control and linear feedback, Zhu and Cao realized the complete synchronization of chaotic Cohen-Crossberg neural networks with mixed time delays [9]. Without solving any linear matrix inequality, the author investigated the adaptive synchronization of different Cohen-Grossberg chaotic neural networks [10]. Shi and Zhu studied the adaptive synchronization problem of Cohen–Grossberg neural networks [11].

However, most of recent work above on the synchronization of chaotic Cohen-Grossberg networks has been restricted to the less general synchronization scheme as

© Springer International Publishing Switzerland 2015
X. Hu et al. (Eds.): ISNN 2015, LNCS 9377, pp. 19–27, 2015.
DOI: 10.1007/978-3-319-25393-0_3

complete synchronization, adaptive synchronization, etc. Function projective synchronization which is characterized by drive and response systems that can be synchronized up to a scaling function instead of a constant, is the extension of projective synchronization [12-13]. Because the unpredictability of the scaling function can additionally enhance the security of communication [14], function projective synchronization of neural networks is necessary to probe into. In [15-16], the authors investigated the function projective synchronization of some simple neural network models. In hybrid function projective synchronization, the drive and response systems could be synchronized to a desired scaling function matrix, which means the different variables in a node system could be synchronized to different scaling functions. However, to the best of our knowledge, there are very few or even no results on the function projective synchronization of Cohen-Grossberg neural networks, not mention the hybrid function projective synchronization.

Noise is ubiquitous in the real systems [17-18]. Therefore the effect of noise on synchronization has been well studied by many researchers. In addition, most of the work on the synchronization of chaotic Cohen-Grossberg neural networks assumes that the chaotic systems' parameters are known in advance. But in many practical situations, the values of some system's parameters cannot be exactly known beforehand [11-12]. Motivated by the above discussions, in this paper, we are concerned with the hybrid function projective synchronization of unknown Cohen-Grossberg neural networks with time delays and noise perturbation.

2   Model Description and Preliminaries

Consider the delayed Cohen-Grossberg neural network which can be called drive system and expressed as follows:

$$dx(t) = -D(x(t))[\Lambda(x(t)) - Bf(x(t) - Cg(x(t-\tau)) + J]dt \tag{1}$$

where $x(t) = [x_1(t), x_2(t), ..., x_n(t)]^T \in R^n$ is the state vector associated with n neurons. $D(x(t)) = diag[d_1(x(t)), d_2(x(t)), ..., d_n(x(t))]$ are amplification functions. $\Lambda(x(t)) = [\alpha_1(x_1(t)), \alpha_2(x_2(t)), ..., \alpha_n(x_n(t))]$ are appropriately behaved functions such that the solutions of model (1) remain bounded. $B = (b_{ij})_{n \times n}$, $C = (c_{ij})_{n \times n}$ are the connection weight matrix and the delayed connection weight matrix respectively. $J = [J_1, J_2, ..., J_n]^T$ denotes a constant external input vector. f and g represent the neuron activation functions and $f(x(t)) = [f_1(x_1(t)), f_2(x_2(t)), ..., f_n(x_n(t))]^T \in R^n$, $g(x(t-\tau)) = [g_1(x_1(t-\tau)), g_2(x_2(t-\tau)), ..., g_n(x_n(t-\tau))]^T \in R^n$, where $\tau > 0$ represents the transmission delay. For the drive system (1), the unknown response system with noise perturbation and time delay is given as follows:

$$dy(t) = \left\{ -D(y(t))[\Lambda(y(t)) - \hat{B}f(y(t)) - \hat{C}g(y(t-\tau)) + J] + U \right\}dt$$
$$+ H(t, y(t) - M(t)x(t), y(t-\tau) - M(t-\tau)x(t-\tau))dW \tag{2}$$

where $\hat{B} = B + \Delta B, \hat{C} = C + \Delta C$, $\Delta B = (\Delta b_{ij})_{n \times n}, \Delta C = (\Delta c_{ij})_{n \times n}$. $W = [W_1, W_2, \ldots, W_n]^T$ is the n-dimensional Brownian motion, $H(t, u, v) = [h_1(t, u, v), h_2(t, u, v), \ldots, h_n(t, u, v)]^T$ is called the noise intensity matrix, $M(t) = diag(m_1(t), m_2(t), \ldots, m_n(t))$ is a scaling function matrix, and $U = [u_1, u_2, \ldots, u_n]$ is a controller vector.

Definition 1. (HFPS) For the drive Cohen-Grossberg neural network (1) and the response Cohen-Grossberg neural network (2), it is said that they are hybrid function projective synchronized, if there exists a continuously differentiable scaling function matrix $M(t)$ such that $\lim_{t \to \infty} \|e(t)\| = \lim_{t \to \infty} \|y(t) - M(t)x(t)\| = 0$.

Remark 1. If the scaling functions $m_i(t)$ are taken as the nonzero constants θ_i, 1 or -1, the synchronization problem will be turned to the projective synchronization, the complete synchronization or the anti-synchronization.

To proceed, the following assumptions and lemmas are given.

Assumption 1. There exist positive constants $\overline{d}_i, N_i > 0$, such that $0 \le d_i(x) \le \overline{d}_i$, $|d_i(y) - d_i(mx)| \le N_i |y - mx|$ for all $x, y, m \in R$, $(i = 1, 2, \ldots, n)$.

Assumption 2. For any $i = 1, 2, \ldots, n$, there exists a positive constant $r_i > 0$, such that
$$\frac{d_i(y)\alpha_i(y) - d_i(mx)\alpha_i(mx)}{y - mx} \ge r_i \text{ for all } x, y, m \in R, \text{ and } y \ne mx.x$$

Assumption 3. For any $i = 1, 2, \ldots, n$, there exist constants $L_i^f > 0$ and $L_i^g > 0$ such that $0 \le \dfrac{f_i(y) - f_i(mx)}{y - mx} \le L_i^f$, $0 \le \dfrac{g_i(y) - g_i(mx)}{y - mx} \le L_i^g$ for all $x, y, m \in R$, and $y \ne mx$.

Assumption 4. For the noise intensity function matrix $H(t, x, y)$, there exists two positive constants p_1, p_2 such that, $trace[H^T(t, x, y)H(t, x, y)] \le p_1 x^T x + p_2 y^T y$, holds for all $(t, x, y) \in R_+ \times R^n \times R^n$. Moreover $H(t, 0, 0) \equiv 0$.

In order to get our main results in the next section, some necessary concepts and a lemma about stochastic differential equations [19] are provided in advance. Consider the following n-dimensional stochastic differential delay equation: $dx(t) = f(x(t), x(t - \tau), t)dt + \sigma(x(t), x(t - \tau), t)dW(t)$. It is known that there exists a unique solution $x(t, \xi)$ on $t \ge 0$ with initial data $\xi \in C_{F_0}^b([-\tau, 0], R^n)$. Moreover, both $f(x, y, t)$ and $\sigma(x, y, t)$ are locally bounded in (x, y) and uniformly bounded in t. For each $V \in C^{2,1}(R^n \times R_+; R_+)$, we define an operator LV from $R^n \times R^n \times R_+$ to R by
$$LV = \partial V / \partial t + \partial V / \partial x \cdot f + 1/2 trace[\sigma^T (\partial^2 V / \partial x_i \partial x_j)\sigma] \tag{3}$$
where $\partial V / \partial z = (\partial V / \partial z_1, \ldots, \partial V / \partial z_n)$.

Lemma 1. (Invariance principle [20]) Assume that there are functions $V \in C^{2,1}(R^n \times R_+; R_+)$, $\beta \in L^1(R_+, R_+)$ and $\omega_1, \omega_2 \in C(R^n, R_+)$ such that $LV(x, y, t) \le \beta(t) - \omega_1(x) + \omega_2(y), (x, y, t) \in R^n \times R^n \times R_+$,

$\omega_1(x,t) \geq \omega_2(x,t+\tau), \quad (x,t) \in R^n \times R_+,$

and $\lim\limits_{\|x\| \to \infty} \inf\limits_{0 \leq t \leq \infty} V(x,t) = \infty$, then, for every $\xi \in C_{F_0}^b([-\tau,0], R^n)$, $\lim\limits_{x \to \infty} x(t;\xi) = 0$ a.s.

Lemma 2. ([9]) For any vectors $x, y \in R^n$ and positive definite matrix $G \in R^{n \times n}$, the following matrix inequality holds : $2x^T y \leq x^T Gx + y^T G^{-1} y$

3   Main Results

In this section, hybrid function projective synchronization between networks (1) and (2) is investigated, and the main results are given in the following theorem.

By defining the synchronization error as $e(t) = y(t) - M(t)x(t)$, one can drive the error system as

$$de(t) = [-(D(y(t))\Lambda(y(t)) - M(t)D(x(t))\Lambda(x(t))) + D(y(t))\hat{B}f(y(t))$$
$$-M(t)D(x(t))Bf(x(t)) + D(y(t))\hat{C}g(y(t-\tau)) - M(t)D(x(t))Cg(x(t-\tau)) \quad (4)$$
$$-[D(y(t)) - M(t)D(x(t))]J - \dot{M}(t)x(t) + U]dt + H(t,e(t),e(t-\tau))dW$$

where $\dot{M}(t) = diag(\dot{m}_1(t), \dot{m}_2(t), \ldots, \dot{m}_n(t))$.

The hybrid controller and parameter update law are designed as follows:

$$U = U_1 + U_2 + U_3 + U_4, \quad U_1 = -Ke(t) \quad K = diag(k_1, k_2, \ldots, k_n) \quad \dot{k}_i = \lambda_i e_i^2(t)$$
$$U_2 = D(M(t)x(t))\Lambda(M(t)x(t)) - M(t)D(x(t))\Lambda(x(t))$$
$$+D(M(t)x(t))J - M(t)D(x(t))J \quad (5)$$
$$U_3 = M(t)D(x(t))Bf(x(t)) - D(y(t))Bf(M(t)x(t)) + \dot{M}(t)x(t)$$
$$U_4 = M(t)D(x(t))Cg(x(t-\tau)) - D(y(t))Cg(M(t-\tau)x(t-\tau))$$

$$\Delta\dot{b}_{ij} = -w_{ij}e_i(t)d_i(y_i(t))f_j(y_j(t)), \quad \Delta\dot{c}_{ij} = -q_{ij}e_i(t)d_i(y_i(t))g_j(y_j(t-\tau)) \quad (6)$$

where U_1 is an adaptive controller and U_2, U_3, U_4 are open-loop controllers.

Thus, under the control input (5), the error system (4) turns out to be the following one

$$de(t) = [-D(y(t))\Lambda(y(t)) + D(M(t)x(t))\Lambda(M(t)x(t))$$
$$+D(y(t))Bf(y(t)) - D(y(t))Bf(M(t)x(t)) + D(y(t))Cg(y(t-\tau))$$
$$-D(y(t))Cg(M(t-\tau)y(t-\tau)) + D(y(t))\Delta Bf(y(t)) + D(y(t))\Delta Cg(y(t-\tau)) \quad (7)$$
$$-(D(y(t)) - D(M(t)x(t)))J - Ke(t)]dt + H(t,e(t),e(t-\tau))dW$$

Theorem 1. Under the Assumption A1-A4, the drive neural network (1) and the response neural network (2) can achieve hybrid function projective synchronization under the control scheme (5) and (6).

Proof Consider the following Lyapunov-Krasovskii function candidate:

$$V(e(t)) = \frac{1}{2}e^T(t)e(t) + \int_{t-\tau}^{t} e^T(s)Qe(s)ds + \frac{1}{2}\sum_{i=1}^{n}\sum_{j=1}^{n}(\frac{1}{w_{ij}}\Delta b_{ij}^{\ 2} + \frac{1}{q_{ij}}\Delta c_{ij}^{\ 2})$$

$$+\frac{1}{2}\sum_{i=1}^{n}\frac{1}{\lambda_i}(k_i - k_i^*)^2 \tag{8}$$

where Q is a positive definite matrix.

Computing LV along the trajectory of error system (7), we have

$$LV(e(t)) = e^T(t)[-D(y(t))A(y(t)) + D(M(t)x(t))A(M(t)x(t)) - K^*e(t)$$
$$+D(y(t))Bf(y(t)) - D(y(t))Bf(M(t)x(t)) + D(y(t))Cg(y(t-\tau))$$
$$-D(y(t))Cg(M(t-\tau)y(t-\tau)) - (D(y(t)) - D(M(t)x(t)))J] + e^T(t)Qe(t) \tag{9}$$
$$-e^T(t-\tau)Qe(t-\tau) + \frac{1}{2}trace[H^T(t,e(t),e(t-\tau))H(t,e(t),e(t-\tau))]$$

where $K^* = diag(k_1^*, k_2^*, ..., k_n^*)$.

According to Assumption1, it follows that

$$LV(e(t)) \le e^T(t)[-D(y(t))A(y(t)) + D(M(t)x(t))A(M(t)x(t))$$
$$+\bar{D}B(f(y(t)) - f(M(t)x(t))) + \bar{D}C(g(y(t-\tau))$$
$$-g(M(t-\tau)y(t-\tau))) + \bar{N}Je(t) - K^*e(t)] + e^T(t)Qe(t) \tag{10}$$
$$-e^T(t-\tau)Qe(t-\tau) + \frac{1}{2}trace[H^T(t,e(t),e(t-\tau))H(t,e(t),e(t-\tau))]$$

where $\bar{D} = diag(\bar{d}_1, \bar{d}_2, ..., \bar{d}_n)$, $\bar{N} = diag(N_1, N_2, ..., N_n)$ and $\bar{J} = diag(J_1, J_2, ..., J_n)$.

By employing Assumption A2-A4, we have

$$LV(e(t)) \le -e^T(t)Re(t) + e^T(t)\bar{D}BL^f e(t) + e^T(t)\bar{D}CL^g e(t-\tau)$$
$$+\bar{N}Je(t) - e(t)K^*e(t) + \frac{1}{2}p_1 e^T(t)e(t) + \frac{1}{2}p_2 e^T(t-\tau)e(t-\tau) \tag{11}$$
$$+e^T(t)Qe(t) - e^T(t-\tau)Qe(t-\tau)$$

where $R = diag(r_1, r_2, ..., r_n)$, $L^f = diag(L_1^f, L_2^f ..., L_n^f)$ and $L^g = diag(L_1^g, L_2^g ..., L_n^g)$.

According to lemma 2, it yields

$$LV(e(t)) \le -e^T(t)Re(t) + e^T(t)\bar{D}BL^f e(t) + \frac{1}{2}e^T(t)Ge(t)$$

$$+\frac{1}{2}e^T(t-\tau)(\bar{D}CL^g)^T \bar{D}CL^g e(t-\tau) + \bar{N}Je(t) - e^T(t)K^*e(t)$$

$$+\frac{1}{2}p_1 e^T(t)e(t) + \frac{1}{2}p_2 e^T(t-\tau)e(t-\tau) + e^T(t)Qe(t) - e^T(t-\tau)Qe(t-\tau) \tag{12}$$

$$\le e^T(t)[-R + \bar{D}BL^f + \frac{1}{2}G + \bar{N}J - K^* + Q + \frac{1}{2}p_1 I_n]e(t)$$

$$+e^T(t-\tau)[\frac{1}{2}(\bar{D}CL^g)^T \bar{D}CL^g - Q + \frac{1}{2}p_2 I_n]e(t-\tau)$$

$$\triangleq -e^T(t)W_1 e(t) + e^T(t-\tau)W_2 e(t-\tau)$$

where $W_1 = -[-R + \bar{D}BL^f + \frac{1}{2}G + \overline{NJ} - K^* + Q + \frac{1}{2}p_1 I_n]$, $W_2 = \frac{1}{2}(\bar{D}CL^g)^T \bar{D}CL^g - Q + \frac{1}{2}p_2 I_n$.

Referring to the above calculation, it can be seen that for a sufficiently large positive constant k_i^*, the following inequality holds: $\omega_1(e) > \omega_2(e), \forall e \neq 0$. Moreover, $\lim_{|e| \to \infty} \inf_{0 \le t \le \infty} V(t,e) = \infty$. From lemma 1, we have $\lim_{t \to \infty} e(t) = 0$, a.s.

4   Illustrative Examples

In this section, one illustrative example and its simulation are presented to demonstrate the effectiveness of the obtained theoretical results.

Consider the system (1) and system (2) with the following parameters:

$$B = \begin{bmatrix} 1.5 & -0.5 \\ -1.2 & 3 \end{bmatrix}, \quad C = \begin{bmatrix} -0.8 & 0.5 \\ -2.2 & -1.2 \end{bmatrix}, \quad J = \begin{bmatrix} 0 \\ 0 \end{bmatrix}, \tau = 1, f(x) = g(x) = \begin{bmatrix} \tanh(x_1) \\ \tanh(x_2) \end{bmatrix}$$

$$D(x(t)) = diag(d_1(x_1(t)), d_2(x_2(t))) = diag(6 + \frac{1}{1 + x_1^2(t)}, 3 - \frac{1}{1 + x_2^2(t)}),$$

$$A(x(t)) = \begin{bmatrix} \alpha_1(x_1(t)) \\ \alpha_2(x_2(t)) \end{bmatrix} = \begin{bmatrix} 1.4x_1(t) + \sin(x_1(t)) \\ 1.6x_2(t) - \cos(x_2(t)) \end{bmatrix}.$$

The drive neural network with above coefficients exhibits a chaotic behavior as shown in Fig. 1, with initial values $x(0) = [1, -1]^T$. The noise perturbation is

$$H(t, e(t), e(t-\tau)) = \sqrt{2} \begin{bmatrix} \|e(t)\| & 0 \\ 0 & \|e(t-\tau)\| \end{bmatrix}, \tag{13}$$

Let the initial conditions of the unknown parameters and feedback strength as $\Delta b_{ij}(0) = \Delta c_{ij}(0) = 0.1, \lambda_i = 10, (i, j = 1, 2)$. We set $w_{ij} = q_{ij} = 5$ and without loss of generality we choose the scaling function $m_i(t) = 1.5\sin(0.5\pi t)$.

According to Theorem 1, the response system and the drive system with the hybrid controller U can be synchronized to a scaling function matrix. Fig. 2 depicts the synchronization errors of state variables between drive and response systems for $m_i(t) = 1.5\sin(0.5\pi t)$. Fig.3 shows the adaptive parameters $\hat{b}_{ij}, \hat{c}_{ij} (i, j = 1, 2)$. It is clear that the parameters $\hat{b}_{ij}, \hat{c}_{ij}$ adapt themselves to the true values b_{ij}, c_{ij} respectively. The numerical simulations clearly verify the effectiveness and feasibility of the proposed hybrid control method.

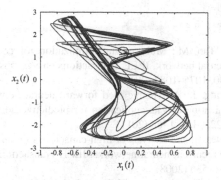

Fig. 1. Chaotic attractor of the drive neural network

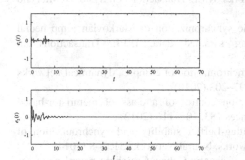

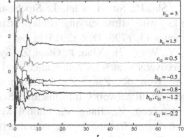

Fig. 2. Time evolution of synchronization errors

Fig. 3. Variation of unknown parameters

5   Conclusions

In this paper, we have dealt with the hybrid function projective synchronization problem for unknown Cohen-Grossberg neural networks with time delays and noise perturbation. Based on the LaSalle-type invariance principle for stochastic differential equations, a hybrid control scheme combining open-loop control and linear feedback control is designed, which can be easily generalized to other simple neural network models. Finally, numerical simulations are provided to show the effectiveness of the main result.

Acknowledgments. This work was supported by the National Natural Science Foundation of China (No. 61374154), National Basic Research Program of China (973 Program) (No. 2013CB430403).

References

1. Hou, Z., Cheng, L., Tan, M.: Multicriteria optimization for coordination of redundant robots using a dual neural network. IEEE Transactions on Systems, Man, and Cybernetics, Part B: Cybernetics 40, 1075–1087 (2010)
2. Han, M., Fan, J., Wang, J.: A dynamic feed forward neural network based on Gaussian particle swarm optimization and its application for predictive control. IEEE Transactions on Neural Networks 22, 1457–1468 (2011)
3. Zeng, Z., Wang, J.: Design and analysis of high-capacity associative memories based on a class of discrete-time recurrent neural networks. IEEE Transactions on Systems, Man, and Cybernetics 38, 1525–1536 (2008)
4. Cohen, M., Grossberg, S.: Absolute stability of global pattern formation and parallel memory storage by competitive neural networks. IEEE Transactions on Systems, Man and Cybernetics 1983, 815–826 (1983)
5. Wu, Z., Shi, P., Su, H., Chu, J.: Stochastic synchronization of Markovian jump neural networks with time-varying delay using sampled data. IEEE Transactions on Cybernetics 43, 1796–1806 (2013)
6. Wang, Y., Wang, H., Xiao, J., Guan, Z.: Synchronization of complex dynamical networks under recoverable attacks. Automatica 46, 197–203 (2010)
7. Wu, A., Wen, S., Zeng, Z.: Synchronization control of a class of memristor-based recurrent neural networks. Information Sciences 183, 106–116 (2012)
8. Chen, J., Zeng, Z., Jiang, P.: Global Mittag-Leffler stability and synchronization of memristor-based fractional-order neural networks. Neural Networks 51, 1–8 (2014)
9. Zhu, Q., Cao, J.: Adaptive synchronization of chaotic Cohen–Crossberg neural networks with mixed time delays. Nonlinear Dynamics 61, 517–534 (2010)
10. Gan, Q.: Adaptive synchronization of Cohen–Grossberg neural networks with unknown parameters and mixed time-varying delays. Communications in Nonlinear Science and Numerical Simulation 17, 3040–3049 (2012)
11. Shi, Y., Zhu, P.: Adaptive synchronization of different Cohen–Grossberg chaotic neural networks with unknown parameters and time-varying delays. Nonlinear Dynamics 73, 1721–1728 (2013)
12. Zhang, R., Yang, Y., Xu, Z., Hu, M.: Function projective synchronization in drive–response dynamical network. Physics Letters A 374, 3025–3028 (2010)
13. Du, H., Shi, P., Lv, N.: Function projective synchronization in complex dynamical networks with time delay via hybrid feedback control. Nonlinear Analysis: Real World Applications 14, 1182–1190 (2013)
14. Liu, S., Zhang, F.: Complex function projective synchronization of complex chaotic system and its applications in secure communication. Nonlinear Dynamics 76, 1087–1097 (2014)
15. Abdurahman, A., Jiang, H., Teng, Z.: Function projective synchronization of impulsive neural networks with mixed time-varying delays. Nonlinear Dynamics 78, 2627–2638 (2014)
16. Cai, G., Ma, H., Gao, X., Wu, X.: Generalized function projective lag synchronization between two different neural networks. In: Guo, C., Hou, Z.-G., Zeng, Z. (eds.) ISNN 2013, Part I. LNCS, vol. 7951, pp. 125–132. Springer, Heidelberg (2013)

17. Wang, Y., Cheng, L., Ren, W., et al.: Seeking consensus in networks of linear agents: communication noises and Markovian switching topologies. IEEE Transactions on Automatic Control 60(5), 1374–1379 (2015)
18. Li, T., Zhang, J.F.: Consensus conditions of multi-agent systems with time-varying topologies and stochastic communication noises. IEEE Transactions on Automatic Control 55(9), 2043–2057 (2010)
19. Sun, Y., Li, W., Ruan, J.: Generalized outer synchronization between complex dynamical networks with time delay and noise perturbation. Communications in Nonlinear Science and Numerical Simulation 18, 989–998 (2013)
20. Mao, X.: LaSalle-type theorems for stochastic differential delay equations. Journal of mathematical analysis and applications 236, 350–369 (1999)

Neural Dynamic Surface Control
for Three-Phase PWM Voltage Source Rectifier*

Liang Diao, Dan Wang**, Zhouhua Peng, and Lei Guo

School of Marine Engineering, Dalian Maritime University, Dalian 116026, PR China
diaoliang678@sina.com, dwangdl@gmail.com

Abstract. In this brief, a neural dynamic surface control algorithm is proposed for three-phase pulse width modulation voltage source rectifier with the parametric variations. Neural networks are employed to approximate the uncertainties, including the parametric variations and the unknown load-resistance. The actual control laws are derived by using the dynamic surface control method. Furthermore, a linear tracking differentiator is introduced to replace the first-order filter to calculate the derivative of the virtual control law. Thus, the peaking phenomenon of the filter is suppressed during the initial phase. The system stability is analyzed by using the Lyapunov theory. Simulation results are provided to validate the efficacy of the proposed controller.

Keywords: PWM rectifier, dynamic surface control, neural network, linear tracking differentiator.

1 Introduction

In recent years, three-phase pulse width modulation (PWM) voltage source rectifier has been widely used in industrial applications such as uninterruptible power supply systems, static synchronous compensator, active power filter, and renewable energies. Their attractive features are low harmonic distortion of the utility currents, bi-directional power flow, and controllable power factor [1].

Various control strategies of three-phase PWM voltage source rectifier have been proposed during the past few years, including the phase and amplitude control [2], direct power control (DPC) [3], model predictive control [4], and voltage-oriented control (VOC) [5] methods. As one of the most popular methods, the VOC can indirectly control the active and reactive powers by controlling the input d-q axis currents. Thus, the dynamic and static performance of power control is affected by the performance of the internal current controller [6]. In

---

* This work was in part supported by the National Nature Science Foundation of China under Grants 61273137, 51209026, 51579023, and in part by the Scientific Research Fund of Liaoning Provincial Education Department under Grant L2013202, and in part by the Fundamental Research Funds for the Central Universities under Grants 3132015021, 3132014321, and in part by the China Postdoctoral Science Foundation under Grant 2015M570247.
** Corresponding author.

X. Hu et al. (Eds.): ISNN 2015, LNCS 9377, pp. 28–35, 2015.
DOI: 10.1007/978-3-319-25393-0_4

the previous study, the proportional integral (PI) controller has been commonly used into the inner current loop since it is convenient to be implemented. However, the main drawback of the PI regulator is that it strongly depends on the operating point and has weak robust to the parametric variations. In [7], a non-linear adaptive backstepping approach is applied to three-phase PWM AC-DC converter, which has achieved a good tracking performance. However, the parametric variations are not considered. In practice, it is difficult to obtain the precise parameters of three-phase PWM voltage source rectifier [8]. Moreover, the parameters are influenced by the environment and operation condition. For example, the resistance and the capacitor may vary with the temperature, and the inductance may vary with the magnetic saturation. On the other hand, a drawback of the backstepping control technique is the problem of "explosion of complexity", which is caused by the repeated differentiations of the virtual control law. In [9], a dynamic surface control (DSC) method is proposed to solve the above algebraic loop problem by introducing a first-order filter in each step of the controller design procedure. Further, neural network (NN) control [10] and fuzzy control [11] are incorporated into the DSC to approximate the system uncertainties. However, the initial value of the virtual control law may not be available in applications. Thus, the derivative of the virtual control law calculated by the first-order filter would inevitably produce a setpoint jump during the initial phase, which is known as the peaking phenomenon. The system stability would be affected if the peaking phenomenon is serious.

In this paper, a neural dynamic surface control (NDSC) strategy is proposed for three-phase PWM voltage source rectifier. In the general synchronously rotating reference frame, the dynamic model of three-phase PWM voltage source rectifier is derived by considering the parametric variations. Neural networks (NNs) are applied to approximate the uncertainties, including the parametric variations and the unknown load-resistance. The actual control laws are derived by using the DSC method. As the peaking phenomenon affects the system stability, the first-order filter is replaced by a linear tracking differentiator (LTD). Thus, the peaking phenomenon of the filter is suppressed during the initial phase. Lyapunov analysis demonstrates that all signals in the closed-loop system are uniformly ultimately bounded, and the tracking errors of the output voltage and the q axis current can converge to a small neighborhood of the origin. Simulation results are provided to validate the efficacy of the proposed controller.

2   Dynamic Model of the Rrectifier Under Study

Fig. 1 represents the topology of three-phase PWM voltage source rectifier. e_a, e_b, e_c are the supply power sources; i_a, i_b, i_c are the input currents; u_a, u_b, u_c are the rectifier input voltages; u_{dc} is the DC-link output voltage; L and R are the line inductance and resistance; C and R_L are the DC-link capacitor and load-resistance. As the d axis is oriented in the direction of the supply voltage vector, the dynamic model of the rectifier under study can be described by

$$\begin{cases} \frac{du_{dc}}{dt} = \frac{3E_m}{2C'u_{dc}}i_d + \delta_{dc} \\ \frac{di_d}{dt} = \omega i_q + \frac{1}{L'}(E_m - u_d) + \delta_d \\ \frac{di_q}{dt} = -\omega i_d - \frac{1}{L'}u_q + \delta_q \end{cases} \tag{1}$$

$$\begin{cases} \delta_{dc} = (-\frac{u_{dc}}{R_L} - \Delta C \frac{du_{dc}}{dt})/C' \\ \delta_d = (\omega \Delta L i_q - R i_d - \Delta L \frac{di_d}{dt})/L' \\ \delta_q = (-\omega \Delta L i_d - R i_q - \Delta L \frac{di_q}{dt})/L' \end{cases} \tag{2}$$

where i_d and i_q are the input currents in d-q axis; u_d and u_q are the rectifier input voltages in d-q axis; ω is the synchronous angular speed; E_m is the amplitude of the supply voltage; δ_{dc}, δ_d, δ_q are the system uncertainties; L' and C' are the measurement values of the line inductance and DC-link capacitor, $L = L' + \Delta L$, $C = C' + \Delta C$.

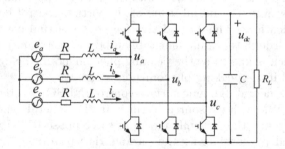

Fig. 1. Three-phase PWM voltage source rectifier.

3   NDSC for PWM Rectifier

3.1   Controller Design

The control objective is to make the output voltage track the reference output voltage. Meanwhile, the q axis input current must be forced to be zero to achieve unity power factor. The design procedure is elaborated in three steps as follows.

Step 1: Define the first surface error as

$$S_1 = u_{dc} - u_{dcr}, \tag{3}$$

where u_{dcr} is the reference output voltage.

The time derivative of S_1 is given by

$$\dot{S}_1 = \frac{3E_m}{2C'u_{dc}}i_d + \delta_{dc}. \tag{4}$$

According to the Stone Weierstrass approximation theorem, there exists an ideal weight W_1 such that δ_{dc} can be represented by an NN as

$$\delta_{dc} = W_1^T h(\bar{u}_{dc}) + \varepsilon_1, \tag{5}$$

where $h(\cdot)$ is a known activation function; $\bar{u}_{dc}$ is the normalized value of u_{dc}; ε_1 is the function reconstruction error satisfying $|\varepsilon_1| \leq \varepsilon_1^*$ with ε_1^* being a positive constant; $\|W_1\| \leq W_1^*$ with W_1^* being a positive constant.

Choose a virtual control law i_d^* as follows

$$i_d^* = \frac{2C' u_{dc}}{3E_m}(-k_1 S_1 - \hat{W}_1^T h(\bar{u}_{dc})), \tag{6}$$

where k_1 is a positive constant to be designed; $\hat{W}_1$ is the estimation of W_1, the update law for $\hat{W}_1$ is given by

$$\dot{\hat{W}}_1 = r_1(h(\bar{u}_{dc})S_1 - \beta_1 \hat{W}_1), \tag{7}$$

where r_1, β_1 are two positive constants to be designed.

Step 2: Define the second surface error as

$$S_2 = i_d - i_d^*. \tag{8}$$

The time derivative of S_2 is given by

$$\dot{S}_2 = \omega i_q + \frac{1}{L'}(E_m - u_d) + \delta_d - \dot{i}_d^*. \tag{9}$$

The derivative of the virtual control law i_d^* can be obtained by using the following LTD.

$$\begin{cases} \dot{x}_1 = x_2, \\ \dot{x}_2 = -\tau_1(x_1 - i_d^*) - \tau_2 x_2, \end{cases} \tag{10}$$

where x_1, x_2 are the estimations of i_d^* and $\dot{i}_d^*$; τ_1, τ_2 are two positive constants to be designed.

There exists an ideal weight W_2 such that δ_d can be represented by an NN as

$$\delta_d = W_2^T h(\bar{i}_d, \bar{i}_q) + \varepsilon_2, \tag{11}$$

where $\bar{i}_d$, $\bar{i}_q$ are the normalized values of i_d and i_q; $|\varepsilon_2| \leq \varepsilon_2^*$ with ε_2^* being a positive constant; $\|W_2\| \leq W_2^*$ with W_2^* being a positive constant.

Select the actual control law u_d^* as

$$u_d^* = E_m + L'(k_2 S_2 + \omega i_q + \hat{W}_2^T h(\bar{i}_d, \bar{i}_q) - x_2), \tag{12}$$

where k_2 is a positive constant to be designed; $\hat{W}_2$ is the estimation of W_2, the update law for $\hat{W}_2$ is given by

$$\dot{\hat{W}}_2 = r_2(h(\bar{i}_d, \bar{i}_q)S_2 - \beta_2 \hat{W}_2), \tag{13}$$

where r_2, β_2 are two positive constants to be designed.

Step 3: Define the third surface error as

$$S_3 = i_q. \tag{14}$$

The time derivative of S_3 is given by

$$\dot{S}_3 = -\omega i_d - \frac{1}{L'}u_q + \delta_q. \tag{15}$$

There exists an ideal weight W_3 such that δ_q can be represented by an NN as

$$\delta_q = W_3^T h(\bar{i}_d, \bar{i}_q) + \varepsilon_3, \tag{16}$$

where $|\varepsilon_3| \leq \varepsilon_3^*$ with ε_3^* being a positive constant; $\|W_3\| \leq W_3^*$ with W_3^* being a positive constant.

Select the actual control law u_q^* as

$$u_q^* = L'(k_3 S_3 - \omega i_d + \hat{W}_3^T h(\bar{i}_d, \bar{i}_q)), \tag{17}$$

where k_3 is a positive constant to be designed; $\hat{W}_3$ is the estimation of W_3, the update law for $\hat{W}_3$ is given by

$$\dot{\hat{W}}_3 = r_3(h(\bar{i}_d, \bar{i}_q)S_3 - \beta_3 \hat{W}_3), \tag{18}$$

where r_3, β_3 are two positive constants to be designed.

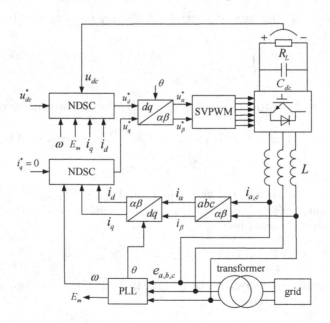

Fig. 2. The block diagram of the proposed control system.

Table 1. Parameter values

Supply's voltage (phase to phase) and frequency	200V(rms), 50Hz
Line's inductance and resistance	2mH, 0.1Ω
DC-link capacitor	5000μF

Table 2. Control design constants

k_1	100	k_2	$1 * 10^3$	k_3	$1 * 10^3$
r_1	$1.1 * 10^4$	r_2	$3 * 10^4$	r_3	$3 * 10^4$
β_1	$9 * 10^{-6}$	β_2	$3.3 * 10^{-7}$	β_3	$3.3 * 10^{-7}$
τ_1	400	τ_2	30		
L'	$2 * 10^{-3}$	C'	$5 * 10^{-3}$		

3.2 Stability Analysis

Theorem 1: Consider the closed-loop system consisting of the system (1), (2) with the virtual control law (6), the actual control laws (12), (17), the NNs (7), (13), (18), and the LTD (10). All signals in the closed-loop system are uniformly ultimately bounded, and the tracking errors of the output voltage and the q axis current can converge to a small neighborhood of the origin.

Proof. Omitted here due to the limited space.

4 Simulation Results and Analysis

In this section, the computer simulations are conducted in order to validate the feasibility and effectiveness of the proposed method. The block diagram of the proposed control system is shown in Fig. 2. A phase locked loop (PLL) is applied to obtain E_m, ω and θ (the angle of the supply voltage vector). The system parameters and design constants are given in Tables 1 and 2, respectively.

The reference output voltage is 330V. At the beginning, the system operates with no-load. The initial value of the output voltage is 283V, which is the three-phase uncontrolled rectifier voltage value. A 150Ω resistance is connected to the DC-link at 1s. To evaluate the robustness of the proposed control strategy, the system parameters are adjusted by increasing R and C to 0.3Ω and 5100μF, and decreasing L to 0.1mH at 2s. The proposed strategy is compared with the PI regulator strategy.

Fig. 3(a) illustrates the the output voltage responses. Compared with the classical PI regulator strategy, the proposed strategy has faster response and less overshoot. As the load variation, the drop voltage and recovery time obtained by the NDSC are smaller and faster. Moreover, the proposed controller has good robustness to the parametric variations. The fluctuations caused by the parametric variations remain almost negligible. Fig. 3(b) shows the supply voltage and the input current of phase-A. The supply voltage and the input current are in phase, which means that the proposed system operates at unity power factor. The estimated values of δ_{dc}, δ_d, δ_q are shown in Fig. 3(c)-(e). The NNs

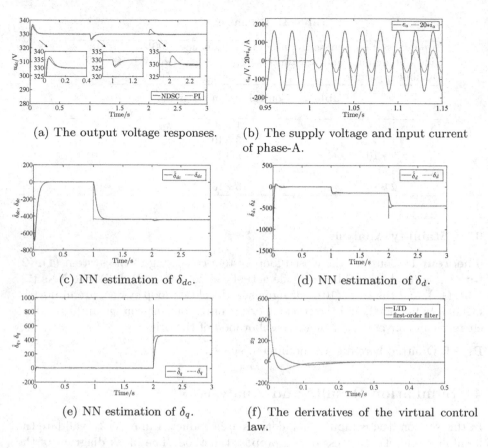

(a) The output voltage responses.

(b) The supply voltage and input current of phase-A.

(c) NN estimation of δ_{dc}.

(d) NN estimation of δ_d.

(e) NN estimation of δ_q.

(f) The derivatives of the virtual control law.

Fig. 3. The simulation results.

can approximate the uncertainties quickly and accurately. Fig. 3(f) shows the derivatives of the virtual control law obtained by the LTD and the first-order filter. It can be noticed that the peaking phenomenon of the filter is suppressed by using the LTD, which can ensure the system stability during the initial phase.

5   Conclusions

In this paper, a NDSC algorithm is proposed for three-phase PWM voltage source rectifier with the parametric variations. The simulation results validate the better performances of the NDSC controller compared with the classical PI regulator. In the future, the challenge will be to reduce the overshoot of the output voltage during the initial phase.

References

1. Choi, D.K., Lee, K.B.: Dynamic performance improvement of AC/DC converter using model preictive direct power control with finite control set. IEEE Transactions on Industrial Electronics 62(2), 757–767 (2015)
2. Wu, R., Dewan, S.B., Slemon, G.R.: Analysis of an ac to dc voltage source converter using PWM with phase and amplitude control. IEEE Transactions on Industry Applications 27(2), 355–364 (1991)
3. Sato, A., Noguchi, T.: Voltage-source PWM rectifier-inverter based on direct power control and its operation characteristic. IEEE Transactions on Power Electronics 26(5), 1559–1567 (2011)
4. Xia, C.L., Wang, M., Song, Z.F., Liu, T.: Robust model predictive current control of three-phase voltage source PWM rectifier with online disturbance observation. IEEE Transactions on Industrial Informatics 8(3), 459–471 (2012)
5. Liang, J.Q., Qiao, W., Hareley, R.G.: Feed-forward transient current control for low-voltage ride-through enhancement of DFIG wind turbines. IEEE Transactions on Energy Conversion 25(3), 836–843 (2010)
6. Yin, Z.G., Liu, J., Zhong, Y.R.: Study and control of three-phase PWM rectifier based on dual single-input single-output model. IEEE Transactions on Industrial Informatics 9(2), 1064–1073 (2013)
7. Allag, A., Hammoudi, M.Y., Mimoune, S.M., Ayad, M.Y.: Adaptive backstepping voltage controller design for an PWM AC-DC converter. International Journal of Electrical and Power Engineering 1(1), 62–69 (2007)
8. Wang, G.D., Wai, R.J., Liao, Y.: Design of backstepping power control for grid-side converter of voltage source converter-based high-voltage dc wind power generation system. IET Renewable Power Generation 7(2), 118–133 (2013)
9. Swaroop, D., Gerdes, J.C., Yip, P.P., Hedrick, J.K.: Dynamic surface control of nonlinear systems. In: Proceedings of the 1997 American Control Conference, Albuquerque, pp. 3028–3034 (1997)
10. Wang, D., Huang, J.: Neural network-based adaptive dynamic surface control for a class of uncertain nonlinear systems in strict-feedback form. IEEE Transactions on Neural Networks 16(1), 195–202 (2005)
11. Han, S.I., Lee, J.M.: Precise positioning of nonsmooth dynamic systems using fuzzy wavelet echo state network and dynamic surface sliding mode control. IEEE Transactions on Industrial Electronics 60(11), 5124–5136 (2013)
12. Guo, B.Z., Han, J.Q., Xi, F.B.: Linear tracking-differentiator and application to online estimation of the frequency of a sinusoidal signal with random noise perturbation. International Journal of Systems Science 33(5), 351–358 (2002)

A Terminal-Sliding-Mode-Based Frequency Regulation

Hong Liu and Dianwei Qian

School of Control and Computer Engineering,
North China Electric Power University, Changping District, Beijing, China
dianwei.qian@ncepu.edu.cn

Abstract. In this paper, a terminal reaching law based sliding mode control (SMC) method for load frequency control (LFC) is investigated in interconnected power systems in the presence of wind turbines and generation rate constraint (GRC). Neural networks are adopted to compensate the entire uncertainties. Simulation results show the validity and robustness of the presented method.

Keywords: load frequency control, terminal reaching law, neural network.

1 Introduction

Load frequency control (LFC) is the most effective way to guarantee the stable operation of power systems [1]. Sliding mode control (SMC) invented by V.I. Utkin [2] is considered as a highly efficient design tool for the LFC problem. For the applications in [3, 4], only conventional power sources have been taken into account. Nowadays, wind power is a power generation way with most promise. However, the randomness and the intermittency of wind plant pose a critical challenge to real-time stability and balancing of power systems [5]. Therefore, the increasing penetration of wind generation in power systems calls for more and more attentions to the LFC problem for the power systems with nonconventional generation systems [6].

Power systems are complicated nonlinear systems [7]. Neural networks (NNs) are able to suppress the nonlinearities and uncertainties with the function approximation [8]. It has been proved that the governor dead band (GDB) nonlinearity can be approximated and compensated by NNs [4]. Further research is needed in processing the generation rate constraint (GRC) nonlinearity by NNs for the LFC problem.

2 System Configuration

2.1 Load Frequency Control System

Fig. 1 is the block diagram of the ith control area in a multi-area power system. Variables $\Delta P_{gi}(t)$, $\Delta X_{gi}(t)$, $\Delta f_i(t)$ and $\Delta P_{tie,i}(t)$ are the incremental changes of generator output, governor valve position, frequency and tie-line active power. ACE_i is area control error. $\Delta P_{Li}(t)$ is load disturbance, $\Delta P_{ci}(t)$ is control input. $\Delta E_i(t)$ is the integral of $ACE_i(t)$ and K_{Ei} is the integral gain. T_{gi}, T_{ti} and T_{pi} are the time constants of governor, turbine and electric system governor. $B_i = 1/R_i + 1/K_{pi}$ is the frequency bias factor where

© Springer International Publishing Switzerland 2015
X. Hu et al. (Eds.): ISNN 2015, LNCS 9377, pp. 36–42, 2015.
DOI: 10.1007/978-3-319-25393-0_5

R_i is adjustment deviation coefficient and K_{pi} is electric system gain. T_{ij} is the synchronizing power coefficient between area i and area j, $i=1,\ldots, N$ and N is the number of areas.

Define $x_i(t)=[\Delta X_{gi}(t), \Delta P_{gi}(t), \Delta f_i(t), \Delta P_{tie,i}(t), \Delta E_i(t)]^T$. $u_i(t)=\Delta P_{ci}(t)$ is control input, $\Delta P_{di}(t)=[\Delta P_{Li}(t), \sum_{\substack{j=1 \\ j\neq i}}^{N} T_{ij}\Delta f_j(t)]^T$ is disturbance vector. Then, the system model (1) can be deduced and employed for the LFC design of the ith control area. And for a nominal system, the detailed expressions of the matrixes A_i, B_i and F_i can be obtained from [9].

$$\dot{x}_i(t) = A_i x_i(t) + B_i u_i(t) + F_i \Delta P_{di}(t) \tag{1}$$

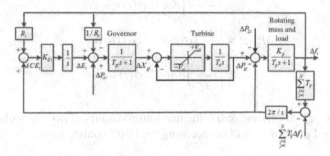

Fig. 1. Dynamic model of the ith control area with GRC

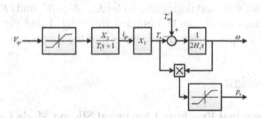

Fig. 2. Structure of a DFIG based wind turbine

2.2 Simplified Wind Turbine Model

Fig. 2 is the simplified model of the doubly-fed induction generator (DFIG) [9]. $V_{qr}(t)$ and $i_{qr}(t)$ are the q-axis components of the rotor voltage and the rotor current. $w(t)$ is the rotational speed, $T_m(t)$ is the mechanical power, H_t is the equivalent inertia constant, $P_e(t)$ is the active power. $X_2=1/R_r$, $X_3=L_m/L_{ss}$, $T_1=L_0/(w_sR_s)$, $L_0=L_{rr}+L_m^2/L_{ss}$, and $L_{ss}=L_s+L_m$, $L_{rr}=L_r+L_m$, here L_m is the magnetizing inductance, R_r and R_s are the rotor and stator resistances, L_r and L_s are the rotor and stator leakage inductances, L_{rr} and L_{ss} are the rotor and stator self-inductances, w_s is the synchronous speed.

Define $x_{wi}(t)=[\Delta X_{gi}(t), \Delta P_{gi}(t), \Delta f_i(t), \Delta P_{tie,i}(t), \Delta i_{qr,i}(t), \Delta w_i(t), \Delta E_i(t)]^T$, $u_i(t) = [\Delta P_{ci}(t), \Delta V_{qr,i}(t)]^T$ and $\Delta P_{di}(t)=[\Delta P_{Li}(t), \sum_{\substack{j=1 \\ j\neq i}}^{N} T_{ij}\Delta f_j(t), \Delta T_{mi}(t)]^T$. Then, the following state space equation can be deduced. (2) will be employed for the LFC design for wind turbines. The details about A_{wi}, B_{wi} and F_{wi} are available in [9].

$$\dot{x}_{wi}(t) = A_{wi}x_{wi}(t) + B_{wi}u_{wi}(t) + F_{wi}\Delta P_{wdi}(t) \tag{2}$$

2.3   Analysis about Nonlinearities

The GRC nonlinearity limits the rate of the generating power change. And (3) formulates the constraint relationship.

$$\Delta P_{gi}(t) = \begin{cases} -\dfrac{1}{T_{gi}}\int\delta dt & \sigma(t) < -\delta \\ \dfrac{1}{T_{gi}}\int\sigma(t)dt & |\sigma(t)| < \delta \\ \dfrac{1}{T_{gi}}\int\delta dt & \sigma(t) > \delta \end{cases} \tag{3}$$

Here $\sigma(t) = \Delta X_{gi}(t) - \Delta P_{gi}(t)$ and $\delta > 0$ is the maximum output of GRC. Models (1) and (2) can be described uniformly as (4) concerning the GRC nonlinearity.

$$\dot{x}(t) = (A' + \Delta A)x(t) + (B' + \Delta B)u(t) + (F' + \Delta F)\Delta P(t) + \phi(t) \tag{4}$$

In (4), $\phi(t)$ indicates the uncertainties due to GRC. A', B' and F' stand for the nominal constant matrices which can be obtained from Fig. 1 and Fig. 2. $\Delta Ax(t)$, $\Delta Bu(t)$ and $\Delta F\Delta P(t)$ denote the parameter uncertainties and the modelling errors.

3   Control Design

3.1   Design of Terminal Reaching Law based Sliding Mode Control

Assumption 1: $\|d(t)\| \le \bar{d}_0$, here $\|\cdot\|$ denotes Euclidean norm, $\bar{d}_0$ is constant but unknown and $d(t)$ satisfies

$$d(t) = F'\Delta P(t) + \Delta Ax(t) + \Delta Bu(t) + \Delta F\Delta P(t) + \phi(t) \tag{5}$$

The switching surface $s(t)$ is designed to satisfy $s(t) = Cx(t)$ where C is the switching gain matrix. In order to make the system reach the sliding surface in a finite time, the terminal reaching law is adopted as

$$\dot{s}(t) = -\alpha s(t) - \beta \| s(t) \|^{q/p-1} s(t) \tag{6}$$

In (6), α and β are both positive constants, p and q ($q < p < 2q$) are positive odd numbers.

When the system trajectories reach the predefined sliding surface $s(t)$ and keep a sliding motion thereafter, there is $s(t) = \dot{s}(t) = 0$. Then the control law based on the terminal reaching law can be deduced as

$$u(t) = -(CB)^{-1}[\alpha s(t) + \beta \parallel s(t) \parallel^{q/p-1} s(t) + CAx(t) + \parallel C \parallel \overline{d}_0 \, \text{sgn}(s(t))] \qquad (7)$$

3.2 Design of Radial Basis Function (RBF) Neural Networks

RBF neural networks can be employed to adaptively learn the upper bound of system uncertainties according to the characteristics of NNs.

For the LFC problem, the system state vector $x(t)$ is picked up as the network input. The boundary value of the system uncertainties is the network output y calculated by

$$y = \hat{\overline{d}}_0(x, \omega) = \hat{\omega}^T h(x) \qquad (8)$$

where $\hat{\omega}$ is the connection weight and $h(x)$ is the Gaussian function vector defined by

$$h_p(x) = \exp\left(-\frac{\parallel x - c_p \parallel^2}{2b_p^2}\right) \quad p = 1, 2, 3..., l \qquad (9)$$

In (9), c_p is the center vector of the pst neuron and b_p is the width of the pst neuron. Adopting the RBF approximation technology, the control law (7) becomes

$$u(t) = -(CB)^{-1}[\alpha s(t) + \beta \parallel s(t) \parallel^{q/p-1} s(t) + CAx(t) + \parallel C \parallel \hat{\overline{d}}_0 \, \text{sgn}(s(t))] \qquad (10)$$

3.3 Stability Analysis

Assumption 2: $\mid \omega^{*T}h(x) - \overline{d}_0 \mid = \varepsilon(x) < \varepsilon_1$, ω^* is the optimal weight vector of NNs.

Assumption 3: $\overline{d}_0 - \parallel d \parallel > \varepsilon_0 > \varepsilon_1$

Adopt adaptive algorithm to adjust weights online with Assumptions 1, 2 and 3 holding true. Adopt the update law of the network weight vector satisfying

$$\dot{\hat{\omega}} = \xi \parallel s \parallel_1 \cdot \parallel C \parallel h(x) \qquad (11)$$

where $\xi = \parallel C \parallel (\varepsilon_0 - \varepsilon_1) > 0$ is constant.

Define Lyapunov candidate function as the following equation

$$V = \frac{1}{2}s^T s + \frac{1}{2}\xi^{-1}\tilde{\omega}^T \tilde{\omega} \qquad (12)$$

where $\tilde{\omega} = \omega^* - \hat{\omega}$. The inequality $\mid \varepsilon(x) \mid - (\overline{d}_0 - \parallel d \parallel) < \varepsilon_1 - \varepsilon_0$ exists according to Assumptions 2 and 3. Differentiate V with respect to time t and the derivative of V can be formulated by

$$\dot{V} = s^T \dot{s} - \xi^{-1} \tilde{\omega}^T \dot{\hat{\omega}}$$
$$= s^T(-\alpha s - \beta \parallel s \parallel^{q/p-1} s - \parallel C \parallel \hat{\bar{d}}_0 \operatorname{sgn}(s) + Cd) - \xi^{-1} \tilde{\omega}^T \dot{\hat{\omega}}$$
$$\leq (-\alpha \parallel s \parallel^2 - \beta \parallel s \parallel^{q/p+1}) - \parallel s \parallel_1 \cdot \parallel C \parallel \varepsilon(x) - \parallel s \parallel \cdot \parallel C \parallel (\bar{d}_\theta - \parallel d \parallel) \qquad (13)$$
$$< (-\alpha \parallel s \parallel^2 - \beta \parallel s \parallel^{q/p+1}) - \parallel s \parallel_1 \cdot \parallel C \parallel (\varepsilon_0 - \varepsilon_1)$$
$$= -\alpha \parallel s \parallel^2 - \beta \parallel s \parallel^{q/p+1} - \xi \parallel s \parallel_1$$

Because $\alpha > 0$, $\beta > 0$ and $\xi > 0$, $V \geq 0$ and $\dot{V} < 0$ can be seen from (12) and (13). The control law (10) can drive the system to the sliding stage and remain on the sliding surface.

4   Simulation Results

Consider a two-area interconnected power system. Each control area has a 400 MVA-scale conventional generating unit with GRC in Fig. 1 and a 400 MVA-scale wind turbine in Fig. 2. The parameters are given in Tables 1 and 2. In addition, X_m is the magnetizing reactance. The integral gains are selected as $K_{E1} = K_{E2} = 1$.

Considering the conventional power units, $C = [1\ 2\ 3\ 4\ 5]^T$, adopt the RBF NNs with 5-6-1 structures, select the initial weight vectors as $[0.1\ 0.1\ 0.1\ 0.1\ 0.1\ 0.1]^T$ and the widths of the Gaussian function vectors as $[0.2\ 0.2\ 0.2\ 0.2\ 0.2\ 0.2]^T$. For wind turbines, $C = [1.1\ 2.2\ 3.3\ 3.4\ 2.5\ 2.6\ 1.1; 1.1\ 2.2\ 3.3\ 3.7\ 2.1\ 3.3\ 1.1]^T$. The RBF NNs have 7-8-1 structures. The initial weight vectors are $[0.1\ 0.1\ 0.1\ 0.1\ 0.1\ 0.1\ 0.1\ 0.1]^T$ and $b = [0.2\ 0.2\ 0.2\ 0.2\ 0.2\ 0.2\ 0.2\ 0.2]^T$. All the centers of the four networks take random numbers between -1 and 1. Other parameters are $\alpha = 10$, $\beta = 2$; $p = 5$, $q = 3$; $\varepsilon_{0r} = 0.002$, $\varepsilon_{1r} = 0.001$.

In order to verify the effectiveness of the proposed control method, $\Delta P_{L1} = \Delta P_{L2} = 1\%$ pu and $\Delta T_{m1} = \Delta T_{m2} = 1\%$ pu are applied to the system at $t = 5$s simultaneously. It can be seen from Fig. 3 that the Δf and ACE are damped to zero with small oscillations. These results embody the performances against load disturbance and wind power fluctuation of the designed terminal reaching law based SMC scheme. From the comparisons in Fig. 3, the performances of the frequency regulation with RBF NNs (blue curves) is superior to those without RBF NNs (red curves) in term of settling time. From Fig. 4, the control inputs of the load frequency control system with RBF NNs are much smoother while those without RBF NNs chatter severely. It can thus be seen the designed RBF NNs effectively reduce chattering and improve response speed. Moreover, the outputs of the RBF NNs are illustrated in Fig. 5 reflecting their convergences.

Table 1. Parameters and data of two control areas

Area	D (pu/Hz)	$2H$ (pu.s)	R (Hz/pu)	T_g (s)	T_t (s)	T_{ij}	B
Area1	0.015	0.1667	3.00	0.08	0.40	0.2	0.425
Area2	0.016	0.2017	2.73	0.06	0.44	0.2	0.425

Table 2. Wind turbine parameters at 247 MW operating point

R_r(pu)	R_s(pu)	X_{lr}(pu)	X_{ls}(pu)	X_m(pu)	H_t(pu)
0.00552	0.00491	0.1	0.09273	3.9654	4.5

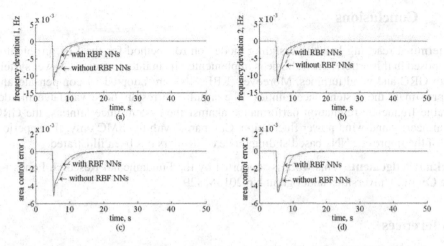

Fig. 3. Simulation results of the considered power system with and without RBF NNs. **a** frequency deviation Δf_1; **b** frequency deviation Δf_2; **c** area control error ACE_1; **d** area control error ACE_2

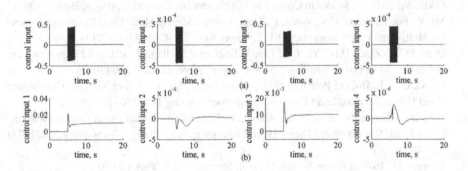

Fig. 4. Simulation results of control inputs. **a** without RBF NNs; **b** with RBF NNs

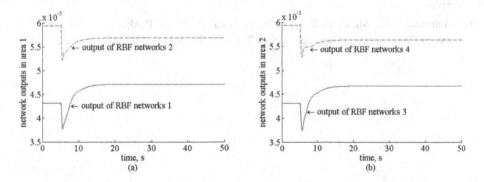

Fig. 5. Simulation results of RBF network outputs. **a** in area1; **b** in area 2

5   Conclusions

A terminal reaching law based sliding mode control method for the LFC problem is proposed in this article. The scheme is implemented in an interconnected power system with GRC and wind turbines. Moreover, RBF NNs are adopted to compensate and approximate the system uncertainties. The simulation results have validated the desirable frequency regulation performance against the system uncertainties, the GRC nonlinearity and wind power fluctuation. Compared with the SMC only, the superiority of the improved NNs-based sliding mode controllers has been illustrated.

Acknowledgements. This work is supported by the Fundamental Research Funds for the Central Universities under grant No.2015MS29.

References

1. Hajian, M., Foroud, A.A., Abdoos, A.A.: New Automated Power Quality Recognition System for Online/offline Monitoring. Neurocomputing 128, 389–406 (2014)
2. Utkin, V.I.: Sliding Modes in Control and Optimization, 2nd edn. Springer, Berlin (1992)
3. Mi, Y., Fu, Y., Wang, C.S., Wang, P.: Decentralized Sliding Mode Load Frequency Control for Multi-Area Power Systems. IEEE T. Power Syst. 28, 4301–4309 (2013)
4. Qian, D.W., Zhao, D.B., Yi, J.Q., Liu, X.J.: Neural Sliding-Mode Load Frequency Controller Design of Power Systems. Neural Comput. Appl. 22, 279–286 (2013)
5. Li, X., Cao, J., Du, D.: Probabilistic Optimal Power Flow for Power Systems Considering Wind Uncertainty and Load Correlation. Neurocomputing 148, 240–247 (2015)
6. Das, D.C., Sinha, N., Roy, A.K.: Automatic Generation Control of an Organic Rankine Cycle Solar-Thermal/Wind-Diesel Hybrid Energy System. Energy Technology 2, 721–731 (2014)
7. Bevrani, H.: Robust Power System Control. Springer, New York (2009)
8. Cheng, L., Hou, Z.G., Tan, M.: Adaptive Neural Network Tracking Control for Manipulators with Uncertain Kinematics, Dynamics and Actuator Model. Automatica 45, 2312–2318 (2009)
9. Mohamed, T.H., Morel, J., Bevrani, H., Hiyama, T.: Model Predictive Based Load Frequency Control Design Concerning Wind Turbines. Int. J. Elec. Power 43, 859–867 (2012)

A New Discrete-Time Iterative Adaptive Dynamic Programming Algorithm Based on Q-Learning*

Qinglai Wei[1] and Derong Liu[2]

[1] The State Key Laboratory of Management and Control for Complex Systems
Institute of Automation, Chinese Academy of Sciences, Beijing 100190, China
qinglai.wei@ia.ac.cn
[2] School of Automation and Electrical Engineering, University of Science and Technology
Beijing, Beijing 100083, China
derong@ustb.edu.cn

Abstract. In this paper, a novel Q-learning based policy iteration adaptive dynamic programming (ADP) algorithm is developed to solve the optimal control problems for discrete-time nonlinear systems. The idea is to use a policy iteration ADP technique to construct the iterative control law which stabilizes the system and simultaneously minimizes the iterative Q function. Convergence property is analyzed to show that the iterative Q function is monotonically non-increasing and converges to the solution of the optimality equation. Finally, simulation results are presented to show the performance of the developed algorithm.

Keywords: Adaptive critic designs, adaptive dynamic programming, approximate dynamic programming, Q-learning, policy iteration, neural networks, nonlinear systems, optimal control.

1 Introduction

Characterized by strong abilities of self-learning and adaptivity, adaptive dynamic programming (ADP), proposed by Werbos [25, 26], has demonstrated powerful capability to find the optimal control policy by solving the Hamilton-Jacobi-Bellman (HJB) equation forward-in-time and becomes an important brain-like intelligent optimal control method for nonlinear systems [4, 6–9, 12, 17, 23]. Policy and value iterations are basic iterative algorithms in ADP. Value iteration algorithm was proposed in [3]. In [2], the convergence of value iteration was proven. Policy iteration algorithms for optimal control of continuous-time (CT) systems were given in [1]. In [5], policy iteration algorithm for discrete-time nonlinear systems was developed. For many traditional iterative ADP algorithms, they require to build the model of nonlinear systems and then perform the ADP algorithms to derive an improved control policy [11, 16, 18–22, 24, 27, 28]. In contrast, Q-learning, proposed by Watkins [14, 15], is a typical data-based ADP algorithm. In [10], Q-learning was named action-dependent heuristic dynamic programming (AD-HDP). For Q-learning algorithms, Q functions are used instead of value functions in

---

* This work was supported in part by the National Natural Science Foundation of China under Grants 61273140, 61304086, 61374105, and 61233001, and in part by Beijing Natural Science Foundation under Grant 4132078.

© Springer International Publishing Switzerland 2015
X. Hu et al. (Eds.): ISNN 2015, LNCS 9377, pp. 43–52, 2015.
DOI: 10.1007/978-3-319-25393-0_6

the traditional iterative ADP algorithms. Q functions depend on both system state and control, which means that they already include the information about the system and the utility function. Hence, it is easier to compute control policies from Q functions than the traditional performance index functions. Because of this merit, Q-learning algorithms are preferred to unknown and model-free systems to obtain the optimal control.

In this paper, inspired by [5], a novel Q-learning based policy iteration ADP algorithm is developed for discrete-time nonlinear systems. First, the procedure of the Q-learning based policy iteration ADP algorithm is described. Next, property analysis of the Q-learning based policy iteration ADP algorithm is established. It is proven that the iterative Q functions will monotonically non-increasing and converges to the optimal solution of the HJB equation. Finally, simulation results will illustrate the effectiveness of the developed algorithm.

The rest of this paper is organized as follows. In Section 2, the problem formulation is presented. In Section 3, the properties of the developed Q-learning based policy iteration ADP algorithm will be proven in this section. In Section 4, numerical results are presented to demonstrate the effectiveness of the developed algorithm. Finally, in Section 5, the conclusion is drawn.

2   Problem Formulation

In this paper, we will study the following discrete-time nonlinear system

$$x_{k+1} = F(x_k, u_k), \ k = 0, 1, 2, \ldots, \tag{1}$$

where $x_k \in \mathbb{R}^n$ is the state vector and $u_k \in \mathbb{R}^m$ is the control vector. Let x_0 be the initial state and $F(x_k, u_k)$ be the system function. Let $\underline{u}_k = \{u_k, u_{k+1}, \ldots\}$ be an arbitrary sequence of controls from k to ∞. The performance index function for state x_0 under the control sequence $\underline{u}_0 = \{u_0, u_1, \ldots\}$ is defined as

$$J(x_0, \underline{u}_0) = \sum_{k=0}^{\infty} U(x_k, u_k), \tag{2}$$

where $U(x_k, u_k) > 0$, for $x_k, u_k \neq 0$, is the utility function. The goal of this paper is to find an optimal control scheme which stabilizes the system (1) and simultaneously minimizes the performance index function (2). For convenience of analysis, results of this paper are based on the following assumptions.

Assumption 1. *System (1) is controllable and the function $F(x_k, u_k)$ is Lipschitz continuous for x_k, u_k.*

Assumption 2. *The system state $x_k = 0$ is an equilibrium state of system (1) under the control $u_k = 0$, i.e., $F(0, 0) = 0$.*

Assumption 3. *The feedback control $u_k = u(x_k)$ satisfies $u_k = u(x_k) = 0$ for $x_k = 0$.*

Assumption 4. *The utility function $U(x_k, u_k)$ is a continuous positive definite function of x_k and u_k.*

Define the control sequence set as $\underline{\mathfrak{U}}_k = \{\underline{u}_k \colon \underline{u}_k = (u_k, u_{k+1}, \dots), \forall u_{k+i} \in \mathbb{R}^m, i = 0, 1, \dots\}$. Then, for a control sequence $\underline{u}_k \in \underline{\mathfrak{U}}_k$, the optimal performance index function is defined as

$$J^*(x_k) = \min_{\underline{u}_k} \left\{ J(x_k, \underline{u}_k) \colon \underline{u}_k \in \underline{\mathfrak{U}}_k \right\}. \tag{3}$$

According to [14] and [15], the optimal Q function satisfies the Q-Bellman equation

$$Q^*(x_k, u_k) = U(x_k, u_k) + \min_{u_{k+1}} Q^*(x_{k+1}, u_{k+1}). \tag{4}$$

The optimal performance index function satisfies

$$J^*(x_k) = \min_{u_k} Q^*(x_k, u_k). \tag{5}$$

The optimal control law $u^*(x_k)$ can be expressed as

$$u^*(x_k) = \arg\min_{u_k} Q^*(x_k, u_k). \tag{6}$$

From (5), we know that if we obtain the optimal Q function $Q^*(x_k, u_k)$, then the optimal control law $u^*(x_k)$ and the optimal performance index function $J^*(x_k)$ can be obtained. However, the optimal Q function $Q^*(x_k, u_k)$ is generally an unknown and non-analytic function, which cannot be obtained directly by (4). Hence, a discrete-time Q learning algorithm is developed in [15] to solve for the Q function iteratively.

3   Discrete-Time Policy Iteration ADP Algorithm Based on Q-Learning

In this section, the Q-learning based policy iteration ADP algorithm will be developed to obtain the optimal controller for discrete-time nonlinear systems. Convergence and optimality proofs will also be given to show that the iterative Q function will converge to the optimum.

3.1   Derivation of the Discrete-Time Policy Iteration ADP Algorithm Based on Q-Learning

In the developed policy iteration algorithm, the Q function and control law are updated by iterations, with the iteration index i increasing from 0 to infinity. Let $v_0(x_k)$ be an arbitrary admissible control law [5]. For $i = 0$, let $Q_0(x_k, u_k)$ be the initial iterative Q function constructed by $v_0(x_k)$, i.e.,

$$Q_0(x_k, v_0(x_k)) = \sum_{j=0}^{\infty} U(x_{k+j}, v_0(x_{k+j})). \tag{7}$$

Thus, initial iterative Q function satisfies the following generalized Q-Bellman equation

$$Q_0(x_k, u_k) = U(x_k, u_k) + Q_0(x_{k+1}, v_0(x_{k+1})). \tag{8}$$

Then, the iterative control law is computed by

$$v_1(x_k) = \arg\min_{u_k} Q_0(x_k, u_k). \tag{9}$$

For $i = 1, 2, \ldots$, let $Q_i(x_k, u_k)$ be the iterative Q function constructed by $v_i(x_k)$, which satisfies the following generalized Q-Bellman equation

$$Q_i(x_k, u_k) = U(x_k, u_k) + Q_i(x_{k+1}, v_i(x_{k+1})), \tag{10}$$

and the iterative control law is updated by

$$v_{i+1}(x_k) = \arg\min_{u_k} Q_i(x_k, u_k). \tag{11}$$

3.2 Properties of the Policy Iteration Based Deterministic Q-Learning Algorithm

For the policy iteration algorithm of discrete-time nonlinear systems [5], it shows that the iterative value function is monotonically non-increasing and converges to the optimum. In this subsection, inspired by [5], we will show that the iterative Q function will also be monotonically non-increasing and converges to its optimum.

Theorem 1. *For $i = 0, 1, \ldots$, let $Q_i(x_k, u_k)$ and $v_i(x_k)$ be obtained by (8)–(11). If Assumptions 1–4 hold, then the iterative Q function $Q_i(x_k, u_k)$ is monotonically non-increasing and converges to the optimal Q function $Q^*(x_k, u_k)$, as $i \to \infty$, i.e.,*

$$\lim_{i\to\infty} Q_i(x_k, u_k) = Q^*(x_k, u_k), \tag{12}$$

which satisfies the optimal Q-Bellman equation (4).

Proof. The statement can be proven in two steps.
1) Show that the iterative Q function $Q_i(x_k, u_k)$ is monotonically non-increasing as i increases, i.e.,

$$Q_{i+1}(x_k, u_k) \le Q_i(x_k, u_k). \tag{13}$$

According to (11), we have

$$Q_i(x_k, v_{i+1}(x_k)) = \min_{u_k} Q_i(x_k, u_k) \le Q_i(x_k, v_i(x_k)). \tag{14}$$

For $i = 0, 1, \ldots$, define a new iterative Q function $\mathcal{Q}_{i+1}(x_k, u_k)$ as

$$\mathcal{Q}_{i+1}(x_k, u_k) = U(x_k, u_k) + Q_i(x_{k+1}, v_{i+1}(x_{k+1})), \tag{15}$$

where $v_{i+1}(x_{k+1})$ is obtained by (11). According to (14), we can obtain

$$\begin{aligned}
\mathcal{Q}_{i+1}(x_k, u_k) &= U(x_k, u_k) + Q_i(x_{k+1}, v_{i+1}(x_{k+1})) \\
&= U(x_k, u_k) + \min_{u_{k+1}} Q_i(x_{k+1}, u_{k+1}) \\
&\le U(x_k, u_k) + Q_i(x_{k+1}, v_i(x_{k+1})) \\
&= Q_i(x_k, u_k).
\end{aligned} \tag{16}$$

Now we prove inequality (13) by mathematical induction. For $i = 0, 1, \ldots$, as

$$
Q_i(x_{k+1}, v_i(x_{k+1})) - Q_i(x_k, v_i(x_k))
$$
$$
= -U(x_k, v_i(x_k))
$$
$$
< 0, \tag{17}
$$

we have $v_i(x_k)$ is a stable control. Thus, we have $x_{\mathcal{N}} = 0$ for $\mathcal{N} \to \infty$. According to Assumptions 1–4, we have $v_{i+1}(x_{\mathcal{N}}) = v_i(x_{\mathcal{N}}) = 0$, which obtains

$$
Q_{i+1}(x_{\mathcal{N}}, v_{i+1}(x_{\mathcal{N}})) = Q_{i+1}(x_{\mathcal{N}}, v_{i+1}(x_{\mathcal{N}})) = Q_i(x_{\mathcal{N}}, v_i(x_{\mathcal{N}})) = 0, \tag{18}
$$

and

$$
Q_{i+1}(x_{\mathcal{N}-1}, u_{\mathcal{N}-1}) = Q_{i+1}(x_{\mathcal{N}-1}, u_{\mathcal{N}-1}) = Q_i(x_{\mathcal{N}-1}, u_{\mathcal{N}-1}) = U(x_{\mathcal{N}-1}, u_{\mathcal{N}-1}). \tag{19}
$$

Let $k = \mathcal{N} - 2$. According to (11),

$$
\begin{aligned}
Q_{i+1}(x_{\mathcal{N}-2}, u_{\mathcal{N}-2}) &= U(x_{\mathcal{N}-2}, u_{\mathcal{N}-2}) + Q_{i+1}(x_{\mathcal{N}-1}, v_{i+1}(x_{\mathcal{N}-1})) \\
&= U(x_{\mathcal{N}-2}, u_{\mathcal{N}-2}) + Q_i(x_{\mathcal{N}-1}, v_{i+1}(x_{\mathcal{N}-1})) \\
&= Q_{i+1}(x_{\mathcal{N}-2}, u_{\mathcal{N}-2}) \\
&\leq Q_i(x_{\mathcal{N}-2}, u_{\mathcal{N}-2}).
\end{aligned} \tag{20}
$$

So, the conclusion holds for $k = \mathcal{N} - 2$. Assume that the conclusion holds for $k = \ell + 1$, $\ell = 0, 1, \ldots$. For $k = \ell$ we can get

$$
\begin{aligned}
Q_{i+1}(x_\ell, u_\ell) &= U(x_\ell, u_\ell) + Q_{i+1}(x_{\ell+1}, v_{i+1}(x_{\ell+1})) \\
&\leq U(x_\ell, u_\ell) + Q_i(x_{\ell+1}, v_{i+1}(x_{\ell+1})) \\
&= Q_{i+1}(x_\ell, u_\ell) \\
&\leq Q_i(x_\ell, u_\ell).
\end{aligned} \tag{21}
$$

Hence, we can obtain that for $i = 0, 1, \ldots$, the inequality (13) holds, for x_k, u_k. The proof of mathematical induction is completed.

As $Q_i(x_k, u_k)$ is a non-increasing and lower bounded sequence, i.e., $Q_i(x_k, u_k) \geq 0$, the limit of the iterative Q function $Q_i(x_k, u_k)$ exists as $i \to \infty$, i.e.,

$$
Q_\infty(x_k, u_k) = \lim_{i \to \infty} Q_i(x_k, u_k). \tag{22}
$$

2) *Show that the limit of the iterative Q function $Q_i(x_k, u_k)$ satisfies the optimal Q-Bellman equation, as $i \to \infty$.*

According to (21), we can obtain

$$
\begin{aligned}
Q_\infty(x_k, u_k) &= \lim_{i \to \infty} Q_{i+1}(x_k, u_k) \leq Q_{i+1}(x_k, u_k) \leq Q_{i+1}(x_k) \\
&= U(x_k, u_k) + Q_i(x_{k+1}, v_{i+1}(x_{k+1})) \\
&= U(x_k, u_k) + \min_{u_k} Q_i(x_{k+1}, u_{k+1}).
\end{aligned} \tag{23}
$$

Letting $i \to \infty$, we obtain

$$Q_\infty(x_k, u_k) \leq U(x_k, u_k) + \min_{u_{k+1}} Q_\infty(x_{k+1}, u_{k+1}). \tag{24}$$

Let $\zeta > 0$ be an arbitrary positive number. There exists a positive integer p such that

$$Q_p(x_k, u_k) - \zeta \leq Q_\infty(x_k, u_k) \leq Q_p(x_k, u_k). \tag{25}$$

Hence, we can get

$$\begin{aligned}
Q_\infty(x_k, u_k) &\geq Q_p(x_k, u_k) - \zeta \\
&= U(x_k, u_k) + Q_p(x_{k+1}, v_p(x_{k+1})) - \zeta \\
&\geq U(x_k, u_k) + Q_\infty(x_{k+1}, v_p(x_{k+1})) - \zeta \\
&\geq U(x_k, u_k) + \min_{u_{k+1}} Q_\infty(x_{k+1}, u_{k+1}) - \zeta.
\end{aligned} \tag{26}$$

Since ζ is arbitrary, we have

$$Q_\infty(x_k, u_k) \geq U(x_k, u_k) + \min_{u_{k+1}} Q_\infty(x_{k+1}, u_{k+1}). \tag{27}$$

Combining (24) and (27), we obtain

$$Q_\infty(x_k, u_k) = U(x_k, u_k) + \min_{u_{k+1}} Q_\infty(x_{k+1}, u_{k+1}). \tag{28}$$

According to the definition of the optimal Q function in (4), we have $Q_\infty(x_k, u_k) = Q^*(x_k, u_k)$. The proof is completed.

4   Simulation Study

We now examine the performance of the developed policy iteration algorithm in a nonlinear torsional pendulum system [13]. The dynamics of the pendulum is as follows

$$\begin{bmatrix} x_{1(k+1)} \\ x_{2(k+1)} \end{bmatrix} = \begin{bmatrix} 0.1 x_{2k} + x_{1k} \\ -0.49 \sin(x_{1k}) - 0.1 f_d x_{2k} + x_{2k} \end{bmatrix} + \begin{bmatrix} 0 \\ 0.1 \end{bmatrix} u_k, \tag{29}$$

where $f_d = 0.2$ is the rotary inertia and frictional factor. Let the initial state be $x_0 = [1, -1]^T$. The utility function is expressed as $U(x_k, u_k) = x_k^T Q x_k + u_k^T R u_k$, where $Q = I$, $R = I$ and I denotes the identity matrix with suitable dimensions. Choose the critic and action networks as back propagation (BP) networks with the structures of 3–12–1 and 2–12–1, respectively. We randomly choose $p = 20000$ training data to implement the developed algorithm to obtain the optimal control law. For each iteration step, the critic network and the action network are trained for 1000 steps using the learning rate of $\alpha_c = \beta_a = 0.01$ so that the neural network training error becomes less than 10^{-5}. Implementing the developed Q-learning based policy iteration adaptive dynamic programming algorithm for $i = 25$ iterations to reach the computation precision $\varepsilon = 0.01$. The plots of the iterative function $Q_i(x_k, v_i(x_k))$ are shown in Fig. 1.

For nonlinear system (29), the iterative Q function is monotonically non-increasing and converges to its optimum by the Q-learning based policy iteration ADP algorithm. The corresponding iterative trajectories of system states and controls are shown in Figs. 2 and 3, respectively.

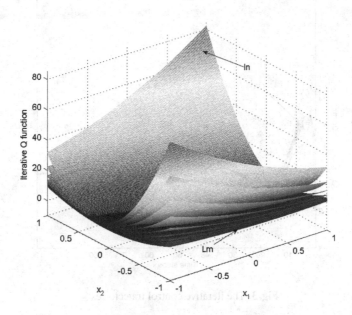

Fig. 1. The plots of the iterative Q function

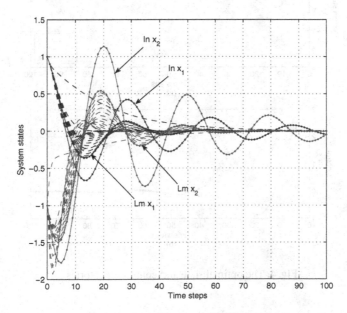

Fig. 2. The iterative state trajectories

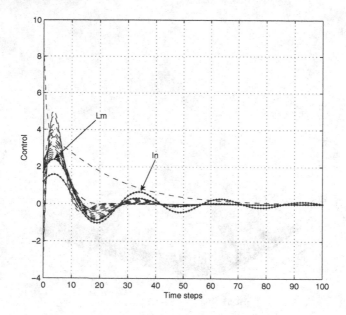

Fig. 3. The iterative control trajectories

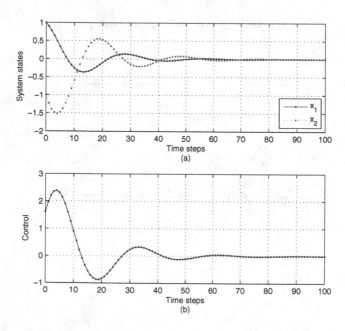

Fig. 4. The optimal state and control trajectories

From Figs. 2 and 3, we can see that the iterative system states and controls are both convergent to their optimal ones. The nonlinear system (29) can be stabilized under an arbitrary iterative control law $v_i(x_k)$, where the stability properties of the developed

Q-learning based policy iteration ADP algorithm can be verified. The optimal states and control trajectories are shown in Fig. 4.

5 Conclusions

In this paper, an effective policy iteration adaptive dynamic programming algorithm based on Q-learning is developed to solve optimal control problems for infinite horizon discrete-time nonlinear systems. The iterative Q functions is proven to be monotonically non-increasing and converges to the optimum as the iteration index increases to infinity. Finally, simulation results are presented to illustrate the performance of the developed algorithm.

References

1. Abu-Khalaf, M., Lewis, F.L.: Nearly optimal control laws for nonlinear systems with saturating actuators using a neural network HJB approach. Automatica 41(5), 779–791 (2005)
2. Al-Tamimi, A., Lewis, F.L., Abu-Khalaf, M.: Discrete-time nonlinear HJB solution using approximate dynamic programming: convergence proof. IEEE Transactions on Systems, Man, and Cybernetics–Part B: Cybernetics 38(4), 943–949 (2008)
3. Bertsekas, D.P., Tsitsiklis, J.N.: Neuro-Dynamic Programming. Athena Scientific, Belmont (1996)
4. Jiang, Y., Jiang, Z.P.: Robust adaptive dynamic programming with an application to power systems. IEEE Transactions on Neural Networks and Learning Systems 24(7), 1150–1156 (2013)
5. Liu, D., Wei, Q.: Policy iteration adaptive dynamic programming algorithm for discrete-time nonlinear systems. IEEE Transactions on Neural Networks and Learning Systems 25(3), 621–634 (2014)
6. Modares, H., Lewis, F.L.: Linear quadratic tracking control of partially-unknown continuous-time systems using reinforcement learning. IEEE Transactions on Automatic Control 59(11), 3051–3056 (2014)
7. Modares, H., Lewis, F.L., Naghibi-Sistani, M.B.: Adaptive optimal control of unknown constrained-input systems using policy iteration and neural networks. IEEE Transactions on Neural Networks and Learning systems 24(10), 1513–1525 (2013)
8. Modares, H., Lewis, F.L.: Optimal tracking control of nonlinear partially-unknown constrained-input systems using integral reinforcement learning. Automatica 50(7), 1780–1792 (2014)
9. Kiumarsi, B., Lewis, F.L., Modares, H., Karimpur, A., Naghibi-Sistani, M.B.: Reinforcement Q-learning for optimal tracking control of linear discrete-time systems with unknown dynamics. Automatica 50(4), 1167–1175 (2014)
10. Prokhorov, D.V., Wunsch, D.C.: Adaptive critic designs. IEEE Transactions on Neural Networks 8(5), 997–1007 (1997)
11. Song, R., Xiao, W., Zhang, H., Sun, C.: Adaptive dynamic programming for a class of complex-valued nonlinear systems. IEEE Transactions on Neural Networks and Learning Systems 25(9), 1733–1739 (2014)
12. Song, R., Lewis, F.L., Wei, Q., Zhang, H., Jiang, Z.-P., Levine, D.: Multiple actor-critic structures for continuous-time optimal control using input-output data. IEEE Transactions on Neural Networks and Learning Systems 26(4), 851–865 (2015)

13. Si, J., Wang, Y.-T.: On-line learning control by association and reinforcement. IEEE Transactions on Neural Networks 12(2), 264–276 (2001)
14. Watkins, C.: Learning from Delayed Rewards. Ph.D. Thesis, Cambridge University, Cambridge (1989)
15. Watkins, C., Dayan, P.: Q-learning. Machine Learning 8(3-4), 279–292 (1992)
16. Wei, Q., Liu, D.: An iterative ϵ-optimal control scheme for a class of discrete-time nonlinear systems with unfixed initial state. Neural Networks 32, 236–244 (2012)
17. Wei, Q., Zhang, H., Dai, J.: Model-free multiobjective approximate dynamic programming for discrete-time nonlinear systems with general performance index functions. Neurocomputing 72(7–9), 1839–1848 (2009)
18. Wei, Q., Liu, D.: A novel iterative θ-adaptive dynamic programming for discrete-time nonlinear systems. IEEE Transactions on Automation Science and Engineering 11(4), 1176–1190 (2014)
19. Wei, Q., Liu, D.: Data-driven neuro-optimal temperature control of water gas shift reaction using stable iterative adaptive dynamic programming. IEEE Transactions on Industrial Electronics 61(11), 6399–6408 (2014)
20. Wei, Q., Liu, D.: Neural-network-based adaptive optimal tracking control scheme for discrete-time nonlinear systems with approximation errors. Neurocomputing 149(3), 106–115 (2015)
21. Wei, Q., Liu, D., Shi, G., Liu, Y.: Optimal multi-battery coordination control for home energy management systems via distributed iterative adaptive dynamic programming. IEEE Transactions on Industrial Electronics (2015) (article in press)
22. Wei, Q., Liu, D., Shi, G.: A novel dual iterative Q-learning method for optimal battery management in smart residential environments. IEEE Transactions on Industrial Electronics 62(4), 2509–2518 (2015)
23. Wei, Q., Liu, D.: Adaptive dynamic programming for optimal tracking control of unknown nonlinear systems with application to coal gasification. IEEE Transactions on Automation Science and Engineering 11(4), 1020–1036 (2014)
24. Wei, Q., Liu, D., Yang, X.: Infinite horizon self-learning optimal control of nonaffine discrete-time nonlinear systems. IEEE Transactions on Neural Networks and Learning Systems 26(4), 866–879 (2015)
25. Werbos, P.J.: Advanced forecasting methods for global crisis warning and models of intelligence. General Systems Yearbook 22, 25–38 (1977)
26. Werbos, P.J.: A menu of designs for reinforcement learning over time. In: Miller, W.T., Sutton, R.S., Werbos, P.J. (eds.) Neural Networks for Control, pp. 67–95. MIT Press, Cambridge (1991)
27. Xu, X., Lian, C., Zuo, L., He, H.: Kernel-based approximate dynamic programming for real-time online learning control: An experimental study. IEEE Transactions on Control Systems Technology 22(1), 146–156 (2014)
28. Zhang, H., Wei, Q., Luo, Y.: A novel infinite-time optimal tracking control scheme for a class of discrete-time nonlinear systems via the greedy HDP iteration algorithm. IEEE Transactions on System, Man, and cybernetics–Part B: Cybernetics 38(4), 937–942 (2008)

Adaptive Neural Network Control for a Class
of Stochastic Nonlinear Strict-Feedback Systems[*]

Zifu Li[1,2] and Tieshan Li[1,**]

[1] Navigation College, Dalian Maritime University, Dalian 116026, China
[2] Navigation College, Jimei University, Xiamen, 361021, China
tieshanli@126.com

Abstract. An adaptive neural network control approach is proposed for a class of stochastic nonlinear strict-feedback systems with unknown nonlinear function in this paper. Only one NN (neural network) approximator is used to tackle unknown nonlinear functions at the last step and only one actual control law and one adaptive law are contained in the designed controller. This approach simplifies the controller design and alleviates the computational burden. The Lyapunov Stability analysis given in this paper shows that the control law can guarantee the solution of the closed-loop system uniformly ultimate boundedness (UUB) in probability. The simulation example is given to illustrate the effectiveness of the proposed approach.

Keywords: adaptive control, neural networks, stochastic nonlinear strict-feedback system.

1   Introduction

Backstepping technique has been a powerful method for synthesizing adaptive controllers for deterministic strict-feedback nonlinear systems, and some useful control schemes have been developed [1-3]. However, little attention has been paid to the stabilization problem for the stochastic nonlinear systems until recently. Efforts toward stabilization of stochastic nonlinear systems have been initiated in the work of Florchinger [4]. By employing the quadratic Lyapunov functions and the Itô differentiation rule Deng and Krstić[5] gave a backstepping design for stochastic strict-feedback system with the form of quartic Lyapunov function.

As well known, both neural network (NN) and fuzzy logic system (FLS) have been found to be particularly useful for controlling nonlinear systems with nonlinearly

[*] This work was supported in part by the National Natural Science Foundation of China (Nos.51179019, 60874056), the Natural Science Foundation of Liaoning Province (No. 20102012), the Program for Liaoning Excellent Talents in University (LNET) (Grant No.LR 2012016) and the Applied Basic Research Program of Ministry of Transport of P. R. China (Nos. 2011-329-225-390 and 2013-329-225-270).
[**] Corresponding author.

© Springer International Publishing Switzerland 2015
X. Hu et al. (Eds.): ISNN 2015, LNCS 9377, pp. 53–61, 2015.
DOI: 10.1007/978-3-319-25393-0_7

parameterized uncertainties. The main advantage is that the unknown nonlinear functions can be approximated by the neural networks [6-8]. For simplifying the complexity of control design and alleviating the computation burden, numerous control approaches have been developed. For instance, Chen [9] and Li [10] introduced the adaptive neural network control schemes to the output-feedback stochastic nonlinear strict-feedback systems, and only an NN to compensate for all upper bounding functions depending on the system output. A novel direct adaptive neural network controller was proposed to control a class of stochastic system with completely unknown nonlinear functions in [11]. For the purpose of solving the problem of the explosion of neural network learning parameters, Yang et al. first solved the problem in their pioneering work [12], where the so-called "minimal learning parameter (MLP)" algorithm containing much less online adaptive parameters were constructed by fusion of traditional backstepping technique and radial-basis-function (RBF) NNs. By combining dynamic surface control (DSC) and MLP techniques, Li et al in [13] first proposed an algorithm which can simultaneously solve both problems of the explosion of learning parameters and the explosion of computation complexity. However, many approximators are still used to construct virtual control laws and actual control law and all the virtual control law also must be actually implemented in the process of controller design. In order to eliminate the complexity growing problem and deduce the computation burden mentioned above completely, Sun et al. proposed a new adaptive control design approach to handle the problems mentioned above [14], only one NN is used to approximate the lumped unknown function of the system.

Motivated by the aforementioned discussion, in this paper, a single neural network approximation based adaptive control approach is presented for the strict-feedback stochastic nonlinear systems. The main contributions lie in the following: (i) only one NN is used to deal with those unknown system functions, those virtual control law are not necessary to be actually implemented in the process of control design; (ii) there is only one adaptive law proposed in this paper, which make the computational burden significantly alleviated and the control scheme more easily implemented in practical applications.

2   Preliminaries and Problem Formulation

Consider an n-dimensional stochastic nonlinear system

$$dx = f(x)dt + \psi(x)dw \tag{1}$$

where $x \in R^n$ is the system state, w is an r-dimensional standard Brownian motion defined on the complete probability space (Ω, F, P) with Ω be a sample space, F being a σ-field. $f(x): R^n \to R^n$, $\psi(x): R^n \to R^{n \times r}$ are locally Lipschitz.

In this paper, the following RBF NN will be used to approximate any unknown continuous function $h(Z)$, namely $h_{nn}(Z) = W^T S(Z)$, where $Z \in \Omega_Z \subset R^q$ is the input vector with q being the input dimension of neural networks,

$W = [w_1,\ w_2, \cdots,\ w_l]^T \in R^l$ is the weight vector, $l > 1$ is the neural networks node number, and $S(Z) = [s_1(Z), \cdots,\ s_l(Z)]^T$ means the basis function vector, $s_i(Z)$ is the Gaussian function of the form $s_i(Z) = \exp\left[-(z - \mu_i)^T(z - \mu_i)/\varsigma^2\right]$, $i = 1,2,\cdots,l$, where $\mu_i = [\mu_{i1},\mu_{i2},\cdots,\mu_{iq}]^T$ is the center of the receptive field and $\varsigma > 0$ are the width of the basis function.

It has been proven that neural network can approximate any continuous function over a compact set $\Omega_z \subset R^q$ to arbitrary any accuracy such as $h(Z) = W^{*T}S(Z) + \delta(Z)$, where W^* is the ideal constant weight vector and $\delta(Z)$ denotes the approximation error and satisfies $|\delta(Z)| \le \varepsilon$.

Assumption 1 [15]. There exist constants b_m and b_M such that for $1 \le i \le n, \forall \overline{x}_i \in R^i$, $0 < b_m \le g_i(\overline{x}_i) \le b_M < \infty$.

Assumption 2. The desired trajectory signal $y_d(t)$ is continuous and bounded, and its time derivatives up to the nth order are also continuous and bounded.

Lemma 1 [16]. Consider the stochastic system (1). If there exists a positive definite, radially unbounded, twice continuously differentiable Lyapunov function $V : R^n \to R$, and constants $a_0 > 0, \gamma_0 \ge 0$, such that

$$LV(x) \le -a_0 V(x) + \gamma_0 \tag{2}$$

Then, the system has a unique solution almost surely, and the system is bounded in probability.

Consider the following stochastic nonlinear strict-feedback system

$$\begin{cases} dx_i = (g(\overline{x}_i)x_{i+1} + f_i(\overline{x}_i))dt + \psi_i(\overline{x}_i)dw \\ dx_n = (g(\overline{x}_n)u + f_n(\overline{x}_n))dt + \psi_n(\overline{x}_n)dw \\ y = x_1 \end{cases} \tag{3}$$

where $x = [x_1, \cdots,\ x_n]^T \in R^n$, $u \in R$ and $y \in R$ are the state variable, the control input, and the system output respectively, $\overline{x}_i = [x_1, \cdots,\ x_i]^T \in R^i$, $f_i(\cdot)$, $g_i(\cdot): R^i \to R$ and $\psi_i(\cdot): R^i \to R^r$, $(i = 1,\cdots,n)$ are unknown smooth nonlinear functions with $f_i(0) = 0$, $\psi_i(0) = 0 (1 \le i \le n)$ $(i = 1,2,\cdots,n)$.

3   Controller Design

Step 1: Define the first error surface as $z_1 = x_1 - y_d$, where y_d is the desired trajectory. Its differential is

$$dz_1 = \left[g_1(x_1)x_2 + f_1(x_1) - \dot{y}_d\right]dt + \psi_1(x_1)dw \tag{4}$$

Define the virtual controller α_2 as follows

$$\alpha_2 = -k_1 z_1 - F_1(x_1, y_d, \dot{y}_d) \tag{5}$$

where $k_1 > 0$ is a positive real design constant, $F_1(x_1, y_d, \dot{y}_d)$ is an unknown smooth function in the following form

$$F_1(x_1, y_d, \dot{y}_d) = \frac{1}{g_1(x_1)}\left[f_1(x_1) - \dot{y}_d + \frac{3}{4}l_1^{-2}z_1 \left\| \varphi(x_1) \right\|^4 \right] \tag{6}$$

where $\varphi(x_1) = \psi(x_1)$. Define the second error surface as $z_2 = x_2 - \alpha_2$. Then, we have

$$z_2 = x_2 - \dot{y}_d + k_1(x_1 - y_d) + F_1^*(x_1, y_d, \dot{y}_d) \tag{7}$$

where $F_1^*(x_1, y_d, \dot{y}_d) = F_1(x_1, y_d, \dot{y}_d) + \dot{y}_d$.

Step i $(2 \le i \le n-1)$ **:** A similar procedure is recursively employed for each step i, from the former step, it can be obtained that

$$z_i = x_i - y_d^{(i-1)} + \sum_{j=1}^{i-1} k_j k_{j+1} \cdots k_{i-1}\left(x_j - y_d^{(j-1)} \right) + F_{i-1}^*\left(\overline{x}_i, y_d, \dot{y}_d, \cdots, y_d^{(i-1)} \right) \tag{8}$$

where

$$F_{i-1}^*\left(\overline{x}_i, y_d, \dot{y}_d, \cdots, y_d^{(i-1)} \right) = k_{i-1} F_{i-2}^*\left(\overline{x}_i, y_d, \dot{y}_d, \cdots, y_d^{(i-2)} \right) + F_{i-1}\left(\overline{x}_i, y_d, \dot{y}_d, \cdots, y_d^{(i-1)} \right) + y_d^{(i-1)} \tag{9}$$

The differential of z_2 is

$$dz_i = \left[g_i(\overline{x}_i)x_{i+1} + \sum_{j=1}^{i-1} \frac{\partial F_{i-1}^*\left(\overline{x}_i, y_d, \dot{y}_d, \cdots, y_d^{(j-1)} \right)}{\partial x_j}\left(g_j(\overline{x}_j)x_{j+1} + f_j(\overline{x}_j) \right) \right.$$

$$+ f_i(\overline{x}_i) - y_d^{(i)} + \sum_{j=1}^{i-1} k_j k_{j+1} \cdots k_{i-1}\left(g_j(\overline{x}_j)x_{j+1} + f_j(\overline{x}_j) - y_d^{(j)} \right)$$

$$\left. + \sum_{j=1}^{i} \frac{\partial F_{i-1}^*\left(\overline{x}_i, y_d, \dot{y}_d, \cdots, y_d^{(j-1)} \right)}{\partial y_d^{(j-1)}} y_d^{(j)} \right] + \varphi_i(\overline{x}_i)dw \tag{10}$$

where

$$\varphi_i(\overline{x}_i) = \psi_i(\overline{x}_i) + \sum_{j=1}^{i-1} k_j k_{j+1} \cdots k_{i-1}\psi_j(\overline{x}_j) + \sum_{j=1}^{i} \frac{\partial F_{i-1}^*\left(\overline{x}_i, y_d, \dot{y}_d, \cdots, y_d^{(j-1)} \right)}{\partial y_d^{(j-1)}}\psi_j(\overline{x}_j)$$

The virtual control law α_{i+1} is chosen as follows:

$$\alpha_{i+1} = -k_i z_i - F_i\left(\overline{x}_i, y_d, \dot{y}_d, \cdots, y_d^{(i)} \right) \tag{11}$$

where k_i is a positive real design constant, $F_i\left(\overline{x}_i, y_d, \dot{y}_d, \cdots, y_d^{(i)} \right)$ is an unknown smooth function in the following form

$$F_i\left(\bar{x}_i, y_d, \dot{y}_d, \cdots, y_d^{(i)}\right) = \frac{1}{g_i\left(\bar{x}_i\right)}\left[\sum_{j=1}^{i-1}\frac{\partial F_{i-1}^*\left(\bar{x}_{i-1}, y_d, \dot{y}_d, \cdots, y_d^{(j-1)}\right)}{\partial x_j}\left(g_j\left(\bar{x}_j\right)x_{j+1} + f_j\left(\bar{x}_j\right)\right)\right.$$

$$+ f_i\left(\bar{x}_i\right) - y_d^{(i)} + \sum_{j=1}^{i-1}k_jk_{j+1}\cdots k_{i-1}\left(g_j\left(\bar{x}_j\right)x_{j+1} + f_j\left(\bar{x}_j\right) - y_d^{(j)}\right)$$

$$\left.+ \sum_{j=1}^{i}\frac{\partial F_{i-1}^*\left(\bar{x}_{i-1}, y_d, \dot{y}_d, \cdots, y_d^{(j-1)}\right)}{\partial y_d^{(j-1)}}y_d^{(j)} + \frac{3}{4}l_i^{-2}z_i\left\|\varphi_i\left(\bar{x}_i\right)\right\|^4\right] \quad (12)$$

Define the $(i+1)$th error surface as $z_{i+1} = x_{i+1} - \alpha_{i+1}$, Substituting α_{i+1} into z_{i+1}, it can be obtained that

$$z_{i+1} = x_{i+1} - y_d^{(i)} + \sum_{j=1}^{i}k_jk_{j+1}\cdots k_i\left(x_j - y_d^{(j-1)}\right) + F_i^*\left(\bar{x}_i, y_d, \dot{y}_d, \cdots, y_d^{(i)}\right) \quad (13)$$

where $F_i^*\left(\bar{x}_i, y_d, \dot{y}_d, \cdots, y_d^{(i)}\right)$ is also an unknown function in the following form

$$F_i^*\left(\bar{x}_i, y_d, \dot{y}_d, \cdots, y_d^{(i)}\right) = k_iF_{i-1}^*\left(\bar{x}_{i-1}, y_d, \dot{y}_d, \cdots, y_d^{(i-1)}\right) + F_i\left(\bar{x}_i, y_d, \dot{y}_d, \cdots, y_d^{(i)}\right) + y_d^{(i)} \quad (14)$$

Step n: The differential of z_n is

$$dz_n = \left[g_n\left(\bar{x}_n\right)u + f_n\left(\bar{x}_n\right) + \sum_{j=1}^{n-1}\frac{\partial F_{n-1}^*\left(\bar{x}_i, y_d, \dot{y}_d, \cdots, y_d^{(j-1)}\right)}{\partial x_j}\left(g_j\left(\bar{x}_j\right)x_{j+1} + f_j\left(\bar{x}_j\right)\right)\right.$$

$$- y_d^{(n)} + \sum_{j=1}^{n-1}k_jk_{j+1}\cdots k_{n-1}\left(g_j\left(\bar{x}_j\right)x_{j+1} + f_j\left(\bar{x}_j\right) - y_d^{(j)}\right)$$

$$\left.+ \sum_{j=1}^{n}\frac{\partial F_{n-1}^*\left(\bar{x}_i, y_d, \dot{y}_d, \cdots, y_d^{(j-1)}\right)}{\partial y_d^{(j-1)}}y_d^{(j)}\right] + \varphi_n\left(\bar{x}_n\right)dw \quad (15)$$

where

$$\varphi_n\left(\bar{x}_n\right) = \psi_n\left(\bar{x}_n\right) + \sum_{j=1}^{n-1}k_jk_{j+1}\cdots k_{n-1}\psi_j\left(\bar{x}_j\right) + \sum_{j=1}^{n}\frac{\partial F_{n-1}^*\left(\bar{x}_i, y_d, \dot{y}_d, \cdots, y_d^{(j-1)}\right)}{\partial y_d^{(j-1)}}\psi_j\left(\bar{x}_n\right)$$

Chose the desired control law as

$$u^* = -k_n\left[x_n - y_d^{(n-1)} + \sum_{j=1}^{n-1}k_jk_{j+1}\cdots k_{n-1}\left(x_j - y_d^{(j-1)}\right)\right] - F_n^*\left(\bar{x}_n, y_d, \dot{y}_d, \cdots, y_d^{(n)}\right) \quad (16)$$

where

$$F_n^*\left(\bar{x}_n, y_d, \dot{y}_d, \cdots, y_d^{(n)}\right) = k_nF_{n-1}^*\left(\bar{x}_{n-1}, y_d, \dot{y}_d, \cdots, y_d^{(n-1)}\right) + F_n\left(\bar{x}_n, y_d, \dot{y}_d, \cdots, y_d^{(n)}\right) + y_d^{(n)}$$ is

an unknown smooth function. Where k_n is a positive real design constant, $F_n\left(\bar{x}_n, y_d, \dot{y}_d, \cdots, y_d^{(n)}\right)$ is an unknown smooth function in the following form

$$F_n\left(\overline{x}_n, y_d, \dot{y}_d, \cdots, y_d^{(n)}\right) = \frac{1}{g_n\left(\overline{x}_n\right)} \left[\sum_{j=1}^{n-1} \frac{\partial F_{n-1}^*\left(\overline{x}_{n-1}, y_d, \dot{y}_d, \cdots, y_d^{(j-1)}\right)}{\partial x_j}\left(g_j\left(\overline{x}_j\right)x_{j+1} + f_j\left(\overline{x}_j\right)\right) \right.$$

$$+ f_n\left(\overline{x}_n\right) - y_d^{(n)} + \sum_{j=1}^{n-1} k_j k_{j+1} \cdots k_{n-1}\left(g_j\left(\overline{x}_j\right)x_{j+1} + f_j\left(\overline{x}_j\right) - y_d^{(j)}\right)$$

$$\left. + \sum_{j=1}^{n} \frac{\partial F_{n-1}^*\left(\overline{x}_{n-1}, y_d, \dot{y}_d, \cdots, y_d^{(j-1)}\right)}{\partial y_d^{(j-1)}} y_d^{(j)} + \frac{3}{4} l_n^{-2} z_n \left\| \varphi_n\left(\overline{x}_n\right) \right\|^4 \right] \quad (17)$$

Since function $F_n^*\left(\overline{x}_n, y_d, \dot{y}_d, \cdots, y_d^{(n)}\right)$ is unknown, an RBF neural network can be used to approximate it. That is

$$F_n^*\left(\overline{x}_n, y_d, \dot{y}_d, \cdots, y_d^{(n)}\right) = W^{*T} S\left(\overline{x}_n, y_d, \dot{y}_d, \cdots, y_d^{(n)}\right) + \varepsilon \quad (18)$$

Then the actual control law u is chosen as follows:

$$u = -k_n\left[x_n - y_d^{(n-1)} + \sum_{j=1}^{n-1} k_j k_{j+1} \cdots k_{n-1}\left(x_j - y_d^{(j-1)}\right)\right] - \hat{W}^T S\left(\overline{x}_n, y_d, \dot{y}_d, \cdots, y_d^{(n)}\right) \quad (19)$$

where $\hat{W}$ is the estimation of W^* and is updated as follows:

$$\dot{\hat{W}} = \Gamma\left(z_1^3 S\left(\overline{x}_n, y_d, \dot{y}_d, \cdots, y_d^{(n)}\right) - \gamma \hat{W}\right) \quad (20)$$

with a constant matrix $\Gamma = \Gamma^T > 0$, and a real scalar $\gamma > 0$.

4   Stability Analysis

Theorem 1. Consider the system (3), and the above closed-loop systems, according to lemma 1, for any initial condition satisfying

$$\Pi = \left\{ \sum_{i=1}^{n} z_i^4(0) + \frac{1}{2}\tilde{W}^T \Gamma^{-1}\tilde{W} < M_0 \right\} \quad (21)$$

where M_0 is any positive constant, then there exist the control parameters k_i, Γ and γ such that all the signals in the closed-loop system are UUB in forth moment. Moreover, the ultimate boundedness of the above closed-loop signals can be tuned arbitrarily small by choosing suitable design parameters.

Proof: Consider the following Lyapunov function candidate

$$V = \frac{1}{4}\sum_{i=1}^{n} z_i^4 + \frac{1}{2}\tilde{W}^T \Gamma^{-1}\tilde{W} \quad (22)$$

According to the Itô's differential rules and Young inequality, together with equations (19) and (20), the differential of the above function V can be found as follows

$$LV \leq -\left[\left(k_1 - \frac{3}{4}\right)b_m - \frac{3}{4}\chi^{\frac{4}{3}}\right]z_1^4 - \sum_{i=2}^{n-1}\left[\left(k_i - \frac{3}{4}\right)b_m - \frac{1}{4}b_M\right]z_i^4 - \frac{1}{2}\gamma\tilde{W}^T\tilde{W} + \frac{1}{2}\|W^*\|^2$$

$$-\left[\left(k_n - \frac{3}{4}\chi^{\frac{4}{3}} - \frac{3}{4}\right)b_m - \frac{1}{4}b_M\right]z_i^4 + \frac{3}{4}\sum_{i=1}^{n}l_i^2 + \frac{1}{4}(1+b_M)\varpi^4 + \frac{1}{4}b_M\|\varepsilon\|^4 \qquad (23)$$

where $\left\|S\left(\overline{x}_n, y_d, \dot{y}_d, \cdots, y_d^{(n)}\right)\right\| \leq \chi^{[17]}$, $\|\tilde{W}\| \leq \varpi$.

Choosing the positive constants as

$$\begin{cases} c_1 = \left(k_1 - \frac{3}{4}\right)b_m - \frac{3}{4}\chi^{\frac{4}{3}} \\ c_i = \left(k_i - \frac{3}{4}\right)b_m - \frac{1}{4}b_M \qquad i = 2, \cdots, n-1 \\ c_n = \left(k_n - \frac{3}{4}\chi^{\frac{4}{3}} - \frac{3}{4}\right)b_m - \frac{1}{4}b_M \end{cases} \qquad (24)$$

Define a positive constant $a_0 = \min\{4c_i, r\}$. It follows from Equation (27) that

$$LV \leq -aV + D \qquad (25)$$

where

$$D = \frac{1}{2}\|W^*\|^2 + + \frac{3}{4}\sum_{i=1}^{n}l_i^2 + \frac{1}{4}(1+b_M)\varpi^4 + \frac{1}{4}b_M\|\varepsilon\|^4 \qquad (26)$$

From Equation (25) we can clearly observe that the first term is negative definite and the second term D is a positive constant.

Furthermore, it follows from (25) that

$$E[V(t)] \leq \left(V(0) - \frac{D}{a}\right)e^{-a(t-t_0)} + \frac{D}{a} \qquad (27)$$

According to lemma 2 the above analysis on the closed-loop system means that all the signals in the system (3) are UUB in the sense of probability. Furthermore, for any $\zeta_1 > \sqrt{D/a_0}$, there exists a constant $T > 0$, such that $|z_1(t)| \leq \zeta_1$ for all $t \geq t_0 + T$. Since $\sqrt{D/a_0}$ can be made arbitrarily small if the design parameters are chosen appropriately, thus, for any given ζ_1, one has $\lim_{t\to\infty} |z_1(t)| \leq \zeta_1$. That is to say, by adjusting the design parameters, the tracking error can be made arbitrarily small.

The proof is thus completed.   □

5   Simulation Example

Consider the following third-order stochastic nonlinear system

$$\begin{cases} dx_1 = \left(1+x_1^2\right)x_2 + x_1 \sin x_1 + x_1^3 dw \\ dx_2 = \left(1+x_2^2\right)x_3 + x_2 e^{-0.5x_1} + x_1 \cos x_2 dw \\ dx_3 = \left(3+\cos\left(x_2 x_3\right)\right)u + x_1 x_2 x_3 + 3x_1 e^{-x_3^2} dw \\ y = x_1 \end{cases} \qquad (28)$$

Based on the adaptive NN controller design proposed in section 3, the true control law are designed and the adaptive law. In applied mathematics, the Wiener process can be described as the integral form of Gauss white noise, which has two main parameters, i.e., mean and variance. We choose the neuron's center and variance as $\{-5,5\} \times \{-5,5\} \times \{-5,5\} \times \{-5,5\} \times \{-5,5\}$ and 1 respectively. If we chose the desired trajectory $y_d = \sin(t)$, the suitable parameters were chosen as $k_1 = 35$, $k_2 = 2.5$, $k_3 = 100$, $\Gamma = 0.05$, $\gamma = 100$, The initial conditions are given by $[x_1(0), x_2(0), x_3(0)]^T = [0.8, 0.4, 0.5]^T$ and the initial weight vector $\hat{W}(0) = 0.5$. The simulation results are shown in Figs.1~4.

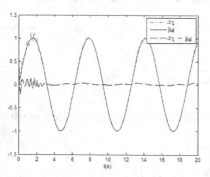

Fig. 1. The output y, the reference signal y_d

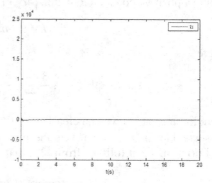

Fig.2. The control input u

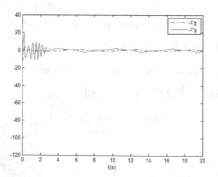

Fig.3. The state of x_2 and x_3

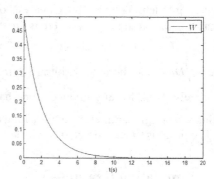

Fig.4. The adaptive law $\hat{W}$

6 Conclusion

In this paper, An adaptive NN controller has been proposed for a class of stochastic nonlinear strict-feedback systems. Using the proposed technique we can alleviate the computational burden and simplify the designed controller. Only one neural network is used to compensate the lumped unknown function at the last step. The closed-loop system has been proved UUB. The effectiveness of the proposed approach has been verified by the simulation example.

References

1. Krstić, M., Kanellakopulos, I., Kocotovic, P.V.: Nonlinear and Adaptive Control Design. Wiley, New York (1995)
2. Ge, S.S., Huang, C.C., Lee, T., Zhang, T.: Stable Adaptive Neural Network Control. Kluwer, Boston (2002)
3. Hua, C., Guan, X., Shi, P.: Robust backstepping control for a class of time delayed systems. IEEE Trans. Automat. Control 50(6), 894–899 (2005)
4. Florchinger, P.: Lyapunov-like techniques for stochastic stability. SIAM J. Contr. Optim. 33, 1151–1169 (1995)
5. Deng, H., Kristić, M.: Stochastic nonlinear stabilization, Part I: a backstepping design. Syst. Control Lett. 32(3), 143–150 (1997)
6. Zhou, J., Meng, J.E., Zurada, J.M.: Adaptive neural network control of uncertain nonlinear systems with nonsmooth actuator nonlinearities. Neurocomputing 70, 1062–1070 (2007)
7. Cheng, L., Hou, Z.G., Tan, M.: Adaptive neural network tracking control for manipulators with uncertain kinematics, dynamics and actuator model. Automatica 45(10), 2312–2318 (2009)
8. Cheng, L., Hou, Z.G., Tan, M., Zhang, W.J.: Tracking control of a closed-chain five-bar robot with two degrees of freedom by integration of approximation-based approach and mechanical design. IEEE Transactions on Systems, Man, and Cybernetics, Part B: Cybernetics 42(5), 1470–1479 (2012)
9. Chen, W.S., Jiao, L.C., Li, J., Li, R.H.: Adaptive NN Backstepping Output-Feedback Control for Stochastic Nonlinear Strict-Feedback Systems with Time-Varying Delays. IEEE Trans. Syst., Man, Cybern. B 40(3), 939–950 (2010)
10. Li, J., Chen, W.S., Li, J.M., Fang, Y.Q.: Adaptive NN output-feedback stabilization for a class of stochastic nonlinear strict-feedback systems. ISA Transactions 48, 468–475 (2009)
11. Wang, H.Q., Chen, B., Lin, C.: Direct adaptive neural control for strict-feedback stochastic nonlinear systems. Nonlinear Dyn. 67, 2703–2718 (2012)
12. Yang, Y.S., Ren, J.S.: Adaptive fuzzy robust tracking controller design via small gain approach and its application. IEEE Trans. Fuzzy Syst. 11(6), 783–795 (2003)
13. Li, T.S., Li, R.H., Wang, D.: Adaptive neural control of nonlinear MIMO systems with unknown time delays. Neurocomputing 78, 83–88 (2012)
14. Sun, G., Wang, D., Li, T.S., Peng, Z.H., Wang, H.: Single neural network approximation based adaptive control for a class of uncertain strict-feedback nonlinear systems. Nonlinear Dyn. 72, 175–184 (2013)
15. Wang, Y.C., Zhang, H.G., Wang, Y.Z.: Fuzzy adaptive control of stochastic nonlinear systems with unknown virtual control gain function. Acta Autom. Sin. 32(2), 170–178 (2006)
16. Psillakis, H.E., Alexandridis, A.T.: NN-based adaptive tracking control of uncertain nonlinear systems disturbed by unknown covariance noise. IEEE Transactions on Neural Networks 18(6), 1830–1835 (2007)
17. Kurdila, A.J., Narcowich, F.J., Ward, J.D.: Persistency of excitation in identification using radial basis function approximants. SIAM Journal on Control and Optimization 33(2), 625–642 (1995)

Event-Triggered H_∞ Control
for Continuous-Time Nonlinear System

Dongbin Zhao[1,*], Qichao Zhang[1], Xiangjun Li[2], and Lingda Kong[2]

[1] State Key Laboratory of Management and Control for Complex Systems,
Institute of Automation, Chinese Academy of Sciences, Beijing 100190, China
[2] China Electric Power Research Institute, Beijing 100192, China
{dongbin.zhao,zhangqichao2014}@ia.ac.cn

Abstract. In this paper, the H_∞ optimal control for a class of continuous-time nonlinear systems is investigated using event-triggered method. First, the H_∞ optimal control problem is formulated as a two-player zero-sum differential game. Then, an adaptive triggering condition is derived for the closed loop system with an event-triggered control policy and a time-triggered disturbance policy. For implementation purpose, the event-triggered concurrent learning algorithm is proposed, where only one critic neural network is required. Finally, an illustrated example is provided to demonstrate the effectiveness of the proposed scheme.

1 Introduction

From the perspective of minmax optimization problem, the H_∞ control problem can be formulated as a two-player zero-sum differential game [1]. In order to obtain a controller that minimizes a cost function in the presence of worst-case disturbances, ones need to find the Nash equilibrium solution by solving the Hamilton-Jacobi-Isaacs (HJI) equation. Several reinforcement learning (RL) methods [2–4] have been successfully applied to solve the HJI equation for discrete-time systems [5] and continuous-time systems [6,7].

Due to the capability of computation efficiency, event-triggered control method has been integrated with the RL approach recently [8,9]. In the event-triggered control method, the controller is updated based on a new sampled state only when an event is triggered at event-triggering instants. This can reduce the communication between the plant and the controller significantly. In [10], an optimal adaptive event-triggered control algorithm was implemented based on an actor-critic structure for continuous-time nonlinear systems. On the other hand, the concurrent learning technique, which can relax the traditional persistency of excitation (PE) condition, was proposed for an uncertain system in [11]. In [12], a related idea called experience replay was adopted in Integral reinforcement learning (IRL) algorithm for constrained-input nonlinear systems.

To the best of our knowledge, there are no results on event-triggered H_∞ control of nonlinear system via concurrent learning. This is the motivation of our research. In this paper, the H_∞ control problem is described as a two-player zero-sum differential game and an online event-triggered concurrent learning (ETCL)

---

* Corresponding author.

© Springer International Publishing Switzerland 2015
X. Hu et al. (Eds.): ISNN 2015, LNCS 9377, pp. 62–70, 2015.
DOI: 10.1007/978-3-319-25393-0_8

algorithm is proposed to approximate the optimal control policy. Simulation results show the effectiveness of the proposed scheme.

2 Problem Statement

Consider the following nonlinear system with external disturbance:

$$\dot{x}(t) = f(x) + g(x)u(t) + k(x)w(t) , \tag{1}$$

where $x \in \mathbf{R}^n$ is the state vector, $u \in \mathbf{R}^m$ is the control input, $w \in \mathbf{R}^q$ is the nonlinear perturbation with $w(t) \in L_2(0, \infty)$. $f(\cdot) \in \mathbf{R}^n$, $g(\cdot) \in \mathbf{R}^{n \times m}$ and $k(\cdot) \in \mathbf{R}^{n \times q}$ are smooth nonlinear dynamics. Assume that $f(x) + g(x)u + k(x)w$ is Lipschitz continuous on a compact set $\Omega \subseteq \mathbf{R}^n$ with $f(0) = 0$. Let $x(0) = x_0$ be the initial state. Assume that $w(0) = 0$, so that $x = 0$ is an equilibrium of system (1). It is assumed that the system (1) is controllable.

Here, we introduce a sampled-data system that is characterized by a monotonically increasing sequence of event-triggering instants $\{\lambda_j\}_{j=0}^\infty$, where λ_j is the jth consecutive sampling instant with $\lambda_j < \lambda_{j+1}$. Define the event-trigger error between the current state $x(t)$ and the sampled state $\hat{x}_j$ as follows

$$e_j(t) = \hat{x}_j - x(t), \forall t \in [\lambda_j, \lambda_{j+1}) . \tag{2}$$

In the event-triggered control mechanism, the event-triggering condition is determined by the event-trigger error and a state-dependent threshold. When the event-triggering condition is not satisfied at $t = \lambda_j$, we say an event is triggered. Then, the system state is sampled that resets the event-trigger error $e_j(t)$ to zero, and the controller $\upsilon(\hat{x}_j)$ is updated based on the new sampled state. Note that $\upsilon(\hat{x}_j)$ is a function of the event-based state vector. The obtained control sequence $\{\upsilon(\hat{x}_j)\}_{j=0}^\infty$ becomes a continuous-time input signal $\upsilon(t) = \{\upsilon(\hat{x}_j, t)\}_{j=0}^\infty$ after using a zero-order hold (ZOH). In order to simplify the expression, we use $\upsilon(\hat{x}_j)$ to represent $\upsilon(\hat{x}_j, t)$ for $t \in [\lambda_j, \lambda_{j+1})$ in the following presentation.

Similar to the traditional H_∞ problem, our primary objective is to find a sequence of control inputs $\{\upsilon(\hat{x}_j)\}_{j=0}^\infty$, which for some prescribed $\gamma > 0$, renders

$$J(x_0, \upsilon(\hat{x}_j), w) = \sum_{\underset{j}{\cup}[\lambda_j, \lambda_{j+1}) = [0, \infty)} \int_{\lambda_j}^{\lambda_{j+1}} r(x, \upsilon(\hat{x}_j), w) dt \tag{3}$$

nonpositive for all $w(t) \in L_2[0, \infty)$ and $x(0) = 0$, where utility $r(x, \upsilon(\hat{x}_j), w) = x^T Q x + \upsilon^T(\hat{x}_j) R \upsilon(\hat{x}_j) - \gamma^2 \|w(t)\|^2$, Q and R are symmetric and positive definite matrices, and $\gamma \geq \gamma^* \geq 0$. Here, γ^* is the smallest γ such that the system (1) is stabilized. The quantity γ^* is known as the H-infinity gain.

3 Event-Triggered Optimal Controller Design

In this section, the H_∞ control problem is formulated as a two-player zero-sum differential game, where the control input u is a minimizing player while the disturbance w is a maximizing one. It is well known that the solution of H_∞ control problem is the zero-sum game theoretic saddle point (u^*, w^*), where u^* and w^* are the optimal control and the worst-case disturbance.

In the time-triggered case, the value function is generally defined as

$$V(u, w) = \int_t^\infty \left(x^T Q x + u^T R u - \gamma^2 \|w\|^2 \right) d\tau .$$ (4)

The corresponding nonlinear zero-sum Bellman equation is

$$r(x, u, w) + (\nabla V)^T (f(x) + g(x)u + k(x)w) = 0 ,$$ (5)

where $\nabla V = \partial V(x)/\partial x$ is the partial derivative of the value function with respect to the state. Then, the two-player zero-sum game has a unique solution if a saddle point (u^*, w^*) exists, that is if the Nash condition holds

$$\min_u \max_w V(u, w) = \max_w \min_u V(u, w) .$$ (6)

Define the Hamiltonian of the time-triggered problem

$$H(x, \nabla V, u, w) = (\nabla V)^T (f + gu + kw) + x^T Q x + u^T R u - \gamma^2 \|w\|^2 .$$ (7)

Then the associated HJI equation can be written as

$$\min_u \max_w H(x, \nabla V^*, u, w) = 0 ,$$ (8)

where the optimal value function V^* is the solution to the HJI equation. The associated control and disturbance policies are given as follows:

$$u^*(t) = -\frac{1}{2} R^{-1} g^T(x) \nabla V^* .$$ (9)

$$w^*(t) = \frac{1}{2\gamma^2} k^T(x) \nabla V^* .$$ (10)

In the event-triggered case, the control input is updated based on the sampled-state information $\hat{x}_j$ instead of the real state $x(t)$. Hence, (9) becomes

$$v^*(\hat{x}_j) = -\frac{1}{2} R^{-1} g^T(\hat{x}_j) \nabla V^*(\hat{x}_j), \forall t \in [\lambda_j, \lambda_{j+1}) ,$$ (11)

where $\nabla V^*(\hat{x}_j) = \partial V^*(\hat{x}_j)/\partial x(t)$. By using (10) and (11), the event-triggered HJI equation can be written as

$$(\nabla V^*)^T f(x) + x^T Q x - \frac{1}{2} (\nabla V^*)^T g(x) R^{-1} g^T(\hat{x}_j) \nabla V^*(\hat{x}_j)$$

$$+ \frac{1}{4} (\nabla V^*(\hat{x}_j))^T g(\hat{x}_j) R^{-1} g^T(\hat{x}_j) \nabla V^*(\hat{x}_j) + \frac{1}{4\gamma^2} (\nabla V^*)^T k(x) k^T(x) \nabla V^* = 0 .$$

(12)

Assumption 1. *The controller $u(x)$ is Lipschitz continuous with respect to the event-trigger error,*

$$\|u(x(t)) - u(\hat{x}_j)\| = \|u(x(t)) - u(x(t) + e_j(t))\| \leq L\|e_j(t)\| , \qquad (13)$$

where L is a positive real constant and $u(\hat{x}_j) = v(\hat{x}_j)$.

Theorem 1. *Suppose that $V^*(x)$ is the solution of the event-triggered HJI equation (12). For $\forall t \in [\lambda_j, \lambda_{j+1}), j = 0, ..., \infty$, the disturbance policy and control policy are given by (10) and (11), respectively. If the triggering condition is defined as follows*

$$\|e_j(t)\|^2 \leq e_T = \frac{(1 - \beta^2)}{L^2\|s\|^2}\underline{\theta}(Q)\|x\|^2$$
$$+ \frac{1}{L^2}\|v(\hat{x}_j)\|^2 - \frac{\gamma^2}{L^2\|s\|^2}\|w(t)\|^2 , \qquad (14)$$

where e_T is the threshold, $\underline{\theta}(Q)$ is the minimal eigenvalue of Q, $\beta \in (0,1)$ is a designed sample frequency parameter and $s^T s = R$. Then the closed-loop system (1) is asymptotically stable.

Remark 1: The event-trigger instants $\{\lambda_j\}_{j=0}^\infty$ is determined by the triggering condition (14). Based on the event-triggered mechanism, an event is generated by the violation of the triggering condition. Note that this method can reduce the communication between the controller and the plant effectively. On the other hand, the sample frequency can be adjusted by the designed parameter β in the triggering condition (14). When β is close to 1 one samples more frequently whereas when β is close to zero, the sampling periods become longer.

4   Online Neuro-Optimal Control Scheme

In this section, an online event-triggered concurrent learning (ETCL) algorithm is proposed, where only one critic neural network is required.

According to the Weierstrass high-order approximation theorem, the value function based on NN can be written as

$$V(x) = W_c^T \phi(x) + \varepsilon , \qquad (15)$$

where $W_c \in \mathbf{R}^N$ and $\phi(x) \in \mathbf{R}^N$ are the critic NN ideal weights and activation function vector, with N the number of hidden neurons, and $\varepsilon \in \mathbf{R}$ the critic NN approximation error.

The derivative of (15) with respect to x can be given by

$$\nabla V(x) = \nabla\phi^T(x)W_c + \nabla\varepsilon . \qquad (16)$$

Then, the zero-sum Bellman equation (5) can be rewritten as

$$x^T Q x + v^T(\hat{x}_j)Rv(\hat{x}_j) - \gamma^2\|w(t)\|^2 + W_c^T\nabla\phi(f(x) + g(x)v(\hat{x}_j) + k(x)w(t)) = \varepsilon_H , \qquad (17)$$

where the residual error is $\varepsilon_H = -(\nabla\varepsilon)^T(f(x) + g(x)\upsilon(\hat{x}_j) + k(x)w(t))$. Under the Lipschitz assumption on the system dynamics, the residual error is bounded locally. It is shown in [7] that this error converges uniformly to zero as the number of hidden-layer units increases. That is, there exists $\varepsilon_{Hmax} > 0$ such that $\|\varepsilon_H\| \leq \varepsilon_{Hmax}$.

Let $\hat{W}_c$ be the estimation of the unknown ideal weight vector W_c. The actual output of critic NN can be presented as

$$\hat{V}(x) = \hat{W}_c^T \phi(x) . \tag{18}$$

Accordingly, the time-triggered disturbance policy (10) and event-triggered control policy (11) can be approximated by

$$\hat{w}(t) = \frac{1}{2\gamma^2} k^T(x)\phi^T(x)\hat{W}_c . \tag{19}$$

$$\hat{\upsilon}(\hat{x}_j) = -\frac{1}{2}R^{-1}g^T(\hat{x}_j)\phi^T(\hat{x}_j)\hat{W}_c(\hat{x}_j) . \tag{20}$$

where $\hat{W}_c(\hat{x}_j)$ is the event-based estimation of ideal weight W_c. Then the closed-loop system dynamics (1) can now be written as

$$\dot{x} = f(x) + g(x)\hat{\upsilon}(\hat{x}_j) + k(x)\hat{w}(t), t \geq 0 . \tag{21}$$

The approximate Hamilton function is

$$\hat{W}_c^T \nabla\phi(x)f + x^T Qx - \frac{1}{2}\hat{W}_c^T \nabla\phi(x)g(x)R^{-1}g^T(\hat{x}_j)\nabla\phi^T(\hat{x}_j)\hat{W}_c(\hat{x}_j) + \frac{1}{4}\hat{W}_c^T(\hat{x}_j)\times$$
$$\nabla\phi(\hat{x}_j)g(\hat{x}_j)R^{-1}g^T(\hat{x}_j)\nabla\phi^T(\hat{x}_j)\hat{W}_c(\hat{x}_j) + \frac{1}{4\gamma^2}\hat{W}_c^T \nabla\phi(x)k(x)k^T(x)\nabla\phi^T(x)\hat{W}_c = e . \tag{22}$$

where e is a residual equation error.

Based on concurrent learning, the critic NN' weights can be updated by recorded data concurrently with current data. Define the residual equation error at time t_k as

$$e(t_k) = r(t_k) + \hat{W}_c^T(t)\sigma_k . \tag{23}$$

where $r(t_k) = x^T(t_k)Qx(t_k) + \hat{\upsilon}^T(\hat{x}_j)R\hat{\upsilon}(\hat{x}_j) - \gamma^2\|\hat{w}(t)\|^2$, $\sigma_k = \nabla\phi(x(t_k))$ $(f(x(t_k))+g(x(t_k))\hat{\upsilon}(\hat{x}_j)+k(x(t_k))\hat{w}(t)))$ are stored data at time $t_k \in [\lambda_j, \lambda_{j+1})$, $k \in \{1, ..., p\}$, $j = 0, 1..., \infty$, and p is the number of stored samples.

Condition 1: Let $M = [\sigma_1, ..., \sigma_p]$ be the recorded data corresponding to the critic NN's weights. Then M contains as many linearly independent elements as the number of corresponding critic NN's hidden neurons, i.e., $\text{rank}(M) = N$.

To derive the minimum value of e, it is desired to choose $\hat{W}_c$ to minimize the corresponding squared residual error $E = \frac{1}{2}e^T e$. Considering the concurrent learning, we develop a novel weight update law for the critic NN

$$\dot{\hat{W}}_c = -\alpha\sigma\left(\sigma^T\hat{W}_c(t) + r(x, \hat{\upsilon}(\hat{x}_j), \hat{w}(t))\right) - \alpha\sum_{k=1}^{p}\sigma_k\left(\sigma_k^T\hat{W}_c(t) + r(t_k)\right) . \tag{24}$$

where $\sigma = \nabla\phi(x)(f(x) + g(x)\upsilon(\hat{x}_j) + k(x)w)$, σ_k is defined in (23), $k \in \{1, ..., p\}$ denote the index of a stored data point, and $\alpha > 0$ denote the learning rate.

Remark 2: The online algorithm presented in this paper dose not rely on traditional PE condition which is difficult to check online. According to [11], the second term in (24) can be utilised to relax the PE condition with Condition 1.

By defining the weight estimation error of the critic NN as $\tilde{W}_c = W_c - \hat{W}_c$ and taking the time derivative one has

$$\dot{\tilde{W}}_c = -\alpha\sigma\left(\sigma^T\tilde{W}_c - \varepsilon_H\right) - \alpha\sum_{k=1}^{p}\sigma_k\left(\sigma_k^T\tilde{W}_c - \varepsilon_H(t_k)\right). \tag{25}$$

Assumption 2. *a. The critic NN activation function and its gradient are bounded, i.e., $\| \phi(x) \| \leq \phi_M$ and $\| \nabla\phi(x) \| \leq \nabla\phi_M$, with ϕ_M, $\nabla\phi_M$ being positive constants.*

b. The system dynamics $g(x)$ and $k(x)$ are upper bounded by positive constants such that $\|g(x)\| \leq g_M$ and $\|k(x)\| \leq k_M$.

Theorem 2. *Consider the nonlinear two-player zero-sum game (1) with the critic neural network (18), the time-triggered disturbance policy (19) and the event-triggered control policy (20). The tuning law based on concurrent learning technique for the continuous-time critic neural network is given by (24). Then the system is asymptotically stable and the critic weight estimation error is guaranteed to be Uniformaly Ultimately Bounded (UUB) if the adaptive triggering condition*

$$\|e_j(t)\|^2 \leq \frac{(1-\beta^2)}{L^2\|s\|^2}\underline{\theta}(Q)\|x\|^2 + \frac{1}{4L^2\|R\|^2}\|g^T(\hat{x}_j)\phi^T(\hat{x}_j)$$
$$\times \hat{W}_c(\hat{x}_j)\|^2 - \frac{1}{4\gamma^2L^2\|s\|^2}\|k^T(x)\phi^T(x)\hat{W}_c(t)\|^2 \tag{26}$$

and the following inequality are satisfied

$$\|\tilde{W}_c\| > \sqrt{\frac{a^2\sum_{k=1}^{p+1}\varepsilon_{Hmax}^2}{4(a-1)\left(\underline{\theta}(M) + \sum_{k=1}^{p}\underline{\theta}(M_k)\right)}} \triangleq B_M \tag{27}$$

for the critic network and $a > 1$.

Remark 3: Note that the triggering condition (26) is adaptive, because the threshold is designed as function of the system state vector and the critic NN weight estimates. The controller is adjusted with events.

Then we give the structure diagram of the online ETCL algorithm for two-player zero-sum game in Fig. 1.

5   Simulation

Consider the continuous-time F16 aircraft plant [7]:

$$\dot{x} = \begin{bmatrix} -1.01887 & 0.90506 & -0.00215 \\ 0.82225 & -1.07741 & -0.17555 \\ 0 & 0 & -1 \end{bmatrix} x + \begin{bmatrix} 0 \\ 0 \\ 1 \end{bmatrix} u + \begin{bmatrix} 1 \\ 0 \\ 0 \end{bmatrix} w$$

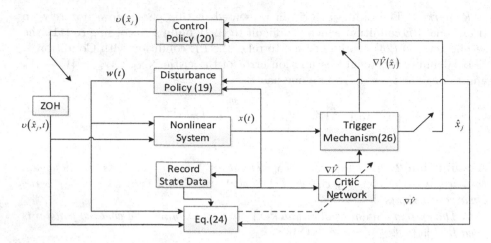

Fig. 1. Structure diagram of the ETCL algorithm for two-player ZS game

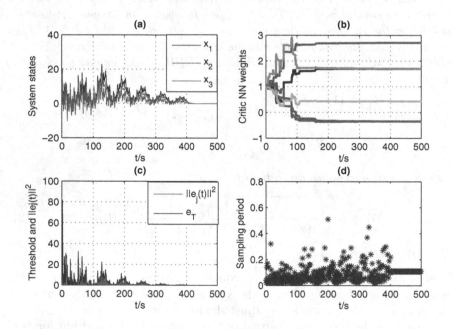

Fig. 2. (a) Evolution of system states. (b) Convergence of the critic parameters. (c) Triggering threshold e_T and $\|e_j(t)\|^2$. (d) Sampling period.

Let Q and R be identity matrices with approximate dimensions, and $\gamma = 5$. Choose the critic NN activation function as $\phi(x) = [x_1^2 \ x_1x_2 \ x_1x_3 \ x_2^2 \ x_2x_3 \ x_3^2]^T$. According to [8], the ideal values of the NN weights are $W_c = [1.6573 \ 2.7908 \ -0.3322 \ 1.6573 \ -0.3608 \ 0.4370]^T$. Select the initial state as $x_0 = [1, -1, 1]^T$, and $\alpha = 15$, $p = 10$, $L = 3$, $\beta = 0.8$. During the learning process, a probing noise is added to the control input and disturbance for the first 400s. Fig. 2(a) presents the evolution of the system states. Fig. 2(b) shows the convergence of the critic parameters. After 100s the critic parameters converged to $\hat{W}_c = [1.6563 \ 2.7788 \ -0.3389 \ 1.6490 \ -0.3615 \ 0.4354]^T$ which are nearly the ideal values above. In Fig. 2(c), one can see that the event-trigger error converges to zero as the states converge to zero. The sampling period during the event-triggered learning process for the control policy is provided in Fig. 2(d). In particular, the event-triggered controller uses 1055 samples of the state while the time-triggered controller uses 50000 samples, which means the event-triggered method improved the learning process.

Select a disturbance signal with $t_0 = 5$ as

$$w(t) = \begin{cases} 8e^{-(t-t_0)} \cos(t - t_0), & t \geq t_0 \\ 0, & t < t_0 \end{cases} \tag{28}$$

Fig. 3 shows the system state trajectories and the event-triggered control input with the H_∞ event-triggered controller. These simulation results verify the effectiveness of the developed control approach.

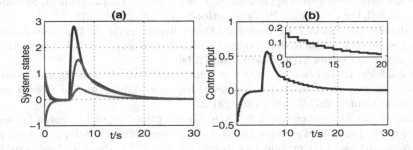

Fig. 3. (a) Closed-loop system states. (b) Event-triggered control input

6   Conclusion

In this paper, we propose an online ETCL algorithm to solve the HJI equation of H_∞ control problem for nonlinear system. The H_∞ control problem is described as a two-player zero-sum game, where the control is a minimizing player and the disturbance is a maximizing one. With an event-triggered control policy and a time-triggered disturbance policy, the online ETCL algorithm is presented. For implementation purpose, only one critic NN is used to approximate the value function, the optimal control and disturbance policies. Furthermore, a novel critic tuning law based on concurrent learning technique is given, which can relax the traditional PE condition. In our future work, we will develop an online ETCL algorithm for the unknown two-player zero-sum game system.

Acknowledgments. This work is supported in part by National Natural Science Foundation of China (NSFC) under Grant No. 61273136 and No. 61473176, Beijing Nova Program under Grant Z141101001814094, and Science and Technology Foundation of SGCC under Grant DG71-14-032.

References

1. Basar, T., Olsder, G.J., Clsder, G.J., et al.: Dynamic noncooperative game theory. Academic Press, London (1995)
2. Zhao, D., Zhu, Y.: MEC—A Near-Optimal Online Reinforcement Learning Algorithm for Continuous Deterministic Systems. IEEE Transactions on Neural Networks and Learning Systems 26(2), 346–356 (2015)
3. Zhao, D., Xia, Z., Wang, D.: Model-Free Optimal Control for Affine Nonlinear Systems With Convergence Analysis. IEEE Transactions on Automation Science and Engineering (2015), doi:10.1109/TASE.2014.2348991
4. Alippi, C., Ferrero, A., Piuri, V.: Artificial intelligence for instruments and measurement applications. IEEE Instrumentation & Measurement Magazine 1(2), 9–17 (1998)
5. Al-Tamimi, A., Abu-Khalaf, M., Lewis, F.L.: Adaptive critic designs for discrete-time zero-sum games with application to control. IEEE Transactions on Systems, Man, and Cybernetics, Part B: Cybernetics 37(1), 240–247 (2007)
6. Abu-Khalaf, M., Lewis, F.L., Huang, J.: Policy iterations on the Hamilton-Jacobi-Isaacs equation for state feedback control with input saturation. IEEE Transactions on Automatic Control 51(12), 1989–1995 (2006)
7. Vamvoudakis, K.G., Lewis, F.L.: Online solution of nonlinear two-player zero-sum games using synchronous policy iteration. International Journal of Robust and Nonlinear Control 22(13), 1460–1483 (2012)
8. Sahoo, A., Xu, H., Jagannathan, S.: Event-based optimal regulator design for nonlinear networked control systems. In: 2014 IEEE Symposium on Adaptive Dynamic Programming and Reinforcement Learning, pp. 1–8. IEEE Press, Orlando (2014)
9. Zhong, X., Ni, Z., He, H., Xu, X., Zhao, D.: Event-triggered reinforcement learning approach for unknown nonlinear continuous-time system. In: 2014 International Joint Conference on Neural Networks, pp. 3677–3684. IEEE Press, Beijing (2014)
10. Vamvoudakis, K.G.: Event-triggered optimal adaptive control algorithm for continuous-time nonlinear systems. IEEE/CAA Journal of Automatica Sinica 1(3), 282–293 (2014)
11. Chowdhary, G., Johnson, E.: Concurrent learning for convergence in adaptive control without persistency of excitation. In: 49th IEEE Conference on Decision and Control (CDC), pp. 3674–3679. IEEE Press, Atlanta (2010)
12. Modares, H., Lewis, F.L., Naghibi-Sistani, M.B.: Integral reinforcement learning and experience replay for adaptive optimal control of partially-unknown constrained-input continuous-time systems. Automatica 50(1), 193–202 (2014)

Adaptive Control of a Class of Nonlinear Systems with Parameterized Unknown Dynamics

Jing-Chao Sun[1,2], Ning Wang[1,*], and Yan-Cheng Liu[1]

[1] Marine Engineering College, Dalian Maritime University, Dalian 116026, China
n.wang.dmu.cn@gmail.com
[2] Dalian Electric Traction R&D Center, China CNR Corporation Limited,
Dalian 116022, P. R. China
sunjingchaofirst@126.com

Abstract. In this paper, an observer-based adaptive control scheme for a class of nonlinear systems with parametric uncertainties is proposed. The adaptive observers using parameter estimates ensure the identification errors of system states are convergent to zero, and force the parameter estimates approach to the true values especially if the observer gains are selected large enough. By combining the Lyapunov synthesis with backstepping framework, the global asymptotical stability and bounded signals of the resulting closed-loop system can be ensured. A numerical example is employed to demonstrate the effectiveness of the proposed adaptive control scheme.

Keywords: Adaptive Control, Uncertain Nonlinear Systems, Unknown Dynamics, Nonlinear Observer.

1 Introduction

Adaptive control of nonlinear systems with uncertainties and/or unknown dynamics has always been a challenging issue in the field of control community. In the early stage, the feedback linearization technique plays a significant role in the control of nonlinear systems which are feedback linearizable, and thereby leading to an iterative design called backstepping [1]. Instead of the feedback cancellation, a domination approach, called adding a power integrator [2], is developed to stabilize nonlinear systems which are unnecessarily feedback linearizable. In this context, the development of backstepping-based and adding-a-power-integrator-based techniques provides powerful tools for control synthesis of nonlinear systems excluding uncertainties and unknown dynamics. It should be noted that the foregoing methods require exact knowledge of system dynamics or bounding functions of homogeneous conditions.

However, it is impossible to acquire exact model dynamics of any nonlinear system. In this context, uncertainties and unknown dynamics need to be allowed in the controller design for nonlinear systems. Actually, system uncertainties

---

* Corresponding author.

© Springer International Publishing Switzerland 2015
X. Hu et al. (Eds.): ISNN 2015, LNCS 9377, pp. 71–80, 2015.
DOI: 10.1007/978-3-319-25393-0_9

and/or unknown dynamics are usually assumed to be linearly or nonlinearly parameterized, and even totally unknown continuous functions.

For linearly parameterized uncertainties, unknown parameters are updated by various adaptive laws derived from Lyapunov approach. It should be noted that the adaptation of parameters towards the true values would be stopped once the system states are stabilized, and cannot ensure the convergence of estimate error to the origin. Under some restrictions, nonlinearly parameterized problems can be transformed into linearly parameterized cases which can be solved by various backstepping-based or adding-a-power-integrator-based frameworks together with adaptive laws for unknown parameters [3]. To be specific, a parameter separation principle is proposed to realize smooth [4] and nonsmooth [5] solutions for adaptive control of nonlinear systems with nonlinearly parameterized dynamics. However, the centralized parameter would inevitably suffer from conservative upper bound, and thereby resulting in high-gain feedback control signals. Although a universal adaptive control scheme for a triangular system satisfying linear growth condition is proposed in [6], a monotone non-decreasing function is required to significantly suppress the unknown growth rate, and thereby always leading to conservatively large control signals and non-zero estimate errors of unknown constant.

If uncertainties are considered as totally unknown nonlinear functions, linear-in-parameter or nonlinear-in-parameter approximators, e.g. fuzzy logic systems [7,8], fuzzy/neural networks [9,10,11,12], are usually used to identify unknown dynamics by employing linear or nonlinear parameters with adaptive laws. In this context, the foregoing case would fall into linearly or nonlinearly parameterized dynamics together with approximation residuals. It still suffers from the deficiency that the convergence of adaptive parameters to the origin can not be guaranteed, and only uniformly ultimately bounded regulation or tracking errors can be obtained.

In this paper, auxiliary observers of system states are designed by employing adaptive parameters, and are incorporated into adaptive controller synthesis for a class of nonlinear systems with parameterized uncertainties. In this context, estimate errors of system states can render the convergence of parameter estimates approach to the true values especially if observer gains are selected large enough. By using the Lyapunov approach, global asymptotical stability and bounded signals of the resulting closed-loop system can be guaranteed.

2   Problem Formulation

We consider a class of nonlinear system with parameterized unknown dynamics as follows:

$$\dot{x}_i = x_{i+1} + \boldsymbol{\theta}_i^T \mathbf{f}_i(x_1, \cdots, x_i), \ i = 1, 2, \cdots, n$$
$$y = x_1 \tag{1}$$

where $\mathbf{x} = [x_1, \cdots, x_n]^T \in \mathbb{R}^n$ are the states, $u \triangleq x_{n+1} \in \mathbb{R}$ is the control input, $\mathbf{f}_i(\cdot) = [f_{i,1}, \cdots, f_{i,s_i}]^T \in \mathbb{R}^{s_i}$ is a vector of any continuous nonlinearities, and $\boldsymbol{\theta}_i = [\theta_{i,1}, \cdots, \theta_{i,s_i}]^T \in \mathbb{R}^{s_i}$ is a vector of unknown constant parameters.

Remark 1. In the nonlinear system (1), unknown dynamics are assumed to be parameterized by $\boldsymbol{\theta}_i^T \mathbf{f}_i(x_1, \cdots, x_i)$. Actually, this point is reasonable in practice since unknown dynamics and/or uncertainties can be decomposed into a set of weighted basis nonlinearities. Moreover, from the viewpoint of approximation theory, any continuous function can be approximated by a polynomial with high enough order to any accuracy.

Remark 2. For nonlinear systems with unknown dynamics and/or uncertainties like system (1), there still exists an open issue that parameter estimates $\widehat{\boldsymbol{\theta}}_i$ are unnecessarily convergent to the true values $\boldsymbol{\theta}_i$ although the output $y = x_1$ is stabilized to the origin.

Example 1. Considering a simple nonlinear system as follows:

$$\dot{x} = u + \theta x^2 \tag{2}$$

where θ is an unknown parameter. A traditional adaptive controller can be obtained as follows:

$$u = -x - \widehat{\theta} x^2 \tag{3}$$

$$\dot{\widehat{\theta}} = x^3 \tag{4}$$

and can stabilize x to the origin. However, the parameter estimate $\widehat{\theta}$ would be denied to update as $x = 0$.

Since unknown parameters $\boldsymbol{\theta}_i \in \mathbb{R}^{s_i}$ can be any real values, we straightforwardly make a minor assumption on basis nonlinearities $\mathbf{f}_i(x_1, \cdots, x_i) \in \mathbb{R}^{s_i}$ as follows:

Assumption 1. *In the nonlinear system (1), basis nonlinearities $\mathbf{f}_i(x_1, \cdots, x_i)$, $i = 1, 2, \cdots, n$ satisfy*

$$\begin{cases} f_{i,j_i}(\overline{\mathbf{x}}_i) > 0, \ \forall\, \overline{\mathbf{x}}_i \in \mathbb{R}^i \setminus 0 \\ f_{i,j_i}(\overline{\mathbf{x}}_i) = 0, \ \overline{\mathbf{x}}_i = 0 \end{cases} \quad j_i = 1, 2, \cdots, s_i \tag{5}$$

where $\overline{\mathbf{x}}_i = [x_1, \cdots, x_i]^T$.

In this context, our objective in this paper is to design a novel adaptive control law for the system (1) such that not only the system (1) can be stabilized but also parameter estimates $\widehat{\boldsymbol{\theta}}_i$ are able to converge to the true values.

3   Main Result

In this section, we present the main result on designing an adaptive controller with parameter estimates converging to the true values for system (1) and corresponding stability analysis of the closed-loop system.

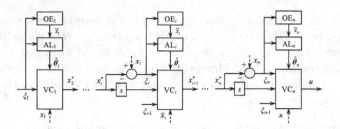

Fig. 1. Block diagram of the observer-based adaptive backstepping.

3.1 Adaptive Controller Design

By recursively combining adaptive observers of system states with backstepping technique, an adaptive controller shown in Fig. 1 for the nonlinear system (1) is explicitly designed as follows.

Step 1: Consider the system (1) with dimension $n = 1$ and design an observer of state x_1 as follows:

$$\dot{\hat{x}}_1 = x_2 + \widehat{\boldsymbol{\theta}}_1^T \mathbf{f}_1(x_1) + \mathbf{k}_1^T \mathbf{f}_1(x_1)(x_1 - \widehat{x}_1) \tag{6}$$

where $\mathbf{k}_1 = [k_{1,1}, \cdots, k_{1,s_1}]^T, k_{1,j_1} > 0$ and $\widehat{\boldsymbol{\theta}}_1 \in \mathbb{R}^{s_1}$ is the estimate of $\boldsymbol{\theta}_1$.

Let $\xi_1 = x_1, \widetilde{x}_1 = x_1 - \widehat{x}_1$ and $\widetilde{\boldsymbol{\theta}}_1 = \boldsymbol{\theta}_1 - \widehat{\boldsymbol{\theta}}_1$, we have

$$\dot{\xi}_1 = x_2 + (\widetilde{\boldsymbol{\theta}}_1 + \widehat{\boldsymbol{\theta}}_1)^T \mathbf{f}_1(x_1) \tag{7}$$

Choose the virtual control (VC) x_2^* as follows:

$$x_2^* = -\lambda_1 \xi_1 - \widehat{\boldsymbol{\theta}}_1^T \mathbf{f}_1(x_1) \tag{8}$$

with the adaptive law (AL) for $\widehat{\boldsymbol{\theta}}_1$ given by

$$\dot{\widehat{\boldsymbol{\theta}}}_1 = \gamma_1 (\xi_1 + \widetilde{x}_1) \mathbf{f}_1(x_1) \tag{9}$$

where $\lambda_1 > 0, \gamma_1 > 0$ are user-defined parameters, and the observer error (OE) $\widetilde{x}_1$ is determined by

$$\dot{\widetilde{x}}_1 = (\widehat{\boldsymbol{\theta}}_1^T - \mathbf{k}_1^T \widetilde{x}_1) \mathbf{f}_1(x_1) \tag{10}$$

$$\dot{\widetilde{\boldsymbol{\theta}}}_1 = -\gamma_1 (\xi_1 + \widetilde{x}_1) \mathbf{f}_1(x_1) \tag{11}$$

where $\xi_1 = x_1$.

Step i (2 ≤ i ≤ n − 1): Consider the observer of state x_i as follows:

$$\dot{\hat{x}}_i = x_{i+1} + \widehat{\boldsymbol{\theta}}_i^T \mathbf{f}_i(x_1, \cdots, x_i) + \mathbf{k}_i^T \mathbf{f}_i(x_1, \cdots, x_i)(x_i - \widehat{x}_i) \tag{12}$$

where $\mathbf{k}_i = [k_{i,1}, \cdots, k_{i,s_i}]^T, k_{i,j_i} > 0$ and $\widehat{\boldsymbol{\theta}}_i \in \mathbb{R}^{s_i}$ is the estimate of $\boldsymbol{\theta}_i$.

Let $\xi_i = x_i - x_i^*$, $\tilde{x}_i = x_i - \hat{x}_i$ and $\tilde{\theta}_i = \theta_i - \hat{\theta}_i$, we have

$$\dot{\xi}_i = x_{i+1} + (\tilde{\theta}_i + \hat{\theta}_i)^T \mathbf{f}_i(x_1, \cdots, x_i) - \dot{x}_i^* \tag{13}$$

Choose the virtual control (VC) x_{i+1}^* as follows:

$$x_{i+1}^* = -\xi_{i-1} - \lambda_i \xi_i - \hat{\theta}_i^T \mathbf{f}_i(x_1, \cdots, x_i) + \dot{x}_i^* \tag{14}$$

with the adaptive law (AL) for $\hat{\theta}_i$ given by

$$\dot{\hat{\theta}}_i = \gamma_i \left(\xi_i + \tilde{x}_i \right) \mathbf{f}_i(x_1, \cdots, x_i) \tag{15}$$

where $\lambda_i > 0$, $\gamma_i > 0$ are user-defined parameters, and the observer error (OE) $\tilde{x}_i$ is determined by

$$\dot{\tilde{x}}_i = (\tilde{\theta}_i^T - \mathbf{k}_i^T \tilde{x}_i) \mathbf{f}_i(x_1, \cdots, x_i) \tag{16}$$

$$\dot{\tilde{\theta}}_i = -\gamma_i \left(\xi_i + \tilde{x}_i \right) \mathbf{f}_i(x_1, \cdots, x_i) \tag{17}$$

where $\xi_i = x_i - x_i^*$.

Step n: Consider the observer of state x_n as follows:

$$\dot{\hat{x}}_n = u + \hat{\theta}_n^T \mathbf{f}_n(x_1, \cdots, x_n) + \mathbf{k}_n^T \mathbf{f}_n(x_1, \cdots, x_n)(x_n - \hat{x}_n) \tag{18}$$

where $\mathbf{k}_n = [k_{n,1}, \cdots, k_{n,s_n}]^T$, $k_{n,j_n} > 0$ and $\hat{\theta}_n \in \mathbb{R}^{s_n}$ is the estimate of θ_n.

Let $\xi_n = x_n - x_n^*$, $\tilde{x}_n = x_n - \hat{x}_n$ and $\tilde{\theta}_n = \theta_n - \hat{\theta}_n$, we have

$$\dot{\xi}_n = u + (\tilde{\theta}_n + \hat{\theta}_n)^T \mathbf{f}_n(x_1, \cdots, x_n) - \dot{x}_n^* \tag{19}$$

Choose the final control u as follows:

$$u = -\xi_{n-1} - \lambda_n \xi_n - \hat{\theta}_n^T \mathbf{f}_n(x_1, \cdots, x_n) + \dot{x}_n^* \tag{20}$$

with the adaptive law (AL) for $\hat{\theta}_n$ given by

$$\dot{\hat{\theta}}_n = \gamma_n \left(\xi_n + \tilde{x}_n \right) \mathbf{f}_n(x_1, \cdots, x_n) \tag{21}$$

where $\lambda_n > 0$, $\gamma_n > 0$ are user-defined parameters, and the observer error (OE) $\tilde{x}_n$ is determined by

$$\dot{\tilde{x}}_n = (\tilde{\theta}_n^T - \mathbf{k}_n^T \tilde{x}_n) \mathbf{f}_n(x_1, \cdots, x_n) \tag{22}$$

$$\dot{\tilde{\theta}}_n = -\gamma_n \left(\xi_n + \tilde{x}_n \right) \mathbf{f}_n(x_1, \cdots, x_n) \tag{23}$$

where $\xi_n = x_n - x_n^*$.

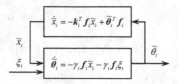

Fig. 2. Block diagram of observer error dynamics in (31) and (32).

3.2   Stability Analysis

The key result for stability analysis of the resulting closed-loop system is stated here.

Theorem 1. *Consider the nonlinear system (1) under Assumption 1, together with the state observers (6), (12) and (18), the adaptive controller (20) and adaptive laws (9), (15) and (21) driven by (10)-(11), (16)-(17) and (22)-(23) respectively, then ξ_i and $\widetilde{x}_i, i = 1, 2, \cdots, n$ globally asymptotically converge to zero, and $\widehat{\boldsymbol{\theta}}_i$ can also converge to the true value $\boldsymbol{\theta}_i$ if $\mathbf{k}_i$ is large enough.*

Proof. Substituting (14) into (13) yields

$$\dot{\xi}_i = -\xi_{i-1} - \lambda_i \xi_i + \xi_{i+1} + \widetilde{\boldsymbol{\theta}}_i^T \mathbf{f}_i(x_1, \cdots, x_i) \tag{24}$$

where $\xi_0 = 0$ and $\xi_{n+1} = 0$.

Together with error dynamics in (16) and (17), we consider the following Lyapunov candidate:

$$V = \frac{1}{2} \sum_{i=1}^{n} \left(\xi_i^2 + \widetilde{x}_i^2 + \gamma_i^{-1} \widetilde{\boldsymbol{\theta}}_i^T \widetilde{\boldsymbol{\theta}}_i \right) \tag{25}$$

Differentiating V along error dynamics in (24), (16) and (17) yeilds

$$\dot{V} = \sum_{i=1}^{n} \xi_i \left(-\xi_{i-1} - \lambda_i \xi_i + \xi_{i+1} + \widetilde{\boldsymbol{\theta}}_i^T \mathbf{f}_i(x_1, \cdots, x_i) \right)$$

$$+ \sum_{i=1}^{n} \widetilde{x}_i (\widetilde{\boldsymbol{\theta}}_i^T - \mathbf{k}_i^T \widetilde{x}_i) \mathbf{f}_i(x_1, \cdots, x_i)$$

$$+ \sum_{i=1}^{n} \widetilde{\boldsymbol{\theta}}_i^T (\xi_i + \widetilde{x}_i) \mathbf{f}_i(x_1, \cdots, x_i)$$

$$= - \sum_{i=1}^{n} \lambda_i \xi_i^2 - \sum_{i=1}^{n} \widetilde{x}_i^2 \mathbf{k}_i^T \mathbf{f}_i(x_1, \cdots, x_i) \tag{26}$$

By Assumption 1, we have

$$\dot{V} \leq -\frac{\alpha}{2} \sum_{i=1}^{n} \left(\xi_i^2 + \widetilde{x}_i^2 \right) \tag{27}$$

where $\alpha = \min_i \left\{ \lambda_i, \mathbf{k}_i^T \mathbf{f}_i(\cdot) \right\}$. It implies that ξ_i and $\widetilde{x}_i, i = 1, 2, \cdots, n$ globally asymptotically converge to zero. Namely, the closed-loop system is globally stabilized and observer errors approach to zero. Together with (17), we have the estimate error $\widetilde{\boldsymbol{\theta}}_i$ is bounded and converge to a constant.

Moreover, by (16), we have

$$\dot{\widetilde{x}}_i = -\mathbf{k}_i^T \mathbf{f}_i(x_1, \cdots, x_i)\widetilde{x}_i + \widetilde{\boldsymbol{\theta}}_i^T \mathbf{f}_i(x_1, \cdots, x_i) \tag{28}$$

Since $\dot{\widetilde{x}}_i, \widetilde{x}_i \to 0$ as $t \to +\infty$, there exists a time instant T such that

$$\dot{\widetilde{x}}_i(t) \not\equiv 0, \text{ and } \widetilde{x}_i(t) \not\equiv 0, \ \forall t \in [0, T] \tag{29}$$

Together with (28), we have

$$\mathbf{f}_i(t) \not\equiv 0, \text{ and } \widetilde{\boldsymbol{\theta}}_i(t) - \widetilde{x}_i(t)\mathbf{k}_i \not\equiv 0, \ \forall t \in [0, T] \tag{30}$$

It implies that $\overline{\mathbf{x}}_i(t) \not\equiv 0$, i.e. system states have not been stabilized. In this context, error dynamics consisting of (16) and (17) can be rewritten as follows:

$$\dot{\widetilde{x}}_i = -\mathbf{k}_i^T \mathbf{f}_i \widetilde{x}_i + \widetilde{\boldsymbol{\theta}}_i^T \mathbf{f}_i \tag{31}$$

$$\dot{\widetilde{\boldsymbol{\theta}}}_i = -\gamma_i \mathbf{f}_i \widetilde{x}_i - \gamma_i \mathbf{f}_i \xi_i \tag{32}$$

with ξ_i being the external input, and the foregoing system (31) and (32) can be depicted in Fig. 2, from which we can see that $\widetilde{\boldsymbol{\theta}}_i$ would be updated to drive $\dot{\widetilde{x}}_i$ towards zero. Once $\dot{\widetilde{x}}_i$ converge to zero, $\widetilde{x}_i$ would keep at the origin. Since $\mathbf{f}_i$ can not be zero before $\widetilde{x}_i$ is stabilized, we have $\widetilde{\boldsymbol{\theta}}_i \to \widetilde{x}_i \mathbf{k}_i$ as $t \to \infty$, and thereby $\widetilde{\boldsymbol{\theta}}_i \to 0$ and $\dot{\widetilde{\boldsymbol{\theta}}}_i \to 0$ as $t \to \infty$. Together with (32), it implies that $\xi_i \to 0$ and $\mathbf{f}_i \to 0$, which yields system state stabilization.

As a consequence, the convergence of ξ_i and $\overline{\mathbf{x}}_i$ induces that of $\widetilde{x}_i$ and $\widetilde{\boldsymbol{\theta}}_i$, and thereby contributing to the convergence of $\widehat{\boldsymbol{\theta}}_i$ to the true value $\boldsymbol{\theta}_i$.

It should be noted that if $\mathbf{k}_i$ is selected large enough, $\widetilde{x}_i$ would converge even faster to zero before $\mathbf{f}_i$ vanishes.

This concludes the proof.

Remark 3. Unlike previous adaptive controllers, the observer errors $\widetilde{x}_i$ are incorporated into adaptive laws (15) for parameter estimates $\widehat{\theta}_i, i = 1, 2, \cdots, n$, and thereby contributing to a real adaptive control scheme with adaptive parameters almost converging to the true values in addition to closed-loop system stabilization.

4 Illustrative Example

In order to demonstrate the effectiveness of the proposed observer-based adaptive control scheme, we consider the following numerical example:

$$\dot{x}_1 = x_2 + \theta_{11}\left(2x_1^2 + 3\sin^2 x_1\right) + \theta_{12}\left(x_1^4 + 2\sin^2 x_1\right)$$
$$\dot{x}_2 = u + \theta_{21}\left((x_1 + 3x_2)^2 + \sin^2(x_1 - x_2)\right) + \theta_{22}x_2^2 \tag{33}$$

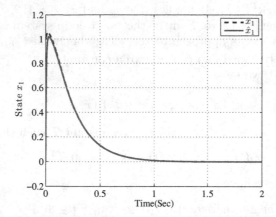

Fig. 3. System state x_1 and estimate $\widehat{x}_1$.

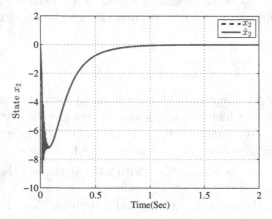

Fig. 4. System state x_2 and estimate $\widehat{x}_2$.

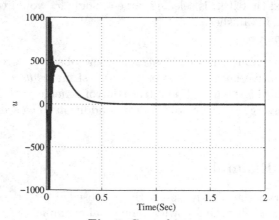

Fig. 5. Control input u.

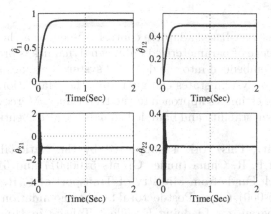

Fig. 6. Adaptive parameters $\widehat{\theta}_{11}$, $\widehat{\theta}_{12}$, $\widehat{\theta}_{21}$ and $\widehat{\theta}_{22}$.

The control law is defined as follows:

$$u = -x_1 - \lambda_2 \left(x_2 + \lambda_1 x_1 + \widehat{\boldsymbol{\theta}}_1^T \mathbf{f}_1 \right) - \widehat{\boldsymbol{\theta}}_2^T \mathbf{f}_2$$
$$- \lambda_1 \dot{x}_1 - \dot{\widehat{\boldsymbol{\theta}}}_1^T \mathbf{f}_1 - \widehat{\boldsymbol{\theta}}_1^T \dot{\mathbf{f}}_1 \qquad (34)$$

where

$$\dot{\widehat{\boldsymbol{\theta}}}_2 = \gamma_2 \left(x_2 + \lambda_1 x_1 + \widehat{\boldsymbol{\theta}}_1^T \mathbf{f}_1 + \tilde{x}_2 \right) \mathbf{f}_2 \qquad (35)$$
$$\dot{\widehat{\boldsymbol{\theta}}}_1 = \gamma_1 \left(x_1 + \tilde{x}_1 \right) \mathbf{f}_1 \qquad (36)$$

with $\tilde{x}_1 = x_1 - \widehat{x}_1$, $\tilde{x}_2 = x_2 - \widehat{x}_2$, and design the observer as follows:

$$\dot{\widehat{x}}_1 = x_2 + \widehat{\boldsymbol{\theta}}_1^T \mathbf{f}_1(x_1) + \mathbf{k}_1^T \mathbf{f}_1(x_1) \left(x_1 - \widehat{x}_1 \right)$$
$$\dot{\widehat{x}}_2 = u + \widehat{\boldsymbol{\theta}}_2^T \mathbf{f}_2(x_1, x_2) + \mathbf{k}_2^T \mathbf{f}_2(x_1, x_2) \left(x_2 - \widehat{x}_2 \right) \qquad (37)$$

and the user-defined parameters are chosen as follows: $\mathbf{k}_1 = [30, 20]^T$, $\mathbf{k}_1 = [10, 20]^T$, $\lambda_1 = 5$, $\lambda_2 = 3$, $\gamma_1 = 1.3$ and $\gamma_2 = 5$. The initial conditions are set as $x_1(0) = 1$, $x_2(0) = -1$, $\widehat{\boldsymbol{\theta}}_1 = \widehat{\boldsymbol{\theta}}_2 = 0$, and $\widehat{x}_1 = \widehat{x}_2 = 0$. The true values of unknown parameters are assumed as $\boldsymbol{\theta}_1 = [0.9, 0.5]^T$ and $\boldsymbol{\theta}_2 = [-1, 0.2]^T$.

Simulation results are shown in Figs. 3–6, from which we can see that system states can be rapidly stabilized in addition that the designed observers are able to accurately identify system states. Due to the rapid update of adaptive parameters $\widehat{\boldsymbol{\theta}}_1$ and $\widehat{\boldsymbol{\theta}}_2$, the resulting control input u shown in Fig. 20 might oscillate at the beginning of the control action. Significantly, as shown in Fig. 6, adaptive parameters $\widehat{\theta}_{11}$, $\widehat{\theta}_{12}$, $\widehat{\theta}_{21}$ and $\widehat{\theta}_{22}$ can eventually estimate corresponding true values of θ_{11}, θ_{12}, θ_{21} and θ_{22} respectively, since the system stabilization can be triggered by the incorporated state observers which are able to render the parameter adaptation not slower than state convergence.

5   Conclusions

In this paper, an observer-based adaptive control scheme for a class of nonlinear systems in the presence of parameterized unknown dynamics is proposed. Adaptive parameters are embedded into the designed system state observers whereby the estimate errors of system states can guarantee the adaptation of parameter identification till the estimates approach to the true values. Moreover, the resulting closed-loop system stability and bounded signals are also eventually ensured.

Acknowledgement. This work is supported by the National Natural Science Foundation of P. R. China (under Grants 51009017 and 51379002), Applied Basic Research Funds from Ministry of Transport of P. R. China (under Grant 2012-329-225-060), China Postdoctoral Science Foundation (under Grant 2012M520629), Program for Liaoning Excellent Talents in University (under Grant LJQ2013055).

References

1. Farrell, J.A., Polycarpou, M., Sharma, M., Dong, W.: Commanded Filtered Backstepping. IEEE Trans. Autom. Control 54, 1391–1395 (2009)
2. Lin, W., Qian, C.: Adding One Power Integrator: A Tool for Global Stabilization of High Order Lower-Triangular Systems. Syst. Control Lett. 39, 339–351 (2000)
3. Seto, D., Annaswamy, A.: Adaptive Control of Nonlinear Systems With a Triangular Structure. IEEE Trans. Autom. Control 39, 1411–1428 (1994)
4. Lin, W., Qian, C.: Adaptive Control of Nonlinearly Parametrized Systems: The Smooth Feedback Framework. IEEE Trans. Automat. Control 47, 1249–1266 (2002)
5. Lin, W., Qian, C.: Adaptive Control of Nonlinearly Parametrized Systems: A Nonsmooth Feedback Framework. IEEE Trans. Automat. Control 47, 757–774 (2002)
6. Lei, H., Lin, W.: Universal Adaptive Control of Nonlinear Systems with Unknown Growth Rate by Output Feedback. Automatica 42, 1783–1789 (2006)
7. Li, T.S., Tong, S.C., Feng, G.: A Novel Robust Adaptive-Fuzzy-Tracking Control for a Class of Nonlinear Multi-Input/Multi-Output Systems. IEEE Trans. Fuzzy Syst. 18, 150–160 (2010)
8. Tong, S.C., Li, Y.M., Feng, G., Li, T.S.: Observer-Based Adaptive Fuzzy Backstepping Dynamic Surface Control for a Class of Mimo Nonlinear Systems. IEEE Trans. Syst. Man Cybern. B, Cybern. 41, 1124–1135 (2011)
9. Wang, N., Er, M.J.: Self-Constructing Adaptive Robust Fuzzy Neural Tracking Control of Surface Vehicles with Uncertainties and Unknown Disturbances. IEEE Trans. Control Syst. Tech. 23, 991–1002 (2015)
10. Ge, S.S., Wang, C.: Direct Adaptive NN Control of a Class of Nonlinear Systems. IEEE Trans. Neural Netw. 13, 214–221 (2002)
11. Chen, M., Ge, S.S., How, B.V.E.: Robust Adaptive Neural Network Control for a Class of Uncertain MIMO Nonlinear Systems with Input Nonlinearities. IEEE Trans. Neural Netw. 21, 796–812 (2010)
12. Liu, Y.J., Tong, S.C., Wang, D., Li, T.S., Chen, C.L.P.: Adaptive Neural Output Feedback Controller Design with Reduced-Order Observer for a Class of Uncertain Nonlinear SISO Systems. IEEE Trans. Neural Netw. 22, 1328–1334 (2011)

H_∞ Control Synthesis for Linear Parabolic PDE Systems with Model-Free Policy Iteration

Biao Luo[1], Derong Liu[2], Xiong Yang[1], and Hongwen Ma[1]

[1] The State Key Laboratory of Management and Control for Complex Systems, Institute of Automation, Chinese Academy of Sciences, Beijing 100190, China
{biao.luo,xiong.yang,mahongwen2012}@ia.ac.cn
[2] School of Automation and Electrical Engineering, University of Science and Technology Beijing, Beijing 100083, China
derong@ustb.edu.cn

Abstract. The H_∞ control problem is considered for linear parabolic partial differential equation (PDE) systems with completely unknown system dynamics. We propose a model-free policy iteration (PI) method for learning the H_∞ control policy by using measured system data without system model information. First, a finite-dimensional system of ordinary differential equation (ODE) is derived, which accurately describes the dominant dynamics of the parabolic PDE system. Based on the finite-dimensional ODE model, the H_∞ control problem is reformulated, which is theoretically equivalent to solving an algebraic Riccati equation (ARE). To solve the ARE without system model information, we propose a least-square based model-free PI approach by using real system data. Finally, the simulation results demonstrate the effectiveness of the developed model-free PI method.

Keywords: Parabolic PDE systems, H_∞ control, model-free, policy iteration, algebraic Riccati equation.

1 Introduction

The control of parabolic partial differential equation (PDE) systems has attracted wide attention in recent years [1–7]. The main feature of parabolic PDEs is that the eigenspectrum of their spatial differential operator can be partitioned into a finite-dimensional slow one and an infinite-dimensional stable fast complement. This motivates applying the reduce-then-design methods for their control synthesis. Model reduction techniques are initially employed for deriving a finite-dimensional system of ordinary differential equation (ODE), which are subsequently used as the basis for the design of the finite-dimensional controllers. Following this framework, a lot of meaningful works have been reported. For example, nonlinear Galerkin's method and the concept of approximate inertial manifold were used to derive a lower-order ODE model [1], which was further used for the synthesis of finite-dimensional nonlinear feedback controllers that enforce stability and output tracking in the closed-loop PDE system. In [2],

© Springer International Publishing Switzerland 2015
X. Hu et al. (Eds.): ISNN 2015, LNCS 9377, pp. 81–90, 2015.
DOI: 10.1007/978-3-319-25393-0_10

Xu et al. considered a class of bilinear parabolic PDE systems and suggested a sequential linear quadratic regulator approach based on an iterative scheme. Neuro-dynamic programming was proposed to solve the optimal control problem of highly dissipative PDE systems [7]. H_∞ fuzzy control methods [5,6] were also used to nonlinear parabolic PDE systems when external disturbances exist. It is noted that most of these approaches are model-based that require the full knowledge of the mathematical system models.

Over the past few decades, reinforcement learning (RL) have been successfully introduced to solve optimal and H_∞ control problems of ODE systems. It is known that the H_∞ control problem of ODE systems can be converted to solve the Hamilton-Jacobi-Isaacs equation (HJIE) [8,9], which is an algebraic Riccati equation (ARE) for linear systems [10]. To solve the HJIE or the ARE, some important works have been developed, such as, synchronous policy iteration method [11], iterative methods [12,13], simultaneous policy update algorithms [14–16] and off-policy RL [17]. Based on the iterative method in [13], Vrabie and Lewis [18] suggested a policy iteration (PI) approach for learning the solution of ARE online. These approaches are either completely model-based that require the full system models, or partially model-based that need system dynamics in part. In contrast, Al-Tamimi et al. [19] proposed a model-free Q-learning algorithm for the discrete H_∞ control problem. Recently, two attractive PI methods have been suggested respectively in [20] and [21], to find adaptive optimal controllers for continuous-time linear systems with completely unknown system dynamics. However, RL approaches are rarely studied for solving the control problems of PDE system till present and only a few works have been reported [22–24]. In [22,23], PI and heuristic dynamic programming were employed to solve the optimal control problem of partially unknown linear hyperbolic PDE systems. To the best of our knowledge, the data-based H_∞ control problem of completely unknown parabolic PDE systems has not yet been addressed with RL, which motivates the present study.

2 Problem Description

We consider the following linear continuous-time parabolic PDE systems:

$$\begin{cases} \dfrac{\partial y(z,t)}{\partial t} = \dfrac{\partial^2 y(z,t)}{\partial z^2} + \overline{A}_1(z)y(z,t) + \overline{B}_1(z)w(t) + \overline{B}_2(z)u(t) \\ y_h(t) = \displaystyle\int_{z_1}^{z_2} H(z)y(z,t)dz \end{cases} \tag{1}$$

subjected to the mixed-type boundary conditions

$$\begin{cases} M_1 y(z_1,t) + N_1 \partial y(z_1,t)/\partial z_1 = d_1 \\ M_2 y(z_2,t) + N_2 \partial y(z_2,t)/\partial z_2 = d_2 \end{cases} \tag{2}$$

and the initial condition

$$y(z,0) = y_0(z) \tag{3}$$

where $z \in [z_1, z_2] \subset \mathbb{R}$ is the spatial coordinate, $t \in [0, \infty)$ is the temporal coordinate, $y(z,t) = [y_1(z,t) \;\cdots\; y_n(z,t)]^T \in \mathbb{R}^n$ is the state, $u(t) \in \mathbb{R}^p$ is the manipulated input, $y_h(z,t) = [y_{h,1}(z,t) \;\cdots\; y_{h,m}(z,t)]^T \in \mathbb{R}^m$ is the objective output, $w(t) \in L_2([0,\infty), \mathbb{R}^p)$ is the exogenous disturbance. $\overline{A}_1(z)$, $\overline{B}_1(z)$ and $\overline{B}_2(z)$ are unknown sufficiently smooth matrix functions of appropriate dimensions, $\overline{B}_1(z)$ and $\overline{B}_2(z)$ describe how disturbance and control actions are distributed in spatial domains respectively, $H(z)$ is a given sufficiently smooth matrix function of appropriate dimension and $y_0(z)$ is a smooth vector function representing the initial state profile.

For convenience, we denote $y(\cdot, t) \triangleq y(z,t)$ and $M(\cdot) \triangleq M(z)$, $z \in [z_1, z_2]$. The H_∞ control problem under consideration is to find a state feedback control law such that the PDE system (1)-(3) is asymptotically stable in the L_2-norm sense, i.e., $\|y(\cdot, t)\|_2 \to 0$ when $t \to 0$, and has L_2-gain less than or equal to γ, that is,

$$\int_0^\infty \left(\|y_h(z,t)\|^2 + \|u(t)\|_R^2 \right) dt \leq \gamma^2 \int_0^\infty \|w(t)\|^2 dt \tag{4}$$

for all $w(t) \in L_2([0,\infty], \mathbb{R}^q), R > 0$ and $\gamma > 0$ is some prescribed level of disturbance attenuation.

3   Model-Free PI Method for H_∞ Control Design

In this section, a model-free PI method is developed for solving the H_∞ control problem in Section 2. Firstly, the H_∞ control problem is reformulated based on the finite-dimensional ODE system. Subsequently, a model-free PI approach is proposed and its implementation is based on a least-square scheme.

3.1   Finite-Dimensional Problem Reformulation

To proceed with the presentation of the SP technique, define the linear parabolic spatial differential operator (SDO) $\mathcal{A}$ of the PDE system (1)-(3) as

$$\mathcal{A}\phi(z) \triangleq \frac{\partial^2 \phi(z)}{\partial z^2} \tag{5}$$

where $\phi(z)$ is a smooth vector function on $[z_1, z_2]$ that satisfies the boundary condition (2). The standard eigenvalue problem is defined as:

$$\mathcal{A}\phi_i(z) = \lambda_i \phi_i(z) \tag{6}$$

where $\lambda_i \in \mathbb{R}$ denotes the ith eigenvalue and $\phi_i(z)$ is its orthogonal eigenfunction; the eigenspectrum of $\mathcal{A}$, $\sigma(\mathcal{A})$ is defined as the set of all eigenvalues of $\mathcal{A}$, i.e., $\sigma(\mathcal{A}) \triangleq \{\lambda_1, \lambda_2, ...\}$.

One important feature of parabolic PDE systems is that the eigenspectrum of its spatial differential operator (SDO) can be partitioned into a finite-dimensional slow one $\sigma_s(\mathcal{A}) \triangleq \{\lambda_1, ..., \lambda_N\}$ and an infinite-dimensional stable fast complement $\sigma_f(\mathcal{A}) \triangleq \{\lambda_{N+1}, \lambda_{N+2}, ...\}$. Accordingly, the eigenfunctions of $\mathcal{A}$ can also be divided into two parts, i.e., $\Phi_s(z) \triangleq [\phi_1(z), \cdots, \phi_N(z)]^T$ and $\Phi_f(z) \triangleq [\phi_{N+1}(z), \cdots, \phi_\infty(z)]^T$.

To simplify the notation, we consider the PDE system (1)-(3) with $n = 1$ without loss of generality. Assume that the PDE state $y(z, t)$ can be represented as an infinite weighted sum of the eigenspectrum, i.e.,

$$y(z, t) = \sum_{i=1}^{\infty} x_i(t) \phi_i(z) \tag{7}$$

where $x_i(t)$ is a time-varying coefficient named the mode of the PDE system. By taking inner product with $\Phi_s(z)$ on both sides of PDE system (1)-(3), we obtain the following finite-dimensional ODE system:

$$\begin{cases} \dot{x}(t) = Ax(t) + B_1 w(t) + B_2 u(t) & x(0) = x_0 \\ y_{hs}(t) = H_s x(t) \end{cases} \tag{8}$$

where

$$x(t) \triangleq \langle y(\cdot, t), \Phi_s(\cdot) \rangle = [x_1(t) \ ... \ x_N(t)]^T \tag{9}$$

$A \triangleq diag(\sigma_s(\mathcal{A})) + \langle \overline{A}_1(\cdot), \Phi_s(\cdot) \rangle$, $B_1 \triangleq \langle \overline{B}_1(\cdot), \Phi_s(\cdot) \rangle$, $B_2 \triangleq \langle \overline{B}_2(\cdot), \Phi_s(\cdot) \rangle$, $H_s \triangleq \int_{z_1}^{z_2} H(z) \Phi_s^T(z) dz$ and $x_0 \triangleq \langle y_0(\cdot), \Phi_s(\cdot) \rangle$. It is noted that the matrices A, B_1 and B_2 are unknown for the matrix functions $\overline{A}_1$, $\overline{B}_1$ and $\overline{B}_2$ are unknown in the original PDE system (1)-(3). Accordingly, the L_2 disturbance attenuation criterion (4) can be given by

$$\int_0^\infty \left(\|x(t)\|_{Q_s}^2 + \|u(t)\|_R^2 \right) dt \leqslant \gamma_s^2 \int_0^\infty \|w(t)\|^2 dt \tag{10}$$

where $Q_s \triangleq H_s^T H_s$ and γ_s is a given positive constant satisfying $0 < \gamma_s < \gamma$. Then, the H_∞ control problem can be reformulated as: Considering the PDE system (1)-(3), design state feedback control based on the finite-dimensional ODE system (9) and the L_2-gain performance (10).

3.2   Model-Free PI Method

If matrices A, B_1 and B_2 are known, the H_∞ control problem of the ODE model (9) with L_2-gain performance (10), can be transformed to solve the following ARE [10]

$$A^T P + PA + Q_s + \gamma_s^{-2} PB_1 B_1^T P - PB_2 R^{-1} B_2^T P = 0 \tag{11}$$

for $P \geqslant 0$. Then, the H_∞ control policy u^* and worst disturbance w^* are given as

$$u^*(t) = K_u^* x(t) \tag{12}$$
$$w^*(t) = K_w^* x(t) \tag{13}$$

with

$$K_u^* \triangleq -R^{-1}B_2^T P \tag{14}$$

$$K_w^* \triangleq \gamma_s^{-2}B_1^T P. \tag{15}$$

Obviously, the H_∞ control policy (12) depends on the solution of the ARE (11). Lanzon and Feng et al. [13] proposed a model-based iterative method, which solves a series of the following H_2-ARE

$$[\widetilde{A}^{(i)}]^T Z^{(i+1)} + Z^{(i+1)} \widetilde{A}^{(i)} - Z^{(i+1)}B_2 R^{-1}B_2^T Z^{(i+1)} + \mathcal{F}_2(P^{(i)}) = 0 \tag{16}$$

for $Z^{(i+1)}$, where

$$\widetilde{A}^{(i)} \triangleq A + B_1 K_w^{(i)} + B_2 K_u^{(i)} \tag{17}$$

with

$$K_u^{(i)} \triangleq -R^{-1}B_2^T P^{(i)} \tag{18}$$

$$K_w^{(i)} \triangleq \gamma_s^{-2}B_1^T P^{(i)}. \tag{19}$$

Then, update $P^{(i+1)}$ with

$$P^{(i+1)} \triangleq P^{(i)} + Z^{(i+1)}. \tag{20}$$

As mentioned in [13], $\mathcal{F}_2(P^{(i)}) \triangleq= \gamma_s^{-2}Z^{(i)}B_1 B_1^T Z^{(i)}$ is a "negative semidefinite quadratic term". The Kleinman's method [25] was then used to solve the H_2-ARE (16) by solving a sequence of Lyapunov matrix equations (LMEs). Thus, we can derive the following algorithm for solving the ARE (11).

Algorithm 1. Model-based iterative method for ARE

▶ *Step 1:* Select $P^{(0)}$ such that $\widetilde{A}^{(0)}$ is Hurwitz, and let $i = 0$;
▶ *Step 2:* Let $Z^{(i+1,0)} = 0$ and $j = 0$; Compute $P^{(i+1,j)}$ with $P^{(i+1,j)} = P^{(i)} + Z^{(i+1,j)}$;
▶ *Step 3:* Solve the following LME for $Z^{(i+1,j+1)}$:

$$[\widetilde{A}^{(i,j)}]^T Z^{(i+1,j+1)} + Z^{(i+1,j+1)} \widetilde{A}^{(i,j)} - Z^{(i+1,j+1)}B_2 R^{-1}B_2^T Z^{(i+1,j+1)}$$
$$+\mathcal{F}_2(P^{(i)}) = 0 \tag{21}$$

where

$$\widetilde{A}^{(i,j)} \triangleq \widetilde{A}^{(i)} - B_2 R^{-1}B_2^T Z^{(i+1,j)} = A + B_1 K_w^{(i)} + B_2 K_u^{(i+1,j)}$$

with

$$K_u^{(i+1,j)} \triangleq -R^{-1}B_2^T P^{(i+1,j)}.$$

▶ *Step 4:* Set $j = j + 1$. If $Z^{(i+1,j)}$ is convergent *w.r.t* (with respect to) j, go to Step 5, else go to Step 3.

▶ *Step 5:* Set $Z^{(i+1)} = Z^{(i+1,j)}$, compute $P^{(i+1)}$ with (20) and let $i = i + 1$. If $Z^{(i+1)}$ is convergent *w.r.t* i, stop and use $P^{(i)}$ as the solution of the ARE (11), else go to Step 2 and continue. □

By completing the square, the LME (21) can be equivalently rewritten as

$$[\tilde{A}^{(i,j)}]^T P^{(i+1,j+1)} + P^{(i+1,j+1)} \tilde{A}^{(i,j)} + Q^{(i,j)} = 0 \tag{22}$$

where

$$Q^{(i,j)} \triangleq Q_s - \gamma_s^{-2} P^{(i)} B_1 B_1^T P^{(i)} + P^{(i+1,j)} B_2 R^{-1} B_2^T P^{(i+1,j)}$$
$$= Q_s - \gamma_s^2 (K_w^{(i)})^T K_w^{(i)} + (K_u^{(i+1,j)})^T R K_u^{(i+1,j)}.$$

Note that Algorithm 1 still requires the model of A, B_1 and B_2. To avoid using system model, we propose a model-free PI approach. Rewrite the finite-dimensional ODE system (8) as

$$\dot{x} = \tilde{A}^{(i,j)} x + B_1 (w - K_w^{(i)} x) + B_2 (u - K_u^{(i+1,j)} x). \tag{23}$$

By using (22) and (23), we have that

$$x^T(t + \Delta t) P^{(i+1,j+1)} x(t + \Delta t) - x^T(t) P^{(i+1,j+1)} x(t)$$
$$= \int_t^{t+\Delta t} d[x^T(\tau) P^{(i+1,j+1)} x(\tau)]$$
$$= 2 \int_t^{t+\Delta t} \dot{x}^T(\tau) P^{(i+1,j+1)} x(\tau) d\tau$$
$$= -\int_t^{t+\Delta t} x^T(\tau) Q^{(i,j)} x(\tau) d\tau$$
$$+ 2 \int_t^{t+\Delta t} \gamma_s^2 [w(\tau) - K_w^{(i)} x(\tau)]^T K_w^{(i+1,j+1)} x(\tau) d\tau$$
$$- 2 \int_t^{t+\Delta t} [u(\tau) - K_u^{(i+1,j)} x(\tau)]^T K_u^{(i+1,j+1)} x(\tau) d\tau \tag{24}$$

where $K_w^{(i+1,j+1)} \triangleq \gamma_s^{-2} B_1^T P^{(i+1,j+1)}$.

Note that the system matrices A, B_1 and B_2 are not involved in the equation (24). Thus, replacing (21) with (24) in Algorithm 1 generates the model-free PI method.

3.3 Implementation

To solve the equation (24) for unknown matrices $P^{(i+1,j+1)}$, $K_w^{(i+1,j+1)}$ and $K_u^{(i+1,j+1)}$, we develop a least-square scheme, which is similar with that in [16, 20, 26]. To avoid repetition, we give the least-square scheme directly as follows:

$$\theta^{(i+1,j+1)} = [(X^{(i+1,j)})^T X^{(i+1,j)}]^T X^{(i+1,j)} \eta^{(i,j)} \tag{25}$$

where $\theta^{(i+1,j+1)} \triangleq [(\overline{p}^{(i+1,j+1)})^T \quad vec((K_w^{(i+1,j+1)})^T) \quad vec((K_u^{(i+1,j+1)})^T)]^T$ is the unknown parameter vector, $X^{(i+1,j)} \triangleq [\Pi \quad -2\gamma_s^2 I_{xw} + 2\gamma_s^2 I_{xx}(I_n \otimes K_w^{(i)T})$ $2I_{xu}(I_n \otimes R) - 2I_{xx}(I_n \otimes K_u^{(i+1,j)T} R)]$ and $\eta^{(i,j)} \triangleq -I_{xx} vec(Q^{(i,j)})$. The notations Π, I_{xx}, I_{xu} and I_{xw} are given by $\Pi \triangleq [\delta_{xx}(0) \quad \delta_{xx}(1) \cdots \delta_{xx}(L-1)]^T$, $I_{xx} \triangleq [\rho_{xx}(0) \quad \rho_{xx}(1) \cdots \rho_{xx}(L-1)]^T$, $I_{xu} \triangleq [\rho_{xu}(0) \quad \rho_{xu}(1) \cdots \rho_{xu}(L-1)]^T$ and $I_{xw} \triangleq [\rho_{xw}(0) \quad \rho_{xw}(1) \cdots \rho_{xw}(L-1)]^T$, where $\delta_{xx}(k)\overline{x} \triangleq (t_{k+1}) - \overline{x}(t_k)$, $\delta_{xx}(k) \triangleq \int_{t_k}^{t_{k+1}} x(\tau) \otimes x(\tau)d\tau$, $\delta_{xw}(k) \triangleq \int_{t_k}^{t_{k+1}} x(\tau) \otimes w(\tau)d\tau$ and $\delta_{xu}(k) \triangleq \int_{t_k}^{t_{k+1}} x(\tau) \otimes u(\tau)d\tau$, with $t_k = k \triangle t, (k = 0, 1, ..., L)$ and L be a positive integer.

Based on Algorithm 1 and the parameter update rule (25), the following implementation procedure for model-free PI method is obtained.

Algorithm 2. Implementation procedure for model-free PI method

▶ *Step 1:* Select initial gain matrices $K_u^{(0)}$ and $K_w^{(0)}$ such that $\widetilde{A}^{(0)}$ is Hurwitz; Let $i = 0$;
▶ *Step 2:* Collect system data from real PDE system for computing Π, I_{xx}, I_{xu} and I_{xw};
▶ *Step 3:* Let $K_u^{(i+1,0)} = K_u^{(i)}$ and $j = 0$;
▶ *Step 4:* Compute $\theta^{(i+1,j+1)}$ with (25);
▶ *Step 5:* If $\overline{\sigma}(K_u^{(i+1,j+1)} - K_u^{(i+1,j)}) \leqslant \zeta_u$ (ζ_u is a small positive real number), let $K_u^{(i+1)} = K_u^{(i+1,j+1)}$ and go to Step 6, else $j = j+1$ and go back to Step 4;
▶ *Step 6:* If $\overline{\sigma}(K_w^{(i+1,j+1)} - K_w^{(i)}) \leqslant \zeta_w$ (ζ_w is a small positive real number), let $K_w^{(i+1)} = K_w^{(i+1,j+1)}$, terminate algorithm and use $K_u^{(i)}$ as the H_∞ control gain, else $i = i+1$ and go back to Step 3. □

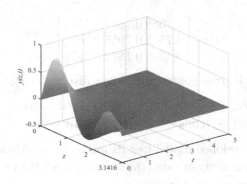

Fig. 1. State profile $y(z, t)$ of the closed-loop PDE system.

4   Simulation Studies

To verify the developed model-free PI method, we consider the following linear parabolic PDE system:

$$\begin{cases} \dfrac{\partial y(z,t)}{\partial t} = \dfrac{\partial^2 y(z,t)}{\partial z^2} + \bar{a}_1(z)y(z,t) + \bar{b}_1(z)w(t) + \bar{b}_2(z)u(t) \\ y_h(t) = \displaystyle\int_{z_1}^{z_2} y(z,t)dz \end{cases} \tag{26}$$

subjected to the Dirichlet boundary conditions

$$y(0,t) = y(\pi,t) = 0 \tag{27}$$

and the initial condition

$$y_0(z) = 0.2\sin(z) + 0.3\sin(2z) + 0.5\sin(3z) \tag{28}$$

where y is the PDE state, $z \in [0,\pi]$. The parameters are given as $\bar{a}_1(z) = 3\cos(z)+3$, $\bar{b}_1 = [H(z-0.8\pi)-H(z-0.9\pi)]$ and $\bar{b}_2 = [H(z-0.4\pi)-H(z-0.5\pi)]$, where $H(\cdot)$ is the standard Heaviside function. The weighting matrix R in (4) is given as $R = 1$. The parameter γ_s in (10) is given as $\gamma_s = 3$, then $\gamma > 3$ for the L_2-gain performance (4).

To verify the effectiveness of the developed model-free PI method, we firstly consider the PDE system model (26)-(28) is known. It is noted that its eigenvalue problem (6) of the SDO can be solved analytically and its solution is of the form $\lambda_i = -i^2$, $\phi_i(z) = \sqrt{2/\pi}\sin(iz)$, $i = 1,2,\dots$. Taking $\Phi_s(z) \triangleq [\phi_1(z) \quad \phi_2(z) \quad \phi_3(z)]^T$, the finite-dimensional ODE system (8) is given by

$$\begin{cases} \dot{x}(t) = \begin{bmatrix} 2 & 1.5 & 0 \\ 1.5 & -1 & 1.5 \\ 0 & 1.5 & -6 \end{bmatrix} x(t) + \begin{bmatrix} 0.8167 \\ -1.4366 \\ 1.7160 \end{bmatrix} w(t) + \begin{bmatrix} 1.7746 \\ 0.5610 \\ -1.5389 \end{bmatrix} u(t) \\ x(0) = [1.7725 \quad 2.6587 \quad 4.4311]^T \end{cases} \tag{29}$$

Then, use the MATLAB command CARE to solve the associated ARE (11), we obtain its solution as

$$P = \begin{bmatrix} 1.7089 & 0.7533 & 0.1762 \\ 0.7533 & 0.7955 & 0.1872 \\ 0.1762 & 0.1872 & 0.1872 \end{bmatrix} \tag{30}$$

From (14) and (15), the true control and worst disturbance gains are given as

$$K_u^* = [-3.1839 \quad -1.4950 \quad -0.2237] \tag{31}$$

$$K_w^* = [0.0684 \quad -0.0229 \quad 0.0101]. \tag{32}$$

Next, we use the developed model-free PI method (i.e., Algorithm 2) to solve the H_∞ control problem without using the system model. Select $K_u^{(0)} = [-20 \quad - 10 \quad 30]$ and $K_w^{(0)} = [0 \quad 0 \quad 0]$. Let parameters $\zeta_u = 10^{-5}$ and $\zeta_u = 10^{-7}$. It can be found that K_u, K_w and P approach to their true values at 16th, 5th and 16th steps, respectively. By using the convergent H_∞ control gain K_u for simulation on the closed-loop PDE system (26)-(28), Fig. 1 shows the state profile.

5 Conclusions

A model-free PI method has been developed in this paper for solving the data-based H_∞ control problem of linear parabolic PDE systems with completely unknown system dynamics. Firstly, the finite-dimensional ODE system is derived, which was used as the basis for H_∞ controller design by solving an ARE. Subsequently, a least square-based model-free PI approach is proposed to solve the ARE by using collected system data instead of mathematical model. Finally, simulation studies were conducted and the achieved results demonstrate the effectiveness of the developed method.

Acknowledgements. This work was supported in part by the National Natural Science Foundation of China under Grants 61233001, 61273140, 61304086, and 61374105, in part by Beijing Natural Science Foundation under Grant 4132078, and in part by the Early Career Development Award of SKLMCCS.

References

1. Baker, J., Christofides, P.D.: Finite-dimensional approximation and control of nonlinear parabolic PDE systems. International Journal of Control 73(5), 439–456 (2000)
2. Xu, C., Ou, Y., Schuster, E.: Sequential linear quadratic control of bilinear parabolic PDEs based on POD model reduction. Automatica 47(2), 418–426 (2011)
3. Luo, B., Wu, H.N.: Approximate optimal control design for nonlinear one-dimensional parabolic PDE systems using empirical eigenfunctions and neural network. IEEE Transactions on Systems, Man, and Cybernetics, Part B: Cybernetics 42(6), 1538–1549 (2012)
4. Wu, H.N., Luo, B.: L_2 disturbance attenuation for highly dissipative nonlinear spatially distributed processes via HJI approach. Journal of Process Control 24(5), 550–567 (2014)
5. Chen, B.S., Chang, Y.T.: Fuzzy state-space modeling and robust observer-based control design for nonlinear partial differential systems. IEEE Transactions on Fuzzy Systems 17(5), 1025–1043 (2009)
6. Chang, Y.T., Chen, B.S.: A fuzzy approach for robust reference-tracking-control design of nonlinear distributed parameter time-delayed systems and its application. IEEE Transactions on Fuzzy Systems 18(6), 1041–1057 (2010)
7. Luo, B., Wu, H.N., Li, H.X.: Adaptive optimal control of highly dissipative nonlinear spatially distributed processes with neuro-dynamic programming. IEEE Transactions on Neural Networks and Learning Systems 26(4), 684–696 (2015)
8. Schaft, A.V.D.: L_2-Gain and Passivity in Nonlinear Control. Springer-Verlag New York, Inc. (1996)
9. Başar, T., Bernhard, P.: H_∞ Optimal Control and Related Minimax Design Problems: A Dynamic Game Approach. Springer (2008)
10. Green, M., Limebeer, D.J.: Linear Robust Control. Prentice-Hall, Englewood Cliffs (1995)
11. Vamvoudakis, K.G., Lewis, F.L.: Online solution of nonlinear two-player zero-sum games using synchronous policy iteration. International Journal of Robust and Nonlinear Control 22(13), 1460–1483 (2012)

12. Feng, Y., Anderson, B., Rotkowitz, M.: A game theoretic algorithm to compute local stabilizing solutions to HJBI equations in nonlinear H_∞ control. Automatica 45(4), 881–888 (2009)
13. Lanzon, A., Feng, Y., Anderson, B.D., Rotkowitz, M.: Computing the positive stabilizing solution to algebraic riccati equations with an indefinite quadratic term via a recursive method. IEEE Transactions on Automatic Control 53(10), 2280–2291 (2008)
14. Wu, H.N., Luo, B.: Neural network based online simultaneous policy update algorithm for solving the HJI equation in nonlinear H_∞ control. IEEE Transactions on Neural Networks and Learning Systems 23(12), 1884–1895 (2012)
15. Luo, B., Wu, H.N.: Computationally efficient simultaneous policy update algorithm for nonlinear H_∞ state feedback control with Galerkin's method. International Journal of Robust and Nonlinear Control 23(9), 991–1012 (2013)
16. Wu, H.N., Luo, B.: Simultaneous policy update algorithms for learning the solution of linear continuous-time H_∞ state feedback control. Information Sciences 222, 472–485 (2013)
17. Luo, B., Wu, H.N., Huang, T.: Off-policy reinforcement learning for H_∞ control design. IEEE Transactions on Cybernetics 45(1), 65–76 (2015)
18. Vrabie, D., Lewis, F.: Adaptive dynamic programming for online solution of a zero-sum differential game. Journal of Control Theory and Applications 9(3), 353–360 (2011)
19. Al-Tamimi, A., Lewis, F.L., Abu-Khalaf, M.: Model-free Q-learning designs for linear discrete-time zero-sum games with application to H-infinity control. Automatica 43(3), 473–481 (2007)
20. Jiang, Y., Jiang, Z.P.: Computational adaptive optimal control for continuous-time linear systems with completely unknown dynamics. Automatica 48(10), 2699–2704 (2012)
21. Lee, J.Y., Park, J.B., Choi, Y.H.: Integral Q-learning and explorized policy iteration for adaptive optimal control of continuous-time linear systems. Automatica 48(11), 2850–2859 (2012)
22. Luo, B., Wu, H.N.: Online policy iteration algorithm for optimal control of linear hyperbolic PDE systems. Journal of Process Control 22(7), 1161–1170 (2012)
23. Wu, H.N., Luo, B.: Heuristic dynamic programming algorithm for optimal control design of linear continuous-time hyperbolic PDE systems. Industrial & Engineering Chemistry Research 51(27), 9310–9319 (2012)
24. Luo, B., Wu, H.N., Li, H.X.: Data-based suboptimal neuro-control design with reinforcement learning for dissipative spatially distributed processes. Industrial & Engineering Chemistry Research 53(29), 8106–8119 (2014)
25. Kleinman, D.L.: On an iterative technique for Riccati equation computations. IEEE Transactions on Automatic Control 13(1), 114–115 (1968)
26. Vrabie, D., Pastravanu, O., Abu-Khalaf, M., Lewis, F.L.: Adaptive optimal control for continuous-time linear systems based on policy iteration. Automatica 45(2), 477–484 (2009)

Exponential Synchronization of Complex Delayed Dynamical Networks with Uncertain Parameters via Intermittent Control

Haoran Zhao and Guoliang Cai[*]

Nonlinear Scientific Research Center, Jiangsu University, Zhenjiang, 212013, PR China
glcai@ujs.edu.cn

Abstract. In this paper, intermittent control scheme is adopted to investigate the exponential synchronization of complex delayed dynamical networks with uncertain parameters. Based on Lyapunov function method and mathematical analysis technique, some novel and useful criteria for exponential synchronization are established. Finally, two numerical simulations are given to illustrate the effectiveness and correctness of the derived theoretical results.

Keywords: exponential synchronization, intermittent control, time-varying delay, uncertain parameters, complex dynamical networks.

1   Introduction

During the last two decades, the investigation on complex networks has attracted a great deal of attentions due to its potential applications in various fields, such as communication, physics, biological, networks and engineering science, and so on [1,2]. The ubiquity of complex networks has naturally resulted in a range of important research problems on the networks structure facilitating and constraining the networks dynamical behaviours. In particular, more attention has been focused on synchronization and control problems of complex dynamical networks. There are many widely effective control schemes including adaptive control [3,4], feedback control [5,6], intermittent control [7,8] and impulsive control [9,10] which have been adopted to drive the networks to achieve synchronization.

As we know, controlling all the nodes of complex dynamical networks, especially those coupled with large number of nodes, is hard to implement and high-cost. To reduce the number of controlled node, pining control, in which controllers are only applied to a small fraction of networks node, has been proposed. In recent years, many results about pinning control and synchronization of complex networks have been proposed. In [11], Cai et al. investigated the synchronization of complex dynamical networks by pinning periodically intermittent control. However, in the real world, the external disturbance, parameter fluctuation and parameter uncertainties which can

---

[*] Corresponding author.

© Springer International Publishing Switzerland 2015
X. Hu et al. (Eds.): ISNN 2015, LNCS 9377, pp. 91–98, 2015.
DOI: 10.1007/978-3-319-25393-0_11

make the synchronization more difficult to realize due to measure errors or may destroy the networks stability are unavoidable.

Motivated by the above discussions, what we are going to investigate in our present study is whether the exponential synchronization can be achieved in complex networks with parameter uncertainties. In order to achieve the exponential synchronization via intermittent control, intermittent controllers are designed. By using the Lyapunov function method, we derived some novel and less conservative exponential synchronization criteria. Finally, two numerical examples are provided to show the effectiveness of the proposed method.

2   Model and Preliminaries

In this paper, we consider delayed complex dynamical networks consisting of N identical nodes with linearly diffusive couplings. The dynamic behaviors of nodes in the presence of uncertain parameters is described as

$$\dot{x}_i(t) = (A + \Delta A(t))x_i(t) + (B + \Delta B(t))f(x_i(t)) + \sum_{j=1}^{N} c_{ij}\Gamma_1 x_j(t) + \sum_{j=1}^{N} d_{ij}\Gamma_2 x_j(t - \tau(t)), \quad t > 0,$$

$$x(t) = \varphi(t), \quad -\tau \le t \le 0, \quad i = 1, 2, 3, \ldots, N.$$

$$(1)$$

where $x_i(t) = [x_{i1}(t), x_{i2}(t), \ldots, x_{in}(t)]^T \in R^n$ is the state vector of the i-th node at time t, $f: R \times R^n \rightarrow R^n$ is a continuously vector-valued function, the time delay $\tau(t)$ may be unknown but is bounded by a known constant. A, B are the nominal constant matrix, $\Delta A(t)$ and $\Delta B(t)$ denote the uncertain parameters Γ_1, Γ_2 are the inner connecting matrix, the coupling matrix $C = (c_{ij})_{n \times n}$ and $D = (d_{ij})_{n \times n}$ represent the topological structure of the whole networks, if there is a connection from node i to node $j (i \neq j)$, $c_{ij} \neq 0$ and $d_{ij} \neq 0$, otherwise $c_{ij} = 0$, $d_{ij} = 0$.

Based on the former research, we propose a general complex networks consisting of N dynamic nodes with parameters as the response networks:

$$\dot{y}_i(t) = (A + \Delta A(t))y_i(t) + (B + \Delta B(t))f(y_i(t)) + \sum_{j=1}^{N} c_{ij}\Gamma_1 y_j(t) + \sum_{j=1}^{N} d_{ij}\Gamma_2 y_j(t - \tau(t)) + u_i(t),$$

$$(2)$$

where $y_i(t) = [y_{i1}(t), y_{i2}(t), \ldots, y_{in}(t)]^T \in R^n$ is the state vector of the i-th node. In order to achieve synchronization between the drive and response systems, intermittent controllers are added to all nodes of networks (1). Here, the intermittent controller $u_i(t)$ is designed as follows:

$$u_i(t) = -k_i(t)(y_i(t) - x_i(t)) \quad i = 1, 2, \ldots, N, \quad (3)$$

where $k_i(t)$ is the intermittent feedback control gain defined as follows:

$$k_i(t) = \begin{cases} k_i, nT \le t < (n+\theta)T \\ 0, (n+\theta T) \le t \le (n+1)T \end{cases} \quad i = 1, 2, \ldots, N, \quad (4)$$

where $k_i > 0$ is a positive constant, $T > 0$ is the control period, θ is control rate and $n = 0$, 1, 2, 3,....

According to the control law (4), the error dynamical system can be derived as

$$\dot{e}_i(t) = (A + \Delta A(t))e_i(t) + (B + \Delta B(t))(f(y_i(t)) - f(x_i(t))) + \sum_{j=1}^{N} c_{ij}\Gamma_1 e_j(t)$$

$$+ \sum_{j=1}^{N} d_{ij}\Gamma_2 e_j(t - \tau(t)), \quad nT \le t < (n+\theta)T, \tag{5}$$

$$\dot{e}_i(t) = (A + \Delta A(t))e_i(t) + (B + \Delta B(t))(f(y_i(t)) - f(x_i(t))) + \sum_{j=1}^{N} c_{ij}\Gamma_1 e_j(t)$$

$$+ \sum_{j=1}^{N} d_{ij}\Gamma_2 e_j(t - \tau(t)) - k_i(t)e_i(t), \quad (n+\theta)T \le t \le (n+1)T,$$

Assumption 1. For any $x(t)$, $y(t) \in R^n$, there exists a positive constant $L > 0$ such that $\|f(y(t)) - f(x(t))\| \le L\|y((t)) - x((t))\|$. The norm $\|\cdot\|$ of a variable is defined as $\|x\| = (x^T x)^{1/2}$.

Assumption 2. The parametric uncertainties $\Delta A(t)$ and $\Delta B(t)$ are in the form of $[\Delta A(t), \Delta B(t)] = EH(t)[G_1, G_2]$, where E and $G_i (i=1, 2)$ are known real constant matrices with appropriate dimensions, and the unknown real time-varying matrix $H(t)$ satisfies the following condition: $H^T(t)H(t) \le I$.

Lemma 1. [12] Suppose that function $y(t)$ is continuous and non-negative when $t \in [-\tau, \infty)$ and satisfies the following conditions:

$$\begin{cases} \dot{y}(t) \le -\gamma_1 y(t) + \gamma_2 \left(\sup_{t-\tau \le s \le t} y(s) \right), nT \le t < (n+\theta)T, \\ \dot{y}(t) \le \gamma_3 y(t) + \gamma_4 \left(\sup_{t-\tau \le s \le t} y(s) \right), (n+\theta)T \le t \le (n+1)T, \end{cases} \tag{6}$$

where $\gamma_1, \gamma_2, \gamma_3, \gamma_4, 0 < \theta < 1$ are constants and $n=0, 1, 2,...,$ if $\delta = \gamma_1 + \gamma_3 > 0$, $\gamma_1 > \gamma^* = \max\{\gamma_2, \gamma_4\} > 0$, and $\eta = \lambda - \delta(1-\theta) > 0$,then $y(t) \le \left(\sup_{-\tau \le s \le 0} y(s) \right) \exp\{-\eta t\}, t \ge 0$, where $\lambda > 0$ is the only positive solution of the equation $\lambda - \gamma_1 + \gamma^* \exp\{\lambda \tau\} = 0$.

3   Main Result

In this section, $\mu_{\min}$ is defined as the minimum eigenvalue of the matrix $(\Gamma_1 + \Gamma_1^T)/2$. We assume that $\mu_{\min} \ne 0$ and $\|\Gamma_1\| = \mu > 0$. Let $\hat{C}^s = (\hat{C} + \hat{C}^T)/2$ where $\hat{C}$ is a modified matrix of C via replacing the diagonal elements C_{ii} by $(\mu_{\min}/\mu)C_{ii}$. We note that generally $\hat{C}$ not possessing the property of zero row sums. In addition, the eigenvalues of C do not have a appropriate relationship with those of $\hat{C}$ for the general matrix C. Let $P = D \otimes \Gamma_2$ where $\otimes$ stands for the kronecker product. For realizing globally exponential synchronization, the suitable $k_i(t)(i=1, 2, ..., N)$, θ and T need to be designed in the following discussions.

Theorem 1. Suppose that Assumption 1 holds. If there exist positive constants β_1, β_2, β_3 and $k_i(i=1, 2, 3, ..., N)$ such that the following conditions hold:

(i) $Q+\mu\hat{C}^s+(\lambda_{max}(\frac{1}{2}PP^T)+\beta_1-k)I_n\leq 0,$

(ii) $Q+\mu\hat{C}^s-(\beta_3-\beta_1-\lambda_{max}(\frac{1}{2}PP^T))I_n\leq 0,$

(iii) $1-2\beta_1<0,$

(iv) $\eta=\lambda-2\beta_3(1-\theta)>0,$

where $Q=A+L(B+EE^T+G_2^TG_2)+EE^T+G_1^TG_1$ and $\lambda>0$ is the unique positive solution of the equation $-2\beta_1+\lambda+\exp\{\lambda\tau\}=0$, then the drive system (1) and the response system (2) can achieve globally exponential synchronization.

Proof. Construct the candidate Lyapunov function as follows:

$$V(t)=\frac{1}{2}e^T(t)e(t)=\frac{1}{2}\sum_{i=1}^{N}e_i^T(t)e_i(t),\tag{7}$$

Calculating the derivative of $V(t)$ along the trajectories of system (5).
  When $nT\leq t<(n+\theta)T$, for $n=0, 1, 2,...$

$$\dot{V}(t)=\sum_{i=1}^{N}e_i^T(t)\left(A+\Delta A(t)\right)e_i(t)+\sum_{i=1}^{N}e_i^T(t)(B+\Delta B(t))\left(f\left(y_i(t)\right)-f\left(x_i(t)\right)\right)$$

$$+\sum_{i=1}^{N}\sum_{j=1}^{N}c_{ij}e_i^T(t)\Gamma_1e_j(t)+\sum_{i=1}^{N}\sum_{j=1}^{N}d_{ij}e_i^T(t)\Gamma_2e_j\left(t-\tau(t)\right),\tag{8}$$

and when $(n+\theta)T\leq t\leq(n+1)T$, for $n=0, 1, 2,...$

$$\dot{V}(t)=\sum_{i=1}^{N}e_i^T(t)\left(A+\Delta A(t)\right)e_i(t)+\sum_{i=1}^{N}e_i^T(t)(B+\Delta B(t))\left(f\left(y_i(t)\right)-f\left(x_i(t)\right)\right)$$

$$+\sum_{i=1}^{N}\sum_{j=1}^{N}c_{ij}e_i^T(t)\Gamma_1e_j(t)+\sum_{i=1}^{N}\sum_{j=1}^{N}d_{ij}e_i^T(t)\Gamma_2e_j\left(t-\tau(t)\right)-\sum_{i=1}^{N}k_i(t)e_i^T(t)e_i(t),\tag{9}$$

By Assumptions 1 and 2 and Lemma 1, we have the following two inequations:

$$\sum_{i=1}^{N}e_i^T(t)\Delta A(t)e_i(t)=\sum_{i=1}^{N}e_i^T(t)EH(t)G_1e_i(t)$$

$$\leq\sum_{i=1}^{N}e_i^T(t)EE^Te_i(t)+\sum_{i=1}^{N}e_i^T(t)G_1^TG_1e_i(t),\tag{10}$$

and

$$\sum_{i=1}^{N} e_i^{\mathrm{T}}(t)\Delta B(t)\big(f\big(y_i(t)\big)-f\big(x_i(t)\big)\big) \leq \sum_{i=1}^{N} e_i^{\mathrm{T}}(t)\Delta B(t)Le_i(t) \tag{11}$$

$$\leq \sum_{i=1}^{N} e_i^{\mathrm{T}}(t)L\big(EE^{\mathrm{T}}+G_2^{\mathrm{T}}G_2\big)e_i(t),$$

Using expressions (8)–(11) and performing more detailed calculations, we obtain: when $nT \leq t < (n+\theta)T$, for $n=0, 1, 2,...$

$$V(t) \leq e^{\mathrm{T}}(t)\left(Q+(\lambda_{\max}(\tfrac{1}{2}PP^{\mathrm{T}})-k)I_n\right)e(t)+ue^{\mathrm{T}}(t)(\hat{C}\otimes I_n)e(t)+\tfrac{1}{2}e^{\mathrm{T}}(t-\tau(t))e(t-\tau(t))$$

$$= e^{\mathrm{T}}(t)\left(Q+u\hat{C}^s+(\lambda_{\max}(\tfrac{1}{2}PP^{\mathrm{T}})+\beta_1-k)I_n\right)e(t)-\beta_1 e^{\mathrm{T}}(t)e(t)+\tfrac{1}{2}e^{\mathrm{T}}(t-\tau(t))e(t-\tau(t)) \tag{12}$$

$$\leq -2\beta_1 V(t)+V(t-\tau(t) \leq -2\beta_1 V(t)+(\sup_{t-\tau\leq s\leq t} V(s)),$$

and when $(n+\theta)T \leq t < (n+1)T$, for $n=0,1,2,...$

$$V(t) \leq e^{\mathrm{T}}(t)\left(Q+\lambda_{\max}(\tfrac{1}{2}PP^{\mathrm{T}})I_n\right)e(t)+ue^{\mathrm{T}}(t)(\hat{C}\otimes I_n)e(t)+\tfrac{1}{2}e^{\mathrm{T}}(t-\tau(t))e(t-\tau(t))$$

$$= e^{\mathrm{T}}(t)\left(Q+u\hat{C}^s-(\beta_3-\beta_1-\lambda_{\max}(\tfrac{1}{2}PP^{\mathrm{T}}))I_n\right)e(t) \tag{13}$$

$$+(\beta_3-\beta_1)e^{\mathrm{T}}(t)e(t)+\tfrac{1}{2}e^{\mathrm{T}}(t-\tau(t))e(t-\tau(t))$$

$$\leq 2(\beta_3-\beta_1)V(t)+V(t-\tau(t)) \leq 2(\beta_3-\beta_1)V(t)+(\sup_{t-\tau\leq s\leq t} V(s)).$$

From the above equation, we have

$$V(t) \leq \left(\sup_{-\tau\leq s\leq 0} V(s)\right)\exp(-\eta t), \quad t\geq 0, \tag{14}$$

Hence, the zero solution of the error dynamical system (5) is globally exponentially stable, the proof of Theorem 1 is completed.

In the following, we will discuss how to achieve synchronization by using simple date and selecting appropriate parameters. According to Theorem 1, letting

$$m_0 = \lambda_{\max}(Q)+u\lambda_{\max}(\hat{C}^s)+\lambda_{\max}(\tfrac{1}{2}PP^{\mathrm{T}}),$$

and selecting $\beta_3=m_0+\beta_1>0$, $\beta_2=\dfrac{1}{2}$. Then condition (ii) in Theorem 1 holds.

Corollary 1. If there exists a positive $\beta_1 > \beta_2$ and k_i being sufficiently large such that

$$(\mathrm{i})\ \lambda_{\max}(\hat{C}^s) < -\frac{k-\beta_1-q}{\mu},$$

$$(\mathrm{ii})\eta = \lambda - 2(m_0+\beta_1)(1-\theta) > 0,$$

where $q = \lambda_{\max}(Q) + \lambda_{\max}(\frac{1}{2}PP^{\mathrm{T}})$, and $\lambda > 0$ is the unique positive solution of the function equation $-2\beta_1 + \lambda + 2\beta_2 \exp\{\lambda\tau\} = 0$, Therefore, the controlled delayed networks (1) with uncertain parameters can be globally exponentially synchronized.

Corollary 2. Assume that β_1 is given as $\beta_1^* > \beta_2$, and k_i being sufficiently large

$$(\mathrm{i})\ \lambda_{\max}(\hat{C}^s) < -\frac{k - \beta_1^* - q}{\mu},$$

$$(\mathrm{ii})\ 1 - \frac{\varphi(\beta_1^*)}{m_0 + \beta_1^*} < \theta < 1,$$

if the above conditions (i) , (ii) hold, the controlled dynamical networks (1) with uncertain parameters can be globally exponentially synchronized.

4   Numerical Example

In order to verify and demonstrate the effectiveness of the proposed approach clearly, in this section, we show that the networks with ten nodes described by

$$\dot{x}_i(t) = (A + \Delta A(t))x_i(t) + (B + \Delta B(t))f(x_i(t)) + \sum_{j=1}^{10} c_{ij}\Gamma_1 x_j(t) + \sum_{j=1}^{10} d_{ij}\Gamma_2 x_j(t - \tau(t)), i = 1,2,...,10. \quad (15)$$

where $x(t) = (x_1(t), x_2(t), x_3(t))^{\mathrm{T}} \in \mathrm{R}^3$ is the state vector, $C = (c_{ij})_{10 \times 10}$ and $D = (d_{ij})_{10 \times 10}$ are configuration coupling matrices.

Example 1. We consider the following Lorenz system:

$$\dot{x} = \begin{pmatrix} a(x_2 - x_1) \\ cx_1 - x_1 x_3 - x_2 \\ -bx_3 + x_1 x_2 \end{pmatrix} = A \begin{pmatrix} x_1 \\ x_2 \\ x_3 \end{pmatrix} + Bf(t,x),$$

where a=10, b=8/3, c=28, Thus $\alpha = \|A\| = 28.0176$, $\beta = \|B\| = 1$.

Here we assume that $\Delta A(t) = \Delta B(t) = 0$, $\tau(t) = \frac{e^t}{1 + e^t}$, $0 < \tau(t) < 1$. the initial values are chosen as $x_i(0)=(0.3+0.1i, 0.3+0.1i, 0.3+0.1i)^{\mathrm{T}}$, $y_i(0)=(2.0+0.7i, 2.0+0.7i, 2.0+0.7i)^{\mathrm{T}}$ (i, j=1, 2,..., 10). In this numerical simulation, We fix the inner coupling matrix $\Gamma_1 = \Gamma_2 = \mathrm{diag}(1, 1, 1)$, and let L=1, T=0.7. Fig.1 shows the synchronization error system is globally stable, and the numerical result means that the exponential synchronization of the two complex networks can be achieved.

Example 2. When $[\Delta A(t), \Delta B(t)] = EH(t)[G_1, G_2]$ with $E = \mathrm{diag}\{0.2, 0.04, 0.5\}$, $H(t) = (\sin(t))I$, $G_1 = -0.4I$, and $G_2 = 0.3I$. In this example, we consider the matrix $A = -I$,

$$B = \begin{bmatrix} -1.14 & -1.3 & -1.3 \\ -1.3 & 1.14 & -3.0 \\ -1.5 & 3.0 & 1.14 \end{bmatrix},$$

$$f\left(x(t)\right) = \left(\frac{1}{2}(|x_1(t)+1|-|x_1(t)-1|), \frac{1}{2}(|x_2(t)+1|-|x_2(t)-1|), \frac{1}{2}(|x_3(t)+1|-|x_3(t)-1|) \right)^T$$

is a nonlinear function. In this example, Fig. 2 describes the synchronization errors $e_{ij}(t)$ ($i=1, 2,\ldots, 10, j=1, 2, 3$), which turn to zero as time goes. it is easy to see that the uncertain complex dynamic networks (1) via intermittent control can achieve globally exponentially synchronization.

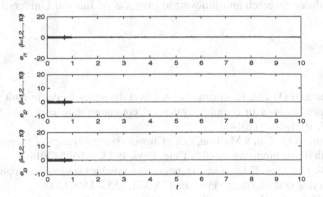

Fig. 1. Synchronization errors for $\Delta A(t) = \Delta B(t) = 0$ of example 1

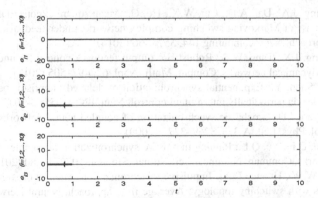

Fig. 2. Synchronization errors for $\Delta A(t) \neq 0$, $\Delta B(t) \neq 0$ of example 2

5   Conclusions

In this letter, a detailed analysis is presented for the exponential synchronization of complex delayed dynamical networks with uncertain parameters. Some useful exponential synchronization criteria for the uncertain complex networks are obtained via utilizing the Lyapunov method and Lipschitz condition. The result shows that the intermittent control remains valid for exponential synchronization even if there exits uncertain parameters in complex dynamical networks. Finally, two numerical simulations have been presented to demonstrate the effectiveness of the theoretical results.

Acknowledgements. This work was supported in part by the National Nature Science foundation of China (Grants 51276081, 11171135), the Society Science Foundation from Ministry of Education of China (Grants 12YJAZH002, 08JA790057), The Priority Academic Program Development of Jiangsu Higher Education Institutions, the Advanced Talents' Foundation of Jiangsu University (Grants 07JDG054,10JDG140), and the Postgraduate research and innovation projects of Jiangsu University (KYXX-0038)

References

1. Hong, S., Yang, H.Q., Zio, E., Huang, N.: A novel dynamics model of fault propagation and equilibrium analysis in complex dynamical communication networks. Appl. Math. Comput. 247, 1021–1029 (2014)
2. Cai, G.L., Jiang, S.Q., Cai, S.M., Tian, L.X.: Cluster synchronization of uncertain complex networks with desynchronizing impulse. Chin. Phys. B 23, 120505 (2014)
3. Zheng, S., Bi, Q.S., Cai, G.L.: Adaptive projective synchronization in complex networks with time-varying coupling delay. Phys. Lett. A 373, 1553–1559 (2009)
4. Yang, X.S., Cao, J.D.: Hybrid adaptive and impulsive synchronization of uncertain complex networks with delays and general uncertain perturbations. Appl. Math. Comput. 227, 480–493 (2014)
5. Wang, X., Fang, J.A., Dai, A.D., Cui, W.X., He, G.: Mean square exponential synchronization for a class of Markovian switching complex networks under feedback control and M-matrix approach. Neurocomputing 144, 357–366 (2014)
6. Li, S.K., Zhang, J.X., Tang, W.S.: Robust H∞ output feedback control for uncertain complex delayed dynamical networks. Comput. Math. Appl. 62, 497–505 (2011)
7. Zhang, G.D., Shen, Y.: Exponential synchronization of delayed memristor-based chaotic neural networks via periodically intermittent control. Neural Netw. 55, 1–10 (2014)
8. Yang, X.S., Cao, J.D.: Stochastic synchronization of coupled neural networks with intermittent control. Phys. Lett. A 373, 3259–3272 (2009)
9. Wu, Z.Y., Liu, D.F., Ye, Q.L.: Pinning impulsive synchronization of complex variable dynamical networks. Commun. Nonlinear Sci. Numer. Simulat. 20, 273–280 (2015)
10. Li, C.J., Yu, W.W., Huang, T.W.: Impulsive synchronization schemes of stochastic complex networks with switching topology: Average time approach. Neural Netw. 54, 85–94 (2014)
11. Cai, S.M., Hao, J.J., He, Q.B., Liu, Z.R.: Exponential synchronization of complex delayed dynamical networks via pinning periodically intermittent control. Phys. Lett. A 375, 1965–1971 (2011)
12. Cai, S.M., Zhou, P.P., Liu, Z.R.: Pinning synchronization of hybrid-coupled directed delayed dynamical network via intermittent control. Chaos 24, 33102 (2014)

Inverse-Free Scheme of G1 Type to Velocity-Level Inverse Kinematics of Redundant Robot Manipulators

Yunong Zhang[1,2,3], Liangyu He[1,2,3], Jingyao Ma[1,2,3], Ying Wang[1,2,3], and Hongzhou Tan[1,2]

[1] School of Information Science and Technology, Sun Yat-sen University (SYSU), Guangzhou 510006, China
[2] SYSU-CMU Shunde International Joint Research Institute, Shunde 528300, China
[3] Key Laboratory of Autonomous Systems and Networked Control, Ministry of Education, Guangzhou 510640, China
zhynong@mail.sysu.edu.cn,ynzhang@ieee.org,jallonzyn@sina.com

Abstract. With the superiority of owning more degrees of freedom than ordinary robot manipulators, redundant robot manipulators have gotten much attention in recent years. In order to control the trajectory of the robot end-effector with a desired velocity, it is very popular to apply the inverse kinematics approaches, such as pseudo-inverse scheme. However, calculating the inverse of Jacobian matrix requires a lot of time. Thus base on gradient neural dynamics (GND), an inverse-free scheme is proposed at the joint-velocity level. The scheme is named G1 type as it uses GND once. In addition, two path tracking simulations based on five-link and six-link redundant robot manipulators illustrate the efficiency and the accuracy of the proposed scheme. What is more, the physical realizability of G1 type scheme is also verified by a physical experiment based on the six-link planar redundant robot manipulator hardware system.

Keywords: redundant robot manipulators, control, inverse-free scheme, gradient neural dynamics, path tracking.

1 Introduction

In recent years, there have been numerous investigations of robot manipulators [1,2,3,4], especially the redundant robot manipulators. Compared with the ordinary robot manipulators, the redundant one has more degrees of freedom than necessary for position and orientation. It is worth noting that this characteristic improves the kinematic and the dynamic performance of the robot manipulator such as increasing dexterity, avoiding obstacles and singularities, and optimizing joint velocity, which makes redundant manipulators widely applied in the field of robotic manipulator control [5]. As one of the central issues in robot control, path tracking refers to making the end-effector move as expected by controlling the joints of robot manipulators. When the end-effector moves in a desired speed, it is often called path tracking control in the velocity level, which can fit in with

© Springer International Publishing Switzerland 2015
X. Hu et al. (Eds.): ISNN 2015, LNCS 9377, pp. 99–108, 2015.
DOI: 10.1007/978-3-319-25393-0_12

the needs of the operators well with two advantages [6,7]. One is that the high velocity can economize the execution time while the other one is that the low velocity can enhance the objective precision [8]. So the control in the velocity level is particularly suited to the tasks well for machining operations, such as cutting and milling [9].

Robot kinematics which studies the relationships between joint space and cartesian space, is an very effective way to control the robot manipulator and attracts numberous researchers to study [10,11,12]. The velocities of all joints play a decisive role in realizing the desired speed of the end-effector. So the most fundamental problem is how to use the kinematic equations to calculate the homologous joint velocities, which is also called inverse kinematics. However, because most kinematic equations involve complex (inverse) trigonometric functions, the inverse kinematics mapping has no closed-form solutions for most manipulators and animation figures [13]. Conventionally, most of researchers exploit pseudo-inverse approaches to obtain a simple general-form solution [14,15]. However, these approaches not only need expensive time in calculating the inverse of Jacobian matrix, but also require the Jacobian matrix to be of full rank which may be away from the reality. Thus various approaches have been proposed, investigated and developed to avoid the calculation of the inverse of Jacobian matrix, such as the quadratic programming method [16].

Gradient neural dynamics (GND) [17,18] is a significant neural dynamic method which attracts many researchers to investigate and develop it. Now GND method is proved to be useful and effective, and it is widely acknowledged in scientific and engineering field, thus generalizing such a GND method has become the primary work [16,19,20,21,22]. In this paper, based on the advantage of GND method that it can help find a minimum of a nonnegative objective function effectively, we propose and investigate a scheme named G1 type for path tracking in the redundant robot manipulator at the joint-velocity level. Besides, in the framework of Zhang-gradient (ZG) method, G1 type scheme is a special Z0G1 situation which only uses the GND method once. The ZG method is an effective method built by combining Zhang neural dynamics and GND to solve the tracking-control and singularity problems [23,24].

The rest of this paper is organized into the following sections. The inverse scheme formulation is presented in Section 2. In Section 3, the G1 type scheme is proposed and analyzed for the redundant robot manipulators at the joint-velocity level. Section 4 illustrates the effectiveness and the accuracy through two simulations, and Section 5 further illustrates the effectiveness and physical realizability of G1 type scheme based on a six-link planar redundant robot manipulator hardware system. Finally, Section 6 concludes the whole work with final remarks. Before ending this introductory section, the main contributions of this paper are listed as follows.

1) By exploiting the GND method, a G1 type scheme in an inverse-free manner is proposed and investigated at the joint-velocity level.
2) The proposed G1 type scheme can solve the inverse kinematics problem effectively but avoid calculating the inverse of Jacobian matrix.

3) The path-tracking simulations and the physical experiment are conducted to further illustrate the effectiveness, the high accuracy and the physical realizability of the G1 type scheme.

2   Inverse Scheme Formulation

For a redundant robot manipulator with n joints, the end-effector pose (or position in this paper) vector $r \in \mathbb{R}^m$ can be described by the following equation:

$$r = f(\theta), \tag{1}$$

where $\theta \in \mathbb{R}^n$ refers to the variables (or angles in this paper) of the n joints, and $f(\cdot)$ is a differentiable nonlinear function with a known structure and parameters for a given manipulator. Then by differentiating (1), the end-effector velocity is

$$\dot{r} = J\dot{\theta}, \tag{2}$$

where $\dot{r} \in \mathbb{R}^m$ refers to the end-effector velocity, and $\dot{\theta} \in \mathbb{R}^n$ refers to the velocities of all joints. Note that $J \in \mathbb{R}^{m \times n}$ is the Jacobian matrix defined as $J = \partial f(\theta)/\partial \theta$. According to (2), for tracking the desired path $r_d \in \mathbb{R}^m$ via the desired speed $\dot{r}_d \in \mathbb{R}^m$, the velocities of all joints can be obtained by the following equation if J is a square matrix and of full rank:

$$\dot{\theta} = J^{-1}\dot{r}_d, \tag{3}$$

If J is rectangular, the velocities of joints may be computed by the following equation of generalized inverse [6]:

$$\dot{\theta} = J^+\dot{r}_d, \tag{4}$$

where J^+ denotes the pseudoinverse of Jacobian matrix J. Note that, the solution obtained by using (4) is a least square solution. However, the inverse scheme theoretically requires the Jacobian matrix to be of full rank, which, in a real world application, may be unavailable sometimes in practice. What is more, the expensive calculating time of Jacobian matrix is also not suited in industry.

3   G1 Type Scheme Formulation

In this section, we generalize the GND method to obtain an inverse-free scheme at the joint-velocity level. By following the GND method, the design procedure of such an inverse-free scheme can be presented detailedly in the following steps.

Firstly, to monitor and control the process of solving the time-varying inverse kinematics problem of redundant manipulators, we define a scalar-valued norm-based energy function according to (1):

$$E = \|r_d - f(\theta)\|_2^2/2, \tag{5}$$

where $\| \cdot \|_2$ denotes the two-norm of a vector.

Secondly, a computational rule is designed to evolve along a descent direction of this energy function until the minimum point is reached. The typical descent direction is the negative gradient of E, i.e.,

$$-\partial E/\partial \boldsymbol{\theta} = J^{\mathrm{T}}(\boldsymbol{r}_{\mathrm{d}} - f(\boldsymbol{\theta})). \tag{6}$$

Then we combine the aforementioned negative gradient (6) and the following GND design formula [17,18]:

$$\dot{\boldsymbol{\theta}} = -\alpha \partial E/\partial \boldsymbol{\theta}, \tag{7}$$

where the design parameter $\alpha > 0$ is used to scale the convergence rate of the GND method.

Finally, we thus have the following generalized G1 type scheme for solving the time-varying inverse kinematics problem of redundant robot manipulators:

$$\dot{\boldsymbol{\theta}} = \alpha J^{\mathrm{T}}(\boldsymbol{r}_{\mathrm{d}} - f(\boldsymbol{\theta})). \tag{8}$$

Evidently, the above scheme (8) does not require the Jacobian inversion appearing in (3) and (4). Besides, since scheme (8) is obtained by applying the GND method only once, it is named G1 type (or Z0G1 type in the ZG framework).

4   Simulations

In this section, the corresponding path-tracking simulations (square path tracking and "Z" path tracking) are performed on five-link and six-link redundant robot manipulators respectively to illustrate the effectiveness and the accuracy of the proposed G1 type scheme. Note that, when we apply such a scheme to solving the time-varying inverse kinematics problem in an inverse-free manner, the design parameter $\alpha = 10^5$ is used throughout this section. For visualized reading, two error functions are defined as follows:

$$\begin{aligned} \varepsilon_{\mathrm{X}} &= r_{\mathrm{dX}} - p_{\mathrm{X}}, \\ \varepsilon_{\mathrm{Y}} &= r_{\mathrm{dY}} - p_{\mathrm{Y}}, \end{aligned} \tag{9}$$

where r_{dX} and r_{dY} refer to the desired positions in the X direction and the Y direction respectively, and p_{X} and p_{Y} refer to the actual positions in the X direction and the Y direction respectively. Accordingly, $\dot{\varepsilon}_{\mathrm{dX}}$ represents the velocity error in the X direction, while $\dot{\varepsilon}_{\mathrm{dY}}$ represents the velocity error in the Y direction.

4.1   Square Path Tracking

In this first example, G1 type scheme (8) is applied to tracking a square path via a desired velocity. The side length of the square path is 0.8 m, and the path tracking is simulated on a five-link redundant robot manipulator, with an initial state $\boldsymbol{\theta}(0)$ being $[\pi/3, \pi/3, \pi/2, -\pi/4, \pi/4]^{\mathrm{T}}$ rad. The corresponding simulation

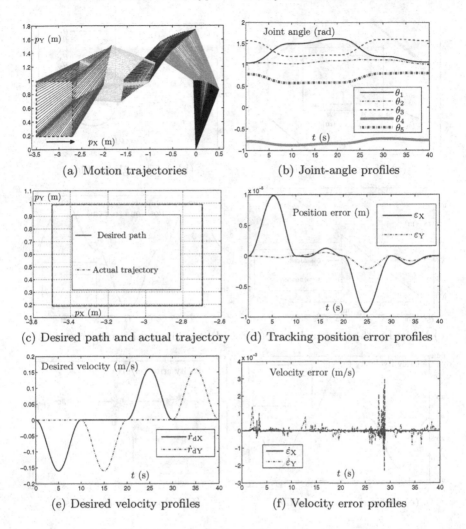

(a) Motion trajectories

(b) Joint-angle profiles

(c) Desired path and actual trajectory

(d) Tracking position error profiles

(e) Desired velocity profiles

(f) Velocity error profiles

Fig. 1. Simulation results of the five-link redundant robot manipulator tracking the given square path synthesized by G1 type scheme (8).

results are shown in Fig. 1, which illustrate the effectiveness and high accuracy of the proposed G1 type scheme (8) for solving the time-varying inverse kinematics of robot manipulators.

Specifically, the results can be seen from Fig. 1(a) and Fig. 1(b), from which we can see how the manipulator tracks the desired path with five joints. It is easily found that the actual end-effector trajectory coincides with the desired square path. Besides, Fig. 1(c) and Fig. 1(d) show us more details about the trajectory. That is, Fig. 1(c) presents the desired path and the actual trajectory, which illustrates the effectiveness and the accuracy intuitively. The corresponding errors shown in Fig. 1(d) are all less than 1×10^{-5} m. This implies that

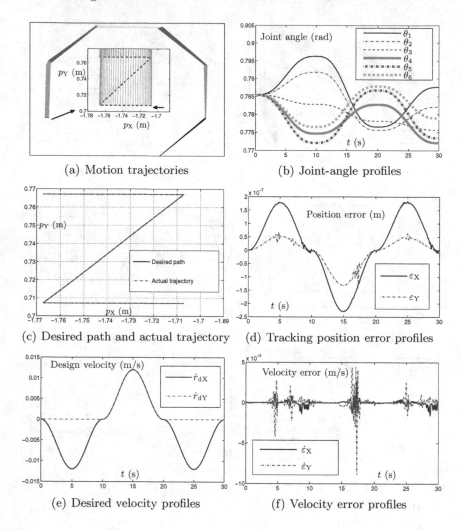

(a) Motion trajectories

(b) Joint-angle profiles

(c) Desired path and actual trajectory

(d) Tracking position error profiles

(e) Desired velocity profiles

(f) Velocity error profiles

Fig. 2. Simulation results of the six-link redundant robot manipulator tracking the given "Z" path synthesized by G1 type scheme (8).

the five-link robot manipulator completes the given square-path tracking task well. In addition, the desired velocities and the corresponding velocity errors are shown in Fig. 1(e) and Fig. 1(f). We see that the velocity errors are less than 4×10^{-3} m/s, validating the high accuracy of such an inverse-free type scheme.

4.2 "Z" Path Tracking

In this (second) example, G1 type scheme (8) is applied to tracking another path task, i.e., a "Z" path task with the side length being 0.8 m. The corresponding simulation results based on a six-link redundant robot manipulator are shown

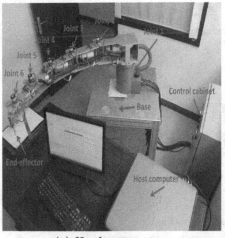

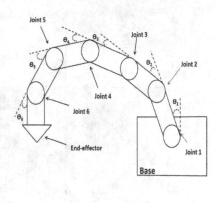

(a) Hardware system   (b) Manipulator-joint structure

Fig. 3. The hardware system of the six-link planar redundant robot manipulator with its structure planform.

in Fig. 2, where an initial state $\theta(0)$ is selected as $[\pi/4, \pi/4, \pi/4, \pi/4, \pi/4, \pi/4]^{\mathrm{T}}$ rad. These results illustrate once more the effectiveness and the high accuracy of the proposed G1 type scheme (8) for solving the time-varying inverse kinematics of robot manipulators.

Specifically, the results can be seen from Fig. 2(a) and Fig. 2(b), from which we can see how the manipulator tracks the desired path with six joints. It is easily found that the actual end-effector trajectory coincides well with the desired "Z" path. Especially, Fig. 2(c) and Fig. 2(d) show us more details about the actual trajectory and corresponding errors. These imply that the six-link robot manipulator completes the given "Z" path tracking task well. Besides, the desired velocities and the corresponding velocity errors are shown in Fig. 2(e) and Fig. 2(f). We can find that any directional velocity error is less than 5×10^{-3} m/s, which validates the high accuracy of such an inverse-free type scheme.

5   Physical Experiment

To verify the physical realizability of the proposed G1 type scheme (8), a six-link planar redundant robot manipulator hardware system is developed, investigated and shown. The whole manipulator system is mainly composed of a robot manipulator, a control cabinet and a host computer. Specifically, Fig. 3(a) shows this planar robot hardware system, and Fig. 3(b) depicts its manipulator-joint structure including a base and an end-effector.

For this experiment, in order to verify the proposed G1 type scheme (8) at the joint-velocity level, the end-effector is expected to move along a "Z" path with the length of 4.5 cm and an initial state $\theta(0) = [\pi/12, \pi/12, \pi/12, \pi/12, \pi/12, \pi/12]^{\mathrm{T}}$

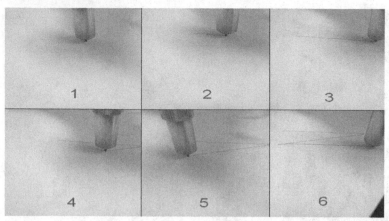

(a) Snapshots of task execution

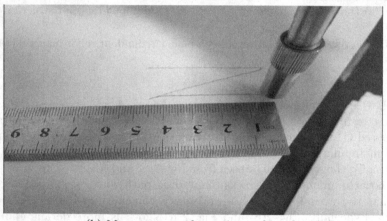

(b) Measurement of experimental result

Fig. 4. The "Z" path tracking experiment of the six-link redundant redundant robot manipulator synthesized by G1 type scheme (8) at the joint-velocity level.

rad, and the design parameter α is also set as 10^5. The task execution can be seen from Fig. 4, i.e., the end-effector of the manipulator moves smoothly and draws a "Z" path precisely. Besides, the video of the process takes 24 seconds. Thus, this experiment illustrates well that the proposed G1 type scheme (8) is effective on the redundant robot manipulator's inverse-free redundancy resolution (or say, motion planning and control).

6   Conclusion

In this paper, to solve the time-varying inverse kinematics problem for redundant robot manipulators with high efficiency and high accuracy, a special type of inverse-free scheme named G1 type scheme has been proposed and investigated at the joint-velocity level. This scheme can avoid the Jacobian inversion in traditional pseudo-inverse methods which not only costs expensive calculating time but also encounters many difficulties in practice. Besides, the corresponding path-tracking simulations have been performed on five-link and six-link redundant robot manipulators using such an inverse-free scheme. The simulation results have illustrated the effectiveness and the accuracy of the aforementioned scheme for solving the time-varying inverse kinematics problem of redundant robot manipulators in an inverse-free manner. In addition, the physical realizability of G1 type scheme has been verified further based on a six-link planar redundant robot manipulator hardware system. In the future, the scheme may be applied in the 3 dimensional case or even with kinematics uncertainties.

Acknowledgments. This work is supported by the National Natural Science Foundation of China (with number 61473323), by the Foundation of Key Laboratory of Autonomous Systems and Networked Control, Ministry of Education, China (with number 2013A07), and also by the Science and Technology Program of Guangzhou, China (with number 2014J4100057).

References

1. Van, M., Kang, H.-J., Ro, Y.-S.: A robust fault detection and isolation scheme for robot manipulators based on neural networks. In: Huang, D.-S., Gan, Y., Bevilacqua, V., Figueroa, J.C. (eds.) ICIC 2011. LNCS, vol. 6838, pp. 25–32. Springer, Heidelberg (2011)
2. Lee, S.C., Ahn, H.S.: Multiple manipulator cooperative control using disturbance estimator and consensus algorithm. In: American Control Conference, pp. 4002–4007. IEEE Press, San Francisco (2011)
3. Kim, C.S., Mo, E.J., Jie, M.S., Hwang, S.C., Lee, K.-W.: Image-based robust control of robot manipulators under jacobian uncertainty. In: Huang, D.-S., Heutte, L., Loog, M. (eds.) ICIC 2007. LNCS, vol. 4681, pp. 502–510. Springer, Heidelberg (2007)
4. Zhou, J., Kang, H., Ro, Y.: Comparison of the observability Indices for robot calibration considering joint stiffness parameters. In: Huang, D.-S., McGinnity, M., Heutte, L., Zhang, X.-P. (eds.) CCIS 2010, CCIS. vol. 93, pp. 372–380. Springer, Heidelberg (2010)
5. Liu, Y., Yu, Y., Jiang, C.: The Development of Redundant Robot Manipulators. Machine Design and Research 19(1), 24–27 (2003)
6. Dubey, R., Luh, J.: Redudant Robot Control for Higher Flexibility. In: IEEE International Conference on Robotics and Automation, pp. 1066–1072. IEEE Press, Raleigh (1987)
7. Zhang, Y., Wang, J.: A Dual Neural Network for Redundancy Resolution of Kinematically Redundant Manipulators Subject to Joint Limits and Joint Velocity Limits. IEEE Transactions on Neural Networks 13(3), 658–667 (2003)
8. Bianco, C., Ghilardelli, F.: Real-Time Planner in the Operational Space for the Automatic Handling of Kinematic Constraints. IEEE Transactions on Automation Science and Engineering 11(3), 730–739 (2014)

9. Moreno, J., Kelly, R.: Manipulator velocity field control with dynamic friction compensation. In: 42nd IEEE Conference on Decision and Control, pp. 3834–3839. IEEE Press, Hawaii (2003)
10. Cheng, L., Hou, Z., Tan, M.: Adaptive Neural Network Tracking Control for Manipulators with Uncertain Kinematics, Dynamics and Actuator Model. Automatica 45(10), 2312–2318 (2009)
11. Cheah, C., Kawamura, S., Arimoto, S.: Stability of Hybrid Position and Force Control for Robotic Manipulator with Kinematics and Dynamics Uncertainties. Automatica 39(5), 847–855 (2003)
12. Cheng, L., Hou, Z., Tan, M., Lin, Y., Zhang, W.: Neural-Network-Based Adaptive Leader-Following Control for Multiagent Systems with Uncertainties. IEEE Transactions on Neural Network 21(8), 1351–1358 (2010)
13. Tarokh, M., Kim, M.: Inverse Kinematics of 7-DOF Robots and Limbs by Decomposition and Approximation. IEEE Transactions on Robotics 23(3), 595–600 (2007)
14. Falco, P., Natale, C.: On the Stability of Closed-Loop Inverse Kinematics Algorithms for Redundant Robots. IEEE Transactions on Robotics 27(3), 780–784 (2011)
15. Tchon, K.: Optimal Extended Jacobian Inverse Kinematics Algorithms for Robotic Manipulators. IEEE Transactions on Robotics 24(6), 1440–1445 (2008)
16. Jin, L., Zhang, Y.: G2-Type SRMPC Scheme for Synchronous Manipulation of Two Redundant Robot Arms. IEEE Transactions on Cybernetics 45(2), 153–164 (2015)
17. Zhang, Y., Yi, C., Guo, D., Zheng, J.: Comparison on Zhang Neural Dynamics and Gradient-Based Neural Dynamics for Online Solution of Nonlinear Time-Varying Equation. Neural Computing and Applications 20(1), 1–7 (2011)
18. Zhang, Y.-N., Yi, C., Ma, W.: Comparison on gradient-based neural dynamics and zhang neural dynamics for online solution of nonlinear equations. In: Kang, L., Cai, Z., Yan, X., Liu, Y. (eds.) ISICA 2008. LNCS, vol. 5370, pp. 269–279. Springer, Heidelberg (2008)
19. Bowling, M., Veloso, M.: Convergence of gradient dynamics with a variable learning rate. In: 18th International Conference on Machine Learning, pp. 27–34. Morgan Kaufmann, Williamstown (2001)
20. Ma, X., Elia, N.: A distributed continuous-time gradient dynamics approach for the active power loss minimizations. In: 51th Annual Allerton Conference, pp. 100–106. IEEE Press, Illinois (2013)
21. Taylor, P., Day, T.: Evolutionary Stability under the Replicator and the Gradient Dynamics. Evolutionary Ecology 11(5), 579–590 (1997)
22. Zhang, Y., Li, Z., Guo, D., Ke, Z., Peng, C.: Discrete-Time ZD, GD and NI for Solving Nonlinear Time-Varying Equations. Numerical Algrithms 64, 721–740 (2013)
23. Zhang, Y., Luo, F., Yin, Y., Liu, J., Yu, X.: Singularity-conquering ZG controller for output tracking of a class of nonlinear systems. In: 32nd Chinese Control Conference, pp. 477–482. IEEE Press, Xi'an (2013)
24. Zhang, Y., Yin, Y., Wu, H., Guo, D.: Dynamics and gradient dynamics with tracking-control application. In: 5th International Symposium on Computational Intelligence and Design, pp. 235–238. IEEE Press, Hangzhou (2013)

Design of Fuzzy-Neural-Network-Inherited Backstepping Control for Unmanned Underwater Vehicle

Yuxin Fu[1], Yancheng Liu[1,*], Siyuan Liu[1], Ning Wang[1], and Chuan Wang[1]

Maritime Engineering College, Dalian Maritime University, Dalian, China
{dlmu_fyx,liuyc3}@126.com,
{dmu.s.y.liu,n.wang.dmu.cn}@gmail.com, chuanwang0101@163.com

Abstract. This paper presents a closed-loop trajectory tracking controller for an Unmanned Underwater Vehicle(UUV) with five degrees of freedom. A backstepping control (BSC) methodology combined with Lyapunov theorem is adopted to design the controller of trajectory tracking. Then an online-tuning fuzzy neural network (FNN) framework is chosen to inherit the conventional BSC law. Moreover, the adaptive parameters tuning laws are derived in the sense of Lyapunov stability theorem and projection algorithm to ensure the network convergence as well as stable control performance. Finally, the simulation results on UUV verify that an excellent performance of the proposed controller can be obtained.

Keywords: trajectory tracking controller, Unmanned Underwater Vehicle (UUV), backstepping control (BSC), fuzzy neural network (FNN).

1 Introduction

The UUV is widely used in risky missions such as underwater exploration, oceanic observations, military applications, etc., [1]. Considering the UUV spatial kinematics model which is complex and highly coupled nonlinear system, it presents a challenging control problem to establish an appropriate mathematical model for the design of a model-based control system. So the high performance tracking control method of the UUV is being studied.

The current trend of control approaches focuses on integrating conventional tracking control techniques such as adaptive control and backstepping control(BSC) [2]. However, the stability of backstepping control system will be damaged while a sudden tracking error happens. R. Fierro and F. L. Lewis [3] utilized a backstepping control method for nonholonomic mobile robot. Based on these conventional control methods, some intelligent schemes such as fuzzy theory and neural network are widely used to improve the performance of classical controllers in various aspects in order to solve the defects of traditional control techniques. A neural network (NN) has an inherent ability to learn and approximate a nonlinear function, which is utilized in

---

* Corresponding author.

© Springer International Publishing Switzerland 2015
X. Hu et al. (Eds.): ISNN 2015, LNCS 9377, pp. 109–118, 2015.
DOI: 10.1007/978-3-319-25393-0_13

UUV controller [4] and even to prevent actuator saturation without prior knowledge of system parameters. S.Cong and Y.Liang [5] utilized a neural network to compensate for unstructured uncertainties in order to improve the robustness of the control system. Recently, the concept of incorporating fuzzy logic into an NN has grown into a popular research topic [6] because the integrated fuzzy neural network (FNN) system possesses the merits of both fuzzy systems and NNs. The fuzzy logic is employed in UUV controllers to provide human logical-thinking capabilities for achieving better control performance [7]. But how to establish some suitable fuzzy rules and guarantee system stability is also a challenging problem to be solved [8], [9]. The notion of robustness has been a core subject in intelligent strategies for UUV trajectory tracking control. Chen et al. [10] considered an adaptive fuzzy control for a class of nonlinear systems by using fuzzy logic systems. Ning Wang, and Meng Joo Er [11], [12] utilized self-constructing adaptive robust fuzzy neural control and adaptive robust online constructive fuzzy control to improve the robustness of the uncertainties and unknown disturbances system.

In this paper, the strategy of trajectory tracking is established by using fuzzy neural network (FNN) to approximate the conventional backstepping control. Furthermore, an adaptive control is designed combining with the Lyapunov functions and projection algorithm. The stability of the control system and the efficiency of this controller can be guaranteed as the simulation results showed.

2   UUV Dynamic Model

In this paper, the UUV model has the following characteristics.

1) Treat the actual situation of the UUV in underwater movement as the rigid body moving in the fluid.

2) Ignore the rolling impact on three dimensional motion of UUV based on the inside structure of ballast water tank in unmanned underwater vehicle.

3) Treat the actual shape of the unmanned underwater vehicle as spheroidicity.

4) The propulsion system of the unmanned underwater vehicle consists of five thrusters, i.e., a main thruster, a left bow horizontal thruster, a right stern horizontal thruster, a down vertical thruster located at the right side of bow and a up vertical thruster located at the left side of stern, so that the motion control, i.e., surge, sway, heave, pitch and yaw, can be realized.

The relationship of the velocity and angular velocity of the UUV in geodetic coordinate system and moving coordinate system can be expressed in the following form:

$$\dot{\eta} = J(\eta)v \tag{1}$$

where $\eta = [x, y, z, \theta, \psi]^{\mathrm{T}}$ is the position and direction vector in geodetic coordinate system; $v = [u, v, w, q, r]^{\mathrm{T}}$ is the velocity and angular velocity vector in moving coordinate system; $J(\eta)$ is the rotation matrix from the moving coordinate system to geodetic coordinate system.

Based on the Newton-Euler kinematical equation, the five degree of freedom (DOF) nonlinear dynamic equations of motion in moving coordinate system can be conveniently expressed as [13]

$$M\dot{v} + C(v)v + D(v)v + g(\eta) = \tau + f \tag{2}$$

where $M=M_{RB}+M_A$ is system inertia matrix (including added mass); $C(v)=C_{RB}(v)+C_A(v)$ is Coriolis-centripetal matrix(including added mass); $D(v)$ is damping matrix; $g(\eta)$ is vector of gravitational/ buoyancy forces; τ is vector of control inputs; f is vector of environmental disturbances(wind, waves and currents).

Simultaneous equations (1) (2), the five DOF kinetic equation in geodetic coordinate system can be expressed as

$$M^*\ddot{\eta} + C^*\dot{\eta} + D^*\dot{\eta} + g^* = \tau + f \tag{3}$$

where $M^* = MJ^{-1}; C^* = (C - MJ^{-1}\dot{J})J^{-1}; D^* = DJ^{-1}; g^* = g$

3   Trajectory Tracking Controller

3.1   Backstepping Control

The first step: Define a tracking error vector and its derivative as $z_1 = \eta - \eta_r$ and $\dot{z}_1 = \dot{\eta} - \dot{\eta}_r$. Where η_r and $\dot{\eta}_r$ are reference trajectory vector and velocity vector in geodetic coordinate system, respectively. The term $\dot{\eta}(t)$ can be viewed as a first virtual control input. Define the first stabilizing function as $\alpha_1 = -k_1 z_1 + \dot{\eta}_r$, where $k_1 \in R^{5 \times 5}$ is a positive-definite diagonal matrix.

The first Lyapunov function is chosen as

$$V_1 = \tfrac{1}{2} z_1^T z_1 \tag{4}$$

Define $z_2 = \dot{\eta} - \alpha_1$; the derivative of V_1 can be represented as

$$\dot{V}_1 = z_1^T(\dot{\eta} - \dot{\eta}_r) = z_1^T(z_2 + \alpha_1 - \dot{\eta}_r) = z_1^T z_2 - z_1^T k_1 z_1 \tag{5}$$

The second step: The derivative of z_2 can be written as

$$\dot{z}_2 = M^{*-1}\tau + M^{*-1}f - M^{*-1}C^*\dot{\eta} - M^{*-1}D^*\dot{\eta} - M^{*-1}g^* - \dot{\alpha}_1 \tag{6}$$

The BSC law is assumed as the follow form

$$\tau_{BSC} = -k_2^{-1}z_1 - f_b \, \mathrm{sgn}(z_2^T k_2)^T + C^*\dot{\eta} - D^*\dot{\eta} + g^* + M^*\dot{\alpha}_1 - z_2 \tag{7}$$

The second Lyapunov function is chosen as

$$V_2 = V_1 + \tfrac{1}{2} z_2^T k_2 M^* z_2 \tag{8}$$

where $k_2 \in R^{5 \times 5}$ is a positive-definite diagonal matrix.

The derivative of V_2 can be represented via (5) (7) (8) as

$$\dot{V}_2 = \dot{V}_1 + z_2^T k_2 M^* \dot{z}_2 = -z_1^T k_1 z_1 - z_2^T k_2 z_2 - z_2^T k_2 [f_b \, \text{sgn}(z_2^T k_2)^T - f]$$
$$\leq -z_1^T k_1 z_1 - z_2^T k_2 z_2 \leq 0 \tag{9}$$

Thereby, it confirms that the vectors z_1, z_2 converge to zero asymptotically if the condition of $\|f\|_1 \leq f_b$ holds.

3.2 Fuzzy-Neural-Network-Inherited Backstepping Control

An FNNIBSC system for UUV trajectory tracking is shown in Fig. 1. Moreover, a four-layer FNN framework is shown in Fig. 2.

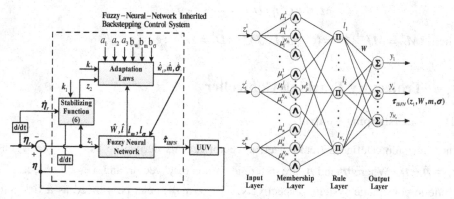

Fig. 1. Framework of the FNNIBSC system   **Fig. 2.** Structure of the four-layer FNN

Each layer of the Structure of the four-layer FNN will be introduced as follows.

1) The input vector z_1^i $(i = 1, ..., n)$ of the input layer is tracking error vector.

2) The membership layer represents the input values with the following Gaussian membership functions:

$$\mu_i^j = \exp[-(z_1^i - m_i^j)^2 / (\sigma_i^j)^2] \tag{10}$$

where m_i^j and σ_i^j $(i = 1, ..., n; j = 1, ..., N_{pi})$ are the mean and standard deviation of the Gaussian function in the jth term of the ith input variable z_i to the node, respectively.

3) The rule layer implements the fuzzy inference mechanism. The output of this layer is described as follows

$$l_k = \prod_{i=1}^n \mu_i^j (z_1^i) \tag{11}$$

where l_k $(k=1, ..., N_y)$ represents the kth output of this layer.

4) The output vector of the output layer is expressed as

$$y_e = \sum_{k=1}^{N_y} w_k^e l_k \tag{12}$$

where y_e $(e=1, ..., N_o)$ represents the eth output of this layer. It can be rewritten in the following vector form

$$\tau = [y_1 \quad y_2 \quad \cdots \quad y_{N_o}]^{\mathrm{T}} = Wl \equiv \tau_{IBFN}(z_1, W, m, \sigma) \tag{13}$$

$$W = \begin{pmatrix} w_1^1 & w_2^1 & \cdots & w_{N_y}^1 \\ w_1^2 & w_2^2 & \cdots & w_{N_y}^2 \\ \vdots & \vdots & \ddots & \vdots \\ w_1^{N_o} & w_2^{N_o} & \cdots & w_{N_y}^{N_o} \end{pmatrix} = \begin{pmatrix} w_1 \\ w_2 \\ \vdots \\ w_{N_o} \end{pmatrix} \tag{14}$$

$$l = [l_1 \quad l_2 \quad \cdots \quad l_{N_y}]^{\mathrm{T}} \tag{15}$$

$$w_i = [w_1^i \quad w_1^i \quad \cdots \quad w_{N_y}^i] \tag{16}$$

To show the effectiveness of the proposed FNNIBSC system, the FNN structure has 5, 15, 243 and 5 neurons at the input, membership, rule and output layers, respectively. It can be regarded that the associated fuzzy sets with the Gaussian function for each input signal are divided into negative small, zero, positive small. That is, $n = 5$, $N_{pi} = 3(i=1,\ldots, n), N_y = \prod_{i=1}^{n} N_{pi} = 243$ and $N_o = 5$.

The optimal FNNIBSC learning the conventional BSC can be written as follow

$$\tau_{BSC} = \tau_{IBFN}^*(z_1, W^*, m^*, \sigma^*) + \varepsilon = W^* l^* + \varepsilon \tag{17}$$

where ε is a minimum reconstructed-error vector; and W^*, m^*, σ^* are optimal parameters of W, m, σ in the FNN, respectively.

The control law of the FNNIBSC scheme will be taken as the following form

$$\tau = \hat{\tau}_{IBFN}(z_1, \hat{W}, \hat{m}, \hat{\sigma}) = \hat{W}\hat{l} \tag{18}$$

where $\hat{W}, \hat{m}$ and $\hat{\sigma}$ are some estimates of the optimal parameters, as provided by tuning algorithms to be introduced later. Subtracting (18) from (17), an approximation error is defined as

$$\tilde{\tau} = \tau_{BSC} - \hat{\tau}_{IBFN} = W^* l^* + \varepsilon - \hat{W}\hat{l} = \tilde{W}l^* + \hat{W}\tilde{l} + \varepsilon \tag{19}$$

where $\tilde{W} = W^* - \hat{W}, \tilde{l} = l^* - \hat{l}$. A Taylor series expansion is utilized to transform the membership functions into partially linear form. Where $\tilde{l}$ can be rewritten as

$$\tilde{l} = l_m \tilde{m} + l_\sigma \tilde{\sigma} + o_{nv} \tag{20}$$

where $\tilde{m} = m^* - \hat{m}$, $\tilde{\sigma} = \sigma^* - \hat{\sigma}$; m^* and σ^* are the optimal parameters of m and σ, respectively; $\hat{m}$ and $\hat{\sigma}$ are the estimates of m^* and σ^*, respectively. o_{nv} is a vector of higher order terms; $l_m = [\frac{\partial l_1}{\partial m} \frac{\partial l_2}{\partial m} \cdots \frac{\partial l_{N_y}}{\partial m}]^{\mathrm{T}}\Big|_{m=\hat{m}} \in R^{243\times 15}$ and $l_\sigma = [\frac{\partial l_1}{\partial \sigma} \frac{\partial l_2}{\partial \sigma} \cdots \frac{\partial l_{N_y}}{\partial \sigma}]^{\mathrm{T}}\Big|_{\sigma=\hat{\sigma}} \in R^{243\times 15}$; simultaneous equations (19) (20), it is revealed that

$$\begin{aligned} \tilde{\tau} &= W^* l^* + \varepsilon - \hat{W}\hat{l} = (\tilde{W} + \hat{W})(\hat{l} + l_m \tilde{m} + l_\sigma \tilde{\sigma} + o_{nv}) + \varepsilon - \hat{W}\hat{l} \\ &= \tilde{W}\hat{l} + \hat{W}l_m \tilde{m} + \hat{W}l_\sigma \tilde{\sigma} + y' \end{aligned} \tag{21}$$

where $y' = \tilde{W}l_m \tilde{m} + \tilde{W}l_\sigma \tilde{\sigma} + \varepsilon + W^* o_{nv}$.

Theorem: For the five DOF kinetic equation of UUV that are represented by (3), the FNNIBSC law is designed as (18), and the corresponding adaptation laws for the estimates of the optimal parameters in the FNN are designed as (22)-(24) so that all the tracking error states(z_1 and z_2) converge to zero as time tends to infinity. Then, the stability of the proposed controller can be proved by the Lyapunov theorem.

$$\dot{\hat{w}}_i = \begin{cases} -a_1 z_2^i k_2^i \hat{l}^T & \text{if} (\|\hat{w}_i\| < b_w) \text{ or } (\|\hat{w}_i\| = b_w \text{ and } z_2^i k_2^i \hat{w}_i \hat{l} \geq 0) \\ -a_1 z_2^i k_2^i \hat{l}^T + a_1 z_2^i k_2^i \hat{l}^T \hat{w}_i^{\ T} \hat{w}_i / \|\hat{w}_i\|^2 & \text{if} (\|\hat{w}_i\| = b_w \text{ and } z_2^i k_2^i \hat{w}_i \hat{l} < 0) \end{cases} \tag{22a,22b}$$

$$\dot{\hat{m}} = \begin{cases} -a_2 (z_2^T k_2 \hat{W} l_m)^T & \text{if} (\|\hat{m}\| < b_m) \text{ or } (\|\hat{m}\| = b_m \text{ and } z_2^T k_2 \hat{W} l_m \hat{m} \geq 0) \\ -a_2 (z_2^T k_2 \hat{W} l_m)^T + a_2 (z_2^T k_2 \hat{W} l_m \dfrac{\hat{m}\hat{m}^T}{\|\hat{m}\|^2})^T & \text{if} (\|\hat{m}\| = b_m \text{ and } z_2^T k_2 \hat{W} l_m \hat{m} < 0) \end{cases} \tag{23a,23b}$$

$$\dot{\hat{\sigma}} = \begin{cases} -a_3 (z_2^T k_2 \hat{W} l_\sigma)^T & \text{if} (\|\hat{\sigma}\| < b_\sigma) \text{ or } (\|\hat{\sigma}\| = b_\sigma \text{ and } z_2^T k_2 \hat{W} l_\sigma \hat{\sigma} \geq 0) \\ -a_3 (z_2^T k_2 \hat{W} l_\sigma)^T + a_3 (z_2^T k_2 \hat{W} l_\sigma \dfrac{\hat{\sigma}\hat{\sigma}^T}{\|\hat{\sigma}\|^2})^T & \text{if} (\|\hat{\sigma}\| = b_\sigma \text{ and } z_2^T k_2 \hat{W} l_\sigma \hat{\sigma} < 0) \end{cases} \tag{24a,24b}$$

where $z_2^i (i = 1,\ldots,5)$ is the element of the vector z_2; k_2^i is the diagonal element of the matrix k_2; $\|\cdot\|$ denotes the Euclidean norm; a_1, a_2 and a_3 are positive learning rates; b_w, b_m and b_σ are given positive parameter bounds; $\hat{w}_i$ is the estimate of w_i^*; w_i^* is the optimal parameter vector of w_i.

Proof: The Lyapunov function is chosen as

$$V = \tfrac{1}{2} z_1^T z_1 + \tfrac{1}{2} z_2^T k_2 M z_2 + \tfrac{1}{2a_1} \text{tr}(\tilde{W}\tilde{W}^T) + \tfrac{1}{2a_2} \tilde{m}^T \tilde{m} + \tfrac{1}{2a_3} \tilde{\sigma}^T \tilde{\sigma} \tag{25}$$

Subtracting (7) (21) into (6), one can obtain

$$\begin{aligned} \dot{z}_2 &= -M^{*-1}C^* \eta - M^{*-1}D^* \eta - M^{*-1}g^* + M^{*-1}f + M^{*-1}(\tau_{BSC} - \tilde{\tau}) - \dot{\alpha}_1 \\ &= -M^{*-1}[\tilde{W}\hat{l} + \hat{W} l_m \tilde{m} + \hat{W} l_\sigma \tilde{\sigma} + z_2 + k_2^{-1} z_1 + f_b \, \text{sgn}(z_2^T k_2)^T + f_n] \end{aligned} \tag{26}$$

where the uncertainty term $f_n = y' - f$ is also assumed to be bounded by $\|f_n\|_1 < f_b$.

Differentiating (25) and using (26), it is concluded that

$$\begin{aligned} \dot{V} &= z_1^T \dot{z}_1 + z_2^T k_2 M^* \dot{z}_2 - \tfrac{1}{a_1} \text{tr}(\tilde{W}\dot{\hat{W}}^T) - \tfrac{1}{a_2} \dot{\hat{m}}^T \tilde{m} - \tfrac{1}{a_3} \dot{\hat{\sigma}}^T \tilde{\sigma} \\ &= -[\tfrac{1}{a_1} \text{tr}(\tilde{W}\dot{\hat{W}}^T) + z_2^T k_2 \tilde{W}\hat{l}] - (\tfrac{1}{a_2} \dot{\hat{m}}^T + z_2^T k_2 \hat{W} l_m)\tilde{m} - \\ &\quad (\tfrac{1}{a_3} \dot{\hat{\sigma}}^T + z_2^T k_2 \hat{W} l_\sigma)\tilde{\sigma} - z_2^T k_2 [f_b \, \text{sgn}(z_2^T k_2)^T + f_n] - z_1^T k_1 z_1 - z_2^T k_2 z_2 \\ &= -V_w - V_m - V_\sigma - z_2^T k_2 [f_b \, \text{sgn}(z_2^T k_2)^T + f_n] - z_1^T k_1 z_1 - z_2^T k_2 z_2 \end{aligned} \tag{27}$$

where $V_w = \frac{\text{tr}(\tilde{W}\dot{\hat{W}}^T)}{a_1} + z_2^T k_2 \tilde{W}\hat{l}$, $V_m = (\frac{\dot{\hat{m}}^T}{a_2} + z_2^T k_2 \hat{W} l_m)\tilde{m}$, $V_\sigma = (\frac{\dot{\hat{\sigma}}^T}{a_3} + z_2^T k_2 \hat{W} l_\sigma)\tilde{\sigma}$.

If the adaptation law for the output weight in the FNN is designed as (22), then V_w can be expressed as follows.

By (22a)

$$V_w = [\tfrac{\text{tr}(\tilde{W}\dot{\hat{W}}^T)}{a_1} + z_2^T k_2 \tilde{W}\hat{l}] = -\sum_{i=1}^{n} z_2^i k_2^i \tilde{w}_i \hat{l} + z_2^T k_2 \tilde{W}\hat{l} = 0 \tag{28}$$

By (22b)

$$V_w = [\frac{\mathrm{tr}(\tilde{W}\dot{\hat{W}}^T)}{a_1} + z_2^T k_2 \tilde{W} \hat{l}] = -\sum_{i=1}^{n} z_2^i k_2^i \tilde{w}_i \hat{l}(I - \frac{\hat{w}_i^T \hat{w}_i}{\|\hat{w}_i\|^2}) + z_2^T k_2 \tilde{W} \hat{l}$$

$$= \sum_{i=1}^{n} z_2^i k_2^i \tilde{w}_i (\hat{w}_i^T \hat{w}_i / \|\hat{w}_i\|^2) \hat{l} \tag{29}$$

The condition $\tilde{w}_i \hat{w}_i^T = \frac{1}{2}(\|w_i^*\|^2 - \|\hat{w}_i\|^2 - \|\hat{w}_i - w_i^*\|^2) < 0$ will hold if the corresponding situations are $\|\hat{w}_i\| = b_w$ and $z_2^i k_2^i \hat{w}_i \hat{l} < 0$ are met because of $\|w_i^*\| < b_w$. As a result from (28) and (29), one can draw a conclusion that $V_w \geq 0$.

If the adaptation law for the means of Gaussian functions in the FNN is designed as in (23), then V_m can be expressed as follows.

By (23a)

$$V_m = (\dot{\hat{m}}^T / a_2 + z_2^T k_2 \hat{W} l_m) \tilde{m} = 0 \tag{30}$$

By (23b)

$$V_m = z_2^T k_2 \hat{W} l_m \hat{m} \frac{\hat{m}^T \tilde{m}}{\|\hat{m}\|^2} \tag{31}$$

The condition $\hat{m}^T(m^* - \hat{m}) = \frac{1}{2}(\|m^*\|^2 - \|\hat{m}\|^2 - \|m^* - \hat{m}\|^2) < 0$ will hold if the corresponding situations $\|m^*\| = b_m$ and $z_2^T k_2 \hat{W} l_m \hat{m} < 0$ are met, because of $\|m^*\| < b_m$. As a result from (30) and (31), one can draw a conclusion that $V_m \geq 0$. According to the method of proving that $V_m \geq 0$, it is easy to draw a conclusion that $V_o \geq 0$.

By combining all terms together, $\dot{V}$ can be analyzed as follow

$$\dot{V} \leq -\|z_2^T k_2\|_1 (f_b - \|f_n\|_1) - z_1^T k_1 z_1 - z_2^T k_2 z_2 \leq -z_1^T k_1 z_1 - z_2^T k_2 z_2 \leq 0 \tag{32}$$

It is noted that the design of the FNN to approximate the BSC law in this paper is just to use the intelligent ability of the FNN without the requirement of system information and auxiliary compensated control for maintaining the robust characteristic of the BSC law. It also confirms that the state variables of trajectory tracking error will converge to zero asymptotically.

4   Simulation Studies

A set of precise UUV trajectory tracking simulation experiment is designed to verify the correctness and robustness of space trajectory tracking controller. The system parameters of UUV are given as

$$m = 390\mathrm{kg}, I_y = 305.67kg \cdot m^2, I_z = 305.67kg \cdot m^2, X_u = -20, Y_v = -200, Z_w = -200$$

$$M_q = -200, N_r = -200, X_{\dot{u}} = -49.12, Y_{\dot{v}} = -311.52, Z_{\dot{w}} = -311.52, M_{\dot{q}} = -87.63$$

$$N_{\dot{r}} = -87.63.$$

The planning trajectories of simulation and the initial position and direction are given as

$x_d = \sin(0.02t), y_d = \cos(0.01t), z_d = \sin(0.01t) + \cos(0.01t).$

$\theta_d = -\cos(0.01t) + \sin(0.01t), \psi_d = 2\cos(0.02t) - \sin(0.01t).$

$x(1) = -0.1, y(1) = 0.8, z(1) = 1.5, \theta(1) = -0.01, \psi(1) = 0.01.$

The parameters of FNNIBSC controller are given as
$k_1 = \text{diag}[3.3, 50, 60, 100, 100], k_2 = \text{diag}[406, 150, 100, 650, 650], a_1=1, a_2=0.1, a_3=0.2.$

The tracking performance in three-dimensional space with BSC controller and FNNIBSC controller are shown as Fig. 3 and Fig.4, respectively. The tracking results of Y axis with BSC controller and FNNIBSC controller are shown as Figs. 5 and 6, respectively, and the tracking errors of η with BSC controller and FNNIBSC controller are shown as Figs. 7 and 8, respectively.

From Figs. 3-8, one can see that both BSC system and the FNNIBSC system can track the desired trajectory. Although the actual initial conditions are different from those of reference trajectory, the FNNIBSC system can approach the reference trajectory with a rapid transient response from Figs. 3-6. As can be seen from Figs. 7 and 8, one can see that the tracking errors of FNNIBSC system are much smaller than those of BSC system. Therefore, the correctness and effectiveness can be demonstrated and it is concluded that the FNNIBSC system can track the reference trajectory with a high steady-state accuracy.

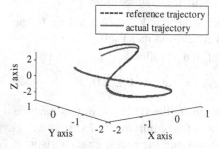

Fig. 3. Curves of UUV trajectory tracking in three-dimensional space with BSC

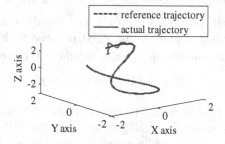

Fig. 4. Curves of UUV trajectory tracking in three-dimensional space with FNNIBSC

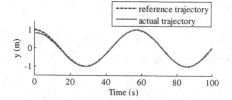

Fig. 5. Curves of UUV trajectory tracking for y with BSC

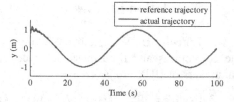

Fig. 6. Curves of UUV trajectory tracking for y with FNNIBSC

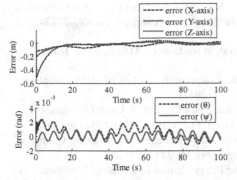

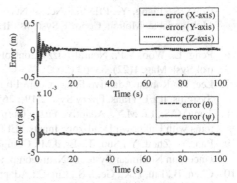

Fig. 7. Curves of UUV trajectory tracking error for X-axis, Y-axis, Z-axis, pitch and yaw with BSC

Fig. 8. Curves of UUV trajectory tracking error for X-axis, Y-axis, Z-axis, pitch and yaw with FNNIBSC

5   Conclusion

In this paper, an adaptive FNN control approach is proposed to approximate the conventional backstepping control. By using the corresponding adaptation laws, the estimates of the optimal parameters of the FNN can be retrieved by the online training methodology. The FNNIBSC method is proved to provide a robust tracking performance of UUV. The correctness and the effectiveness of the proposed FNNIBSC scheme are also confirmed by the simulation results compared with BSC system.

Acknowledgement. This work is supported by the National Natural Science Foundation of PR China (under Grants 51479018, 51379002 and 51009017), the General Project of Liaoning Provincial Department of Education (under Grants L2014209), Fundamental Research Funds for the Central Universities of PR China (under Grants 313201432 and 3132015028), Applied Basic Research Funds from Ministry of Transport of P. R. China (under Grant 2012-329-225-060), and Program for Liaoning Excellent Talents in University (under Grant LJQ2013055).

References

1. Yuh, J.: Design and Control of Autonomous Underwater Robots: A survey. Int. J. Control 8, 7–24 (2000)
2. Repoulias, F., Papadopoulos, E.: Planar Trajectory Planning and Tracking Control Design for Underactuated AUVs. Ocean Eng. 11(34), 1650–1667 (2007)
3. Fierro, R., Lewis, L.F.: Control of a Nonholonomic Mobile Robot Backstepping Kinematics into Dynamics. J. Robot. Syst. 3(14), 149–163 (1997)
4. Gao, W., Selmic, R.R.: Neural Network Control of a class of Nonlinear Systems with Actuator Saturation. IEEE Trans. Neural Netw. 1(17), 147–156 (2006)

5. Cong, S., Liang, Y.: PID-like Neural Network Nonlinear Adaptive Control for Uncertain Multivariable Motion Control Systems. IEEE Trans. Ind. Electron. 10(56), 3872–3879 (2009)
6. Peng, L., Woo, Y.P.: Neural-Fuzzy Control System for Robotic Manipulators. IEEE Control Syst. Mag. 1(22), 53–63 (2002)
7. Lee, H.: Robust Adaptive Fuzzy Control by Backstepping for a class of MIMO Nonlinear Systems. IEEE Trans. Fuzzy Syst. 2(19), 265–275 (2011)
8. Tong, C.S., Li, M.Y.: Adaptive Fuzzy Output Feedback Control of MIMO Nonlinear Systems with Unknown Dead-zone Inputs. IEEE Trans. Fuzzy Syst. 1(21), 134–146 (2013)
9. Pan, Y., Zhou, Y., Sun, T., Er, J.M.: Composite Adaptive Fuzzy H_∞ Tracking Control of Uncertain Nonlinear Systems. Neurocomputing 1(99), 15–24 (2013)
10. Chen, B., Liu, P.X., Ge, S.S., Lin, C.: Adaptive Fuzzy Control of a class of Nonlinear Systems by Fuzzy Approximation Approach. IEEE Trans. Fuzzy Syst. 6(20), 1012–1021 (2012)
11. Wang, N., Er, J.M.: Self-Constructing Adaptive Robust Fuzzy Neural Tracking Control of Surface Vehicles with Uncertainties and Unknown Disturbances. IEEE Transactions on Control Systems Technology 23(3), 991–1002 (2015)
12. Wang, N., Er, J.M., Sun, J.C., Liu, Y.C.: Adaptive Robust Online Constructive Fuzzy Control of a Complex Surface Vehicle System. IEE Trans. on Cybernetics (2015). doi:10.1109/TCYB.2451116
13. Fossen, T.I.: Marine control systems: guidance, navigation, and control of ships, rigs an underwater vehicles. In: Marine Cybernetics AS, Trondheim (2002)

Neurodynamics Analsysis

A New Sampled-Data State Estimator for Neural Networks of Neutral-Type with Time-Varying Delays*

Xianyun Xu, Changchun Yang, Manfeng Hu, Yongqing Yang, and Li Li

School of Science, Jiangnan University, Wuxi 214122, PR China

Abstract. This paper is concerned with the sampled-data state estimation problem for neural networks of neutral-type with time-varying delays. A new state estimator was designed based on the sampled measurements. The sufficient condition for the existence of state estimator is derived by using the Lyapunov functional method. A numerical example is given to show the effectiveness of the proposed estimator.

Keywords: state estimation, sampled measurements, neutral-type, neural network, delay.

1 Introduction

Neural networks have received much attention in the past decades due to their extensive applications in signal processing, pattern classification, optimization and associative memory[1,2,3]. It is known that time delays often occur due to the finite switching speeds of the amplifiers or the finite signal propagation time. So, the delayed neural networks have been considered as viable network models[4,5,6,7,8]. In applications, the neuron states are not often fully available from the network outputs, the neuron state estimation problem becomes precursor and was widely studied. For example, Liu et al. studied H_∞ state estimation of static neural networks with time-varying delays[9]. The state estimation was investigated for nonlinear networked control systems with limited capacity channel in [10]. For some Markovian jumping systems, the robust state estimation and exponential state estimation were discussed[11,12]. The state estimator was designed for discrete-time neural networks and fuzzy neural networks in [13,14,15,16]. In addition, Liu et al. investigated state estimation for the complex networked systems with randomly occurring nonlinearities and randomly missing measurements[17]. Park et al. discussed the state estimation problems of neural networks with neutral-type delay and interval time-varying delays[22,23].

Recently, with the development of computer hardware technology, the sampled-data state estimations have received constant attentions both from academic research and industrial application. Via the output sampled measurement,

* This work was jointly supported by the National Natural Science Foundation of China under Grant 11226116, the Fundamental Research Funds for the Central Universities JUSRP51317B.

© Springer International Publishing Switzerland 2015
X. Hu et al. (Eds.): ISNN 2015, LNCS 9377, pp. 121–128, 2015.
DOI: 10.1007/978-3-319-25393-0_14

state estimation of the neuron needs less information from the network outputs, which can lead to the reduction of the network communication burden. Some sampled-data state estimator have been designed for Markovian jumping neural networks[18,21]. Using discontinuous Lyapunov functional approach, the sampled-data state estimation of some neural networks were investigated[19,20]. To the best of our knowledge, state estimators were usually designed as the same formation with the estimated system. Moreover, it is still scarce on the results that the sampled-data state estimators were designed based on partial information of the estimated systems.

Motivated by the previous discussion, the purpose of this paper is to investigate the sampled-data state estimation problem for delayed neural networks of neutral-type. The main contributions of this paper are twofold. First, a new sampled-data state estimator is designed based on the partial information of the neural networks. Second, The estimator gain matrix can be obtained by solving the linear matrix inequalities.

The organization of the paper is as follows: in section 2, the problem is formulated and some lemmas are introduced. In section 3, some sufficient conditions are given to guarantee the existence of the sampled-data estimator. An examples is given to exemplify the usefulness of our theoretical results in section 4. And in the last section: section 5, we give some conclusions.

2   Problem Formulation

Consider the following neural networks with neutral time-varying delay:

$$
\begin{aligned}
\dot{x}(t) &= -Ax(t) + W_0 f(x(t)) + W_1 g(x(t - \tau(t))) + V\dot{x}(t - d(t)), \\
y(t) &= Cx(t),
\end{aligned}
\tag{1}
$$

where $\tau(t) > 0$ and $d(t) > 0$, correspond to finite speed of axonal signal transmission delay satisfying the following:

$$
\begin{aligned}
0 < \tau(t) \leq \bar{\tau}, \quad \dot{\tau}(t) \leq \mu, \\
0 < d(t) \leq \bar{d}, \quad \dot{d}(t) \leq d_D < 1.
\end{aligned}
$$

Assumption 1. The neuron activation function $f(\cdot)$, $g(\cdot)$ satisfy :

$$
\begin{aligned}
[f(x) - U_1 x]^T [f(x) - U_2 x] \leq 0, \\
[g(x) - U_3 x]^T [g(x) - U_4 x] \leq 0,
\end{aligned}
\tag{2}
$$

The aim of this paper is to propose a new estimation algorithm to observe the neuron states from the available network sampling output. The measurement output is sampled as follows:

$$
\bar{y}(t) = Cx(t_k), \quad t \in [t_k, t_{k+1}),
\tag{3}
$$

where $\bar{y}(t) \in \mathbb{R}^m$ is the actual output, and t_k denotes the sampling instant satisfying $\lim_{k \to \infty} t_k = \infty$.

Based on the available sampled measurement $\bar{y}(t)$, the state estimator is adopted:

$$\dot{\hat{x}}(t) = -A\hat{x}(t) + V\dot{\hat{x}}(t - d(t)) + K[\bar{y}(t) - \hat{y}(t)],$$
$$\hat{y}(t) = C\hat{x}(t), \tag{4}$$

Assumption 2. For $k \geq 0$, there exists a positive constant $\bar{h}$ such that the sampling instant t_k satisfies $t_{k+1} - t_k \leq \bar{h}$.

Define the error vector to be $e(t) = x(t) - \hat{x}(t)$. Let $h(t) = t - t_k, t_k \leq t < t_{k+1}$, the error dynamical system can be expressed by:

$$\begin{aligned}
\dot{e}(t) &= -Ae(t) + V\dot{e}(t - d(t)) + W_0 f(x(t)) + W_1 g(x(t - \tau(t))) - K\left[Cx(t_k) - C\hat{x}(t)\right] \\
&= -\left[A + KC\right]e(t) + V\dot{e}(t - d(t)) + W_0 f(x(t)) + W_1 g(x(t - \tau(t))) \\
&\quad + KCx(t) - KCx(t - h(t)).
\end{aligned} \tag{5}$$

From (1) and (5), we have the following augmented system:

$$\dot{\bar{e}}(t) = \bar{A}\bar{e}(t) + \bar{B}\bar{e}(t - h(t)) + \bar{V}\dot{\bar{e}}(t - d(t)) + \bar{W}_0 f(H\bar{e}(t)) + \bar{W}_1 g(H\bar{e}(t - \tau(t))), \tag{6}$$

where

$$\bar{e}(t) = \begin{bmatrix} x(t) \\ e(t) \end{bmatrix}, \quad \bar{A} = \begin{bmatrix} -A & 0 \\ KC & -(A + KC) \end{bmatrix}, \quad \bar{B} = \begin{bmatrix} 0 & 0 \\ -KC & 0 \end{bmatrix},$$
$$\bar{V} = \begin{bmatrix} V & 0 \\ 0 & V \end{bmatrix}, \quad \bar{W}_0 = \begin{bmatrix} W_0 \\ W_0 \end{bmatrix}, \quad \bar{W}_1 = \begin{bmatrix} W_1 \\ W_1 \end{bmatrix}, \quad H = \begin{bmatrix} I & 0 \end{bmatrix}.$$

The following lemmas will be used in deriving the main results.

Lemma 1. *([21]) (Schur Completement) Given constant matrices Ω_1, Ω_2, and Ω_3, where $\Omega_1 = \Omega_1^T$ and $\Omega_2 > 0$, then*

$$\Omega_1 + \Omega_3^T \Omega_2^{-1} \Omega_3 < 0, \tag{7}$$

if and only if

$$\begin{bmatrix} \Omega_1 & \Omega_3^T \\ \Omega_3 & -\Omega_2 \end{bmatrix} < 0. \tag{8}$$

Lemma 2. *([22]) (Jensen's inequality) For any constant matrix $Q \in \mathbb{R}^{n \times n}, Q = Q^T > 0$, scalar $b > 0$, and vector function $x : [0, b] \to \mathbb{R}^n$, one has*

$$-\int_0^b x^T(s)Qx(s)ds \leq -\frac{1}{b}\left[\int_0^b x(s)ds\right]^T Q\left[\int_0^b x(s)ds\right].$$

3 Main Results

In this section, we derive a delay-dependent criterion for exponential stability of the error system (6) by the Lyapunov method.

Theorem 1. *For given matrices U_i $(i = 1, 2, 3, 4)$ and ρ, the error system (6) is exponentially stable if there exist matrices $P = diag\{P_{11}, P_{22}\} > 0, Q_1 > 0, Q_2 > 0, Q_3 > 0, Z_1 > 0, Z_2 > 0, Z_3 > 0, Y$ and scalars $\lambda_1 > 0, \lambda_2 > 0$ such that $\Upsilon > 0$ and*

$$\Sigma = \begin{bmatrix} \Sigma_1 & \Sigma_2^T P \\ & -2\rho P + \rho^2 \Upsilon \end{bmatrix} < 0, \tag{9}$$

where

$$\Sigma_1 = \begin{bmatrix} \Lambda_{11} & P\bar{B} + \frac{1}{h}Z_1 & 0 & \Lambda_{14} & 0 & P\bar{V} & P\bar{W}_0 - \lambda_1 N_1 & P\bar{W}_1 \\ -\frac{2}{h}Z_1 & & & & & & & \\ * & \frac{1}{h}Z_1 & 0 & 0 & 0 & 0 & 0 & 0 \\ * & * & -Q_1 - \frac{1}{h}Z_1 & 0 & 0 & 0 & 0 & 0 \\ * & * & * & \Lambda_{44} & \frac{1}{\bar{\tau}}Z_2 & 0 & 0 & -\lambda_2 N_2 \\ * & * & * & * & -Q_2 - \frac{1}{\bar{\tau}}Z_2 & 0 & 0 & 0 \\ * & * & * & * & * & -(1-d_D)R & 0 & 0 \\ * & * & * & * & * & * & -\lambda_1 I & 0 \\ * & * & * & * & * & * & * & -\lambda_2 I \end{bmatrix},$$

$\Sigma_2 = \begin{bmatrix} \bar{A} & \bar{B} & 0 & 0 & 0 & \bar{V} & \bar{W}_0 & \bar{W}_1 \end{bmatrix}$,

$\Upsilon = R + \bar{h}Z_1 + \bar{\tau}Z_2 + \bar{\tau}Z_3$,
$\Lambda_{11} = P\bar{A} + \bar{A}^T P + Q_1 + Q_2 + Q_3 - \frac{1}{\bar{\tau}}[Z_2 + (1 - \mu)Z_3] - \frac{1}{h}Z_1 - \lambda_1 M_1$,
$\Lambda_{14} = \frac{1}{\bar{\tau}}[Z_2 + (1 - \mu)Z_3]$,
$\Lambda_{44} = -\frac{1}{\bar{\tau}}[Z_2 + (1 - \mu)Z_3] - \frac{1}{\bar{\tau}}Z_2 - (1 - \mu)Q_3 - \lambda_2 M_2$,
$M_1 = \frac{H^T U_1^T U_2 H + H^T U_2^T U_1 H}{2}$, $\qquad M_2 = \frac{H^T U_3^T U_4 H + H^T U_4^T U_3 H}{2}$,
$N_1 = -\frac{H^T U_1^T + H^T U_2^T}{2}$, $\qquad N_2 = -\frac{H^T U_3^T + H^T U_4^T}{2}$.
Furthermore, if the LMIs given above are solvable, the desired estimator parameters are given as $K = P_{22}^{-1} X$.

Proof. Consider the following Lyapunov functional:

$$V(t) = V_1(t) + V_2(t) + V_3(t) + V_4(t) \tag{10}$$

where

$V_1(t) = \bar{e}^T(t)P\bar{e}(t)$,
$V_2(t) = \int_{t-\bar{h}}^t \bar{e}^T(s)Q_1\bar{e}(s)ds + \int_{t-\bar{\tau}}^t \bar{e}^T(s)Q_2\bar{e}(s)ds + \int_{t-\tau(t)}^t \bar{e}^T(s)Q_3\bar{e}(s)ds$,
$V_3(t) = \int_{t-d(t)}^t \dot{\bar{e}}^T(s)R\dot{\bar{e}}(s)ds$,
$V_4(t) = \int_{-\bar{h}}^0 \int_{t+\theta}^t \dot{\bar{e}}^T(s)Z_1\dot{\bar{e}}(s)dsd\theta + \int_{-\bar{\tau}}^0 \int_{t+\theta}^t \dot{\bar{e}}^T(s)Z_2\dot{\bar{e}}(s)dsd\theta + \int_{-\tau(t)}^0 \int_{t+\theta}^t \dot{\bar{e}}^T(s)Z_3\dot{\bar{e}}(s)dsd\theta$.

Calculate the derivative of V_i $(i = 1, 2, 3, 4)$ along the trajectories of the system (6), we have

$$\dot{V}_1(t) = 2\bar{e}^T(t)P[\bar{A}\bar{e}(t) + \bar{B}\bar{e}(t - h(t)) + \bar{V}\dot{\bar{e}}(t - d(t)) + \bar{W}_0 f(H\bar{e}(t)) + \bar{W}_1 g(H\bar{e}(t - h(t)))], \tag{11}$$

$$\dot{V}_2(t) \leq \bar{e}^T(t)[Q_1 + Q_2 + Q_3]\bar{e}(t) - \bar{e}^T(t - \bar{e}^T(t - \bar{\tau})Q_2\bar{e}(t - \bar{\tau}) \\ - \bar{h})Q_1\bar{e}(t - \bar{h}) - (1 - \mu)\bar{e}^T(t - \tau(t))Q_3\bar{e}(t - \tau(t)) \tag{12}$$

$$\dot{V}_3(t) \le \dot{\bar{e}}^T(t)R\dot{\bar{e}}(t) - (1-d_D)\dot{\bar{e}}^T(t-d(t))R\dot{\bar{e}}(t-d(t)), \tag{13}$$

$$\dot{V}_4(t) \le \dot{\bar{e}}^T(t) \left[\hbar Z_1 + \bar{\tau} Z_2 + \bar{\tau} Z_3\right] \dot{\bar{e}}(t) - \int_{t-\hbar}^t \dot{\bar{e}}^T(s)Z_1\dot{\bar{e}}(s)ds - \int_{t-\bar{\tau}}^t \dot{\bar{e}}^T(s)Z_2\dot{\bar{e}}(s)ds \\ - (1-\mu)\int_{t-\tau(t)}^t \dot{\bar{e}}^T(s)Z_3\dot{\bar{e}}(s)ds. \tag{14}$$

By Lemma 2, we can obtain

$$-\int_{t-\hbar}^t \dot{\bar{e}}^T(s)Z_1\dot{\bar{e}}(s)ds \le -\tfrac{1}{\hbar}\left[\bar{e}(t)-\bar{e}(t-h(t))\right]^T Z_1 \left[\bar{e}(t)-\bar{e}(t-h(t))\right] \\ -\tfrac{1}{\hbar}\left[\bar{e}(t-h(t))-\bar{e}(t-\hbar)\right]^T Z_1 \left[\bar{e}(t-h(t))-\bar{e}(t-\hbar)\right], \tag{15}$$

$$-\int_{t-\bar{\tau}}^t \dot{\bar{e}}^T(s)Z_2\dot{\bar{e}}(s)ds - (1-\mu)\int_{t-\tau(t)}^t \dot{\bar{e}}^T(s)Z_3\dot{\bar{e}}(s)ds \\ \le -\tfrac{1}{\bar{\tau}}\left[\bar{e}(t)-\bar{e}(t-\tau(t))\right]^T \left[Z_2+(1-\mu)Z_3\right]\left[\bar{e}(t)-\bar{e}(t-\tau(t))\right] \\ -\tfrac{1}{\bar{\tau}}\left[\bar{e}(t-\tau(t))-\bar{e}(t-\bar{\tau})\right]^T Z_2 \left[\bar{e}(t-\tau(t))-\bar{e}(t-\bar{\tau})\right]. \tag{16}$$

Note that (2) implies

$$\begin{bmatrix} \bar{e}(t) \\ f(H\bar{e}(t)) \end{bmatrix}^T \begin{bmatrix} M_1 & N_1 \\ & I \end{bmatrix} \begin{bmatrix} \bar{e}(t) \\ f(H\bar{e}(t)) \end{bmatrix} \le 0, \tag{17}$$

$$\begin{bmatrix} \bar{e}(t-\tau(t)) \\ g(H\bar{e}(t-\tau(t))) \end{bmatrix}^T \begin{bmatrix} M_2 & N_2 \\ & I \end{bmatrix} \begin{bmatrix} \bar{e}(t-\tau(t)) \\ g(H\bar{e}(t-\tau(t))) \end{bmatrix} \le 0, \tag{18}$$

where M_1, M_2, N_1 and N_2 are defined in Theorem 1.

Therefore, for any scalars $\lambda_1 > 0$ and $\lambda_2 > 0$, and introducing the new variable $P_{22}K = Y$, after some matrix manipulations together with (11)—(18), we have

$$\begin{aligned} \dot{V}(t) &= \sum_{i=1}^4 \dot{V}_i(t) \\ &\le \bar{e}^T(t)\left[P\bar{A}+\bar{A}^T P+Q_1+Q_2+Q_3-\tfrac{1}{\bar{\tau}}[Z_2+(1-\mu)Z_3]-\tfrac{1}{\hbar}Z_1\right]\bar{e}(t) \\ &\quad + 2\bar{e}^T(t)\left[P\bar{B}+\tfrac{1}{\hbar}Z_1\right]\bar{e}(t-h(t))-\bar{e}^T(t-h(t))\left[\tfrac{2}{\hbar}Z_1\right]\bar{e}(t-h(t)) \\ &\quad + 2\bar{e}^T(t)\left[P\bar{W}_0\right]f(H\bar{e}(t))+2\bar{e}^T(t)\left[P\bar{W}_1\right]g(H\bar{e}(t-\tau(t))) \\ &\quad + 2\bar{e}(t)\left[\tfrac{1}{\bar{\tau}}[Z_2+(1-\mu)Z_3]\right]\bar{e}(t-\tau(t))+2\bar{e}^T(t)\left[P\bar{V}\right]\dot{\bar{e}}(t-d(t))) \\ &\quad - \dot{\bar{e}}^T(t-d(t))\left[(1-d_D)R\right]\dot{\bar{e}}(t-d(t))+2\bar{e}^T(t-h(t))\left[\tfrac{1}{\hbar}Z_1\right]\bar{e}(t-\hbar) \\ &\quad - \bar{e}^T(t-\hbar)\left[Q_1+\tfrac{1}{\hbar}Z_1\right]\bar{e}(t-\hbar)+2\bar{e}^T(t-\tau(t))\left[\tfrac{1}{\bar{\tau}}Z_2\right]\bar{e}(t-\bar{\tau}) \\ &\quad - \bar{e}^T(t-\tau(t))\left[\tfrac{1}{\bar{\tau}}[Z_2+(1-\mu)Z_3]+\tfrac{1}{\bar{\tau}}Z_2+(1-\mu)Q_3\right]\bar{e}(t-\tau(t)) \\ &\quad - \bar{e}^T(t-\bar{\tau})\left[Q_2+\tfrac{1}{\bar{\tau}}Z_2\right]\bar{e}(t-\bar{\tau})+\dot{\bar{e}}^T(t)\left[R+\hbar Z_1+\bar{\tau}Z_2+\bar{\tau}Z_3\right]\dot{\bar{e}}(t) \\ &\quad - \lambda_1 \begin{bmatrix} \bar{e}(t) \\ f(H\bar{e}(t)) \end{bmatrix}^T \begin{bmatrix} M_1 & N_1 \\ & I \end{bmatrix} \begin{bmatrix} \bar{e}(t) \\ f(H\bar{e}(t)) \end{bmatrix} \\ &\quad - \lambda_2 \begin{bmatrix} \bar{e}(t-\tau(t)) \\ g(H\bar{e}(t-\tau(t))) \end{bmatrix}^T \begin{bmatrix} M_2 & N_2 \\ & I \end{bmatrix} \begin{bmatrix} \bar{e}(t-\tau(t)) \\ g(H\bar{e}(t-\tau(t))) \end{bmatrix} \\ &= \xi^T(t)\left[\Sigma_1+\Sigma_2^T\Upsilon\Sigma_2\right]\xi(t), \end{aligned} \tag{19}$$

Using the fact $-P\Upsilon^{-1}P \le -2\rho P+\rho^2 R$ resulted from $(P-\rho\Upsilon)\Upsilon^{-1}(P-\rho\Upsilon) \ge 0$, it is clear that LMIs (9) can guarantee the following inequality:

$$\begin{bmatrix} \Sigma_1 & \Sigma_2^T P \\ & -P\Upsilon^{-1}P \end{bmatrix} < 0. \tag{20}$$

Pre- and post multiplying $diag\{I,I,I,I,I,I,I,I,\Upsilon P^{-1}\}$ and $diag\{I,I,I,I,$ $I,I,I,I,P^{-1}\Upsilon\}$, respectively, we have

$$\begin{bmatrix} \Sigma_1 & \Sigma_2^T\Upsilon \\ & -\Upsilon \end{bmatrix} < 0. \tag{21}$$

By Lemma 1, the inequality (21) is equivalent to the inequality $\Sigma_1 + \Sigma_2^T\Upsilon\Sigma_2 < 0$. This implies that the error dynamic (6) is exponentially stable by the Lyapunov theory. This completes the proof of the theorem.

4   Simulation Example

In this section, a numerical example with simulation results is employed to demonstrated the effectiveness of the proposed method.

Example 1. Consider the neural networks of neutral-type (1) with the following parameters:

$$A = \begin{bmatrix} 1.5 & 0 \\ 0 & 1.2 \end{bmatrix}, \quad W_0 = \begin{bmatrix} 0.3 & -0.2 \\ -0.2 & 0.3 \end{bmatrix}, \quad W_1 = \begin{bmatrix} 0.3 & 0.3 \\ 0.3 & 0.3 \end{bmatrix},$$
$$V = \begin{bmatrix} 0.3 & 0.1 \\ 0.1 & 0.3 \end{bmatrix}, \quad U_1 = \begin{bmatrix} -0.2 & 0 \\ 0.3 & 0.85 \end{bmatrix}, \quad U_2 = \begin{bmatrix} -0.5 & 0 \\ 0.3 & 0.3 \end{bmatrix},$$
$$U_3 = \begin{bmatrix} 0.5 & 0 \\ -0.3 & -0.3 \end{bmatrix}, \quad U_4 = \begin{bmatrix} 0.2 & 0 \\ -0.3 & -0.85 \end{bmatrix}, \quad C = [0.7 \ 0.8],$$
$$f(x) = 0.5tanh(x), \quad g(x(t-h(t))) = tanh(0.4*x(t-h(t))),$$
$$h(t) = 0.14(1-sin(t)), \quad d(t) = 0.45(1-sin(t)), \quad \bar{\tau} = 0.36.$$

From the parameters above, for given $\rho = 25$, The conditions of Theorem 1 can be satisfied by using Matlab LMI Toolbox. Let $x(0) = [-0.36 \ -0.35]^T$, $\hat{x}(0) = [0.45 \ 0.54]^T$. The simulation results are displayed in Fig.1.

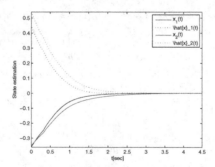

Fig.1: The true state $x(t)$ and its estimate $\hat{x}(t)$.

5   Conclusions

In this paper, we studied the sampled-data state estimation problem for delayed neural networks of neutral-type. By using a delayed-input approach, the sampling period is converted equivalently into a bounded time-varying delay. Based on the available sampled measurement, we construct the state estimator by utilizing partial information of the neural networks. By employing a suitable Lyapunov functional, and combining with Jensen integral inequality, a sufficient condition for the existence of state estimator is derived in terms of linear matrix inequalities (LMIs). Finally, a illustrative example is exploited to show the effectiveness of the proposed method.

References

1. Elanayar, V., Shin, Y.: Radial basis function neural network for approximation and estimation of nonlinear stochastic dynamic systems. IEEE Transaction on Neural Networks 5, 594–603 (1994)
2. Joya, G., Atencia, M., Sandoval, F.: Hopfield neural network for optimization: study of different dynamic. Neurocomputing 43, 219–237 (2002)
3. Cao, J., Liang, J.: Boundedness and stability for Cohen-Grossberg neural network with time-varying delays. Journal of Mathematical Analysis and Applications 296, 665–685 (2004)
4. Cao, J., Zhou, D.: Stability analysis of delayed cellular neural networks. Neural Networks 11, 1601–1605 (1998)
5. Liu, X., Teo, K., Xu, B.: Exponential stability of impulsive high-order Hopfield-type neural networks with time-varying delays. IEEE Transaction on Neural Networks 16, 1329–1339 (2005)
6. Liang, T., Yang, Y., Liu, Y., Li, L.: Existence and global exponential stability of almost periodic solutions to Cohen-Grossberg neural networks with distributed delays on time scales. Neurocomputing 123, 207–215 (2014)
7. Park, J., Park, C., Kwon, O., Lee, S.: A new stability criterion for bidirectional associative memory neural networks of neutral-type. Applied Mathematics and Computation 199, 716–722 (2008)
8. Li, L., Yang, Y., Liang, T., Hu, M.: The exponential stability of BAM neural networks with leakage time-varying delays and sampled-data state feedback input. Advances in Difference Equations 39 (2014)
9. Liu, Y., Lee, S., Kwon, O., Park, J.: A study on H_∞ state estimation of static neural networks with time-varying delays. Applied Mathematics and Computation 226, 589–597 (2014)
10. Liu, M., Wang, Q., Li, H.: State estimation and stabilization for nonlinear networked control systems with limited capacity channel. Journal of the Franklin Institute 348(8), 1869–1885 (2011)
11. Zhang, D., Yu, L.: Exponential state estimation for Markovian jumping neural networks with time-varying discrete and distributed delays. Neural Networks 35, 103–111 (2012)
12. Li, W., Jia, Y., Du, J., Zhang, J.: Robust state estimation for jump Markovian linear systems with missing measurements. Journal of the Franklin Institute 350(6), 1476–1487 (2013)
13. Liu, Y., Wang, Z., Liu, X.: State estimation for discrete-time Markovian jumping neural networks with mixed mode-dependent delays. Physics Letters A 372, 7147–7155 (2008)

14. Balasubramaniam, P., Vembarasan, V., Rakkiyappan, R.: Delay-dependent robust exponential state estimation of Markovian jumping fuzzy Hopfield neural networks with mixed random time-varying delays. Communications in Nonlinear Science and Numerical Simulation 16, 2109–2129 (2011)
15. Balasubramaniam, P., Kalpana, M., Rakkiyappan, R.: State estimation for fuzzy cellular neural networks with time delay in the leakage term, discrete and unbounded distributed delays. Computers and Mathematics with Applications 62, 3959–3972 (2011)
16. Duan, Q., Park, J., Wu, Z.: Exponential state estimator design for discrete-time neural networks with discrete and distributed time-varying delays. Complexity (2014). doi:10.1002/cplx.21494
17. Liu, J., Cao, J., Wu, Z., Qi, Q.: State estimation for complex systems with randomly occurring nonlinearities and randomly missing measurements. International Journal of Systems Science, 1–11 (2014)
18. Rakkiyappan, R., Sakthivel, N., Park, J., Kwon, O.: Sampled-data state estimation for Markovian jumping fuzzy cellular neural networks with mode-dependent probabilistic time-varying delays. Applied Mathematics and Computation 221, 741–769 (2013)
19. Lakshmanan, S., Park, J., Rakkiyappan, R., Jung, H.: State estimator for neural networks with sampled data using discontinuous Lyapunov functional approach. Nonlinear Dynamic Systems 73, 509–520 (2013)
20. Rakkiyappan, R., Zhu, Q., Radhika, T.: Design of sampled data state estimator for Markovian jumping neural networks with leakage time-varying delays and discontinuous Lyapunov functional approach. Nonlinear Dynamic Systems 73, 1367–1383 (2013)
21. Hu, J., Li, N., Liu, X., Zhang, G.: Sampled-data state estimation for delayed neural networks with Markovian jumping parameters. Nonlinear Dynamic Systems 73, 275–284 (2013)
22. Park, J., Kwon, O.: Further results on state estimation for neural networks of neutral-type with time-varying delay. Applied Mathematics and Computation 208, 69–75 (2009)
23. Park, J., Kwon, O., Lee, S.: State estimation for neural networks of neutral-type with interval time-varying delays. Applied Mathematics and Computation 203, 217–223 (2008)
24. Yang, C., Yang, Y., Hu, M., Xu, X.: Sampled-data state estimation for neural networks of neutral type. Advances in Difference Equations, 138 (2014)

Exponential Lag Synchronization for Delayed Cohen-Grossberg Neural Networks with Discontinuous Activations

Abdujelil Abdurahman, Cheng Hu, and Haijun Jiang

College of Mathematics and System Sciences, Xinjiang University,
Urumqi, 830046, Xinjiang, P.R. China
jianghaijunxju@163.com

Abstract. In this paper, we investigate the exponential lag synchronization of delayed Cohen-Grossberg neural networks with discontinuous activation functions. By employing the analysis technique and theory of the differential equations with discontinuous right-hand side, some novel lag synchronization criteria have been obtained. Finally, an example is given to illustrate the effectiveness of the obtained results.

Keywords: Cohen-Grossberg neural network, Exponential lag synchronization, Discontinuous, Time-delay.

1 Introduction and Preliminaries

In the past few years, many considerable efforts have been devoted to investigate the neural network system with discontinuous activation functions due to the fact that neural networks with discontinuous (or non-Lipschitz, or nonsmooth) neuron activations, have been found useful to address a number of interesting engineering tasks, such as dry friction, impacting machines, systems oscillating under the effect of an earthquake, power circuits, switching in electronic circuits, linear complementarity systems, and many others [1-3].

Since Pecora and Carrol [4] first proposed a method to synchronize two identical systems with different initial values, the problem of synchronization in chaotic systems has been extensively investigated over the past few decades owing to their potential applications in many engineering areas, ranging from secure communications to modeling brain activity, even to optimization of nonlinear system performance [5]. Meanwhile, a number of methods have been developed for the synchronization of chaotic systems which include complete synchronization [6], lag synchronization [7], impulsive synchronization [8], phase synchronization [9], projective synchronization [10], function projective synchronization [11], etc.

On the other hand, it has been shown that the complete synchronization of chaos is practically impossible in the remote communication systems due to finite transmission speed of signals or memory effects. For example, in the telephone communication system, the voice one hears on the receiver side at time $t+\sigma$ is the voice from the transmitter side at time t. Hence, it is reasonable to require one

© Springer International Publishing Switzerland 2015
X. Hu et al. (Eds.): ISNN 2015, LNCS 9377, pp. 129–137, 2015.
DOI: 10.1007/978-3-319-25393-0_15

neural network to synchronize the other neural network at a constant time lag [12]. Lag synchronization appears as a coincidence of shifted-in-time states of two systems $y(t) \to x(t - \sigma)$, $t \to \infty$, with a propagation delay $\sigma > 0$ [13, 14]. Also, compared with complete synchronization, lag synchronization may be a more appropriate technique to clearly indicate the fragile nature of neuron systems. Thus, it is of great importance to study lag synchronization of discontinuous neural networks, but till now, there are very few or even no published results in this research area.

Motivated by above analysis, in this paper, we consider the exponential lag synchronization of following delayed Cohen-Grossberg neural networks (CGNNs) model with discontinuous activation functions

$$\dot{x}_i(t) = a_i(x_i(t))\Big[- d_i x_i(t) + \sum_{j=1}^{n} b_{ij} f_j(x_j(t)) + \sum_{j=1}^{n} c_{ij} f_j\left(x_j(t - \tau_j)\right) + I_i\Big], \quad (1)$$

where $i \in \mathscr{I} \triangleq \{1, 2, \cdots, n\}$, $n \geq 2$ denotes the number of neurons in the neural network; $a_i(x_i(t))$ represents an amplification function, $d_i > 0$ represents the rate with which the ith neuron will reset its potential to the resting state when disconnected from the network and external inputs, $f_j(x_j(t))$ is the activation function, $\tau_j \geq 0$ corresponds to the transmission delay. Concerning coefficients b_{ij} and c_{ij} denote the synaptic connection weights, and I_i denotes the external bias on the ith unit.

Throughout this paper, for the model (1), we introduce the following assumptions

A$_1$: $a_i \in C(R, R^+)$ and there exist positive constants $\underline{a_i}$ and $\bar{a}_i$ such that

$$\underline{a_i} \leq a_i(u) \leq \bar{a}_i, \quad \text{for all } u \in R, \quad i \in \mathscr{I} .$$

A$_2$: For each i, $f_i(\cdot)$ is continuous on $\mathbb{R}$ expect a countable set of isolate points ρ_k^i, where there exist finite right and left limits $f_i^+(\rho_k^i)$ and $f_i^-(\rho_k^i)$, respectively. Moreover, f_i has a finite number of discontinuous points on any compact interval of $\mathbb{R}$.

A$_3$: For each i, there exist a nonnegative constants L_i and N_i such that

$$\sup_{\xi_i \in K[f_i(u)], \gamma_i \in K[f_i(v)]} |\xi_i - \gamma_i| \leq L_i|u - v| + N_i, \quad u, v \in \mathbb{R},$$

where $K[f_i(s)] = \big[\min\{f_i^-(s), f_i^+(s)\}, \max\{f_i^-(s), f_i^+(s)\}\big]$ for $s \in \mathbb{R}$.

Let $\tau = \max_{j \in \mathscr{I}}\{\tau_j\}$ and $C = C([-\tau, 0], \mathbb{R}^n)$ denotes the Banach space of all continuous functions mapping $[-\tau, 0]$ into $\mathbb{R}^n$ with the norm $\|\varphi\|_c = \sup_{-\tau \leq s \leq 0} \|\varphi(s)\|$. If for $L \in (0, +\infty]$, $x(t) : [-\tau, L) \to \mathbb{R}^n$ is continuous, then $x_t \in C$ is defined by $x_t(\theta) = x(t + \theta), \theta \in [-\tau, 0]$ for any $t \in [0, L)$.

Definition 1. *For the system* $\dot{x}(t) = f(t, x_t)$, *where* $x_t(\cdot)$ *denotes the history of the state from time* $t - \tau$, *up to the present time* t; $\dot{x}(t)$ *denotes the time derivative of* x *and* $f : \mathbb{R} \times C \to \mathbb{R}^n$ *is measurable and essentially locally bounded. Then, the Flippov set-valued map* $F(t, x_t) : \mathbb{R} \times C \to 2^{\mathbb{R}^n}$ *is defined as follows:*

$$\mathbb{F}(t, x_t) = \bigcap_{r>0} \bigcap_{\mu(N)=0} K[f(\mathscr{B}(x_t, r) \backslash N)],$$

where $K[E]$ is the closure of the convex hull of set E; intersection is taken over all sets N of Lebesgue measure zero and over all $r > 0$; $\mathscr{B}(x_t, r) := \{x_t^* : \|x_t^* - x_t\|_c \leq r\}$, and $\mu(N)$ is Lebesgue measure of set N.

Definition 2. *A function $x : [0, L) \to \mathbb{R}^n$, $L \in (0, +\infty]$, is called a state solution of (1) on $[0, L)$ if*

(i) *x is absolutely continuous on $[0, L)$;*
(ii) *there exists a measurable function $\gamma = (\gamma_1, \gamma_2, ..., \gamma_n)^T : [0, L) \to \mathbb{R}^n$ such that $\gamma(t) \in K[f(x(t))]$ for a.a.t $\in [0, L)$ and*

$$\dot{x}_i(t) = a_i(x_i(t))\Big[-d_i x_i(t) + \sum_{j=1}^{n} b_{ij}\gamma_j(t) + \sum_{j=1}^{n} c_{ij}\gamma_j(t - \tau_j) + I_i\Big]. \quad (2)$$

Any function γ satisfying (2) is called an output solution associated to the state x. With this definition it turns out that the state x is a solution of (1.1) in the sense of Fillipov since it satisfies

$$\dot{x}_i(t) \in a_i(x_i(t))\Big[-d_i x_i(t) + \sum_{j=1}^{n} b_{ij}K[f_j(x_j(t))] + \sum_{j=1}^{n} c_{ij}K[f_j(x_j(t - \tau_j))] + I_i\Big]. \quad (3)$$

Definition 3. *(IVP) For a continuous function $\varphi(\theta) = (\varphi_1(\theta), \cdots, \varphi_n(\theta))^T$ and a measurable function $\phi(\theta) = (\phi_1(\theta), \cdots, \phi_n(\theta))^T \in K[f(\varphi(\theta))]$ for a.a. $\theta \in [-\tau, 0]$, a continuous functions $x(t) = x(t, \varphi, \phi) = (x_1(t), \cdots, x_n(t))^T$ associated with a measurable function $\gamma(t) = (\gamma_1(t), \cdots, \gamma_n(t))^T$ is said to be a solution of the Cauchy problem for system (3) on $[-\tau, L)$ ($L > 0$ might be $+\infty$) with initial condition $[\varphi(\theta), \psi(\theta)]$, $\theta \in [-\tau, 0]$, if $x(t)$ is absolutely continuous on any compact interval of $[0, L)$, and*

$$
\begin{cases}
\dot{x}_i(t) = a_i(x_i(t))\big[-d_i x_i(t) + \sum_{j=1}^{n} b_{ij}\gamma_j(t) + \sum_{j=1}^{n} c_{ij}\gamma_j(t - \tau_j) + I_i\big], & \text{a.a. } t \in [0, L), \\
\gamma_j(t) \in K[f_j(x_j(t))], & \text{a.a. } t \in [0, L), \\
x_i(\theta) = \varphi_i(\theta), & \theta \in [-\tau, 0] \\
\gamma_i(\theta) = \psi_i(\theta), \quad \psi_i(\theta) \in K[f_i(\varphi_i(\theta))], & \text{a.a. } \theta \in [-\tau, 0].
\end{cases} \quad (4)
$$

Remark 1. Suppose that the conditions $\mathbf{A}_1 - \mathbf{A}_3$ are satisfied, then the growth condition (8) in Theorem 1 in [15] holds. Therefore, any IVP for (1) has at least one solution x on $[0, +\infty)$.

Consider the neural network model (1) as the driver system, the controlled response system can be described as follows:

$$\dot{y}_i(t) = a_i(x_i(t))\Big[-d_i y_i(t) + \sum_{j=1}^{n} b_{ij}f_j(y_j(t)) + \sum_{j=1}^{n} c_{ij}f_j(x_j(t - \tau_j)) + I_i\Big] + u_i(t), \quad (5)$$

for $i \in \mathscr{I}$ and $t \geq \sigma \geq 0$, y_i denote the state of the slave system, $u_i(t)$ is the discontinuous state-feedback controller given by

$$u_i(t) = -\alpha_i(y_i(t) - x_i(t - \sigma)) - k_i sign(y_i(t) - x_i(t - \sigma)), \quad i \in \mathscr{I}, \qquad (6)$$

where α_i and k_i are positive constants determined in later.

According to Definition 3 and Remark 1, we can obtain the initial value problem (IVP) of response system (5) as follows:

$$\begin{cases} \dot{y}_i(t) = a_i(y_i(t))\Big[-d_i y_i(t) + \sum_{j=1}^{n} b_{ij}\gamma_j^*(t) + \sum_{j=1}^{n} c_{ij}\gamma_j^*(t - \tau_j) + I_i \Big] + u_i^*(t), \\ \qquad\qquad\qquad\qquad\qquad\qquad\qquad\qquad\qquad\qquad\quad \text{for a.a. } t \in [\sigma, +\infty), \\ \gamma_j^*(t) \in K[f_j(y_j(t))], \qquad\qquad\qquad\qquad\qquad\quad\ \text{a.a. } t \in [\sigma, +\infty), \\ y_i(\theta) = \phi_\sigma^i(\theta), \quad \phi_\sigma^i(\theta) = \phi_i(\theta + \sigma), \qquad\qquad\ \theta \in [-\tau, 0], \\ \gamma_i^*(\theta) = \xi_i(\theta), \quad \xi_i(\theta) \in K[f_i(\phi_i(\theta)], \qquad\qquad \text{a.a. } \theta \in [-\tau + \sigma, \sigma]. \end{cases} \qquad (7)$$

where

$$u_i^*(t) \in K[u_i(t)] = -\alpha_i(y_i(t) - x_i(t - \sigma)) - k_i\vartheta_i(t)$$

with

$$\vartheta_i(t) = K[sign(y_i(t) - x_i(t - \sigma))] = \begin{cases} -1 & \text{if } y_i(t) - x_i(t - \sigma) < 0, \\ [-1, 1] & \text{if } y_i(t) - x_i(t - \sigma) = 0, \\ 1 & \text{if } y_i(t) - x_i(t - \sigma) > 0. \end{cases} \qquad (8)$$

Definition 4. *Drive-response systems (1) and (5) are said to be exponentially lag synchronized, if there exist $M \geq 1$ and $\lambda > 0$ such that*

$$\|y(t) - x(t - \sigma)\| \leq M\|\phi_\sigma - \varphi\|_c e^{-\lambda(t - \sigma)}$$

for any $t \geq \sigma$. Here λ is called the degree of exponential lag synchronization.

2   Main Results

In this section, we consider the global exponential lag synchronization of delayed CGNNs with discontinuous activations by using discontinuous state-feedback controller. Based on extended Filippov-framework and some analytic techniques, we propose a series of new criteria for synchronization which are different from those of the existing literature.

Theorem 1. *Let $\mathbf{A}_1 - \mathbf{A}_3$ hold. If the control strengths α_i and k_i of (6) satisfy the following inequalities*

$$\Gamma_i = -(d_i + \frac{\alpha_i}{\bar{a}_i} - b_{ii}^+) + \sum_{j=1, j \neq i}^{n} |b_{ij}|L_j + \sum_{j=1}^{n} |c_{ij}|L_j < 0,$$

$$\Upsilon_i = b_{ii}^+ N_i + \sum_{j=1, j \neq i}^{n} |b_{ij}|N_j + \sum_{j=1}^{n} |c_{ij}|N_j - \frac{k_i}{\bar{a}_i} < 0, \quad i \in \mathscr{I}, \qquad (9)$$

where $b_{ii}^+ = \max\{0, b_{ii}\}$. Then then drive-response systems (1) and (5) are exponentially lag synchronized under the controller (6).

Proof. Suppose $\varphi, \phi_\sigma \in C$ and let $[x(t), \gamma(t)]$ and $[y(t), \gamma^*(t)]$ are solutions of systems (1) and (5) with different initial values $[\varphi, \psi]$ and $[\phi_\sigma, \xi]$, respectively, where $\psi \in K[f(\varphi(s))], \xi \in K[f(\phi_\sigma(s))]$, and $\phi_\sigma(s) = \phi(\sigma + s)$ for all $s \in [-\tau, 0]$. Further, let $e_i(t) = y_i(t) - x_i(t - \sigma)$ be the synchronization error between the states of the drive system (1) and the response system (3), and set

$$\bar{e}_i(t) = \left| \int_{x_i(t-\sigma)}^{y_i(t)} \frac{ds}{a_i(s)} \right|. \tag{10}$$

Then from $\mathbf{A}_1$, we have

$$\frac{1}{\bar{a}_i} |e_i(t)| \le \bar{e}_i(t) \le \frac{1}{\underline{a}_i} |e_i(t)|. \tag{11}$$

Thus from Lemma 2.4 in [1], for $t \ge \sigma$, we have

$$\begin{aligned}
\frac{d\bar{e}_i(t)}{dt} &= \partial \left| \int_{x_i(t-\sigma)}^{y_i(t)} \frac{ds}{a_i(s)} \right| \left\{ \frac{\dot{y}_i(t)}{a_i(x_i(t))} - \frac{\dot{x}_i(t-\sigma)}{a_i(x_i(t-\sigma))} \right\} \\
&= v_i(t) \left\{ -d_i e_i(t) + \sum_{j=1}^n b_{ij} \eta_j(t) + \sum_{j=1}^n c_{ij} \eta_j(t-\tau_j) - \frac{\alpha_i e_i(t) + k_i \vartheta_i(t)}{a_i(y_i(t))} \right\}.
\end{aligned} \tag{12}$$

where $\eta_j(t) = \gamma_j^*(t) - \gamma(t-\sigma)$, $\eta_j(t-\tau_j) = \gamma_j^*(t-\tau_j) - \gamma(t-\tau_j-\sigma)$, $v_i(t) = \text{sign}\{\int_{x_i(t-\sigma)}^{y_i(t)} \frac{ds}{a_i(s)}\} = \text{sign}[y_i(t) - x_i(t-\sigma)] = \text{sign}(e_i(t))$, if $e_i(t) \neq 0$; while $v_i(t)$ can be arbitrarily chosen in $[-1, 1]$, if $e_i(t) = 0$. In particular, we choose $v_i(t)$ as follows

$$v_i(t) = \begin{cases} 0, & \text{if } e_i(t) = 0 \text{ and } \eta_i(t) = 0, \\ \text{sign}(\eta_i(t)), & \text{if } e_i(t) = 0 \text{ and } \eta_i(t) \neq 0, \\ \text{sign}(e_i(t)), & \text{if } e_i(t) \neq 0. \end{cases} \tag{13}$$

Thus, we have

$$v_i(t) e_i(t) = |e_i(t)|, \quad v_i(t) \eta_i(t) = |\eta_i(t)|.$$

Constructing a Lyapunov functional $V(t)$ by

$$V(t) = \sum_{i=1}^n \bar{e}_i(t) e^{\lambda(t-\sigma)} + \sum_{i=1}^n \sum_{j=1}^n |c_{ij}| \int_{t-\tau_j}^t |\eta_j(s)| e^{\lambda(s+\tau_j-\sigma)} ds. \tag{14}$$

In view of the chain rule in Lemma 2.4 in [1], calculate the time derivative of $V(t)$ along the solutions of systems (1) and (5) along the sense of Eqs. (4) and (7), then for a.a. $t \geq \sigma$ we have

$$
\begin{aligned}
\frac{dV(t)}{dt} &= \sum_{i=1}^{n} \lambda \bar{e}_i(t) e^{\lambda(t-\sigma)} + \sum_{i=1}^{n} v_i(t) \Big\{ -d_i e_i(t) + \sum_{j=1}^{n} b_{ij} \eta_j(t) + \sum_{j=1}^{n} c_{ij} \eta_j(t - \tau_j) \\
&\quad - \frac{\alpha_i e_i(t) + k_i \vartheta_i(t)}{a_i(y_i(t))} \Big\} e^{\lambda(t-\sigma)} + \sum_{i=1}^{n} \sum_{j=1}^{n} e^{\lambda(t-\sigma)} \Big\{ |c_{ij}| |\eta_j(t)| e^{\lambda \tau_j} - |c_{ij}| |\eta_j(t - \tau_j)| \Big\} \\
&\leq \sum_{i=1}^{n} \Big\{ -(d_i + \frac{\alpha_i}{\bar{a}_i} - \frac{\lambda}{\underline{a}_i}|) |e_i(t)| + b_{ii}^{+} |\eta_i(t)| + \sum_{j=1,j\neq i}^{n} |b_{ij}| |\eta_j(t)| - \frac{k_i}{\bar{a}_i} \Big\} e^{\lambda(t-\sigma)} \\
&\quad + \sum_{i=1}^{n} \sum_{j=1}^{n} |c_{ij}| |\eta_j(t)| e^{\lambda(t+\tau_j-\sigma)} \Big\} \\
&\leq \sum_{i=1}^{n} \Big\{ (\frac{\lambda}{\underline{a}_i} - d_i - \frac{\alpha_i}{\bar{a}_i} + b_{ii}^{+}) |e_i(t)| + b_{ii}^{+} N_i + \sum_{j=1,j\neq i}^{n} |b_{ij}| [L_j |e_j(t)| + N_j] \Big\} e^{\lambda(t-\sigma)} \\
&\quad - \frac{k_i}{\bar{a}_i} e^{\lambda(t-\sigma)} + \sum_{i=1}^{n} \sum_{j=1}^{n} |c_{ij}| [L_j |e_j(t)| + N_j] e^{\lambda(t+\tau_j-\sigma)} \\
&\leq \max_{i \in \mathscr{I}} \Big\{ (\frac{\lambda}{\underline{a}_i} - d_i - \frac{\alpha_i}{\bar{a}_i} + b_{ii}^{+}) + \sum_{j=1,j\neq i}^{n} |b_{ij}| L_j + \sum_{j=1}^{n} |c_{ij}| L_j e^{\lambda \tau_j} \Big\} \sum_{i=1}^{n} |e_i(t)| e^{\lambda(t-\sigma)} \\
&\quad + \sum_{i=1}^{n} \Big\{ b_{ii}^{+} N_i + \sum_{j=1,j\neq i}^{n} |b_{ij}| N_j + \sum_{j=1}^{n} |c_{ij}| N_j e^{\lambda \tau_j} - \frac{k_i}{\bar{a}_i} \Big\}.
\end{aligned}
$$

From (9), for a small enough λ, we can obtain that

$$
\begin{aligned}
(\frac{\lambda}{\underline{a}_i} - d_i - \frac{\alpha_i}{\bar{a}_i} + b_{ii}^{+}) + \sum_{j=1,j\neq i}^{n} |b_{ij}| L_j + \sum_{j=1}^{n} |c_{ij}| L_j e^{\lambda \tau_j} &< 0, \\
b_{ii}^{+} N_i + \sum_{j=1,j\neq i}^{n} |b_{ij}| N_j + \sum_{j=1}^{n} |c_{ij}| N_j e^{\lambda \tau_j} - \frac{k_i}{\bar{a}_i} &< 0.
\end{aligned}
\tag{15}
$$

Thus we have

$$
\frac{dV(t)}{dt} \leq 0,
$$

which, together with (12) and (14), leads to

$$
e^{\lambda(t-\sigma)} \sum_{i=1}^{n} \bar{e}_i(t) \leq V(t) \leq V(\sigma),
\tag{16}
$$

where

$$
V(\sigma) \leq \sum_{i=1}^{n} \Big[\Big(\frac{1}{\underline{a}_i} + \sum_{j=1}^{n} |c_{ij}| L_j \tau_j e^{\tau_j} \Big) \|\phi_\sigma - \varphi\|_c + \sum_{j=1}^{n} |c_{ij}| N_j \tau_j e^{\tau_j} \Big] \triangleq N^*.
$$

Then there exists a positive constant M^* such that

$$
N^* \leq M^* \|\phi_\sigma - \varphi\|_c.
\tag{17}
$$

Thus we obtain that the following inequality holds.

$$
\sum_{i=1}^{n} \frac{1}{\underline{a}_i} |e_i(t)| \leq \sum_{i=1}^{n} \bar{e}_i(t) \leq M^* \|\phi_\sigma - \varphi\|_c e^{-\lambda(t-\sigma)}.
$$

That is

$$\|y(t) - x_i(t - \sigma)\| = \sum_{i=1}^{n} |e_i(t)| \leq M \|\phi_\sigma - \varphi\|_c e^{-\lambda(t-\sigma)}, \qquad (18)$$

where $M = (M^*) \max_{1 \leq i \leq n} \bar{a}_i$. The proof of Theorem 2 is completed. □

3 Numerical Simulations

For $n = 2$, consider the following discontinuous CGNNs

$$\dot{x}_i(t) = a_i(x_i(t)) \left[-d_i x_i(t) + \sum_{j=1}^{2} b_{ij} f_j(x_j(t)) + \sum_{j=1}^{2} c_{ij} f_j(x_j(t - \tau_j)) \right], \qquad (19)$$

where $i = 1, 2$, $f_1(u) = f_2(u) = \tanh(u) - 0.06\,\mathrm{sign}(x)$, $d_1 = 1.1$, $d_2 = 1.2$, $b_{11} = 2$, $b_{12} = -0.08$, $b_{21} = -5$, $b_{22} = 4$, $c_{11} = -1.5$, $c_{12} = -0.1$, $c_{21} = -0.2$, $c_{22} = -4$ and $a_1(u) = 0.8 + 0.1/(1 + u^2)$, $a_2(u) = 0.8 - 0.1/(1 + u^2)$, $\tau_1 = \tau_2 = 1$.

Obviously, $0.8 \leq a_1(u) \leq 0.9$ $0.7 \leq a_2(u) \leq 0.8$, thus $\underline{a}_1 = 0.8$, $\underline{a}_2 = 0.7$, $\bar{a}_1 = 0.9$ and $\bar{a}_2 = 0.8$.

The numerical simulation of system (19) with the initial values $x_1(s) = 0.4$ and $x_2(s) = -0.9$ for $s \in [-1, 0]$ is represented in Fig. 1, which shows that system (19) has a chaotic attractor.

In the following, we consider the lag synchronization of drive system (19) and response system described by

$$\dot{y}_i(t) = a_i(y_i(t)) \left[-d_i y_i(t) + \sum_{j=1}^{2} b_{ij} f_j(y_j(t)) + \sum_{j=1}^{2} c_{ij} f_j(y_j(t - \tau_{ij})) \right] + u_i(t), \qquad (20)$$

for $t \geq \sigma$ and $i = 1, 2$, where a_i, d_i, b_{ij}, c_{ij}, f_j and τ_j are the same as defined in system (22) and $u_i(t)$ is given by (6).

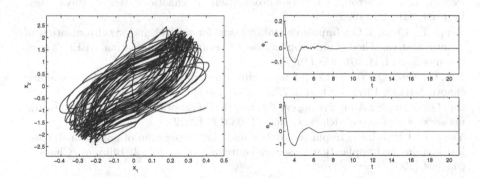

Fig. 1. The chaotic attractor of system (19).

Fig. 2. The evaluation of lag synchronization error e_i.

By choosing $\alpha_1 = 2.4$, $\alpha_2 = 9.7$ and $k_1 = 0.22$, $k_2 = 0.64$, then by simple computation, we obtain that $L_1 = L_2 = 1$, $N_1 = N_2 = 0.06$, $\Gamma_1 = -0.0867$, $\Gamma_2 = -0.1250$,

$\Upsilon_1 = -0.0236$ and $\Upsilon_2 = -0.0080$. Thus, all the conditions of Theorem 1 are satisfied, hence the drive-response systems (19) and (20) are exponentially lag synchronized under the controller (6) and above parameters. Taking $\sigma = 3$ and denoting $e_i(t) = y_i(t) - x_i(t - \sigma)$, the time evolution of synchronization errors between systems (19) and (20) are given in Figs. 2 .

4   Conclusion

In this paper, we study the exponential lag synchronization of delayed CGNNs with discontinuous activation functions. By employing the analysis technique and theory of the differential equations with discontinuous right-hand side, some novel lag synchronization criteria have been obtained. Finally, an example is given to demonstrate the effectiveness of the proposed synchronization method.

Acknowledgments. This work was supported by the National Natural Science Foundation of People's Republic of China (Grant No. 61164004). The authors are grateful to the Editor and anonymous reviewers for their helpful comments.

References

1. Guo, Z., Huang, L.: LMI conditions for global robust stability of delayed neural networks with discontinuous neuron activations. Appl. Math. Comput. 215, 889–900 (2009)
2. Liu, J., Liu, X., Xie, W.: Global convergence of neural networks with mixed time-varying delays and discontinuous neuron activations. Inf. Sci. 183, 92–105 (2012)
3. Lu, W., Chen, T.: Almost periodic dynamics of a class of delayed neural networks with discontinuous activations. Neural Comput. 20, 1065–1090 (2008)
4. Pecora, L.M., Carroll, T.L.: Synchronization in chaotic systems, Phys. Rev. Lett. 64, 821–824 (1990)
5. Yang, T., Chua, L.O.: Impulsive stabilization for control and synchronization of chaotic system: Theory and application to secure communication. IEEE Trans. Circuits Syst. I 44, 976–988 (1997)
6. Yao, C., Zhao, Q., Yu, J.: Complete synchronization induced by disorder in coupled chaotic lattices. Phys. Lett. A 377, 370–377 (2013)
7. Liu, Q., Zhang, S.: Adaptive lag synchronization of chaotic Cohen-Grossberg neural networks with discrete delays. Chaos 22, 033123 (2012)
8. Yang, T., Chua, L.O.: Impulsive control and synchronization of nonlinear dynamical systems and application to secure communication. Int. J. Bifurcat. Chaos 7, 645–664 (1997)
9. Rosenblum, M.G., Pikovsky, A.S., Kurths, J.: From phase to lag synchronization in coupled chaotic oscillators. Phys. Rev. Lett. 78, 4193 (1997)
10. Mainieri, R., Rehacek, J.: Projective synchronization in three-dimensional chaotic systems. Phys. Rev. Lett. 82, 304 (1999)
11. Abdurahman, A., Jiang, H., Teng, Z.: Function projective synchronization of impulsive neural networks with mixed time-varying delays. Nonlinear Dyn. 78, 2627–2638 (2014)

12. Hu, C., Yu, J., Jiang, H., Teng, Z.: Exponential lag synchronization for neural networks with mixed delays via periodically intermittent control. Chaos 20, 023108 (2010)
13. Shahverdiev, E.M., Sivaprakasam, S., Shore, K.A.: Lag synchronization in time-delayed systems. Phys. Lett. A 292, 320–324 (2002)
14. Sun, Y., Cao, J.: Adaptive lag synchronization of unknown chaotic delayed neural networks with noise perturbation. Phy. Lett. A 364, 277–285 (2007)
15. Liu, X., Chen, T., Cao, J., Lu, W.: Dissipativity and quasi-synchronization for neural networks with discontinuous activations and parameter mismatches. Neural Networks 24, 1013–1021 (2011)

Mean Square Exponential Stability of Stochastic Delayed Static Neural Networks with Markovian Switching*

He Huang

School of Electronics and Information Engineering, Soochow University,
Suzhou 215006, P. R. China
hhuang@suda.edu.cn

Abstract. This paper is concerned with globally exponential stability in the mean square of stochastic static neural networks with Markovian switching and time delay. Firstly, the mathematical model of this kind of recurrent neural networks is established by taking information latching and noise disturbance into consideration. Then, a stability condition, which is dependent on both time delay and system mode, is presented in terms of linear matrix inequalities. Based on it, the maximum value of the exponential decay rate can be efficiently found by solving a convex optimization problem.

Keywords: Static neural networks, stability, time delay, Markovian switching, exponential decay rate, convex optimization.

1 Introduction

According to the basic variables adopted in the modeling process, recurrent neural networks can be classified into local field neural networks and static neural networks [12]. In general, only when some strict preconditions are satisfied, the two kinds of recurrent neural networks are equivalent. While, a practical example was presented in [6] to verify that these preconditions were not always met. It means that the stability conditions for local field neural networks are not applicable to static neural networks. On the other hand, comparing with local field neural networks, much less attention has been paid to static neural networks. This motivates the study of static neural networks.

It is well known that the information latching phenomenon is frequently encountered in recurrent neural networks. One of promising ways to resolve it is to extract a finite state representation from the considered neural network [11]. In practice, the switching between the finite states can be well modeled by introducing a Markov chain. As a result, the so-called stochastic recurrent neural

---

* This work was jointly supported by the National Natural Science Foundation of China under Grant Nos. 61273122 and 61005047, and the Natural Science Foundation of Jiangsu Province of China under Grant No. BK2010214.

© Springer International Publishing Switzerland 2015
X. Hu et al. (Eds.): ISNN 2015, LNCS 9377, pp. 138–145, 2015.
DOI: 10.1007/978-3-319-25393-0_16

networks with Markovian switching are proposed. Recently, the stability analysis of this kind of recurrent neural networks has become an active research topic. Many excellent stability conditions have been reported in the literature (see, e.g., [3,4,9]). In [9], the authors discussed the stochastic stability for a class of discrete-time neural networks with Markovian switching and time-varying delays. The parameter uncertainties were also taken into account. In [4], the stochastic stability analysis problem was studied for neutral-type neural networks with Markovian jumping parameters and mode-dependent mixed delays.

It should be pointed out that the above-mentioned results are on local field neural networks with Markovian switching. As discussed before, these criteria can not be applied to judge the stability of stochastic static neural networks. Recently, the authors initially considered the H_∞ filter design in [8] for delayed static neural networks with Markovian jumping parameters. By employing Wirtinger inequality [7], a design criterion was provided and formulated by means of linear matrix inequalities (LMIs). In [10], the stochastic stability was investigated for stochastic Markovian jumping static neural networks with asynchronous mode-dependent delays. To our knowledge, stochastic stability of delayed static neural networks with Markovian switching has not yet been fully investigated. This motivates the current study.

In this paper, our attention focuses on the stability analysis of a class of stochastic delayed static neural networks with Markovian jumping parameters. By constructing a suitable stochastic Lyapunov functional with a tripe integral term, a delay and mode dependent condition is derived under which the considered neural network is globally exponentially stable in the mean square. One of the advantages of our result lies in that the maximum allowable value of the exponential decay rate can be easily obtained by solving a convex optimization problem subject to some LMI-based constraints.

The notations used in this paper are the same as those in [3].

2 Mathematical Model and Problem Formulation

It is known that a delayed static neural network with n neurons can be represented by

$$\dot{x}(t) = -Ax(t) + f(Wx(t - \tau) + J) \tag{1}$$

where $x(t) = [x_1(t), x_2(t), \ldots, x_n(t)]^T \in \mathbb{R}^n$ is the state vector, $W = [w_{ij}]_{n \times n}$ is a connection weight matrix with w_{ij} being the connection weight between neurons j and i, $A = \text{diag}\{a_1, a_2, \ldots, a_n\}$ is a diagonal matrix with positive entries a_i $(i = 1, 2, \ldots, n)$, $f(x) = [f_1(x_1), f_2(x_2), \ldots, f_n(x_n)]^T$ is an activation function, J is an external input, and τ is a constant time delay.

In this study, the activation function $f(\cdot)$ is supposed to be bounded and globally Lipschitz continuous. That is, for each $k \in \{1, 2, \ldots, n\}$ and any $a, b \in \mathbb{R}$, there exists a Lipschitz constant L_k such that

$$|f_k(a) - f_k(b)| \leq L_k|a - b|. \tag{2}$$

By Brouwer's fixed point theorem, when the activation function $f(\cdot)$ is continuous and bounded, the delayed static neural network (1) has at least one equilibrium point. Especially, in some practical applications (e.g., combinatorial optimization and associated memory, etc), it is required that the designed neural network has a unique and stable equilibrium point. Without loss of generality, it is assumed that $x^* = [x_1^*, x_2^*, \ldots, x_n^*]^T$ is an equilibrium point of (1) (that is, x^* satisfies $-Ax^* + f(Wx^* + J) = 0$). For stability analysis, it is necessary to transform the equilibrium point x^* to the origin. This can be easily achieved by making a transformation $u(t) = x(t) - x^*$. Then, (1) can be rewritten as

$$\dot{u}(t) = -Au(t) + g(Wu(t - \tau)) \tag{3}$$

with $g(Wu(t - \tau)) = f(Wx(t - \tau) + J) - f(Wx^* + J)$.

When both the environmental noise and information latching phenomenon are taking into account in (3), the so-called stochastic delayed static neural networks with Markovian switching can be established. The mathematical model of this kind of recurrent neural networks is expressed by a stochastic functional differential equation

$$du(t) = \Big[- A(r(t))u(t) + g(W(r(t))u(t - \tau)) \Big] dt$$
$$+ h\Big(u(t), u(t - \tau), t, r(t) \Big) dw(t) \tag{4}$$

where $w(t)$ is an m-dimension Browian motion defined on a probability space $(\Omega, \mathcal{F}, \mathcal{P})$, $r(t)$, independent of $w(t)$, is a Markov chain whose value is taken from a finite set $\mathcal{N} = \{1, 2, \ldots, N\}$, and $h\Big(u(t), u(t - \tau), t, r(t) \Big) : \mathbb{R}^n \times \mathbb{R}^n \times \mathbb{R}^+ \times \mathcal{N} \to \mathbb{R}^{n \times m}$ is a noise disturbance.

It is assumed that the transition probability matrix $\Pi = [\pi_{ij}]_{N \times N}$ of $r(t)$ is given by

$$\Pr\{r_{t+h} = j | r_t = i\} = \begin{cases} \pi_{ij} h + o(h), & i \neq j \\ 1 + \pi_{ii} h + o(h), & i = j \end{cases}$$

where $h > 0, \lim_{h \to 0+} o(h)/h = 0$, $\pi_{ij} \geq 0$ for $j \neq i$ is the transition rate from mode i at time t to mode j at time $t + h$, and for each $i \in \mathcal{N}$,

$$\pi_{ii} = -\sum_{j=1, j \neq i}^{N} \pi_{ij}.$$

For simplicity, for each $r(t) = i \in \mathcal{N}$, matrix $\mathcal{M}(r(t))$ is denoted by $\mathcal{M}_i$. For examples, $A(r(t))$ and $W(r(t))$ in (4) are respectively written as A_i and W_i. In addition, for each $r(t) = i \in \mathcal{N}$, the noise disturbance is supposed to satisfy

$$\text{Trace}(h^T(u(t), u(t - \tau), t, i) h(u(t), u(t - \tau), t, i))$$
$$\leq \begin{bmatrix} u(t) \\ u(t - \tau) \end{bmatrix}^T \begin{bmatrix} X_{1i} & X_{2i} \\ X_{2i}^T & X_{3i} \end{bmatrix} \begin{bmatrix} u(t) \\ u(t - \tau) \end{bmatrix}, \tag{5}$$

with $\begin{bmatrix} X_{1i} & X_{2i} \\ X_{2i}^T & X_{3i} \end{bmatrix} > 0.$

To ensure the existence and uniqueness of solution of (4), it is required that $h(u(t), u(t-\tau), t, i)$ is locally Lipschitz continuous and linearly increasing. In this situation, for any initial value $\psi \in \mathcal{C}_{\mathcal{F}_0}^b([-\tau, 0]; \mathbb{R}^n)$ and initial mode $r_0 \in \mathcal{N}$, (4) has a unique continuous solution $u(t; \psi)$.

Definition 1. *For any initial condition $\psi \in \mathcal{C}_{\mathcal{F}_0}^b([-\tau, 0]; \mathbb{R}^n)$ and initial mode $r_0 \in \mathcal{N}$, the solution $u(t; \psi)$ of (4) is said to be mean square globally exponentially stable if there are positive scalars $\alpha > 1$ and $\lambda > 0$ such that*

$$\mathbb{E}|x(t; \psi)|^2 \le \alpha e^{-\lambda t} \mathbb{E}\|\psi\|^2.$$

Here, λ is the decay rate, and is used to characterize the transient process of the underlying system.

3   Mean Square Exponential Stability Criterion

In this section, a delay and mode dependent condition is presented by means of LMIs to check mean square globally exponential stability of the stochastic delayed static neural network with Markovian switching (4).

Theorem 1. *For given scalars $\tau > 0$ and $\lambda > 0$, the stochastic delayed static neural network (4) is globally exponentially stable in the mean square with a decay rate λ if there are real scalars $\alpha_i > 0, \beta_i > 0$ and real matrices $P_i > 0, Q_i > 0, R_i > 0, S > 0, X_{1i} > 0, X_{2i}, X_{3i} > 0$ such that the LMIs*

$$P_i \le \beta_i I, \tag{6}$$

$$e^{\lambda \tau} \sum_{j=1}^{N} \pi_{ij} Q_j \le R_i, \tag{7}$$

$$e^{\lambda \tau} \sum_{j=1}^{N} \pi_{ij} R_j \le S, \tag{8}$$

$$\begin{bmatrix} X_{1i} & X_{2i} \\ X_{2i}^T & X_{3i} \end{bmatrix} > 0, \tag{9}$$

$$\begin{bmatrix} \Phi_i & \beta_i X_{2i} & P_i & 0 \\ \beta_i X_{2i}^T & -Q_i + \beta_i X_{3i} & 0 & \alpha_i W_i^T L \\ P_i^T & 0 & -\alpha_i I & 0 \\ 0 & \alpha_i L^T W_i & 0 & -\alpha_i I \end{bmatrix} < 0, \tag{10}$$

are held for any $i = 1, 2, \ldots, N$, where

$$\Phi_i = \lambda P_i - P_i A_i - A_i^T P_i + \beta_i X_{1i} + \sum_{j=1}^{N} \pi_{ij} P_j$$

$$+ e^{\lambda \tau} Q_i + \frac{e^{\lambda \tau} - 1}{\lambda} R_i + \frac{e^{\lambda \tau} - \lambda \tau - 1}{\lambda^2} S,$$

$$L = diag\{L_1, L_2, \ldots, L_n\}.$$

Proof. Since $f_k(\cdot)$ satisfies (2) and $g(W_i u(t-\tau)) = f(W_i x(t-\tau)+J) - f(W_i x^* + J)$, one has

$$g^T(W_i u(t-\tau))g(W_i u(t-\tau)) = \sum_{k=1}^{n} |g_k(W_i^k u(t-\tau))|^2$$

$$\leq \sum_{k=1}^{n} L_k^2 |W_i^k u(t-\tau)|^2$$

$$= u^T(t-\tau)W_i^T L^2 W_i u(t-\tau), \qquad (11)$$

where W_i^k is the k-th row of the matrix W_i. Then, for any scalar $\alpha_i > 0$,

$$\alpha_i g^T(W_i u(t-\tau))g(W_i u(t-\tau)) \leq \alpha_i u^T(t-\tau)W_i^T L^2 W_i u(t-\tau). \qquad (12)$$

It follows from (6) that

$$\mathrm{Trace}(h^T(u(t), u(t-\tau), t, i)P_i h(u(t), u(t-\tau), t, i))$$

$$\leq \beta_i \mathrm{Trace}(h^T(u(t), u(t-\tau), t, i)h(u(t), u(t-\tau), t, i))$$

$$\leq \beta_i \begin{bmatrix} u(t) \\ u(t-\tau) \end{bmatrix}^T \begin{bmatrix} X_{1i} & X_{2i} \\ X_{2i}^T & X_{3i} \end{bmatrix} \begin{bmatrix} u(t) \\ u(t-\tau) \end{bmatrix}. \qquad (13)$$

For $\lambda > 0$ and $R_i > 0$, it is obvious that

$$e^{\lambda t} \int_{t-\tau}^{t} u^T(s)R_i u(s)ds \geq \int_{t-\tau}^{t} e^{\lambda s}u^T(s)R_i u(s)ds. \qquad (14)$$

Then, by noting (7), it yields

$$\sum_{j=1}^{N} \pi_{ij} \int_{t-\tau}^{t} e^{\lambda(s+\tau)}u^T(s)Q_j u(s)ds - e^{\lambda t}\int_{t-\tau}^{t} u^T(s)R_i u(s)ds$$

$$\leq \int_{t-\tau}^{t} e^{\lambda s}u^T(s)\left(e^{\lambda \tau}\sum_{j=1}^{N}\pi_{ij}Q_j\right)u(s)ds - \int_{t-\tau}^{t} e^{\lambda s}u^T(s)R_i u(s)ds$$

$$\leq 0. \qquad (15)$$

Similarly, one can also verify from (8) that

$$\sum_{j=1}^{N} \pi_{ij} \int_{-\tau}^{0}\int_{t+\theta}^{t} e^{\lambda(s-\theta)}u^T(s)R_j u(s)ds d\theta$$

$$-e^{\lambda t}\int_{-\tau}^{0}\int_{t+\theta}^{t} u^T(s)S u(s)ds d\theta \leq 0. \qquad (16)$$

In addition, by the well-known Schur complement [1], it is known that the LMI (10) is equivalent to

$$\Omega_i = \begin{bmatrix} \Phi_i & \beta_i X_{2i} & P_i \\ \beta_i X_{2i}^T & \Psi_i & 0 \\ P_i^T & 0 & -\alpha_i I \end{bmatrix} < 0 \qquad (17)$$

with $\Psi_i = -Q_i + \beta_i X_{3i} + \alpha_i W_i^T L^2 W_i$.

To show the mean square exponential stability of (4), for each $i \in \mathcal{N}$, we consider a stochastic Lyapunov functional

$$V(t, u_t, i) = e^{\lambda t} u^T(t) P_i u(t) + \int_{t-\tau}^{t} e^{\lambda(s+\tau)} u^T(s) Q_i u(s) ds$$

$$+ \int_{-\tau}^{0} \int_{t+\theta}^{t} e^{\lambda(s-\theta)} u^T(s) R_i u(s) ds d\theta$$

$$+ \int_{-\tau}^{0} \int_{\theta}^{0} \int_{t+\delta}^{t} e^{\lambda(s-\delta)} u^T(s) S u(s) ds d\delta d\theta. \tag{18}$$

By calculating the infinitesimal operator $\mathcal{L}$ of $V(t, u_t, i)$, one gets

$$\mathcal{L}V(t, u_t, i) = \lambda e^{\lambda t} u^T(t) P_i u(t) + 2e^{\lambda t} u^T(t) P_i \Big[-A_i u(t) + g(W_i u(t-\tau)) \Big]$$

$$+ e^{\lambda t} \text{Trace}\left(h^T\Big(u(t), u(t-\tau), t, i\Big) P_i h\Big(u(t), u(t-\tau), t, i\Big) \right)$$

$$+ e^{\lambda t} u^T(t) \left(\sum_{j=1}^{N} \pi_{ij} P_j \right) u(t) + e^{\lambda(t+\tau)} u^T(t) Q_i u(t)$$

$$- e^{\lambda t} u^T(t-\tau) Q_i u(t-\tau) + \sum_{j=1}^{N} \pi_{ij} \int_{t-\tau}^{t} e^{\lambda(s+\tau)} u^T(s) Q_j u(s) ds$$

$$+ \frac{e^{\lambda \tau} - 1}{\lambda} e^{\lambda t} u^T(t) R_i u(t) - e^{\lambda t} \int_{t-\tau}^{t} u^T(s) R_i u(s) ds$$

$$+ \sum_{j=1}^{N} \pi_{ij} \int_{-\tau}^{0} \int_{t+\theta}^{t} e^{\lambda(s-\theta)} u^T(s) R_j u(s) ds d\theta$$

$$+ \frac{e^{\lambda \tau} - \lambda \tau - 1}{\lambda^2} e^{\lambda t} u^T(t) S u(t) + e^{\lambda t} \int_{-\tau}^{0} \int_{t+\theta}^{t} u^T(s) S u(s) ds d\theta. \tag{19}$$

Let $\xi_i(t) = \Big[u^T(t), u^T(t-\tau), g^T(W_i u(t-\tau)) \Big]^T$. By combining (12), (13), (15), (16) and (19) together, one can deduce from (17) that for any nonzero $\xi_i(t)$,

$$\mathcal{L}V(t, u_t, i) \le \xi_i^T(t) \Omega_i \xi_i(t) < 0. \tag{20}$$

Now, by the generalized Itô's formula and (20), one can have

$$\mathbb{E}V(t, u_t, r(t)) \le \mathbb{E}V(0, \psi, r_0). \tag{21}$$

On the other hand, it immediately derives from (18) that

$$\mathbb{E}V(t, u_t, r(t)) \ge \min_{i \in \mathcal{N}} \{\lambda_{\min}(P_i)\} e^{\lambda t} \mathbb{E}|u(t)|^2. \tag{22}$$

This together with (21) gives

$$\mathbb{E}|u(t)|^2 \leq \frac{1}{\min_{i\in\mathcal{N}}\{\lambda_{\min}(P_i)\}}e^{-\lambda t}\mathbb{E}V(0,\phi,r_0). \tag{23}$$

Let

$$\mu = \max_{i\in\mathcal{N}}\{\lambda_{\max}(P_i)\} + \frac{e^{\lambda\tau}-1}{\lambda}\max_{i\in\mathcal{N}}\{\lambda_{\max}(Q_i)\}$$

$$+\frac{e^{\lambda\tau}-\lambda\tau-1}{\lambda^2}\max_{i\in\mathcal{N}}\{\lambda_{\max}(R_i)\}$$

$$+\frac{1}{\lambda^3}\left(e^{\lambda\tau}-\lambda^2\tau^2/2-\lambda\tau-1\right)\lambda_{\max}(S).$$

It is not difficult to deduce that

$$\mathbb{E}V(0,\phi,r_0) \leq \mu\mathbb{E}\|\psi\|^2. \tag{24}$$

Therefore, one has

$$\mathbb{E}|u(t)|^2 \leq \frac{\mu}{\min_{i\in\mathcal{N}}\{\lambda_{\min}(P_i)\}}e^{-\lambda t}\mathbb{E}\|\phi\|^2. \tag{25}$$

According to Definition 1, the stochastic delayed static neural network with Markovian switching (4) is mean square globally exponentially stable with a decay rate λ. This completes the proof.

Remark 1. *In Theorem 1, a sufficient condition guaranteeing mean square exponential stability of the stochastic delayed static neural network (4) is derived in terms of LMIs. It can be found that the Lyapunov matrices P_i, Q_i and R_i are dependent on system mode i. That is, distinct P_i, Q_i and R_i can be chosen for different modes. In this circumstance, the choice of these Lyapunov matrices in Theorem 1 is of high flexibility. It is thus believed that Theorem 1 presents a less conservative stability condition. Of course, with more computation, the conservatism of Theorem 1 can be further reduced by employing the delay partition technique, free-weighting matrices based approach and Wirtinger integral inequality. This can be easily obtained by following the similar lines in [2, 5, 7].*

Remark 2. *It is easily shown that the functions $e^{\lambda\tau}$, $\frac{e^{\lambda\tau}-1}{\lambda}$ and $\frac{e^{\lambda\tau}-\lambda\tau-1}{\lambda}$ are strictly monotonically increasing with respect to the decay rate λ. It means that the left-hand sides of the LMIs (7), (8) and (10) are also strictly monotonically increasing with respect to λ. As a result, the maximum value of the decay rate λ allowed by Theorem 1 can be efficiently obtained by solving the LMIs (6)-(10).*

Remark 3. *Comparing with the most recent work [10], there are two advantages for our result. One is that more Lyapunov matrices are mode-dependent in our approach. Specifically, besides P_i and Q_i, R_i in the double integral term is also mode-dependent. The other is that the exponential decay rate can be easily found by Theorem 1, which was not addressed in [10].*

Remark 4. *As seen in Section II, the time delay considered in this study is constant. In fact, it is not difficult to extend our approach to handle time-varying and/or mode-dependent delays. Furthermore, one can also discuss the case that the information of the transition probability matrix Π is partially known.*

4 Conclusion

In this paper, the globally exponential stability in the mean square has been studied for a class of stochastic static neural networks with Markovian switching and time delay. By constructing a suitable stochastic Lyapunov functional with a triple integral term, a stability condition has been derived by means of a set of LMIs. It has been shown that the established LMIs are strictly monotonically increasing with respect to the decay rate. Therefore, the maximum allowable value of the decay rate can be efficiently calculated such that the transient process of the underlying recurrent neural network is explicitly described.

References

1. Boyd, S., EI Ghaoui, L., Feron, E., Balakrishnan, V.: Linear Matrix Inequalities in System and Control Theory. SIAM, Philadelphia (1994)
2. He, Y., Wang, Q.-G., Wu, M., Lin, C.: Delay-dependent state estimation for delayed neural networks. IEEE Trans. Neural Netw. 17, 1077–1081 (2006)
3. Huang, H., Huang, T., Chen, X.: Global exponential estimates of delayed stochastic neural networks with Markovian switching. Neural Netw. 36, 136–145 (2012)
4. Liu, Y., Wang, Z., Liu, X.: Stability analysis for a class of neutral-type neural networks with Markovian jumping parameters and mode-dependent mixed delays. Neurocomputing 94, 46–53 (2012)
5. Mou, S., Gao, H., Lam, J., Qiang, W.: A new criterion of delay-dependent asymptotic stability for Hopfield neural networks with time delay. IEEE Trans. Neural Netw. 19, 532–535 (2008)
6. Seung, H.S.: How the brain keeps the eye still. Proc. Natl. Acad. Sci. USA 93, 13339–13344 (1996)
7. Seuret, A., Gouaisbaut, F.: Wirtinger-based integral inequality: application to time-delay systems. Automatica 49, 2860–2866 (2013)
8. Shao, L., Huang, H., Zhao, H., Huang, T.: Filter design of delayed static neural networks with Markovian jumping parameters. Neurocomputing 153, 126–132 (2015)
9. Syed Ali, M., Marudaib, M.: Stochastic stability of discrete-time uncertain recurrent neural networks with Markovian jumping and time-varying delays. Math. Comput. Model. 54(9), 1979–1988 (2011)
10. Tan, H., Hua, M., Chen, J., Fei, J.: Stability analysis of stochastic Markovian switching static neural networks with asynchronous mode-dependent delays. Neurocomputing 151, 864–872 (2015)
11. Tino, P., Cernansky, M., Benuskova, L.: Markovian architectural bias of recurrent neural networks. IEEE Trans. Neural Netw. 15, 6–15 (2004)
12. Xu, Z.-B., Qiao, H., Peng, J., Zhang, B.: A comparative study of two modeling approaches in neural networks. Neural Netw. 17, 73–85 (2004)

Robust Multistability and Multiperiodicity
of Neural Networks with Time Delays

Lili Wang

School of Mathematics, Shanghai University of Finance and Economics,
Shanghai, P.R. China
wang.lili@mail.shufe.edu.cn

Abstract. In this paper, we are concerned with the robust multistability and multiperiodicity of delayed neural networks. A set of sufficient conditions ensuring
the coexistence of 2^n periodic solutions and their local stability are presented.
And the attraction basin of each periodic solution can be enlarged by rigorous
analysis.

Keywords: Neural networks, periodic solution, multistability, multiperiodicity,
robust stability.

1   Introduction

Neural networks have been extensively studied in recent years due to their potential
applications in classification, associative memory, parallel computation and other fields.
As a prerequisite, the theoretical studies on the dynamical properties of neural networks
are of great importance and have attracted considerable research attentions.

Recently, Refs. [1]-[4] studied the stability, periodic oscillations and other dynamical
behaviors of neural networks, and some sufficient conditions ensuring the global stability of these systems were derived. On the other hand, in an earlier paper [2], the authors
have reported that some one-neuron neural network model can have three equilibrium
points, two of them are locally stable, and one is unstable. Furthermore, it is also pointed
out that the n-neuron neural networks may exhibit more than one equilibrium point or
periodic solution, which is called as multistability or multiperiodicity of systems, see
[5]-[10]. Note that the parameter fluctuation in neural network implementation on very
large scale integration chips is unavoidable, which may lead to some deviations in the
values of system parameters. In this case, it is of great importance to reveal the robust
stability of the systems with respect to the uncertainties in the design and applications
of neural networks [3].

Consider the neural networks described by the following equations

$$\frac{dx_i(t)}{dt} = -d_i(t)x_i(t) + \sum_{j=1}^{n} a_{ij}(t)g_j(x_j(t)) + \sum_{j=1}^{n} b_{ij}(t)g_j(x_j(t - \tau_{ij}(t))) + I_i(t),$$

$$i = 1, \cdots, n, \text{ (1)}$$

where $x_i(t)$ represents the state of the i-th neuron at time t; $d_i(t) > 0$ denotes the rate
with which the i-th unit will reset its potential to the resting state in isolation when

© Springer International Publishing Switzerland 2015
X. Hu et al. (Eds.): ISNN 2015, LNCS 9377, pp. 146–153, 2015.
DOI: 10.1007/978-3-319-25393-0_17

disconnected from the network and external inputs at time t; $a_{ij}(t)$, $b_{ij}(t)$ correspond to the connection weight and the delayed connection weight, respectively; $\tau_{ij}(t) \geq 0$ denotes the transmission delay; $g_j(\cdot)$ is the activation function; and $I_i(t)$ stands for the external input at time t.

Here, we make the following assumptions:

Assumption 1. Suppose that $d_i(t)$, $a_{ij}(t)$, $b_{ij}(t)$, $\tau_{ij}(t)$, $I_i(t)$, $i, j = 1, \cdots, n$, are all continuous periodic functions with period $\omega > 0$.

Assumption 2. Suppose that there exist constants $p_i < 0 < q_i$, $m_i < 0 < M_i$, $i = 1, \cdots, n$, such that

$$g_i(x) = \begin{cases} m_i & -\infty < x < p_i, \\ \frac{M_i - m_i}{q_i - p_i}(x - p_i) + m_i & p_i \leq x \leq q_i, \\ M_i & q_i < x < +\infty. \end{cases} \qquad (2)$$

Assumption 3. Suppose that, for all $t \in [0, \omega]$, there hold that

$$0 < \underline{d}_i \leq d_i(t) \leq \bar{d}_i, \quad \underline{a}_{ij} \leq a_{ij}(t) \leq \bar{a}_{ij}, \quad \underline{b}_{ij} \leq b_{ij}(t) \leq \bar{b}_{ij}, \quad \underline{I}_i \leq I_i(t) \leq \bar{I}_i.$$

Consider the system (1) with initial state $x(\theta) = \phi(\theta)$, for $\theta \in [-\bar{\tau}, 0]$, where $\bar{\tau} = \max\limits_{t \in [0,\omega]} \{\max\limits_{i,j} \tau_{ij}(t)\}$, $\phi(\theta) = [\phi_1(\theta), \cdots, \phi_n(\theta)]^T$, and $\phi_i \in C([-\bar{\tau}, 0])$, $i = 1, \ldots, n$. In the following, we will explore the robust multistability and multiperiodicity of the system and the attraction basin of each periodic orbit under Assumptions 1-3.

2   Main Results

Theorem 1. Under Assumptions 1-3, if there hold that

$$\begin{cases} -\bar{d}_i p_i + \underline{a}_{ii} m_i + \sum\limits_{j \neq i} \max(\underline{a}_{ij} m_j, \underline{a}_{ij} M_j, \bar{a}_{ij} m_j, \bar{a}_{ij} M_j) \\ \qquad + \sum\limits_{j=1}^{n} \max(\underline{b}_{ij} m_j, \underline{b}_{ij} M_j, \bar{b}_{ij} m_j, \bar{b}_{ij} M_j) + \bar{I}_i < 0, \\ -\bar{d}_i q_i + \underline{a}_{ii} M_i + \sum\limits_{j \neq i} \min(\underline{a}_{ij} m_j, \underline{a}_{ij} M_j, \bar{a}_{ij} m_j, \bar{a}_{ij} M_j) \\ \qquad + \sum\limits_{j=1}^{n} \min(\underline{b}_{ij} m_j, \underline{b}_{ij} M_j, \bar{b}_{ij} m_j, \bar{b}_{ij} M_j) + \underline{I}_i > 0, \end{cases} \qquad (3)$$

$i, j = 1, \cdots, n$, then for all $d_i(t), a_{ij}(t), b_{ij}(t), I_i(t)$, there are 2^n periodic solutions of system (1), and each of them is locally exponentially stable.

Proof. Pick a subset region of R^n arbitrarily such as

$$\Omega_1 = \prod\limits_{k \in N_1} (-\infty, p_k] \times \prod\limits_{k \in N_2} [q_k, +\infty),$$

where N_1, N_2 are subsets of $\{1, 2, \cdots, n\}$, and $N_1 \bigcup N_2 = \{1, 2, \cdots, n\}$, $N_1 \bigcap N_2 = \emptyset$. In the following, we will study the existence, uniqueness of periodic solution and its local stability in Ω_1.

First, we claim that Ω_1 is invariant with respect to the solution with initial state in it. That is, if $\phi(\theta) \in \Omega_1, \theta \in [-\bar{\tau}, 0]$, then the solution $x(t)$ $(t \geq 0)$ of the dynamical system (1) will stay in Ω_1. In fact, if for some $t_1 > 0$, for some index i, such that $x(t) \in \Omega_1$ for $t \leq t_1$ while $x_i(t_1) = p_i$ or $x_i(t_1) = q_i$, then, we have

$$\left. \frac{dx_i(t)}{dt} \right|_{t=t_1} = -d_i(t_1)p_i + a_{ii}(t_1)m_i + \sum_{j \neq i} a_{ij}(t_1)g_j(x_j(t_1))$$

$$+ \sum_{j=1}^{n} b_{ij}(t_1)g_j(x_j(t_1 - \tau_{ij}(t_1))) + I_i(t_1)$$

$$\leq -d_i(t_1)p_i + a_{ii}(t_1)m_i + \sum_{j \neq i} \max(a_{ij}(t_1)m_j, a_{ij}(t_1)M_j)$$

$$+ \sum_{j=1}^{n} \max(b_{ij}(t_1)m_j, b_{ij}(t_1)M_j) + I_i(t_1)$$

$$\leq -\bar{d}_i p_i + \underline{a}_{ii} m_i + \sum_{j \neq i} \max(\underline{a}_{ij} m_j, \underline{a}_{ij} M_j, \bar{a}_{ij} m_j, \bar{a}_{ij} M_j)$$

$$+ \sum_{j=1}^{n} \max(\underline{b}_{ij} m_j, \underline{b}_{ij} M_j, \bar{b}_{ij} m_j, \bar{b}_{ij} M_j) + \bar{I}_i < 0, \tag{4}$$

$$\left. \frac{dx_i(t)}{dt} \right|_{t=t_1} = -d_i(t_1)q_i + a_{ii}(t_1)M_i + \sum_{j \neq i} a_{ij}(t_1)g_j(x_j(t_1))$$

$$+ \sum_{j=1}^{n} b_{ij}(t_1)g_j(x_j(t_1 - \tau_{ij}(t_1))) + I_i(t_1)$$

$$\geq -d_i(t_1)q_i + a_{ii}(t_1)M_i + \sum_{j \neq i} \min(a_{ij}(t_1)m_j, a_{ij}(t_1)M_j)$$

$$+ \sum_{j=1}^{n} \min(b_{ij}(t_1)m_j, b_{ij}(t_1)M_j) + I_i(t_1)$$

$$\geq -\bar{d}_i q_i + \underline{a}_{ii} M_i + \sum_{j \neq i} \min(\underline{a}_{ij} m_j, \underline{a}_{ij} M_j, \bar{a}_{ij} m_j, \bar{a}_{ij} M_j)$$

$$+ \sum_{j=1}^{n} \min(\underline{b}_{ij} m_j, \underline{b}_{ij} M_j, \bar{b}_{ij} m_j, \bar{b}_{ij} M_j) + \underline{I}_i > 0, \tag{5}$$

which implies the solution $x(t)$ will never go out of Ω_1 for all $t > 0$.

Then, we are to show that the solution $x(t)$ of system (1) with initial state $\phi(\theta) \in \Omega_1, \theta \in [-\bar{\tau}, 0]$, is bounded.

In fact, if we denote

$$E_i = \underline{d}_i^{-1} \left[\sum_{j=1}^{n} \max(\underline{a}_{ij} m_j, \underline{a}_{ij} M_j, \bar{a}_{ij} m_j, \bar{a}_{ij} M_j) \right.$$

$$+ \sum_{j=1}^{n} \max(\underline{b}_{ij}m_j, \underline{b}_{ij}M_j, \bar{b}_{ij}m_j, \bar{b}_{ij}M_j) + \max(\underline{I}_i, \bar{I}_i) + 1 \Big],$$

for $i = 1, \cdots, n$, and suppose that there exist $t_2 > 0$ and some index i such that $|x_i(t_2)| \geq E_i$, then, we have

$$\frac{d|x_i(t)|}{dt}\Big|_{t=t_2} = -d_i(t_2)|x_i(t_2)| + sign\{x_i(t_2)\}\Big\{ \sum_{j=1}^{n} a_{ij}(t_2)g_j(x_j(t_2))$$

$$+ \sum_{j=1}^{n} b_{ij}(t_2)g_j(x_j(t_2 - \tau_{ij}(t_2))) + I_i(t_2) \Big\}$$

$$\leq -\underline{d}_i E_i + \sum_{j=1}^{n} \max(\underline{a}_{ij}m_j, \underline{a}_{ij}M_j, \bar{a}_{ij}m_j, \bar{a}_{ij}M_j)$$

$$+ \sum_{j=1}^{n} \max(\underline{b}_{ij}m_j, \underline{b}_{ij}M_j, \bar{b}_{ij}m_j, \bar{b}_{ij}M_j) + \max(\underline{I}_i, \bar{I}_i) < 0, \quad (6)$$

which follows that $|x_i(t)|$ is bounded by E_i, $i = 1, \cdots, n$.

Now, define $\tilde{\Omega} = \Omega_1 \bigcap \{x : |x_i| \leq E_i\}$ and a map T on $\tilde{\Omega}$ such that

$$T : \phi(\theta) \rightarrow x(\theta + \omega, \phi)$$

where $x(t) = x(t, \phi)$ is the solution of the system (1) with the initial state $x_i(\theta) = \phi_i(\theta)$, for $\theta \in [-\bar{\tau}, 0]$ and $i = 1, \cdots, n$.

From the analysis above, we know that $T\tilde{\Omega} \subset \tilde{\Omega}$. By Brouwers fixed point theorem, there exists $\phi^* \in \tilde{\Omega}$ such that $T\phi^* = \phi^*$. Hence $x(t, \phi^*) = x(t, T\phi^*)$, i.e.,

$$x(t, \phi^*) = x(t + \omega, \phi^*), \quad (7)$$

which is an ω-periodic solution of the system (1), denoted as $x^*(t)$.

Next, we will address the local stability of $x^*(t)$. Due to the positive invariance of Ω_1, we can rewrite the solution $x(t)$ with initial state $\phi(\theta) \in \Omega_1, \theta \in [-\bar{\tau}, 0]$ as

$$\frac{dx_i(t)}{dt} = -d_i(t)x_i(t) + \sum_{j \in N_1}^{n} (a_{ij}(t) + b_{ij}(t))m_j + \sum_{j \in N_2}^{n} (a_{ij}(t) + b_{ij}(t))M_j + I_i(t). \quad (8)$$

Denote $u(t) = x(t) - x^*(t)$, then, we have

$$\frac{d|u_i(t)|}{dt} = -d_i(t)|u_i(t)| \leq -\underline{d}_i|u_i(t)|, \quad (9)$$

which implies that $x^*(t)$ is exponentially stable in Ω_1, so that it is unique meanwhile.

Therefore, we get that there exists a unique periodic solution in Ω_1 and it is locally exponentially stable. On the other hand, it is easy to see that there are 2^n Ω_1-type subsets of R^n in all, so the same method can be applied and the same result can be derived. To sum up, there are 2^n periodic solutions of system (1) under conditions (3), and each of them is locally exponentially stable. $\qquad \square$

Remark 1. From Theorem 1, we can see that the system (1) is multistable for all $d_i(t), a_{ij}(t), b_{ij}(t), I_i(t)$ with Assumption 3, that is, the system is robust multistable.

Corollary 1. Let $d_i(t) = d_i, a_{ij}(t) = a_{ij}, b_{ij}(t) = b_{ij}, \tau_{ij}(t) = \tau_{ij}, I_i(t) = I_i$. Then, under Assumption 2 and condition (3), for all $0 < \underline{d}_i \leq d_i \leq \bar{d}_i$, $\underline{a}_{ij} \leq a_{ij} \leq \bar{a}_{ij}$, $\underline{b}_{ij} \leq b_{ij} \leq \bar{b}_{ij}$, $\underline{I}_i \leq I_i \leq \bar{I}_i$, the system (1) is robust multistable.

Theorem 2. Suppose that Assumptions 1-3 and (3) hold for $i, j = 1, \cdots, n$. Denote

$$\Gamma_i \triangleq \sup_t \left\{ -\int_0^t e^{-\int_0^s (-d_i(u)+a_{ii}(u)l_i)du} \left(\sum_{j\neq i} \min(a_{ij}(s)m_j, a_{ij}(s)M_j) \right. \right.$$

$$\left. \left. + \sum_{j=1}^n \min(b_{ij}(s)m_j, b_{ij}(s)M_j) + a_{ii}(s)c_i + I_i(s) \right) ds \right\}, \tag{10}$$

$$\Psi_i \triangleq \inf_t \left\{ -\int_0^t e^{-\int_0^s (-d_i(u)+a_{ii}(u)l_i)du} \left(\sum_{j\neq i} \max(a_{ij}(s)m_j, a_{ij}(s)M_j) \right. \right.$$

$$\left. \left. + \sum_{j=1}^n \max(b_{ij}(s)m_j, b_{ij}(s)M_j) + a_{ii}(s)c_i + I_i(s) \right) ds \right\}, \tag{11}$$

where $l_i = \frac{M_i - m_i}{q_i - p_i}$, $c_i = m_i - \frac{p_i(M_i - m_i)}{q_i - p_i}$. Then, for the solution $x(t)$:
(i) if $\Gamma_i < x_i(0) < q_i$, then, $x_i(t)$ will go into $[q_i, +\infty)$ when t is big enough;
(ii) if $p_i < x_i(0) < \Psi_i$, then, $x_i(t)$ will go into $(-\infty, p_i]$ when t is big enough.
Proof. Firstly, we show that $p_i < \Psi_i$, $\Gamma_i < q_i$.
In fact, from condition (3), we can easily find a positive constant ϵ small enough that

$$-d_i(t)p_i + a_{ii}(t)l_ip_i + \sum_{j\neq i} \max(a_{ij}(t)m_j, a_{ij}(t)M_j)$$

$$+ \sum_{j=1}^n \max(b_{ij}(t)m_j, b_{ij}(t)M_j) + a_{ii}(t)c_i + I_i(t) < -\epsilon < 0,$$

$$-d_i(t)q_i + a_{ii}(t)l_iq_i + \sum_{j\neq i} \min(a_{ij}(t)m_j, a_{ij}(t)M_j)$$

$$+ \sum_{j=1}^n \min(b_{ij}(t)m_j, b_{ij}(t)M_j) + a_{ii}(t)c_i + I_i(t) > \epsilon > 0,$$

which obviously imply that $-d_i(t) + a_{ii}(t)l_i > 0$ for all $t \in [0, \omega]$.
Denote $\alpha_i \triangleq \sup_{t\in[0,\omega]} (-d_i(t) + a_{ii}(t)l_i)$, $\beta_i \triangleq \inf_{t\in[0,\omega]} (-d_i(t) + a_{ii}(t)l_i)$, and for any $t > 0$, denote $t = n\omega + t_0$, where n is an integer, $t_0 \in [0, \omega)$, it holds that

$$-\int_0^t e^{-\int_0^s (-d_i(u)+a_{ii}(u)l_i)du}\Big(\sum_{j\neq i}\min(a_{ij}(s)m_j,a_{ij}(s)M_j)$$

$$+\sum_{j=1}^n \min(b_{ij}(s)m_j,b_{ij}(s)M_j)+a_{ii}(s)c_i+I_i(s)\Big)ds$$

$$=-\sum_{k=1}^n\int_{(k-1)\omega}^{k\omega}e^{-\int_0^s(-d_i(u)+a_{ii}(u)l_i)du}\Big(\sum_{j\neq i}\min(a_{ij}(s)m_j,a_{ij}(s)M_j)$$

$$+\sum_{j=1}^n \min(b_{ij}(s)m_j,b_{ij}(s)M_j)+a_{ii}(s)c_i+I_i(s)\Big)ds$$

$$-\int_{n\omega}^{n\omega+t_0}e^{-\int_0^s(-d_i(u)+a_{ii}(u)l_i)du}\Big(\sum_{j\neq i}\min(a_{ij}(s)m_j,a_{ij}(s)M_j)$$

$$+\sum_{j=1}^n \min(b_{ij}(s)m_j,b_{ij}(s)M_j)+a_{ii}(s)c_i+I_i(s)\Big)ds$$

$$=-\sum_{k=1}^n\int_0^\omega e^{-\int_0^{s+(k-1)\omega}(-d_i(u)+a_{ii}(u)l_i)du}\Big(\sum_{j\neq i}\min(a_{ij}(s)m_j,a_{ij}(s)M_j)$$

$$+\sum_{j=1}^n \min(b_{ij}(s)m_j,b_{ij}(s)M_j)+a_{ii}(s)c_i+I_i(s)\Big)ds$$

$$-\int_0^{t_0}e^{-\int_0^{s+n\omega}(-d_i(u)+a_{ii}(u)l_i)du}\Big(\sum_{j\neq i}\min(a_{ij}(s)m_j,a_{ij}(s)M_j)$$

$$+\sum_{j=1}^n \min(b_{ij}(s)m_j,b_{ij}(s)M_j)+a_{ii}(s)c_i+I_i(s)\Big)ds$$

$$=-\sum_{k=1}^n e^{-(k-1)\int_0^\omega(-d_i(u)+a_{ii}(u)l_i)du}\int_0^\omega e^{-\int_0^s(-d_i(u)+a_{ii}(u)l_i)du}\Big(\sum_{j\neq i}\min(a_{ij}(s)m_j,a_{ij}(s)M_j)$$

$$+\sum_{j=1}^n \min(b_{ij}(s)m_j,b_{ij}(s)M_j)+a_{ii}(s)c_i+I_i(s)\Big)ds$$

$$-e^{-n\int_0^\omega(-d_i(u)+a_{ii}(u)l_i)du}\int_0^{t_0} e^{-\int_0^s(-d_i(u)+a_{ii}(u)l_i)du}\Big(\sum_{j\neq i}\min(a_{ij}(s)m_j,a_{ij}(s)M_j)$$

$$+\sum_{j=1}^n \min(b_{ij}(s)m_j,b_{ij}(s)M_j)+a_{ii}(s)c_i+I_i(s)\Big)ds$$

$$\leq \sum_{k=1}^n e^{-(k-1)\int_0^\omega(-d_i(u)+a_{ii}(u)l_i)du}\int_0^\omega e^{-\int_0^s(-d_i(u)+a_{ii}(u)l_i)du}\Big((-d_i(u)+a_{ii}(u)l_i)q_i-\epsilon\Big)ds$$

$$+e^{-n\int_0^\omega(-d_i(u)+a_{ii}(u)l_i)du}\int_0^{t_0} e^{-\int_0^s(-d_i(u)+a_{ii}(u)l_i)du}\Big((-d_i(u)+a_{ii}(u)l_i)q_i-\epsilon\Big)ds$$

$$= q_i\sum_{k=1}^n e^{-(k-1)\int_0^\omega(-d_i(u)+a_{ii}(u)l_i)du}(1-e^{-\int_0^\omega(-d_i(u)+a_{ii}(u)l_i)du})$$

$$+q_i e^{-n\int_0^\omega(-d_i(u)+a_{ii}(u)l_i)du}(1-e^{-\int_0^{t_0}(-d_i(u)+a_{ii}(u)l_i)du})$$

$$-\epsilon\sum_{k=1}^n e^{-(k-1)\int_0^\omega(-d_i(u)+a_{ii}(u)l_i)du}\int_0^\omega e^{-\int_0^s(-d_i(u)+a_{ii}(u)l_i)du}ds$$

$$-\epsilon e^{-n\int_0^\omega(-d_i(u)+a_{ii}(u)l_i)du}\int_0^{t_0} e^{-\int_0^s(-d_i(u)+a_{ii}(u)l_i)du}ds$$

$$\leq q_i-\frac{(1-e^{-\alpha_i\omega})\epsilon}{\alpha_i}.$$

So we know that

$$\sup_t\{-\int_0^t e^{-\int_0^s(-d_i(u)+a_{ii}(u)l_i)du}\Big(\sum_{j\neq i}\min(a_{ij}(s)m_j,a_{ij}(s)M_j)+a_{ii}(s)c_i+I_i(s)\Big)ds\}$$

$$\leq q_i-\frac{(1-e^{-\alpha_i\omega})\epsilon}{\alpha_i}<q_i,\quad i.e.,\ \Gamma_i<q_i.$$

Similarly, we can get that $\Psi_i>p_i$, $i=1,\cdots,n$.

Then, consider the solution $x(t)$ with initial states $\phi(\theta),\theta\in[-\bar\tau,0]$. If for some i, it holds $\Gamma_i<x_i(0)<q_i$, then we claim that $x_i(t)$ will go into the interval $[q_i,+\infty)$. In fact, when it stays in (p_i,q_i), there holds

$$\frac{dx_i(t)}{dt}=-d_i(t)x_i(t)+a_{ii}(t)l_ix_i(t)+\sum_{j\neq i}a_{ij}(t)g(x_j(t))+\sum_{j=1}^n b_{ij}(t)g(x_j(t-\tau_{ij}(t)))$$

$$+a_{ii}(t)c_i+I_i(t)$$

$$\geq(-d_i(t)+a_{ii}(t)l_i)x_i(t)+\sum_{j\neq i}\min(a_{ij}(t)m_j,a_{ij}(t)M_j)$$

$$+\sum_{j=1}^n\min(b_{ij}(t)m_j,b_{ij}(t)M_j)+a_{ii}(t)c_i+I_i(t),$$

which follows that

$$x_i(t)\geq e^{\int_0^t(-d_i(u)+a_{ii}(u)l_i)du}\Big\{x_i(0)+\int_0^t e^{-\int_0^s(-d_i(u)+a_{ii}(u)l_i)du}\Big(\sum_{j\neq i}\min(a_{ij}(s)m_j,a_{ij}(s)M_j)$$

$$+\sum_{j=1}^n\min(b_{ij}(s)m_j,b_{ij}(s)M_j)+a_{ii}(s)c_i+I_i(s)\Big)ds\Big\}$$

$$\geq e^{\beta_it}\Big\{x_i(0)+\int_0^t e^{-\int_0^s(-d_i(u)+a_{ii}(u)l_i)du}\Big(\sum_{j\neq i}\min(a_{ij}(s)m_j,a_{ij}(s)M_j)$$

$$+\sum_{j=1}^n\min(b_{ij}(s)m_j,b_{ij}(s)M_j)+a_{ii}(s)c_i+I_i(s)\Big)ds\Big\}$$

Hence, $x_i(t)$ will exceed q_i when t is big enough. Then, by the proof of Theorem 1, we know that $x_i(t)$ will stays in $[q_i,+\infty)$ afterwards. Similar proof can be applied to the case $\Psi_i<x_i(0)<p_i$ and get the conclusion.   □

Remark 2. From Theorem 2, we can see that the attraction basin of each periodic solution can be extended to $(\Gamma_i,+\infty)$ or $(-\infty,\Psi_i)$, from $[q_i,+\infty)$ or $(-\infty,p_i]$ in previous literatures, respectively.

3   Conclusions

In this paper, the robust multistability and multiperiodicity of neural networks are addressed. Under some mild conditions, the coexistence of 2^n periodic solution and their local stability are established. Furthermore, the attraction basin of each periodic solution can be enlarged from the subset it locates.

Acknowledgments. This work is jointly supported by the National Natural Sciences Foundation of China under Grant No. 61101203, the Innovation Program of Shanghai Municipal Education Commission under Grant No. 14ZZ076, and the Foundation of SUFE under Grant No. 2015110273.

References

1. Arik, S., Tavsanoglu, V.: On the global asymptotic stability of delayed cellular neural networks. IEEE Transactions on Circuits and Systems-I: Fundamental Theory and Applications 47(4), 571–574 (2000)
2. Chen, T., Amari, S.: New theorems on global convergence of some dynamical systems. Neural Networks 14, 251–255 (2001)
3. Chen, T., Rong, L.: Robust global exponential stability of Cohen-Grossberg neural networks with time delays. IEEE Transactions on Neural Networks 15(1), 203–206 (2004)
4. Lu, W., Chen, T.: Global exponential stability of almost periodic solution for a large class of delayed dynamical systems. Science in China Series A-Mathematics 48(8), 1015–1026 (2005)
5. Zeng, Z., Wang, J.: Multiperiodicity and exponential attractivity evoked by periodic external inputs in delayed cellular neural networks. Neural Computation 18(4), 848–870 (2006)
6. Cheng, C., Lin, K., Shih, C.: Multistability in recurrent neural networks. SIAM Journal on Applied Mathematics 66(4), 1301–1320 (2006)
7. Zhang, L., Yi, Z., Yu, J.: Multiperiodicity and attractivity of delayed recurrent neural networks with unsaturating piecewise linear transfer functions. IEEE Transactions on Neural Networks 19(1), 158–167 (2008)
8. Wang, L., Lu, W., Chen, T.: Multistability and new attraction basins of almost-periodic solutions of delayed neural networks. IEEE Transactions on Neural Networks 20(10), 1581–1593 (2009)
9. Wang, L., Lu, W., Chen, T.: Coexistence and local stability of multiple equilibria in neural networks with piecewise linear nondecreasing activation functions. Neural Networks 23, 189–200 (2010)
10. Wang, L., Chen, T.: Multiple μ-stability of neural networks with unbounded time-varying delays. Neural Networks 53, 109–118 (2014)

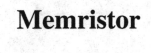
Memristor

A Novel Four-Dimensional Memristive Hyperchaotic System with Its Analog Circuit Implementation

Guoqi Min, Lidan Wang[*], and Shukai Duan

School of Electronic and Information Engineering,
Southwest University, Chongqing, 400715, China
ldwang@swu.edu.cn

Abstract. A novel memristor-based hyperchaotic system is proposed and studied in this paper. The memristor is nonlinear memory element intrinsically, which has the potential application for generating complex dynamics in nonlinear circuit to reduce system power consumption and circuit size. As the non-linear part of a system, the HP memristor is introduced to a four-dimensional system. Chaotic attractors, Lyapunov exponent spectrum, Lyapunov dimension, power spectrum, Poincaré map and bifurcation with respect to various circuit parameter, are considered and observed, which together demonstrate the rich chaotic dynamical behaviors of the system. Finally, the circuit in SPICE are designed for the proposed memristive hyperchaotic system. The SPICE experimental results are consistent with the numerical simulation results, which verifies the feasibility of the memristor hyperchaotic system.

Keywords: memristor, hyperchaotic system, dynamics behavior, circuit implementation.

1   Introduction

The concept of memristor was proposed by Leon O. Chua in 1971 firstly [1]. Stan Williams and his team at HP lab fabricated a solid state implementation of the memristor in 2008, which cemented its place as the fourth circuit element [2-4] and garnered extensive interest from both academic and industrial communities immediately.

Chaos is a very interesting complex nonlinear phenomenon, which has been studied intensively in the last four decades within science, mathematics and engineering communities. Theoretical design and circuit implementation of various chaotic generators have been a key focus of nonlinear science. Due to the nonlinearity of memristor element, memristor-based circuits can generate a chaotic signal easily. The first memristor-based chaotic circuit was proposed by Itoh and Chua [5]. In their paper, they used a piecewise linear nonlinear memristor to replace Chua's diode in Chua's oscillator, based on Chua's circuit, they put forward several memristor-based chaotic

---

[*] Corresponding author.

© Springer International Publishing Switzerland 2015
X. Hu et al. (Eds.): ISNN 2015, LNCS 9377, pp. 157–165, 2015.
DOI: 10.1007/978-3-319-25393-0_18

oscillation circuits. Recently, lots of research efforts on memristive chaotic circuits have been reported [6-8].

This paper investigates a new hyperchaotic system based on HP memristor. Section 2 introduces the memristor chaotic system, calculates the Lyapunov exponent spectrum, Lyapunov dimension and power spectrum, and further analyzes the parameter of chaotic attractor and bifurcation diagram. Section 3 presents a basic analog circuit realization of the memristive hyperchaotic system, and simulates the chaotic attractor of the circuit by SPICE. Finally, conclusions are drawn in section 4.

2   The Memristor as Nonlinear Part in a Hyperchaotic System

2.1   A New Memristor Hyperchaotic System

Firstly, we take the following hyperchaotic system based on memristor into consideration.

$$\begin{cases} \dot{x} = ay, \\ \dot{y} = w - bx, \\ \dot{z} = xy - cz, \\ \dot{w} = 1 - z^2 + 10^4 f(-|x|). \end{cases} \tag{1}$$

where x, y, z and w are the states of the system; a, b and c are parameters. The function $f(-|x|)$ codifies the charge of the memristor [7,8], precisely expressed as follows:

$$f(-|x|) = \begin{cases} \dfrac{x - c_1}{R_{OFF}} & x < \varphi_{low} \\[3mm] \dfrac{\sqrt{2kx + M(0)^2} - M(0)}{k} & \varphi_{low} < x < 0 \\[3mm] \dfrac{\sqrt{-2kx + M(0)^2} - M(0)}{k} & 0 \le x < -\varphi_{low} \\[3mm] \dfrac{-x - c_1}{R_{OFF}} & x \ge -\varphi_{low} \end{cases} \tag{2}$$

Where,

$$c_1 = -(R_{OFF} - M(0))^2 / 2k, \quad \varphi_{low} = (R_{OFF}{}^2 - M(0)^2) / 2k, \quad k = \frac{(R_{ON} - R_{OFF})u_v R_{ON}}{D^2}$$

x is the flux through the memristor; R_{OFF}=20$k\Omega$, R_{ON}=100Ω, $M(0)$= 16$k\Omega$, D=10nm, u_v=10$^{-14}m^2s^{-1}v^{-1}$, and we choose the initial conditions $(x, y, z, w) = (0,1,0,0)$.

2.2   Hyperchaotic Attractors Incorporating with a Memristor

The parameter values of the system which can yield chaotic dynamics are a=5, b=10, c=7. A three-dimensional and three two-dimensional attractors are generated by means of numerical integration, which is illustrated in Fig.1.

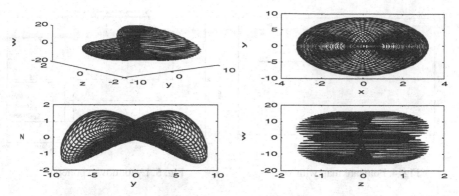

Fig. 1. Various projections of the chaotic attractor codified by system (1).

2.3 Dynamical Behaviors of the System

Typical time-domain waveforms for the four dynamical states, namely $x(t)$, $y(t)$, $z(t)$ and $w(t)$ are illustrated in Fig.2. As we can see, they are non-cyclical, and have the unique state transferring behavior so that the jump behavior can increase the complexity of chaos.The power spectrum of memristive hyperchaotic system is shown in Fig.3.

A basic dynamical behavior is further explored by calculating its Poincaré diagram. The Poincaré section of the system is given in Fig.4. We can draw the conclusion that the system has extremely rich dynamics because the section shows a continuous curve.

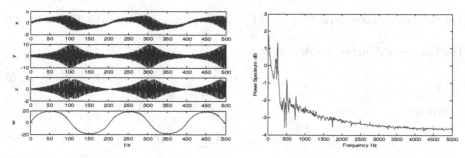

Fig. 2. Time-domain wave form of each state variable.   **Fig. 3.** Power spectrum

The existence of dynamic behavior in the system can be observed from whether the largest Lyapunov exponent is larger than zero; the larger the exponent, the more evident chaotic characteristics will be, as well as the higher degree of chaos. In order to present the chaotic motion state more intuitively, the system's Lyapunov exponent spectrum, varying with time, is drawn in Fig.5, which indicates that the chaotic characteristic is obvious.

Take t=2000 in numerical simulation, the Lyapunov exponent are λ_1= 0.4080, λ_2= 0.1601, λ_3=0.001, λ_4= -0.8893. Obviously, the first two exponents are larger than zero, the third exponent can be equal to zero in the range of allowable error, and the last exponent is less than zero. So we can conclude the attractor is in the hyperchaos state.

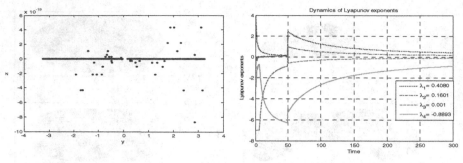

Fig. 4. Poincaré map with $x=0$ **Fig. 5.** Lyapunov exponents

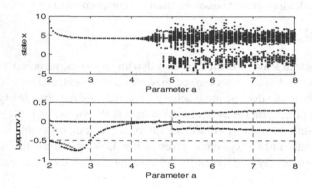

Fig. 6. Bifurcation diagram and maximum Lyapunov exponent spectrum with the changes of a.

The Lyapunov dimension is defined by:

$$D_\lambda = j + \frac{S_j}{\left|\lambda_{j+1}\right|} \tag{5}$$

Where

$$S_j = \sum_{i=1}^{i=j} \lambda_i \geq 0, \quad S_{j+1} = \sum_{i=1}^{i=j+1} \lambda_i \leq 0 \tag{6}$$

and j is a integer that satisfies the above principle.

The four Lyapunov exponents show that j is equal to 3 here. The Lyapunov dimension of the system (1) is $D_\lambda = 3 + (\lambda_1 + \lambda_2 + \lambda_3)/|\lambda_4| \approx 3.6388$, namely $D_\lambda > 3$, which indicates the attractor has a type of complex structure.

Next, bifurcation theory has investigated the topic that bears on chaotic dynamics intensively. In the memristive circuit, a bifurcation of a dynamical system is a qualitative change in its dynamics produced by varying parameters, and the general state of the ensuing series of mutation processes, such as period-doubling route to chaos road. Fig.6 shows the bifurcation diagram generated by system (1), with the corresponding Lyapunov exponent spectrum shown on the bottom. Choose the parameter a as the bifurcation parameter in the range(2, 8). The bifurcation diagram shows that there are many real numbers a for the solution of memristive hyperchaotic system.

By comparison, it can be observed that the bifurcation diagram coincides with the spectrum of the Lyapunov exponents well. When $a < 4.2$, the system has no positive Lyapunov exponents, thus it is in single cycle. With the increasing value of a, the first and second Lyapunov exponents begin to increase, and are more than zero, while the third Lyapunov exponent decreases to less than zero. The above analysis shows that the system is in the hyperchaotic state.

From the phase trajectory, time-domain waveform, Power spectrum, Poincaré map, Lyapunov exponent and dimension, and the bifurcation diagram with the change of parameters, we can conclude that the system is a hyperchaotic system.

So far, several numerical calculations and simulations have demonstrated the basic dynamical behaviors of memristive system. However, more likely scenarios are worthy of further study in the future. In the next section, the memristive hyperchaotic circuit is designed and realized based on OrCAD.

3 Analog Implementation of Memristive Hyperchaotic Attractor

In order to validate hyperchaotic behaviors of the system (1), a circuit is designed, which consists of four analog operation circuits, and the voltages at the nodes are labeled as v_x, v_y, v_z and v_w corresponding to the states of system (1). The circuit is composed of the memristor, diodes, capacitors, resistors, multiplier and operational amplifiers. SPICE memristor simulation model have been built [3]. The operational amplifiers are all LM675 type, which are powered with VCC = +30V and VEE =⁻30V. And the circuit [7,8] is shown in Fig.7. Voltage node Vx, Vy and Vz of the SPICE circuit represent state variables x, y and z of the chaotic system.

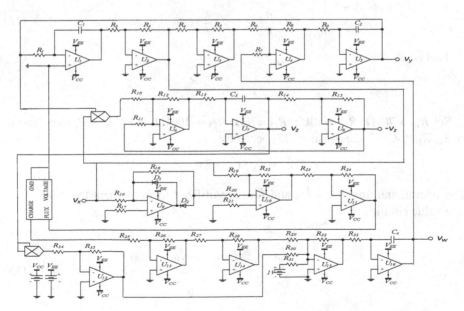

Fig. 7. Analog SPICE implementation of the memristive hyperchaotic system.

The operational amplifiers U_1 and U_2 are used to achieve the following formula:

$$v_x = -\frac{R_3}{R_2}\int -\frac{1}{R_1 C_1} v_y dt = \frac{R_3}{R_1 R_2 C_1}\int v_y dt \tag{7}$$

Or equivalently,

$$\dot{v}_x = \frac{R_3}{R_1 R_2 C_1} v_y \tag{8}$$

Compare with system (1) and set $R_1 = 1M\Omega$, $R_2 = 1k\Omega$, $R_3 = 5k\Omega$ $C_1 = 1\mu F$, then lead to $a=R_3/R_1 R_2 C_1=5$, namely,

$$\dot{v}_x = 5v_y \tag{9}$$

The operational amplifiers U_3 and U_4 are reverse amplifiers while U_5 is realized as an integrator. The following holds:

$$\dot{v}_y = -\frac{R_5 R_8}{R_4 R_6 R_9 C_2} v_x + \frac{R_8}{R_7 R_9 C_2} v_w \tag{10}$$

Set $R_4 = 1k\Omega$, $R_5 = 10k\Omega$, $R_6 = R_7 = R_8 = 1k\Omega$, $R_9 = 1M\Omega$ and $C_2 = 1\mu F$, then lead to $b=R_5 R_8/R_4 R_6 R_9 C_2=10$, the equation (10) becomes:

$$\dot{v}_y = -10v_x + v_w \tag{11}$$

Similarly available:

$$v_z = -\frac{1}{R_{13} C_3}\int(-\frac{R_{12}}{R_{10}} v_x \cdot v_y - \frac{R_{12}}{R_{11}}(-v_z))dt \tag{12}$$

That is:

$$\dot{v}_z = \frac{R_{12}}{R_{10} R_{13} C_3} v_x \cdot v_y - \frac{R_{12}}{R_{11} R_{13} C_3} v_z \tag{13}$$

Set $R_{10} = 70k\Omega$, $R_{11} = 10k\Omega$, $R_{12} = 7k\Omega$, $R_{13} = 100k\Omega$ and $C_3 = 1\mu F$, then lead to $c=R_{12}/R_{11} R_{13} C_3=7$, the equation (13) becomes:

$$\dot{v}_z = v_x \cdot v_y - 7v_z \tag{14}$$

The operational amplifiers U_9 and U_{10} are utilized to implement the absolute value circuit:

$$v_{U10} = \begin{cases} -\dfrac{R_{22}}{R_{19}} v_x + \dfrac{R_{18} R_{22}}{R_{16} R_{20}} v_x, & v_x \geq 0 \\[3mm] -\dfrac{R_{22}}{R_{19}} v_x, & v_x < 0 \end{cases} \tag{15}$$

When $R_{16} = R_{18} = R_{19} = R_{22} = 1k\Omega$, $R_{17} = R_{20} = 500\Omega$, the output voltage of U_{10} is $v_{U10}=|v_x|$, and the output voltage of U_{11} is $v_{U11}=-(R_{24}/R_{23})$ $|v_x|$. Set $R_{23} = R_{24} = 1k\Omega$, and it becomes $v_{U11}=-|v_x|$. The voltage v_{U11} is the input to flux terminal of the flux-controlled memristor's SPICE model. The charge terminal of the memristor is connected with operational amplifiers U_{13}. The memristor can implement the function $f(\bullet)$ in system(2). The memristor parameters are $R_{ON}=100\Omega$, $R_{OFF}=20k\Omega$, $M(0)=16k\Omega$, $D=10nm$ and $\mu_V=10^{-14}m^2s^{-1}V^{-1}$.

The operational amplifiers U_{13} and U_{14} are rephrase amplifiers whose gain are all set to 100. After a two-stage amplification, the charge of the memristor is amplified by 10000 times. Set $R_{25}= R_{27}=100\Omega$, $R_{26}= R_{28}= 10k\Omega$, we can get:

$$v_{U14} = \frac{R_{26}R_{28}}{R_{25}R_{27}} f(-|x|) = 10000 f(-|x|) \tag{16}$$

The operational amplifier U_{12} is to achieve an inverse ratio. While operational amplifiers U_{15} and U_{16} are applied as an adder and an integrator, we get:

$$v_{U12} = -\frac{R_{35}}{R_{34}}v_z \cdot v_z \tag{17}$$

$$v_{U15} = -\frac{R_{32}}{R_{29}}v_{U14} - \frac{R_{32}}{R_{30}}v_{U12} - \frac{R_{32}}{R_{31}}v_1 \tag{18}$$

$$v_w = v_{U16} = -\frac{1}{R_{33}C_4}\int v_{U15}dt = -\frac{1}{R_{33}C_4}\int(-\frac{R_{32}}{R_{29}}v_{U14} - \frac{R_{32}}{R_{30}}v_{U12} - \frac{R_{32}}{R_{31}}v_1)dt \tag{19}$$

Or equivalently,

$$\dot{v}_w = v_{U16} = \frac{R_{32}}{R_{29}R_{33}C_4}v_{U14} + \frac{R_{32}}{R_{30}R_{33}C_4}v_{U12} + \frac{R_{32}}{R_{31}R_{33}C_4}v_1 \tag{20}$$

Substitute v_{U12}, v_{U14} and $v_1=1v$, we can yield:

$$\dot{v}_w = v_{U16} = \frac{R_{26}R_{28}R_{32}}{R_{25}R_{27}R_{29}R_{33}C_4} f(-|x|) - \frac{R_{32}R_{35}}{R_{30}R_{33}R_{34}C_4}v_z \cdot v_z + \frac{R_{32}}{R_{31}R_{33}C_4} \tag{21}$$

Set $R_{25}= R_{27}= 100\Omega$, $R_{26}= R_{28}=10k\Omega$, $R_{29}= R_{30}= R_{31}= R_{32}=1k\Omega$, $R_{33}=1000k\Omega$, $R_{34}= R_{35}=20k\Omega$ and $C_4= 1\mu F$, we get:

$$\dot{v}_w = 1-v_z \cdot v_z +10000 f(-|x|) \tag{22}$$

The SPICE simulation time interval ranges from 0 sec to 500 sec and the maximum step size is set to 0.001 second. Fig.8 shows the t-z time domain graph, x-y, y-z and z-w phase diagram of the analog realization of the chaotic memristive system by SPICE. A comparison between Fig.2, Fig.1 and Fig.8 reveals a good qualitative agreement between numeric (Fig.2, Fig.1).

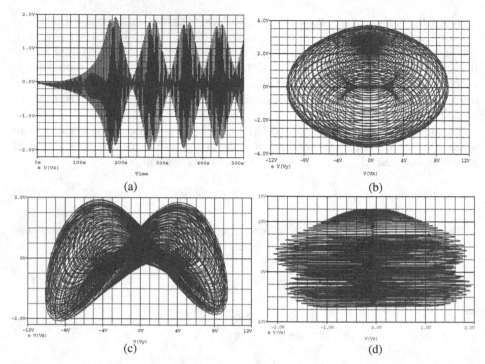

Fig. 8. SPICE simulated results of the state variables (a) t-z. Phase diagrams of (b) x-y, (c) y-z, (d) z-w.

4   Conclusion

The memristor is the latest component, which has the unique properties of not only nanoscale size, low power consumption, but also special components, and can be achieved by using a common electronic device. So it has high value in the fields of chaos circuit confidential communications and information encryption, and electronic measurement. This paper presents a novel hyperchaotic system based on HP memristor. A series of computer simulations have been performed, including the Lyapunov exponent spectrum, Lyapunov dimension, Power spectrum, bifurcation diagram, which verified the complex chaotic dynamics of the new memristive system. In addition, an analog circuit of the memristive hyperchaotic system has been presented by SPICE simulations. The use of memristors assures that the proposed memristive hyperchaotic system has higher perspective than other common chaotic systems.

Acknowledgments. The work was supported by Program for New Century Excellent Talents in University, National Natural Science Foundation of China (Grant Nos. 61372139, 61101233, 60972155, 61374078), "Spring Sunshine Plan" Research Project of Ministry of Education of China (Grant No. z2011148), Technology Foundation for Selected Overseas Chinese Scholars, Ministry of Personnel in China (Grant No. 2012-186), University Excellent Talents Supporting Foundations in of Chongqing

(Grant No. 2011-65), University Key Teacher Supporting Foundations of Chongqing (Grant No. 2011-65), Fundamental Research Funds for the Central Universities (Grant Nos. XDJK2012A007, XDJK2013B011).

References

1. Chua, L.O.: Memristor-the missing circuit element. IEEE Transactions on ircuit Theory 18(5), 507–519 (1971)
2. Strukov, D.B., Snider, G.S., Stewart, D.R., Williams, R.S.: The missing memristor found. Nature 453(7191), 80–83 (2008)
3. Biolek, Z., Biolek, D., Biolkova, V.: SPICE model of memristor with nonlinear dopant drift. Radioengineering 18(2), 210–214 (2009)
4. Chua, L.O., Kang, S.M.: Memristive devices and systems. Proceedings of the IEEE 64(2), 209–223 (1976)
5. Barboza, R., Chua, L.O.: The four-element Chua's circuit. International Journal of Bifurcation and Chaos 18(04), 943–955 (2008)
6. Wang, L., Duan, S.: A chaotic attractor in delayed memristive system. In: Abstract and Applied Analysis. Hindawi Publishing Corporation, November 2012
7. Wang, L., Drakakis, E., Duan, S., He, P., Liao, X.: Memristor model and its application for chaos generation. International Journal of Bifurcation and Chaos 22(08) (2012)
8. Hu, X., Chen, G., Duan, S., Feng, G.: A memristor-based chaotic system with boundary conditions. In: Memristor Networks, pp. 351–364. Springer International Publishing (2014)

Memristor Crossbar Array for Image Storing

Ling Chen[1], Chuandong Li[1,*], Tingwen Huang[2], Shiping Wen[3],
and Yiran Chen[4]

[1] The College of Electronic and Information Engineering,
Southwest University, Chongqing 400715, China
[2] Texas A & M University at Qatar, Doha 5825, Qatar
[3] School of Automation, Huazhong University of Science and Technology,
Wuhan 430074, China
[4] Electrical and Computer Engineering, University of Pittsburgh, PA15261, USA
cdli@swu.edu.cn

Abstract. This letter uses image overlay technique on memristor cross-bar array (MCA) structure for image storing. Different programming circuits with time slot techniques are designed for the MCA consisting of the nonlinear HP memristor (HPMCA) and the MCA composed of the piece-wise linear threshold memristor (TMCA). The experiment results indicate that the HPMCA has a better performance, the TMCA is more practical in the industrial implementation. As a conclusion, the MCA made up of the memristor with both the nonlinear drift boundary property and the threshold property is preferred for image overlay.

Keywords: Memristor, CMOS Unit, Time Slot, Image Overlay.

1 Introduction

Memristor is a nano scale device that has an ability to remember its history via the modulation of the memristance[1, 2]. Memristor-based systems have been applied in many practical fields such as artificial neural network [3–6], image processing[7–9], and neuromorphic circuits [10, 11]. However its properties are greatly affected by the fabricating material and fabricating process[12]. A lot of mathematical models are proposed to describe the behavior of memristors [11, 13–15], two most classic ones are the HP memristor (HPM) model [14] and the threshold model (TM)[11]. The dynamic of the HPM is described as:

$$r_m(t) = r_{off} + (r_{on} - r_{off}) x(t), \quad x \in (0, 1), \tag{1}$$

$$\dot{x}(t) = -\frac{\mu_v r_{on}}{D^2} i(t) f(x) = li(t) f(x), \quad l = -\mu_v r_{on}/D^2, \tag{2}$$

where r_m is the memristance, r_{off} is the maximum memristance, r_{on} is the minimum memristance. x, corresponding to the memristance according to eq.(1), is the position of the drift boundary. D is the length of the device, μ_v is the drift

---

* Corresponding author.

© Springer International Publishing Switzerland 2015
X. Hu et al. (Eds.): ISNN 2015, LNCS 9377, pp. 166–173, 2015.
DOI: 10.1007/978-3-319-25393-0_19

factor, and i is the current through the memristor, $f(x)$ is the window function. The HPM has an advantage in describing the nonlinear behavior of memristor due to the existence of the window function, but it does not capture the threshold characteristic of memristor. The TM can make up for the deficiency [11]:

$$\dot{r}_m(t) = (\beta v_m + 0.5(\alpha - \beta)[|v_m + v_t| - |v_m - v_t|]) \theta (r_m/r_{on} - 1) \theta (1 - r_m/r_{off}),$$
(3)

where $\theta(\cdot)$ is the step function that promises the memristance changing between r_{on} and r_{off}, α and β are the variation rates of memristance at $|v_m| \leq v_t$ and $|v_m| > v_t$ respectively, v_m is the voltage dropping on memristor, v_t is the threshold voltage.

One valuable application of memristor is to serve as the memory device [6, 16, 17]. However with the increasing storage levels, it's getting more difficult to program the memristor into the expected value. In [18], the colorful image is stored in the MCAs by direct pulse signal, but it didn't build the circuit protecting solutions for the nano scale environment. In [16], a functional hybrid MCA/CMOS system is succeed in storing binary images, and achieved a ten-level color image storing by connecting different resistor in series. But the half selected memristors (memristors share the same row line or column line with the selected memristors) will enlarge the series resistor in the real circuit, and lead to a large noise. Besides, the inconvenience of changing the series resistors makes it difficult to program the large scale MCA. In this letter, we choose the controlled pulse circuits, which is ease of integration, to grogram the MCAs, and designed the time slot based circuit to protect the circuit, further combined the image overlay technique in the writing algorithm, to decrease the noise caused by the sneak path(the unexpected circuits)[19]. This method is tested on different MCAS (here are the HPMCA and the TMCA) and especially performs well on the linear processing HPMCA. The experiment results indicate that the crossbar array consisting of nonlinear memristor with threshold property will be the best choice for the image storing by this method.

2   Memristor Crossbar Array for Image Overlay

A MCA comprises two sets of conductive parallel wires intersecting each other perpendicularly. The intersections (or crosspoints) are separated by memristor. The memristance has an counterpart with the pixel value. For binary images, the memristance is divided into the high part and the low part, which correspond to "1" and "0" respectively. For 8-bit gray images, the memristance is divided into 256 intervals to indicate 256 gray levels.

Here we take the MCA in Fig.1(a) for image storing. Set the row wire as the positive electrode, and the column wire as the negative electrode. Different programming circuits are designed for different MCAs, the HMCA in Fig.1(b) and the TMCA in Fig.1(d). Each row connected to a selecting circuit for selecting the writing row, each column connected to a CMOS unit for programming the memristor. The MCA is operated row by row, so for one image, the more rows it has, the more time it needs. The number of the CMOS unit is the same with the column, and thus the decrease of the columns will lead to the decrease of the CMOS units. Note that the function of the row and the column can be reversed. Due to the single selected row, each CMOS unit processes one memristor each time. When reading the stored image, a read voltage is applied to the selected row, and other unselected rows are connected to ground. When storing the image, different signals are applied in the selected row and unselected rows. Thus, the image reading and image writing can be distinguished by signals.

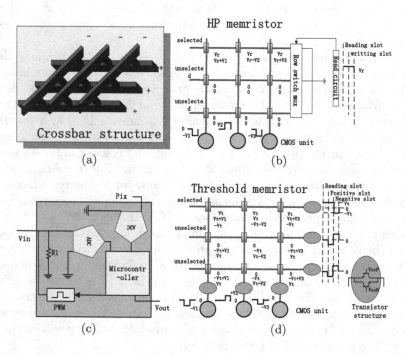

Fig. 1. A new structure combined the memristor crossbar arrays with CMOS units for image storing.

The CMOS unit is the most complex part of the system. It mainly includes the Analog To Digital Converter (ADC), pulse generator, and the micro-controller as shown in Fig.1(c). It receives the input pixel signal p_{in} and obtains the input pixel value. At the mean time, it obtains the memristance r_m through the reading current i_{pix}, and gets the stored pixel value through a floor function $p_m = \lfloor i_{pix} \rfloor$. In the next step, the CMOS unit codes the pulse signal v_b in the column by the image overlay technique, and programs the MCA to the mean value between the input image and the stored image. According to the image overlay theory, the more time one image is programmed, the higher storing accuracy it will get. That is $\Delta r_m(n) = (f(p_{in}) - r_m)/(n+1)$, $f(p_{in})$ is the corresponding memristance of the input pixel, n is the programming times, $n+1$ means that the initial memristance is caculated, and it is obvious that $\lim_{n \to \infty} \Delta r_m(n) = 0$.

In the following we will present the detail programming operations of the HPMCA and the TMCA. Fig.1(b) exhibits the programming structure of the HPMCA. The CMOS unit in each column programs the memristor every $2T$ s, T is one slot time of the pulse signal. The first time slot with a zero amplitude voltage is designed for reading. The second is used for writing, and the amplitude is calculated according to the mean error between the input pixel value and the stored pixel value. Row switch mux is used for selecting the target row. In the selected row, a read voltage signal v_r is always applied, and all the unselected rows are grounded. The final voltage v_m dropping on each memristor in each time slot is clear marked in the MCA. The parameters of the memristor model in the HPMCA are set as: $f(x) = 1 - \left[(x-0.5)^2 + 0.75\right]^p$ [13], $D = 10$nm, $\mu_v = e - 14 m^2 s^{-1} v^{-1}$, $r_{on} = 0.1$kΩ, $r_{off} = 1.6$kΩ, $p = 20$. The read voltage is $v_r = 0.1$V, $T = 0.01s$. For computing convenience, turn the nonlinear HP memristor model into a computable model by assuming $f(x) = 1$. However, due to $f(x) \leq 1$ in the physical memristor, the feedback voltage should be larger than the calculated result. We twice the calculated value to make sure the memristance can change enough or more than Δr_m. This is the linearizing process of the HP memristor. So here the programming pulse signal v_b in the column is:

$$v_b(n) = \Delta r_m (\Delta r_m + 2r_m(n))/(lT(r_{on} - r_{off})). \tag{4}$$

The programing operation of the HPMCA is theoretically practical, in the real nano scale circuit, the unselected nano-scale memristor will be affected by the electric field, and the system would be destroyed by the sneak path. For protecting the circuit, the threshold property of memristor should be utilized. The programming structure of the TMCA is shown in Fig.1(d). Each column pulse signal has three time slots: the reading slot(RS) for reading memristance, the positive slot(PS) for increasing memristance, and the negative slot(NS) for decreasing memristance (if consider the effect of reading voltage, a fourth time slot can be added as a negative reading voltage slot, here we ignore it). Each time slot has a different voltage polarity, a zero voltage is coded in the RS, a negative voltage is coded in the PS to increase memristance, and a positive voltage is coded in the NS to decrease memristance. The selected and unselected rows are also coded with different voltage signals as shown in Fig.1(d). In the selected row, the signal in the RS is v_r, in the PS is v_t, in the NS is $-v_t$. In the unselected row, the signal in the RS is 0V, in the PS is $-v_t$, in the NS is v_t. The separated signal in each row or column is below the threshold, but the combination of the row and the column signals will keep the voltage dropping on the selected memristors above the threshold while others are below the threshold, and that means the selected memristors are programmed while the unselected memristors are protected. This is also clearly proved by the marked voltage v_m in each time slot for each memristor in the TMCA in Fig.1(d).

For suppressing the sneak path, a two-transistor structure is added in each row and column as shown in the inset of Fig.1(d). In the positive time slot, the transistors VccP in the rows and the VccN in the columns are activated for passing the positive current. In the negative time slot, the transistors VccN in the rows and the VccP in the columns are activated for passing the negative current. All four transistors in the selected row and column are activated in the reading slot. And in other cases, all transistors are inactivated. In this way, the sneak path is largely restrained.

The operation of the CMOS unit in the TMCA is similar with the HPMCA, but the period of the CMOS unit is $3T$ s. The parameters of the TM are set as: $\alpha = 0$, $\beta = 1275\text{k}\Omega/(\text{V} \cdot \text{s})$, $v_t = 2\text{V}$, $r_{on} = 0.1\text{k}\Omega, r_{off} = 25.7\text{k}\Omega$. Each interval is 100Ω. The threshold voltage is the read voltage. The backward pulse v_b can be observed by

$$v_b = -\Delta r_m/\beta T. \tag{5}$$

3   Simulation on the Image Storing

In this section, simulink and numerical simulations are taken to present the storing process. Because of the limitation of condition, the memristor and the CMOS unit are emulated by the MATLAB. Take three memristors in the first row of the MCA in Fig.1 as the selected memristors. Train each memristor in the selected row 20 times and the input pixel value sequences is:

$$\begin{bmatrix} 255 & 0 & 255 & 0 & 255 & 0 & 255 & 0 & 255 & 0 & 255 & 0 & 255 & 0 & 255 & 0 & 255 & 0 & 255 & 0 \\ 76 & 12 & 25 & 211 & 178 & 82 & 244 & 9 & 113 & 98 & 196 & 204 & 48 & 126 & 115 & 166 & 182 & 194 & 71 & 175 \\ 108 & 108 & 108 & 108 & 108 & 108 & 108 & 108 & 108 & 108 & 108 & 108 & 108 & 108 & 108 & 108 & 108 & 108 & 108 & 108 \end{bmatrix}^{\text{T}}$$

For the comparison, set the three initial memristances in TMCA as M11 = 100ohm, M12 = 7700ohm, M13 is random, and in HPMCA, M12 = 4838ohm, M11, M13 is random (Mij means the memristor in the i-th row and j-th column). From Fig.2(a)and 2(b), it is easy to find that all the memristances are convergent to the mean values, but different values of the HPMCA and the TMCA is convergent to. The TMCA is completely matching with the mean values, but the HPMCA is not. Turn memristance of M12 into the corresponding pixel value as shown in Fig.2(c), it is obvious that the TMCA is highly consistent with the calculated results but the HPMCA is always slightly fluctuated (sometimes the error is as high as 34 gray levels). The error of the HPMCA comes from the linearized calculation of the HPM, and this error can be reduced by increasing the writing operation. Both in the HPMCA and the TMCA, the stored pixel in M13 gets closer to 108 along with the increasing writing times. But the M13 in the HPMCA convergents to 107 more quickly than the TMCA. These phenomena not only verify the effectiveness of the noise reduction function of the image overlay technique, but also indicate that the image overlay technique is more effective for the memristor with a complex nonlinear behavior. In Fig. 2(d), the time slot character of the pulse signal is clear exhibited, and in different time slot, voltage with different amplitude is also generated in the CMOS unit. Because of the time slot character, a high requirement for time control and device accuracy is need in the industry implementation. Thus the CMOS unit is a big

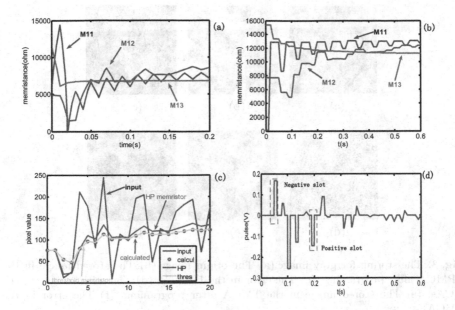

Fig. 2. The simulation of the system. (a) The memristance curves of the HPMCA. (b) The memristance curves of the TMCA. (c) The comparison between input pixels, the stored pixels and the calculated results of M12 (d) The column pulse signals in the CMOS unit for M12 in the TMCS.

challenge in the implementation of this system, and a further optimization is still need.

Fig. 3 is the numerical simulation result of the image storing in the MCA with a random initial memristance. Fig.3(a) is the image to be stored, Fig.3(d) is the image transformed from the initial memristances in the HPMCA, Fig.3(b) and 3(e) is the image stored in the HPMCA and the TMCA after programming 5 times. We can see that the image stored in HPMCA is obvious better than the image in TMCA. Fig.3(c) and 3(f) are the error between Fig. 3(a) and Fig. 3(b) and Fig. 3(e). The error is mapping into a gray image, the lighter the color is, the larger the error is. The maximum error in the HPMCA and the TMCA are 68 (only one pixel) and 37 gray levels respectively, and the rate of pixel which has an error over 5 gray levels is 0.07% and 40.22%. The numerical simulation results further confirmed that the programming method proposed in this letter has a better image storing performance for the HPMCA.

However different from the image storing performance, the circuit designed for the TMCA is more practical in the industry. Considering the advantage of each MCA, a physical memristor combined the complex nonlinear drift boundary behavior and the threshold character [12] should be taken in the physical implementation. Thus, in the physical MCA, the programming circuit for the TMCA would be adopted, but a good image storing effect like the HPMCA would be gotten.

(a)   (b)   (c)

(d)   (e)   (f)

Fig. 3. The storing for gray image.(a) The origin image.(b) The stored image in the HPMCA after programing. (c) The error in the HPMCA.(d) The initial image in the MCAs. (e) The stored image in the TMCA after programing. (f) The error in the TMCA.

4   Conclusions

Different programing operations are designed for the reading and writing of the crossbar arrays consisting of the HP memristor and the threshold memristor. The time slot based on-chip pulse and the image overlay technique are introduced in the CMOS units to control the programming of memristor. The new method has a better noise tolerance ability for image storing in the HPMCA. Both the simulink and numerical simulations demonstrate the efficiency of this system in the image storing.

Acknowledgements. This publication was made possible by NPRP grant NPRP 4-1162-1-181 from the Qatar National Research Fund (a member of Qatar Foundation). This work was supported by NSF-1202225 from US National Science Foundation, Natural Science Foundation of China (grant no: 61374078) at the same time. It was also supported by the Fundamental Research Funds for the Central Universities under Grant XDJK2015C079 and Grant SWU115015.

References

1. Chua, L.O.: Memristor: the missing circuit element. IEEE Transactionson Circuit Theory 18, 507–519 (1971)
2. Strukov, D.B., Snider, G.S., Stewart, D.R., Williams, R.S.: The missing memriastor found. Nature 452, 80–83 (2008)

3. Wen, S.P., Zeng, Z.G., Huang, T.W., Zhang, Y.D.: Exponential lag adaptive synchronization of memristive neural networks and applications in Pseudo-random generators. IEEE Transactions on Fuzzy Systems 22(6), 1704–1713 (2014)
4. Sun, J.W., Shen, Y.: Quasi-ideal memory system (2014)
5. Bao, H.B., Cao, J.D.: Projective synchronization of fractional-order memristor-based neural networks. Neural Networks 63, 1–9 (2015)
6. Dong, Z., Duan, S., Hu, X., Wang, L., Li, H.: A Novel Memristive Multilayer Feedforward Small-World Neural Network with Its Applications in PID Control. The Scientific World Journal (2014)
7. Prodromakis, T., Toumazou, C.: A review on memristive devices and applications. In: 2010 17th IEEE International Conference on Electronics, Circuits, and Systems, Athens, pp. 934–937 (2010)
8. Wen, S.P., Zeng, Z.G., Huang, T.W., Meng, Q.G.: Lag synchronization of switched neural networks via neural activation function and applications in image encryption. IEEE Transactions on Neural Networks and Learning Systems (2014). doi:10.1109/TNNLS.2014.2387355
9. Chen, L., Li, C.D., Huang, T.W., Chen, Y.R., Wang, X.: Memristor crossbar-based unsupervised image learning. Neural Computing and Applications 25(2), 393–400 (2014)
10. Wen, S.P., Huang, T.W., Zeng, Z.G., Chen, Y.R., Li, P.: Circuit design and exponential stabilization of memristive neural networks. Neural Networks 63, 48–56 (2015)
11. Pershin, Y.V., Ventra, M.D.: Experimental demonstration of associative memory with memristive neural networks. Neural Networks 23, 881–886 (2010)
12. Yang, J.J., Strukov, D.B., Stewart, D.R.: Memristive devices for computing. Nature Nanotechnology (2012). doi:10.1038/NNANO.2012.240
13. Kvatinsky, S., Friedman, E.G., Kolodny, A., Weiser, U.C.: TEAM: ThrEshold Adaptive Memristor Model. IEEE Transactions on Circuits and Systems-I 60, 211–221 (2012)
14. Biolek, Z., Biolek, D., Biolkova, V.: SPICE model of memristor with nonlinear dopant drift. Radio Engeering 18, 210–214 (2009)
15. Chen, L., Li, C., Huang, T., Chen, Y., Wen, S., Qi, J.: A synapse memristor model with forgetting effect. Physics Letters A 377(45), 3260–3265 (2013)
16. Kim, K.H., Gaba, S., Wheeler, D., Cruz-Albrecht, J.M., Hussain, T., Srinivasa, N., Lu, W.: A Functional Hybrid Memristor Crossbar-Array/CMOS System for Data Storage and Neuromorphic Applications. Nano Lett. 12, 389–395 (2011)
17. Bayat, F.M., Shouraki, S.B.: Programming of memristor crossbars by using genetic algorithm. Procedia Computer Science 3, 232–237 (2011)
18. Hu, X., Duan, S., Wang, L., Liao, X.: Memristive crossbar array with applications in image processing. Science China Information Sciences 55(2), 461–472 (2012)
19. Li, H.Q., Liao, X.F., Li, C.D., Huang, H.Y., Li, C.J.: Edge detection of noisy images based on cellular neural networks. Commun. Nonlinear Sci. Numer. Simul. 16(9), 3746–3759 (2011)

Lagrange Stability for Memristor-Based Neural Networks with Time-Varying Delay via Matrix Measure

Sanbo Ding[1], Linlin Zhao[2], and Zhanshan Wang[1,*]

[1] School of Information Science and Engineering,
Northeastern University, Shenyang 110819, China
[2] Liaoning Province Product Quality Supervision Procuratorate,
Shenyang 110004, China
{dingsanbo,zhanshan_wang}@163.com

Abstract. In this paper, we study the global exponential stability in Lagrange sense for memristor-based neural networks (MBNNs) with time-varying delays. Based on the nonsmooth analysis and differential inclusion theory, matrix measure technique is employed to establish some succinct criteria which ensure the Lagrange stability of the considered memristive model. In addition, the new proposed criteria are very easy to verify, and they also enrich and improve the earlier publications. Finally, two example are given to demonstrate the validity of the results.

Keywords: Memristor-based Neural Networks, Lagrange Stability, Matrix Measure.

1 Introduction

Memristive neurodynamic systems have received considerable attention over the last few years. Memristor-based neural networks (MBNNs) have demonstrated high efficiency in numerous applications, and it would be an interesting and important research topic in many fields [1–9]. Differently from traditional neural systems [10, 11, 15], MBNNs are characterized by state-dependent nonlinear system families [3, 5]. Such system family can reveal coexisting solutions, jumped, transient chaos of rich and complex nonlinear behaviors [1]. Thus, it is more difficient to analyze the dynamic of memristor-based system than the traditional ones.

Recently, many researchers focus on the global asymptotic or exponential stability for MBNNs with a unique equilibrium point or unique periodic orbit [6, 9]. In many applications, however, monostable neural networks have been found

* Foundation This work was supported by the National Natural Science Foundation of China (Grant Nos. 61473070, 61433004), the Fundamental Research Funds for the Central Universities (Grant Nos. N130504002 and N130104001), and SAPI Fundamental Research Funds (Grant No. 2013ZCX01).

© Springer International Publishing Switzerland 2015
X. Hu et al. (Eds.): ISNN 2015, LNCS 9377, pp. 174–181, 2015.
DOI: 10.1007/978-3-319-25393-0_20

computationally restrictive and multistable dynamics are essential to handle important neural computations desired [2, 12]. Unlike Lyapunov stability, Lagrange stability refers to the stability of the total system, rather than the stability of equilibria. A Lagrange stable system may have multistable property because the Lagrange stability is considered on the basis of the boundedness of solutions and the existence of global attractive sets. Also, with regard to Lagrange stability, outside the global attractive sets, there is no equilibrium point, periodic state, almost periodic state, or chaos attractor [1, 2, 12, 13]. So, it is a significative and important work to study the Lagrange stability of MBNNs.

Motivated by the above discussions, in this paper, we shall consider the Lagrange stability of MBNNs. The dynamic analysis here adopts theories of differential inclusions and set-valued maps to handle MBNNs with discontinuous right-hand side. The activation functions in this paper are assumed to be bounded. By utilizing matrix measure technique, some new criteria which ensure the lagrange stability of the considered MBNNs are derived. The global exponential attractive sets can be directly derived from the parameters of the MBNNs, and the stability criteria are very easily verified.

Throughout this paper, R^n is the n-dimensional Euclidean space. $co[P]$ denotes the closure of the convex hull of set P, and $co[a, b]$ denotes the closure of the convex hull generated by numbers a and b. $a_{ij}^+ = \max\{a_{ij}^*, a_{ij}^{**}\}$, $a_{ij}^- = \min\{a_{ij}^*, a_{ij}^{**}\}$, $b_{ij}^+ = \max\{b_{ij}^*, b_{ij}^{**}\}$, $b_{ij}^- = \min\{b_{ij}^*, b_{ij}^{**}\}$. $a_{ij} = \max\{|a_{ij}^*|, |a_{ij}^{**}|\}$, $b_{ij} = \max\{|b_{ij}^*|, |b_{ij}^{**}|\}$, $A = (a_{ij})_{n \times n}$, $B = (b_{ij})_{n \times n}$. For a given constant $\mathbf{T} > 0$, $\mathbf{C_T}$ is defined as the subset $\sigma \in \mathbf{T} : ||\sigma|| \leq \mathbf{T}$. Let $\mathbf{C}$ be the set of all nonnegative functionals $\mathbf{K} : \mathbf{C} \to [0, +\infty)$, mapping bounded sets in $\mathbf{C}$ into bounded sets in $[0, +\infty)$. For any initial condition $\varphi \in \mathbf{C}$, the solution of the considered system that starts from the initial condition φ will be denoted by $x(t; \varphi)$.

2   Problem Formulation and Preliminaries

Consider the following MBNNs with time-varying delay:

$$\dot{x}(t) = -Cx(t) + A(x(t))f(x(t)) + B(x(t - \tau(t)))g(x(t - \tau(t))) + I(t), \quad (1)$$

where $x(t) = (x_1(t), x_2(t), \cdots, x_n(t))^T$ denotes the state variable of neurons; $I(t) = (I_1(t), I_2(t), \cdots, I_n(t))^T$ is the external inputs satisfying $|I_i(t)| \leq I_i$; $\tau(t)$ denotes system delay satisfying $0 \leq \tau(t) \leq \tau$; $f(x(t)) = (f_1(x_1(t)), f_2(x_2(t)) \cdots, f_n(x_n(t)))^T$ and $g(x(t - \tau(t))) = (g_1(x_1(t - \tau(t))), g_2(x_2(t - \tau(t))) \cdots, g_n(x_n(t - \tau(t))))^T$ denote the continuous activation functions; $C = diag(c_1, c_2, \cdots, c_n)$ is a diagonal matrix with positive entries; $A(x(t)) = (a_{ij}(x_j(t)))_{n \times n}$ and $B(x(t - \tau(t))) = (b_{ij}(x_j(t - \tau(t))))_{n \times n}$ are the memristive synaptic weight matrices with

$$a_{ij}(x_j(t)) = \begin{cases} a_{ij}^*, & |x_j(t)| \leq \mathcal{X}_j, \\ a_{ij}^{**}, & |x_j(t)| > \mathcal{X}_j, \end{cases} \quad b_{ij}(x_j(t - \tau(t))) = \begin{cases} b_{ij}^*, & |x_j(t - \tau(t))| \leq \mathcal{X}_j, \\ b_{ij}^{**}, & |x_j(t - \tau(t))| > \mathcal{X}_j, \end{cases}$$

in which switching jumps $\mathcal{X}_j > 0$, $c_i > 0$, a_{ij}^*, a_{ij}^{**}, b_{ij}^*, b_{ij}^{**} are constants.

Throughout this paper, we assume that the activation functions $f_i(s)$ and $g_i(s)$ are continuous and bounded, i.e., there exist positive constants F_i and G_i such that

$$|f_i(s)| \le F_i, \quad |g_i(s)| \le G_i, \quad \forall s \in R, \quad i = 1, 2, \cdots, n. \tag{2}$$

Notice that the system (1) is differential equation with discontinuous right-hand sides, and based on the theories of set-valued maps and differential inclusions, if $x(t)$ is the solution of (1) in the sense of Filippov, then

$$\dot{x}(t) \in - Cx(t) + co[A(x(t))]f(x(t)) + co[B(x(t-\tau(t)))]g(x(t-\tau(t))) + I(t), \tag{3}$$

where $co[A(x(t))] = (co(a_{ij}(x_j(t))))_{n \times n}$, $co[B(x(t-\tau(t)))] = (co(b_{ij}(x_j(t-\tau(t)))))_{n \times n}$, with

$$co[a_{ij}(x_j(t))] = \begin{cases} a_{ij}^*, & |x_j(t)| < \mathcal{X}_j, \\ [a_{ij}^-, a_{ij}^+], & |x_j(t)| = \mathcal{X}_j, \\ a_{ij}^{**}, & |x_j(t)| > \mathcal{X}_j, \end{cases}$$

$$co[b_{ij}(x_j(t-\tau(t)))] = \begin{cases} b_{ij}^*, & |x_j(t-\tau(t))| < \mathcal{X}_j, \\ [b_{ij}^-, b_{ij}^+], & |x_j(t-\tau(t))| = \mathcal{X}_j, \\ b_{ij}^{**}, & |x_j(t-\tau(t))| > \mathcal{X}_j. \end{cases}$$

or equivalently, there exist measurable functions $\hat{A}(t) \in co[A(x(t))]$ and $\hat{B}(t) \in co[B(x(t-\tau(t)))]$, such that

$$\dot{x}(t) = - Cx(t) + \hat{A}(t)f(x(t)) + \hat{B}(t)g(x(t-\tau(t))) + I(t) . \tag{4}$$

Now we define some concepts and lemmas that are needed later.

Definition 1. *[1] The trajectory of MBNN (1) is said to be uniformly stable in Lagrange sense (or uniformly bounded), if for any $H > 0$, there exists a constant $K = K(H) > 0$ such that $|x(t; \varphi)| < K$ for all $\varphi \in C_H$ and $t \ge 0$.*

Definition 2. *[1] If there exist a radially unbounded and positive definite function $V(x)$, a functional $K \in C$, positive constants ℓ and α, such that for any solution $x(t) = x(t; \varphi)$ of MBNN (1), $V(x(t)) > \ell$, $t \ge 0$, implies*

$$V(x(t)) - \ell \le K(\varphi)exp\{-\alpha t\},$$

then the trajectory of MBNN (1) is said to be globally exponentially attractive with respect to V, and the compact set $\Omega := x \in R^n | V(x) \le \ell$ is called to be a globally exponentially attractive set of MBNN (1).

Definition 3. *[1] The trajectory of MBNN (1) is called globally exponentially stable in Lagrange sense, if it is both uniformly stable in Lagrange sense and globally exponentially attractive.*

Definition 4. *[14] Suppose $A = (a_{ij})_{n \times n}$ is a real matrix, then the matrix measure of A is defined as*

$$\mu_p(A) = \lim_{h \to 0^+} \frac{\|I + hA\|_p - 1}{h}$$

where I is a $n \times n$ identity matrix, $p = 1, 2, \infty$, and $\| \cdot \|_p$ is the corresponding induced matrix norm.

When the matrix norm $\|A\|_1 = max_j \sum_{i=1}^n |a_{ij}|$, $\|A\|_2 = \sqrt{\lambda_{max}(A^T A)}$, $A_\infty = max_i \sum_{j=1}^n |a_{ij}|$, we can obtain the matrix measure $\mu_1(A) = max_j \{a_{jj} + \sum_{i=1, i \neq j}^n |a_{ij}|\}$, $\mu_2(A) = \frac{1}{2}\lambda_{max}(A^T + A)$, $\mu_\infty(A) = max_i \{a_{ii} + \sum_{j=1, j \neq i}^n |a_{ij}|\}$.

Lemma 1. *[12] Let $G \in \mathbf{C}([t_0, +\infty), R)$, and there exist positive constants κ_1 and κ_2 such that $D^+ G(t) \leq -\kappa_1 G(t) + \kappa_2$, $t \geq t_0$, then*

$$G(t) - \frac{\kappa_2}{\kappa_1} \leq \left(G(t_0) - \frac{\kappa_2}{\kappa_1} \right) exp\{-\kappa_1(t - t_0)\}, \qquad t \geq t_0 .$$

In particular, if $G(t) \geq \kappa_2/\kappa_1$, $t \geq t_0$, then $G(t)$ exponentially approaches κ_2/κ_1 as t increases.

3   Main Results

Theorem 1. *Assume that the activation functions $f(\cdot)$ and $g(\cdot)$ satisfy condition (2), then the trajectory of MBNN (1) is globally exponentially stable in Lagrange sense. In addition, there exists a matrix measure $\mu_p(\cdot)(p = 1, 2, \infty)$ such that global exponential attractive sets $\Omega_p(p = 1, 2, \infty)$ of MBNN (1) can be estimated as follows:*

$$\Omega_p = \left\{ x \in R^n | \|x\|_p \leq \frac{M_p}{L_c} \right\}, p = 1, 2, \infty . \tag{5}$$

where $M_p = \|A\|_p F + \|B\|_p \|G + \|I\|_p$, $L_c = -\mu(-C)$, $F = max_{i=1,2,\cdots,n}\{F_i\}$, $G = max_{i=1,2,\cdots,n}\{G_i\}$.

Proof. We first prove that the trajectory of MBNN (1) is uniformly stable in Lagrange sense.

Consider the following function:

$$V(x(t)) = \|x(t)\|_p. \tag{6}$$

Evaluating the upper right Dini derivative of V along the trajectory of (3) or (4) gives

$$
\begin{aligned}
D^+V(x(t)) =\ & \overline{\lim_{h\to0^+}}\frac{\|x(t+h)\|_p - \|x(t)\|_p}{h} = \overline{\lim_{h\to0^+}}\frac{\|x(t)+h\dot{x}(t)+o(h)\|_p - \|x(t)\|_p}{h} \\
\in\ & \overline{\lim_{h\to0^+}}\frac{1}{h}\Big\{\|x(t)+h[-Cx(t)+co[A(x(t))]f(x(t)) \\
& + co[B(x(t-\tau(t)))]g(x(t-\tau(t)))+I(t)]+o(h)\|_p - \|x(t)\|_p\Big\} \\
=\ & \overline{\lim_{h\to0^+}}\frac{1}{h}\Big\{\|x(t)+h[-Cx(t)+\hat{A}(t)f(x(t)) \\
& + \hat{B}(t)g(x(t-\tau(t)))+I(t)]+o(h)\|_p - \|x(t)\|_p\Big\} \\
\le\ & \overline{\lim_{h\to0^+}}\frac{\|I+h(-C)\|_p - 1}{h}\|x(t)\|_p \\
& + \|\hat{A}(t)f(x(t))\|_p + \|\hat{B}(t)g(x(t-\tau(t)))\|_p + \|I(t)\|_p \\
\le\ & \mu_p(-C)\|x(t)\|_p + \|\hat{A}(t)f(x(t))\|_p + \|\hat{B}(t)g(x(t-\tau(t)))\|_p + \|I(t)\|_p \ .
\end{aligned}
\tag{7}
$$

Notice that the boundedness of activation functions, and $|I_i(t)| \le I_i$. One can obtain from the above inequality that

$$
\begin{aligned}
D^+V(x(t)) \le\ & \mu_p(-C)\|x(t)\|_p + \|A\|_p\|f(x(t))\|_p + \|B\|_p\|g(x(t-\tau(t)))\|_p + \|I(t)\|_p \\
\le\ & \mu_p(-C)\|x(t)\|_p + \|A\|_p F + \|B\|_p\|G + \|I\|_p \\
=\ & -L_c V(x(t)) + M_p \ .
\end{aligned}
\tag{8}
$$

where $L_c = -\mu_p(-C)$, $M_p = \|A\|_p F + \|B\|_p\|G + \|I\|_p$.
By Lemma.1, we have

$$
V(x(t)) - \frac{M_p}{L_c} \le \left[V(x(0)) - \frac{M_p}{L_c}\right]\exp\{-L_c t\}, \quad t \ge 0 \ .
\tag{9}
$$

Then

$$
\|x(t)\|_p - \frac{M_p}{L_c} \le \left[\|x(0)\|_p - \frac{M_p}{L_c}\right]\exp\{-L_c t\} \le \left[\|x(0)\|_p - \frac{M_p}{L_c}\right] \ .
$$

This means $\|x(t)\|_p \le \|x(0)\|_p$. This immediately implies the uniform boundedness of the solutions of MBNN (1). Hence, the trajectory of MBNN (1) is uniformly stable in Lagrange sense.

In addition, noticing that $V(x(t)) - \frac{M_p}{L_c} \le V(x(0)) := \mathbf{K}(\varphi)$, then $\mathbf{K} \in \mathbf{C}$, and from (9) it implies that

$$
V(x(t)) - \frac{M_p}{L_c} \le \mathbf{K}(\varphi)\exp\{-L_c t\}, \quad t \ge 0 \ .
\tag{10}
$$

Through Definition 2, MBNN (1) is globally exponentially attractive and Ω_p is a globally exponentially attractive set. This proves the global exponential stability in Lagrange sense of the trajectory of MBNN (1). The proof is completed.

When $g_i = 0 (i = 1, 2, \cdots, n)$, system (1) changes into:

$$\dot{x}(t) = -Cx(t) + A(x(t))f(x(t)) + I(t), \tag{11}$$

Corollary 1. *Assume that the activation function $g(\cdot)$ satisfies condition (2), then the trajectory of network (11) is globally exponentially stable in Lagrange sense. In addition, there exists a matrix measure $\mu_p(\cdot)(p = 1, 2, \infty)$ such that global exponential attractive sets $\Omega_p(p = 1, 2, \infty)$ of MBNN (1) can be estimated as follows:*

$$\Omega_p = \left\{ x \in R^n \Big| \|x\|_p \leq \frac{\widetilde{M}_p}{L_c} \right\}, p = 1, 2, \infty .$$

where $\widetilde{M}_p = \|A\|_p \|F + \|I\|_p$, $L_c = -\mu(-C)$, $G = \max_{i=1,2,\cdots,n}\{G_i\}$.

System (1) includes a special case of MBNNs as follows:

$$\dot{x}(t) = -Cx(t) + B(x(t - \tau(t)))g(x(t - \tau(t))) + I(t), \tag{12}$$

Corollary 2. *Assume that the activation function $g(\cdot)$ satisfies condition (2), then the trajectory of network (12) is globally exponentially stable in Lagrange sense. In addition, there exists a matrix measure $\mu_p(\cdot)(p = 1, 2, \infty)$ such that global exponential attractive sets $\Omega_p(p = 1, 2, \infty)$ of MBNN (1) can be estimated as follows:*

$$\Omega_p = \left\{ x \in R^n \Big| \|x\|_p \leq \frac{\widehat{M}_p}{L_c} \right\}, p = 1, 2, \infty .$$

where $\widehat{M}_p = \|B\|_p \|G + \|I\|_p$, $L_c = -\mu(-C)$, $G = \max_{i=1,2,\cdots,n}\{G_i\}$.

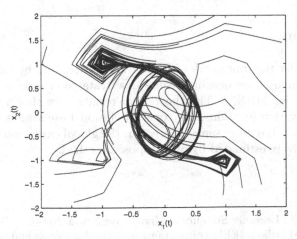

Fig. 1. Transient behaviors of network (13) with 20 initial values.

4   Examples

Example 1. Consider a two-neuron memristor-based neural network model

$$\begin{cases} \dot{x}_1(t) = -x_1(t) + a_{11}(x_1(t))f(x_1(t)) + a_{12}(x_2(t))f(x_2(t)) - sin(t), \\ \dot{x}_2(t) = -0.9x_2(t) + a_{21}(x_1(t))f(x_1(t)) + a_{22}(x_2(t))f(x_2(t)) + sin(t), \end{cases} \quad (13)$$

where $f(s) = tanh(s)$, and

$$a_{11}(x_1(t)) = \begin{cases} 0.1, & |x_1(t)| \le 1, \\ -0.1, & |x_1(t)| > 1, \end{cases} \quad a_{12}(x_2(t)) = \begin{cases} 2, & |x_2(t)| \le 1, \\ -2, & |x_2(t)| > 1, \end{cases}$$

$$a_{21}(x_1(t)) = \begin{cases} -2, & |x_1(t)| \le 1, \\ 2, & |x_1(t)| > 1, \end{cases} \quad a_{22}(x_2(t)) = \begin{cases} 0.1, & |x_2(t)| \le 1, \\ -0.1, & |x_2(t)| > 1. \end{cases}$$

Obviously, the activation function is bounded with $F = 1$. By briefly calculating, we obtain $L_c = -\mu(-C) = 0.9$ $M_1 = M_2 = M_\infty = 3.1$, then the global exponential attractive sets can be estimated as follows:

$$\Omega_p = \{x \in R^2 | \|x\|_p \le 3.4444\}, p = 1, 2, +\infty.$$

The simulation result of network (13) with 20 initial values is shown in Fig. 1.

Example 2. Consider Ikeda-type oscillator with memristor characteristics

$$\dot{x}(t) = -cx(t) + b(x(t - \tau(t)))sin(x(t - \tau(t))) \quad (14)$$

where $\tau(t) = e^t/(1 + e^t)$, $c = 1$ and $b(x(t - \tau(t))) = \begin{cases} -1.3, & |x_1(t)| \le 0.8, \\ 1.3, & |x_1(t)| > 0.8. \end{cases}$

From calculating the parameter $M_p = 1.3$ and $L_c = -\mu_p(-c) = 1$. Then we can obtain the global exponential attractive sets as

$$\Omega_p = \{x \in R | \|x\|_p \le 1.3\}, p = 1, 2, +\infty.$$

5   Conclusion

In this paper, under the framework of Filippovs solution, and by using the matrix measure technique, we obtained some new testable criteria to ensure the lagrange stability for MBNNs. The theoretical results have shown that, under the bounded activation functions, the considered model are always globally exponentially stable in Lagrange sense. Moreover, the global exponential attractive sets can be effectively estimated by our proposed method.

References

1. Wu, A., Zeng, Z.: Lagrange Stability of Memristive Neural Networks With Discrete and Distributed Delays. IEEE Transactions on Neural networks and Learning Systems 25, 690–703 (2014)

2. Zhang, G., Shen, Y., Xu, C.: Global Exponential Stability in a Lagrange Sense for Memristive Recurrent Neural Networks With Time-Varying Delays. Neurocomputing 149, 1330–1336 (2015)
3. Ding, S., Wang, Z.: Stochastic Exponential Synchronization Control of Memristive Neural Networks With Multiple Time-Varying Delays. Neurocomputing 162, 16–25 (2015)
4. Wu, H., Li, R., Zhang, X., Yao, R.: Weak, Modified and Function Projective Synchronization of Chaotic Memristive Neural Networks With Time Delays. Neurocomputing 149, 667–676 (2015)
5. Guo, Z., Wang, J., Yang, Z.: Attractivity Analysis of Memristor-Based Cellular Neural Networks With Time-Varying Delays. IEEE Transactions on Neural Networks and Learning Systems 25, 704–717 (2014)
6. Qi, J., Li, C., Huang, T.: Stability of Delayed Memristive Neural Networks With Time-Varying Impulses. Cogn. Neurodyn. 8, 429–436 (2014)
7. Zhang, G., Shen, Y.: New Algebraic Criteria for Synchronization Stability of Chaotic Memristive Neural Networks With Time-Varying Delays. IEEE Transactions on Neural networks and Learning Systems 24, 1701–1707 (2013)
8. Wen, S., Zeng, Z., Huang, T.: Adaptive Synchronization of Memristor-Based Chua's Circuits. Physics Letters A 376, 2775–2780 (2012)
9. Chen, J., Zeng, Z., Jiang, P.: On the Periodic Dynamics of a Memristor-Based Neural Networks With Time-Varying Delays. Information Science 279, 358–373 (2014)
10. Wang, Z., Zhang, H., Jiang, B.: LMI-based approach for global asymptotic stability analysis of recurrent neural networks with various delays and structures. IEEE Transactions on Neural Networks and Learning Systems 22, 1032–1045 (2011)
11. Wang, Z., Liu, L., Shan, Q., Zhang, H.: Stability criteria for recurrent neural networks with time-varying delay based on secondary delay partitioning method. IEEE Transactions on Neural Networks and Learning Systems (2015). doi:10.1109/TNNLS.2014.2387434
12. Liao, X., Luo, Q., Zeng, Z., Guo, Y.: Global Exponential Stability in Lagrange Sense for Recurrent Neural Networks With Time Delays. Nonlinear Analysis: Real World Applications 9, 1535–1557 (2008)
13. Luo, Q., Zeng, Z., Liao, X.: Global Exponential Stability in Lagrange Sense for Neutral Type Recurrent Neural Networks. Neurocomputing 74, 638–645 (2011)
14. Vidyasagar, M.: Nonlinear System Analysis. Prentice-Hall, Englewood Cliffs (1993)
15. Zhang, S., Xia, Y., Zheng, W.: A complex-valued neural dynamical optimization approach and its stability analysis. Neural Networks 61, 59–67 (2015)

Multistability of Memristive Neural Networks with Non-monotonic Piecewise Linear Activation Functions

Xiaobing Nie and Jinde Cao

Department of Mathematics, Southeast University, Nanjing 210096, China
{xbnie,jdcao}@seu.edu.cn

Abstract. In this paper, a general class of non-monotonic piecewise linear activation functions is introduced and then the coexistence and dynamical behaviors of multiple equilibrium points are studied for a class of memristive neural networks (MNNs). It is proven that under some conditions, such n-neuron MNNs can have 5^n equilibrium points located in $\Re^n$, and 3^n of them are locally exponentially stable, by means of fixed point theorem, nonsmooth analysis theory and rigorous mathematical analysis. The investigation shows that the neural networks with non-monotonic piecewise linear activation functions introduced in this paper can have greater storage capacity than the ones with Mexican-hat-type activation function.

Keywords: Memristive neural networks, Multistability, Non-monotonic piecewise linear activation functions.

1 Introduction

Multistability is necessary whenever neural networks are used for implementing an associative memory or for solving other asks in real time in the field of combinatorial optimization, pattern recognition, image processing, and so on. As a result, multistability of conventional recurrent neural networks has attracted considerable attention from many researchers over the last decade [1–4]. It has been well recognized that multistability analysis of neural networks critically depends upon the type of activation functions. However, most of the activation functions employed in multistability analysis are restricted in sigmoidal activation functions, nondecreasing saturated activation functions, and piecewise linear activation functions, which share the common feature that they are all monotonically increasing. Recently, [5] introduced a class of non-monotonic piecewise linear activation function which is called Mexican-hat-type activation function, and investigated the issue of multistability for Hopfield neural networks.

With the inspiration from Mexican-hat-type activation function and in order to increase the storage capacity of neural networks, in this paper, another class of continuous non-monotonic piecewise linear activation functions is introduced as follows (see Fig. 1):

© Springer International Publishing Switzerland 2015
X. Hu et al. (Eds.): ISNN 2015, LNCS 9377, pp. 182–191, 2015.
DOI: 10.1007/978-3-319-25393-0_21

$$f_i(x) = \begin{cases} m_i, & -\infty < x < p_i, \\ l_{i,1}\,x + c_{i,1}, & p_i \leq x \leq r_i, \\ l_{i,2}\,x + c_{i,2}, & r_i < x < q_i, \\ l_{i,3}\,x + c_{i,3}, & q_i \leq x \leq s_i, \\ M_i, & s_i < x < +\infty, \end{cases} \tag{1}$$

where p_i, r_i, q_i, s_i, m_i, M_i, $l_{i,1}$, $l_{i,2}$, $l_{i,3}$, $c_{i,1}$, $c_{i,2}$, $c_{i,3}$ are constants with $-\infty < p_i < r_i < q_i < s_i < +\infty$, $l_{i,1} > 0$, $l_{i,2} < 0$, $l_{i,3} > 0$, $m_i = f_i(q_i)$ and $M_i > f_i(r_i)$, $i = 1, 2, \cdots, n$. It is easy to see that the activation functions f_i is Lipschitz continuous, i.e., $\forall x, y \in \Re$, there exists positive number $\rho_i = \max\{|l_{i,1}|, |l_{i,2}|, |l_{i,3}|\}$ such that

$$|f_i(x) - f_i(y)| \leq \rho_i |x - y|.$$

Memristor (an abbreviation for memory and resistor), as the fourth fundamental two-terminal circuit element, was first postulated by L. O. Chua [6]. It is well known that the memristor exhibits the feature of pinched hysteresis just as the neurons in the human brain have. Because of this important feature, the memristor can remember its past dynamic history. By replacing the resistors with memristors in conventional neural networks, a new memristive neural networks (MNNs) can be constructed. The great potential for exploiting MNNs will help us build a brain-like neural computer to implement the synapses of biological brains. As a prerequisite, the dynamical analysis of MNNs plays an important role in the design of practical memristive neural networks model. During the last few years there has been an increasing research interest in mono-stability analysis and synchronization control of MNNs [7–11]. However, to the best of the authors' knowledge, the multistability of MNNs is seldom considered. It should be pointed out that MNNs are totally different from conventional recurrent neural networks, since MNNs are state-dependent switching systems which are discontinuous dynamical systems, while conventional recurrent neural networks are continuous dynamical systems. Therefore, the research on multistability of MNNs is more complicated and challenging.

Motivated by the above discussions, our main objective in this paper is to study the multistability of MNNs with activation functions (1). More precisely, the main contributions of this paper lie in the following aspects. Firstly, under the framework of Filippov's solution, we present sufficient condition under which the n-neuron MNNs with activation functions (1) can have 5^n equilibrium points located in $\Re^n$, by applying the known fixed point theorem. Secondly, based on rigorous mathematical analysis and the theories of set-valued maps and differential inclusions, we analyze the dynamical behaviors of MNNs with activation functions (1), and show that the addressed MNNs can have 5^n equilibrium points, and 3^n of them are locally exponentially stable. Thirdly, compared with the neural networks with Mexican-hat-type activation function, the MNNs with activation functions (1) have both more total equilibrium points and more locally stable equilibrium points.

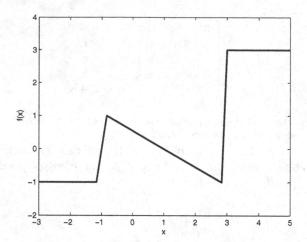

Fig. 1. The configuration of non-monotonic piecewise linear activation functions (1).

2   Model Description and Preliminaries

2.1   Model

In this paper, we consider a class of memristor-based neural networks as follows:

$$\frac{\mathrm{d}x_i(t)}{\mathrm{d}t} = -d_i\,x_i(t) + \sum_{j=1}^{n} a_{ij}\,(x_j(t))\,f_j(x_j(t)) + I_i,\ i = 1, 2, \cdots, n, \qquad (2)$$

where

$$a_{ij}(x_j(t)) = \begin{cases} a_{ij}^*, & |x_j(t)| \le T_j, \\ a_{ij}^{**}, & |x_j(t)| > T_j, \end{cases}$$

in which switching jumps $T_j > 0$, $d_i > 0$, a_{ij}^*, a_{ij}^{**} $(i, j = 1, 2, \cdots, n)$ are all constant numbers, I_i is an external constant input, $f_j(\cdot)$ is the activation function defined in (1).

2.2   Notations

Throughout this paper, solutions of all the systems considered in the following are intended in Filippov's sense. $\dot{x}(t)$ denotes the derivative of $x(t)$. Let $\bar{a}_{ij} = \max\{a_{ij}^*, a_{ij}^{**}\}$, $\underline{a}_{ij} = \min\{a_{ij}^*, a_{ij}^{**}\}$. Given a set $\Omega \subset \Re$, $co[\Omega]$ denotes the closure of the convex hull of Ω. Thus, we have

$$co\,[a_{ij}(x_j(t))] = \begin{cases} a_{ij}^*, & |x_j(t)| < T_j, \\ [\underline{a}_{ij}, \bar{a}_{ij}], & |x_j(t)| = T_j, \\ a_{ij}^{**}, & |x_j(t)| > T_j. \end{cases}$$

$\Re^n$ can be divided into 5^n regions as

$$\Omega = \Bigg\{ \prod_{i=1}^{n} (-\infty, p_i)^{\delta_1^{(i)}} \times [p_i, r_i]^{\delta_2^{(i)}} \times (r_i, q_i)^{\delta_3^{(i)}} \times [q_i, s_i]^{\delta_4^{(i)}} \times (s_i, +\infty)^{\delta_5^{(i)}},$$

$$(\delta_1^{(i)}, \delta_2^{(i)}, \delta_3^{(i)}, \delta_4^{(i)}, \delta_5^{(i)}) = (1,0,0,0,0) \text{ or } (0,1,0,0,0) \text{ or } (0,0,1,0,0)$$

$$\text{or } (0,0,0,1,0) \text{ or } (0,0,0,0,1) \Bigg\}.$$

2.3 Properties and Definitions

By the theories of differential inclusions and set-valued maps [12, 13], the memristor-based neural networks (2) can be written as the following differential inclusion:

$$\frac{dx_i(t)}{dt} \in - d_i\, x_i(t) + \sum_{j=1}^{n} co\, [a_{ij}\, (x_j(t))]\, f_j(x_j(t)) + I_i, \text{ for a.e. } t \geq 0, \quad (3)$$

or equivalently, there exist $\widehat{a}_{ij} \in co\, [a_{ij}\, (x_j(t))]\ (i,j = 1,2,\cdots,n)$ such that

$$\frac{dx_i(t)}{dt} = - d_i\, x_i(t) + \sum_{j=1}^{n} \widehat{a}_{ij}\, f_j(x_j(t)) + I_i, \text{ for a.e. } t \geq 0,\ i = 1,2,\cdots,n. \quad (4)$$

Definition 1. A function $x(t) = (x_1(t), \cdots, x_n(t))^T$ is a solution of (2) in the sense of Filippov, if $x(t)$ is an absolutely continuous function on any compact interval of $[0, +\infty)$ and satisfies the differential inclusion (3).

Definition 2. An equilibrium point of (2) is a constant vector $x^* \in \Re^n$ that satisfies

$$0 \in - d_i\, x_i^* + \sum_{j=1}^{n} co[a_{ij}(x_j^*)]\, f_j(x_j^*) + I_i,\ i = 1,2,\cdots,n, \quad (5)$$

or equivalently, for $i,j = 1,2,\cdots,n$, there exist $\widehat{a}_{ij} \in co[a_{ij}(x_j^*)]$ such that

$$-d_i\, x_i^* + \sum_{j=1}^{n} \widehat{a}_{ij}\, f_j(x_j^*) + I_i = 0,\ i = 1,2,\cdots,n. \quad (6)$$

Lemma 1. Assume that activation functions f_j are Lipschitz continuous on $\Re$ with Lipschitz constants $\rho_j > 0$. If $f_j(\pm T_j) = 0\,(j = 1,2,\cdots,n)$, then

$$|co[a_{ij}(x_j(t))]f_j(x_j(t)) - co[a_{ij}(y_j(t))]f_j(y_j(t))| \leq A_{ij}\,\rho_j\,|x_j(t) - y_j(t)| \quad (7)$$

hold for $i,j = 1,2,\cdots,n$, where $A_{ij} = \max\{|a_{ij}^*|, |a_{ij}^{**}|\}$.

It is obvious that the set-valued map

$$x_i(t) \multimap -d_i\, x_i(t) + \sum_{j=1}^{n} co\, [a_{ij}\, (x_j(t))]\, f_j(x_j(t)) + I_i$$

has nonempty compact convex values. Furthermore, it is upper-semicontinuous. Thus, the local existence of a solution $x(t)$ with initial condition $x(0)$ can be guaranteed from [12]. Moreover, since the activation functions f_j are bounded and Lipschitz continuous, the local solution $x(t)$ can be extended to the interval $[0, +\infty)$.

3   Main Results

In this section, the multistability of MNNs (2) with activation functions (1) is investigated. First of all, we give the following theorem on the coexistence of multiple equilibrium points for MNNs (2) by applying the known fixed point theorem under the framework of Filippov's solution.

Theorem 1. If the following conditions hold for all $i = 1, 2, \cdots, n$:

$$- d_i\, p_i + \max\{\underline{a}_{ii} u_i, \bar{a}_{ii} u_i\} + \sum_{j \neq i, j=1}^{n} \max\{\underline{a}_{ij} u_j, \underline{a}_{ij} v_j, \bar{a}_{ij} u_j, \bar{a}_{ij} v_j\}$$

$$+ I_i < 0, \tag{8}$$

$$- d_i\, r_i + \min\{\underline{a}_{ii} f_i(r_i), \bar{a}_{ii} f_i(r_i)\} + \sum_{j \neq i, j=1}^{n} \min\{\underline{a}_{ij} u_j, \underline{a}_{ij} v_j, \bar{a}_{ij} u_j, \bar{a}_{ij} v_j\}$$

$$+ I_i > 0, \tag{9}$$

$$- d_i\, s_i + \min\{\underline{a}_{ii} v_i, \bar{a}_{ii} v_i\} + \sum_{j \neq i, j=1}^{n} \min\{\underline{a}_{ij} u_j, \underline{a}_{ij} v_j, \bar{a}_{ij} u_j, \bar{a}_{ij} v_j\}$$

$$+ I_i > 0, \tag{10}$$

then MNNs (2) with activation functions (1) can have 5^n equilibrium points located in $\Re^n$.

Proof. Arbitrarily pick a region from Ω as

$$\widetilde{\Omega} = \prod_{i \in N_1} (-\infty, p_i) \times \prod_{i \in N_2} [p_i, r_i] \times \prod_{i \in N_3} (r_i, q_i) \times \prod_{i \in N_4} [q_i, s_i] \times \prod_{i \in N_5} (s_i, +\infty) \subset \Omega,$$

where $N_i\, (i = 1, 2, 3, 4, 5)$ are subsets of $\{1, 2, \cdots, n\}$, and $\bigcup_{i=1}^{5} N_i = \{1, 2, \cdots, n\}$, $N_i \cap N_j = \emptyset (i \neq j,\ i, j = 1, 2, 3, 4, 5)$. We will show that MNNs (2) with activation functions (1) have an equilibrium point located in $\widetilde{\Omega}$.

For any point $(x_1, x_2, \cdots, x_n)^T \in \widetilde{\Omega}$, fix $x_1, \cdots, x_{i-1}, x_{i+1}, \cdots, x_n$ except for x_i, and define

$$F_i(x) = - d_i\, x + \hat{a}_{ii}\, f_i(x) + \sum_{j \neq i, j=1}^{n} \hat{a}_{ij}\, f_j(x_j) + I_i, \ i = 1, 2, \cdots, n, \tag{11}$$

where $\hat{a}_{ij}\, (i, j = 1, 2, \cdots, n)$ are the constants defined in (6). Then there are five possible cases for us to discuss.

Case 1. $i \in N_1$. Note that $u_j \leq f_j \leq v_j$ and $\underline{a}_{ij} \leq \widehat{a}_{ij} \leq \bar{a}_{ij}$, from (8) and (11), we have

$$F_i(p_i) = - d_i \, p_i + \widehat{a}_{ii} \, u_i + \sum_{j \neq i, j=1}^{n} \widehat{a}_{ij} \, f_j(x_j) + I_i$$

$$\leq - d_i \, p_i + \max \left\{ \underline{a}_{ii} \, u_i, \bar{a}_{ii} \, u_i \right\} + \sum_{j \neq i, j=1}^{n} \max \left\{ \underline{a}_{ij} \, u_j, \underline{a}_{ij} \, v_j, \bar{a}_{ij} \, u_j, \bar{a}_{ij} \, v_j \right\}$$

$$+ I_i$$

$$< 0,$$

then due to the continuity of $F_i(x)$ and $\lim_{x \to -\infty} F_i(x) = +\infty$, we can find an $\bar{x}_i \in (-\infty, p_i)$ such that $F_i(\bar{x}_i) = 0$.

Case 2. $i \in N_2$. From (9) and (11), we get

$$F_i(r_i) \geq - d_i \, r_i + \min \left\{ \underline{a}_{ii} \, f_i(r_i), \bar{a}_{ii} \, f_i(r_i) \right\}$$

$$+ \sum_{j \neq i, j=1}^{n} \min \left\{ \underline{a}_{ij} \, u_j, \underline{a}_{ij} \, v_j, \bar{a}_{ij} \, u_j, \bar{a}_{ij} \, v_j \right\} + I_i$$

$$> 0,$$

Then we can find an $\bar{x}_i \in (p_i, r_i)$ such that $F_i(\bar{x}_i) = 0$, due to $F_i(p_i) < 0$.

Case 3. $i \in N_3$. Note that $f_i(q_i) = u_i$ and $p_i < q_i$, it follows from (8) and (11) that

$$F_i(q_i) < F_i(p_i) < 0.$$

Then we can find an $\bar{x}_i \in (r_i, q_i)$ such that $F_i(\bar{x}_i) = 0$, in view of $F_i(r_i) > 0$.

Case 4. $i \in N_4$. By virtue of $f_i(s_i) = v_i$, (10) and (11), we obtain

$$F_i(s_i) \geq - d_i \, s_i + \min \left\{ \underline{a}_{ii} \, v_i, \bar{a}_{ii} \, v_i \right\}$$

$$+ \sum_{j \neq i, j=1}^{n} \min \left\{ \underline{a}_{ij} \, u_j, \underline{a}_{ij} \, v_j, \bar{a}_{ij} \, u_j, \bar{a}_{ij} \, v_j \right\} + I_i$$

$$> 0.$$

Then we can find an $\bar{x}_i \in (q_i, s_i)$ such that $F_i(\bar{x}_i) = 0$, because of $F_i(q_i) < 0$.

Case 5. $i \in N_5$. Note that $F_i(s_i) > 0$ and $\lim_{x \to +\infty} F_i(x) = -\infty$, we can also find an $\bar{x}_i \in (s_i, +\infty)$ such that $F_i(\bar{x}_i) = 0$.

Define a map $H : \widetilde{\Omega} \to \widetilde{\Omega}$ by $H(x_1, x_2, \cdots, x_n) = (\bar{x}_1, \bar{x}_2, \cdots, \bar{x}_n)$. It is clear that the map is continuous. By Brouwer's fixed point theorem, there exists one fixed point $x^* = (x_1^*, \cdots, x_n^*)$ of H in $\widetilde{\Omega}$, which is also the equilibrium point of MNNs (2) in $\widetilde{\Omega}$. As $\Re^n$ is divided into 5^n parts, by arbitrariness of $\widetilde{\Omega}$, MNNs (2) with activation functions (1) can have 5^n equilibrium points located in $\Re^n$.

Denote

$$\Phi = \Big\{ \prod_{i=1}^{n} (-\infty, p_i]^{\delta_1^{(i)}} \times [r_i, q_i]^{\delta_2^{(i)}} \times [s_i, +\infty)^{\delta_3^{(i)}},$$

$$(\delta_1^{(i)}, \delta_2^{(i)}, \delta_3^{(i)}) = (1, 0, 0) \text{ or } (0, 1, 0) \text{ or } (0, 0, 1) \Big\}.$$

It is easy to see that Φ is composed of 3^n regions. We are now ready to analyze the dynamical behavior of the solution $x(t)$ with initial condition $x(0) \in \Phi$.

Theorem 2. Assume that $f_i(\pm T_i) = 0$ $(i = 1, 2, \cdots, n)$ and (8)-(10) hold. Furthermore, if there are positive constants $\xi_1, \xi_2, \cdots, \xi_n$ such that

$$-d_i \xi_i + \sum_{j=1}^{n} \xi_j A_{ij} |l_{j,2}| < 0 \tag{12}$$

hold for all $i = 1, 2, \cdots, n$, where $A_{ij} = \max\{|a_{ij}^*|, |a_{ij}^{**}|\}$, then MNNs (2) with activation functions (1) can have 5^n equilibrium points, and 3^n of which are locally exponentially stable.

Proof. First of all, according to Theorem 1, the coexistence of 5^n equilibrium points for MNNs (2) can be guaranteed under the conditions of Theorem 2. In the following, we will prove the local stability of the 3^n equilibrium points located in Φ in two steps.

Step I. Arbitrarily pick a region from Φ as

$$\widetilde{\Phi} = \prod_{i \in N_1} (-\infty, p_i] \times \prod_{i \in N_3} [r_i, q_i] \times \prod_{i \in N_5} [s_i, +\infty) \subset \Phi,$$

where N_1, N_3, N_5 are subsets of $\{1, 2, \cdots, n\}$, and $N_1 \cup N_3 \cup N_5 = \{1, 2, \cdots, n\}$, $N_i \cap N_j = \emptyset$ $(i \neq j, i, j = 1, 3, 5)$. Let $x(t)$ be a solution of MNNs (2) with initial condition $x(0) \in \widetilde{\Phi}$. Then we claim that $x(t)$ would stay in $\widetilde{\Phi}$ for all $t \geq 0$. If this is not true, then there are three possible cases to discuss.

Case 1. There exists some $i \in N_1$ and $t^* \geq 0$ such that $x_i(t^*) = p_i$, $\dot{x}_i(t^*) > 0$, $x_i(t) \leq p_i$ for $0 \leq t \leq t^*$. Then it follows from (4), (8) and the definition of f_i that

$$\dot{x}_i(t^*) = -d_i x_i(t^*) + \hat{a}_{ii} f_i(x_i(t^*)) + \sum_{j \neq i, j=1}^{n} \hat{a}_{ij} f_j(x_j(t^*)) + I_i$$

$$\leq -d_i p_i + \max\{\underline{a}_{ii} u_i, \bar{a}_{ii} u_i\} + \sum_{j \neq i, j=1}^{n} \max\{\underline{a}_{ij} u_j, \underline{a}_{ij} v_j, \bar{a}_{ij} u_j, \bar{a}_{ij} v_j\} + I_i$$

$$< 0,$$

which is a contradiction.

Case 2. There exists some $i \in N_3$ and $t^* \geq 0$ such that either $x_i(t^*) = r_i$, $\dot{x}_i(t^*) < 0$, $x_i(t) \in [r_i, q_i]$ for $0 \leq t \leq t^*$ or $x_i(t^*) = q_i$, $\dot{x}_i(t^*) > 0$, $x_i(t) \in [r_i, q_i]$ for $0 \leq t \leq t^*$. For the first case, we derive from (4), (9) that

$$\dot{x}_i(t^*) \geq - d_i\, r_i + \min\left\{\underline{a}_{ii}\, f_i(r_i), \bar{a}_{ii}\, f_i(r_i)\right\}$$

$$+ \sum_{j \neq i, j=1}^{n} \min\left\{\underline{a}_{ij}\, u_j, \underline{a}_{ij}\, v_j, \bar{a}_{ij}\, u_j, \bar{a}_{ij}\, v_j\right\} + I_i$$

$$> 0,$$

which is a contradiction. The second case can be proved similarly.

Case 3. There exists some $i \in N_5$ and $t^* \geq 0$ such that $x_i(t^*) = s_i$, $\dot{x}_i(t^*) < 0$, $x_i(t) \geq s_i$ for $0 \leq t \leq t^*$. Then we get

$$\dot{x}_i(t^*) \geq - d_i\, s_i + \min\left\{\underline{a}_{ii} v_i, \bar{a}_{ii} v_i\right\} + \sum_{j \neq i, j=1}^{n} \min\left\{\underline{a}_{ij} u_j, \underline{a}_{ij} v_j, \bar{a}_{ij} u_j, \bar{a}_{ij} v_j\right\} + I_i,$$

$$> 0,$$

which contradicts $\dot{x}_i(t^*) < 0$.

From the above three cases, we know that the solution $x(t)$ will never escape from $\widetilde{\Phi}$ for all $t \geq 0$. That is, $\widetilde{\Phi}$ is positively invariant with respect to the solution $x(t)$ with initial state $x(0) \in \widetilde{\Phi}$.

Step II. We will prove that the equilibrium point x^* of MNNs (2) in $\widetilde{\Phi}$ is locally stable. From (12), there exists a positive constant ε small enough such that

$$(-d_i + \varepsilon)\, \xi_i + \sum_{j=1}^{n} \xi_j\, A_{ij}\, |l_{j,2}| < 0. \tag{13}$$

Let $y_i(t) = x_i(t) - x_i^*$. By the theories of set-valued maps and differential inclusions, (3) and (5), we can get that

$$\dot{y}_i(t) \in - d_i\, y_i(t) + \sum_{j=1}^{n} \left\{co[a_{ij}(x_j(t))] f_j(x_j(t)) - co[a_{ij}(x_j^*)] f_j(x_j^*)\right\}. \tag{14}$$

Note that when $u = (u_1, u_2, \cdots, u_n)^T \in \widetilde{\Phi}$, $f_j(u_j)$ is Lipschitz continuous with Lipschitz constant $|l_{j,2}|$. So it follows from Lemma 1 that

$$|co[a_{ij}(x_j(t))] f_j(x_j(t)) - co[a_{ij}(x_j^*)] f_j(x_j^*)| \leq A_{ij}\, |l_{j,2}|\, |y_j(t)|. \tag{15}$$

Define $z_i(t) = e^{\varepsilon t}\, y_i(t)$ and

$$M(t) = \sup_{s \leq t} \left(\max_{1 \leq i \leq n} \left(\xi_i^{-1} |z_i(s)| \right) \right), \quad t \geq 0. \tag{16}$$

In the following, we claim that $M(t)$ is bounded. More precisely, for all $t \geq 0$, we have $M(t) = M(0)$.

In fact, for any $t_0 \geq 0$, there are two cases:

Case 1. $\max_{1 \leq i \leq n} \left(\xi_i^{-1} |z_i(t_0)| \right) < M(t_0)$. In this case, there exists a $\delta > 0$ such that $\max_{1 \leq i \leq n} \left(\xi_i^{-1} |z_i(t)| \right) < M(t_0)$ for $t \in (t_0, t_0 + \delta)$.

Case 2. $\max_{1 \leq i \leq n} \left(\xi_i^{-1} |z_i(t_0)| \right) = M(t_0)$. In this case, let $i_{t_0} = i_{t_0}(t_0)$ be such an index that

$$\xi_{i_{t_0}}^{-1} |z_{i_{t_0}}(t_0)| = \max_{1 \leq i \leq n} \left(\xi_i^{-1} |z_i(t_0)| \right).$$

By using (13), (14) and (15), we derive that

$$\frac{d|z_{i_{t_0}}(t)|}{dt}\bigg|_{t=t_0} = \text{sign}\left(z_{i_{t_0}}(t_0)\right) \varepsilon e^{\varepsilon t_0} y_{i_{t_0}}(t_0) + \text{sign}\left(z_{i_{t_0}}(t_0)\right) e^{\varepsilon t_0} \dot{y}_{i_{t_0}}(t_0)$$

$$\leq \left(-d_{i_{t_0}} + \varepsilon\right) |z_{i_{t_0}}(t_0)| + \sum_{j=1}^{n} A_{i_{t_0} j} |l_{j,2}| |z_j(t_0)|$$

$$\leq \left[\left(-d_{i_{t_0}} + \varepsilon\right) \xi_{i_{t_0}} + \sum_{j=1}^{n} \xi_j A_{i_{t_0} j} |l_{j,2}| \right] M(t_0)$$

$$\leq 0.$$

Then, there exists a $\delta_1 > 0$ such that $M(t) = M(t_0)$ for $t \in (t_0, t_0 + \delta_1)$.

Therefore, from Case 1-Case 2, we can conclude that $M(t) = M(0)$ for all $t \geq 0$, which implies that

$$\max_{1 \leq i \leq n} \left(\xi_i^{-1} |z_i(t)| \right) \leq M(0),$$

then we can get

$$|y_i(t)| \leq M e^{-\varepsilon t}, \ t \geq 0, \ i = 1, 2, \cdots, n, \tag{17}$$

where $M = M(0) \max_{1 \leq i \leq n} \{\xi_i\}$. That is, x^* is locally exponentially stable in $\widetilde{\Phi}$.

Because $\widetilde{\Phi} \subset \Phi$ is chosen arbitrarily, we conclude that in each subset of Φ, there is a locally exponentially stable equilibrium point. Therefore, MNNs (2) have 3^n locally exponentially stable equilibrium points.

Remark 1. As reported in [5], under some conditions, conventional recurrent neural networks with Mexican-hat-type activation function have at most 3^n equilibrium points and at most 2^n locally stable equilibrium points. Compared with the result in [5], it can be seen from Theorem 2 that MNNs (2) with activation functions (1) now have both more total equilibrium points and more locally stable equilibrium points. This clearly shows that neural networks with activation functions (1) can have greater storage capacity than the ones with Mexican-hat-type activation function.

4 Conclusions

Under the framework of Filippov's solution, the issue of multistability has been studied in this paper for a class of MNNs with non-monotonic piecewise linear activation functions. Some new sufficient conditions have been presented to ensure the coexistence of 5^n equilibrium points and the local stability of 3^n equilibrium points. The obtained results have demonstrated that the non-monotonic activation functions introduced in this paper play a significant role in increasing the storage capacity of neural networks.

Acknowledgements. This work was supported by the National Natural Science Foundation of China under Grant 61203300, the Specialized Research Fund for the Doctoral Program of Higher Education under Grant 20120092120029, the Natural Science Foundation of Jiangsu Province of China under Grant BK2012319, and the China Postdoctoral Science Foundation funded project under Grant 2012M511177.

References

1. Kaslik, E., Sivasundaram, S.: Impulsive Hybrid Discrete-Time Hopfield Neural Networks with Delays and Multistability Analysis. Neural Networks 24, 370–377 (2011)
2. Nie, X., Cao, J.: Multistability of Second-Order Competitive Neural Networks with Nondecreasing Saturated Activation Functions. IEEE Trans. Neural Networks 22, 1694–1708 (2011)
3. Marco, M., Forti, M., Grazzini, M., Pancioni, L.: Limit Set Dichotomy and Multistability for A Class of Cooperative Neural Networks with Delays. IEEE Trans. Neural Networks and Learning Systems 23, 1473–1485 (2012)
4. Huang, Z., Raffoul, Y., Cheng, C.: Scale-Limited Activating Sets and Multiperiodicity for Threshold Networks on Time Scales. IEEE Trans. Cybernetics 44, 488–499 (2014)
5. Wang, L., Chen, T.: Multistability of Neural Networks with Mexican-Hat-Type Activation Functions. IEEE Trans. Neural Networks and Learning Systems 23, 1816–1826 (2012)
6. Chua, L.: Memristor-The Missing Circuit Element. IEEE Trans. Circuit Theory 18, 507–519 (1971)
7. Wu, A., Zeng, Z.: Exponential Stabilization of Memristive Neural Networks with Time Delays. IEEE Trans. Neural Networks and Learning Systems 23, 1919–1929 (2012)
8. Zhang, G., Shen, Y.: New Algebraic Criteria for Synchronization Stability of Chaotic Memristive Neural Networks with Time-Varying Delays. IEEE Trans. Neural Networks and Learning Systems 24, 1701–1707 (2013)
9. Wen, S., Bao, G., Zeng, Z., Chen, Y., Huang, T.: Global Exponential Synchronization of Memristor-Based Recurrent Neural Networks with Time-Varying Delays. Neural Networks 48, 195–203 (2013)
10. Chandrasekar, A., Rakkiyappan, R., Cao, J., Lakshmanan, S.: Synchronization of Memristor-Based Recurrent Neural Networks with Two Delay Components based on Second-Order Reciprocally Convex Approach. Neural Networks 57, 79–93 (2014)
11. Chen, J., Zeng, Z., Jiang, P.: Global Mittag-Leffler Stability and Synchronization of Memristor-Based Fractional-Order Neural Networks. Neural Networks 51, 1–8 (2014)
12. Filippov, A.: Differential Equations with Discontinuous Right-hand Sides. Kluwer, Boston (1988)
13. Clarke, F., Ledyaev, Y., Stem, R., Wolenski, R.: Nonsmooth Analysis and Control Theory. Springer, New York (1998)

Global Exponential Anti-synchronization of Coupled Memristive Chaotic Neural Networks with Time-Varying Delays

Zheng Yan, Shuzhan Bi, and Xijun Xue

Shannon Lab, Huawei Technologies Co., Ltd., Beijing, China
yanzheng@huawei.com

Abstract. This paper investigates the problem of global exponential anti-synchronization of a class of memristive chaotic neural networks with time-varying delays. First, a memrsitive neural network is modeled. Then, considering the state-dependent properties of the memristor, a new fuzzy model employing parallel distributed compensation (PDC) provides a new way to analyze the complicated memristive neural networks with only two subsystems. And the controller is dependent on the output of the system in the case of packed circuits. An illustrative example is also presented to show the effectiveness of the results.

1 Introduction

Through the classical von Neumann bottleneck of conventional digital computers, the sequential processing of fetch, decode, and execution of instructions has resulted in less efficient machines as their eco-systems have grown to be increasingly complex [1]. In order to emulate the brain functionality of animals like a spider, mouse, and cat [2, 3], modern digital computers dissipate a vast amount of energies. For example, to perform certain cortical simulations at the cat scale even at 83 times slower firing rate, a super computer equipped with 147456 CPUs and 144 TBs of main memory, has to be employed by the IBM team [2]. On the other hand, the human brain contains more than 100 billion neurons and each neuron has more than 20000 synapses. Efficient circuit implementation of synapses, therefore, is very essential to build a brain-like machine. However, Since shrinking the current transistor size is very difficult, introducing a more efficient approach is critical for further development of neural network implementations.

In 2008, a successful fabrication of a very compact and non-volatile nano scale memory called the memristor has been announced by the Williams group [4]. It was postulated by Chua [5] as the fourth basic circuit elements in electrical circuits, which is based on the nonlinear characteristics of charge and flux. By supplying a voltage or current to the memristor, its resistance can be altered and stored when the applied voltage is gone. [6]. In this way, the memristor remembers information. Several examples of successful multichip networks of spiking neurons have been recently proposed [7–9]; however there are still a

© Springer International Publishing Switzerland 2015
X. Hu et al. (Eds.): ISNN 2015, LNCS 9377, pp. 192–201, 2015.
DOI: 10.1007/978-3-319-25393-0_22

number of practical problems that hinder the development of truly large-scale, distributed, massively parallel networks of very large scale integration (VLSI) neurons, such as how to set the weight of individual synapses in the network. It is well-known that changing the synaptic connections between neurons are widely believed to contribute to memory storage, and the activity-dependent development of neural networks. These changes are thought to occur through correlated-based, or Hebbian plasticity.

Meanwhile, it is obvious that neural networks have been widely studied in recent years with the immense application prospective [10–14]. Many applications have been developed in different areas such as combinatorial optimization, knowledge acquisition and pattern recognition. Recently, the problem of anti-synchronization of coupled neural networks which is one of hot research fields of complex networks has been a challenging issue due to its potential application such as information science, biological systems and so on [15–19].

However, to the best of the author's knowledge, the research on global exponential anti-synchronization of coupled memristive neural networks is still an open problem that deserves further investigation. To shorten sup gap, we investigate the problem of global exponential anti-synchronization for a class of memristive neural networks with time-varying delays. The main contributions of this paper can be summarized as follows: (i) Based on the circuits design, the model of MNNs is established; (ii) a fuzzy model of memristive neural networks is employed to give a new way to analyze the complicated MNNs with only two subsystems; (iii) a sufficient condition is derived to make the anti-synchronization error system exponentially stable.

2   Problem Formulation

By Kirchoff's current law, the equation of the i-th subsystem of the memristive neural network is presented as follows:

$$\dot{x}_i(t) = -d_i(x_i(t))x_i(t) + \sum_{j=1}^{n} a_{ij} f_j(x_j(t))$$

$$+ \sum_{j=1}^{n} b_{ij} f_j(x_j(t - \tau_j(t))) + s_i, \tag{1}$$

where

$$a_{ij} = \frac{\text{sign}_{ij}}{C_i R_{fij}}, b_{ij} = \frac{\text{sign}_{ij}}{C_i R_{gij}}, \ s_i = \frac{I_i}{C_i},$$

$$d_i(x_i(t)) = \frac{1}{C_i} \Big[\sum_{j=1}^{n} \Big(\frac{1}{R_{fij}} + \frac{1}{R_{gij}} \Big) + W_i(x_i(t)) \Big]$$

$$= \begin{cases} d_{1i}, x_i(t) \leq 0; \\ d_{2i}, x_i(t) > 0. \end{cases}$$

Then

$$\dot{x}(t) = -D(x(t))x(t) + Af(x(t)) + Bf(x(t - \tau(t))) + s, \tag{2}$$

where

$$D(x(t)) = \text{diag}\{d_1(x_1(t)), d_2(x_2(t)), \ldots, d_n(x_n(t))\},$$
$$A = [a_{ij}]_{n \times n}, B = [b_{ij}]_{n \times n}, s = (s_1, s_2, \ldots, s_n)^T,$$
$$f(x(t)) = \left(f_1(x_1(t)), \cdots, f_n(x_n(t))\right)^{\text{T}},$$
$$f(x(t - \tau(t))) = \left(f_1(x_1(t - \tau_1(t))), \cdots, f_n(x_n(t - \tau_n(t)))\right)^{\text{T}}.$$

To solve the problem about nonlinear control, fuzzy logic has attracted much attention as a powerful tool. Among various kinds of fuzzy methods, the Takagi-Sugeno fuzzy systems are widely accepted as a useful tool for design and analysis of fuzzy control system [20–23]. Then, the memristive neural network (1) can be exactly represented by the fuzzy model as follows:

Rule 1: IF $x_i(t)$ is N_{1i}, THEN

$$\dot{x}_i(t) = -d_{1i}x_i(t) + \sum_{j=1}^{n} a_{ij}f_j(x_j(t))$$

$$+ \sum_{j=1}^{n} b_{ij}f_j(x_j(t - \tau_j(t))) + s_i,$$

Rule 2: IF $x_i(t)$ is N_{2i}, THEN

$$\dot{x}_i(t) = -d_{2i}x_i(t) + \sum_{j=1}^{n} a_{ij}f_j(x_j(t))$$

$$+ \sum_{j=1}^{n} b_{ij}f_j(x_j(t - \tau_j(t))) + s_i,$$

where N_{1i} is $x_i(t) \leq 0$, N_{2i} is $x_i(t) > 0$. With a center-average defuzzier, the over fuzzy system is represented as

$$\dot{x}_i(t) = -\sum_{l=1}^{2} \eta_{li}(t)d_{li}x_i(t) + \sum_{j=1}^{n} a_{ij}f_j(x_j(t))$$

$$+ \sum_{j=1}^{n} b_{ij}f_j(x_j(t - \tau_j(t))) + s_i, \tag{3}$$

where

$$\eta_{1i}(t) = \begin{cases} 1, x_i(t) \leq 0, \\ 0, x_i(t) > 0, \end{cases} \quad \eta_{2i}(t) = \begin{cases} 0, x_i(t) \leq 0, \\ 1, x_i(t) > 0. \end{cases}$$

Therefore, system (3) can be represented by

$$\dot{x}(t) = -\sum_{l=1}^{2} \Pi_l(t) D_l x(t) + Af(x(t))$$

$$+ Bf(x(t - \tau(t))) + s, \tag{4}$$

where $\Pi_l(t) = \text{diag}\{\eta_{l1}(t), \cdots, \eta_{ln}(t)\}$, and $\sum_{l=1}^{2} \eta_{li}(t) = 1, i = 1, \cdots, n, l = 1, 2$, and

$$D_l = \text{diag}\{d_{l1}, d_{l2}, \ldots, d_{ln}\}.$$

3 Preliminaries

Denote $u = (u_1, \cdots, u_n)^T$, $|u|$ as the absolute-value vector; i.e., $|u| = (|u_1|, |u_2|, \ldots, |u_n|)^T$, $\|x\|_p$ as the p-norm of the vector x with p, $1 \le p < \infty$. $\|x\|_\infty = \max_{i \in \{1,2,\cdots,n\}} |x_i|$ is the vector infinity norm. Denote $\|D\|_p$ as the p-norm of the matrix D with p. Denote $\mathcal{C}$ as the set of continuous functions.

And the following assumptions will be needed throughout the paper:

A1. For $i \in \{1, 2, \cdots, n\}$, the activation function f_i is Lipschitz continuous; and $\forall r_1, r_2 \in \mathbb{R}$, there exists real number κ_i such that

$$0 \le \frac{f_i(r_1) - f_i(r_2)}{r_1 - r_2} \le \kappa_i.$$

A2. For $i \in \{1, 2, \cdots, n\}$, $\tau_i(t)$ satisfies

$$0 \le \tau_i(t) \le \bar{\tau}_i, \quad \dot{\tau}_i(t) \le \mu_i.$$

In this paper, we consider system (4) as the master system, and through electronic inductors, the values of memristor will be presented in the corresponding slave system, then the slave system is given as:

$$\dot{y}(t) = -\sum_{l=1}^{2} \Pi_l(t) D_l y(t) + Af(y(t))$$

$$+ Bf(y(t - \tau(t))) + s + u(t), t \ge 0, \tag{5}$$

where $y(t) = (y_1(t), y_2(t), \cdots, y_n(t))^T$, $u(t) = (u_1(t), u_2(t), \cdots, u_n(t))^T$ is the control input that will be designed. The initial conditions of system (5) is in the form of $y(t) = \Phi(t) \in \mathcal{C}([-\bar{\tau}, 0], \mathbb{R}^n)$, $\bar{\tau} = \max_{1 \le i \le n} \{\bar{\tau}_i\}$.

In order to derive sufficient conditions for the global exponential anti-synchronization of system (4) with system (5), we will need the following lemmas.

Lemma 1 [24]. It is given any real matrices X, Z, P of appropriate dimensions and a scalar $\varepsilon_0 > 0$, where $P > 0$. Then the following inequality holds:

$$X^T Z + Z^T X \leq \varepsilon_0 X^T P X + \varepsilon_0^{-1} Z^T P^{-1} Z.$$

In particular, if X and Z are vectors, $X^T Z \leq \frac{1}{2}(X^T X + Z^T Z)$.

4   Main Results

As anti-synchronization has been applied in many real applications, the error system can be obtained as

$$
\begin{aligned}
\dot{e}(t) = & - \sum_{l=1}^{2} \Pi_l(t) D_l e(t) + A\Phi(e(t)) \\
& + B\Phi(e(t - \tau(t))) + u(t).
\end{aligned}
\tag{6}
$$

where $e(t) = (e_1(t), e_2(t), \cdots, e_n(t))^T$, is the anti-synchronization error, and $e_i(t) = x_i(t) + y_i(t)$. And the output functions with/without delays are

$$
\begin{aligned}
& \Phi(e(t)) \\
& = (\Phi_1(e_1(t)), \cdots, \Phi_n(e_n(t))) \\
& = f(e(t) - y(t)) + f(y(t)), \\
& \Phi(e(t - \tau(t))) \\
& = (\Phi_1(e_1(t - \tau_1(t))), \cdots, \Phi_n(e_n(t - \tau_n(t))))^T \\
& = f(e(t - \tau(t)) - y(t - \tau(t))) + f(y(t - \tau(t))).
\end{aligned}
$$

As this paper aims to design an output controller

$$u(t) = K\Phi(e(t)).
\tag{7}$$

where $K = (k_{ij})_{n \times n}$ is a constant gain matrix to be determined to anti-synchronize the drive and response systems, $\Phi(e(t))$ is the output function without delays.

With controller (7), the error system (6) is transformed into

$$
\dot{e}(t) = - \sum_{l=1}^{2} \Pi_l(t) D_l e(t) + \hat{A}\Phi(e(t)) + B\Phi(e(t - \tau(t))),
\tag{8}
$$

where $\hat{A} = (a_{ij})_{n \times n} = (a_{ij} + k_{ij})_{n \times n}$.

Theorem 1. Assume the conditions in Theorem 1 hold, then the drive system (4) is globally exponentially anti-synchronized with the response system (5).

Proof. Let $\lambda = \min_{1 \leq l \leq 2} \{\lambda_{\min}\{D_l\}\}$. Obviously, $V(t)$ which is defined in Theorem 1 is a positive definite and radially unbounded Lyapunov functional. A positive number $\epsilon > 0$ is chosen to satisfy

$$\epsilon\alpha - \lambda\alpha + 2\epsilon\|LQ\| + 2\epsilon\bar{\tau}e^{\epsilon\bar{\tau}}\|L^2R\| < 0. \tag{9}$$

It follows that

$$\frac{d}{dt}\{e^{\epsilon t}V(T)\} = e^{\epsilon t}(\epsilon V(t) + \dot{V}(t))$$

$$\leq \sum_{l=1}^{2}\Pi_l(t)e^{\epsilon t}\left(\epsilon\left(\frac{\alpha}{2}e^T(t)e(t) + 2\sum_{i=1}^{n}q_i\int_0^{e_i(t)}\Phi_i(s)ds\right.\right.$$

$$\left.+ \sum_{i=1}^{n}\int_{t-\tau_i(t)}^{t}\Phi_i^2(e_i(s))r_i ds\right) - \frac{\alpha\lambda}{2}e^T(t)e(t)\Bigg)$$

$$\leq \frac{1}{2}\sum_{l=1}^{2}\Pi_l(t)e^{\epsilon t}\left(\epsilon\alpha e^T(t)e(t) - \alpha\lambda e^T(t)e(t)\right.$$

$$\left.+ 4\epsilon\sum_{i=1}^{n}q_i\int_0^{e_i(t)}\Phi_i(s)ds\right)$$

$$+ \epsilon e^{\epsilon t}\sum_{i=1}^{n}\int_{t-\tau_i(t)}^{t}\Phi_i^2(e_i(s))r_i ds.$$

Since

$$\sum_{i=1}^{n}q_i\int_0^{e_i(t)}\Phi_i(s)ds \leq \sum_{i=1}^{n}q_i\int_0^{e_i(t)}\kappa_i s ds \leq \frac{1}{2}e^T(t)LQe(t),$$

we can get

$$\frac{d}{dt}(e^{\epsilon t}V(t)) \leq \frac{1}{2}e^{\epsilon t}\left(\epsilon\alpha - \lambda\alpha + 2\epsilon\|LQ\|\right)e^T(t)e(t)$$

$$+ \epsilon e^{\epsilon t}\sum_{i=1}^{n}\int_{t-\tau_i(t)}^{t}\Phi_i^2(e_i(s))r_i ds. \tag{10}$$

Estimating the second term on the right-hand side of (10) by changing the integrals, we can get

$$\epsilon \int_0^s e^{\epsilon t} \sum_{i=1}^n \int_{t-\tau_i(t)}^t \Phi_i^2(e_i(\varsigma)) r_i d\varsigma dt$$

$$\leq \epsilon \sum_{i=1}^n \int_{-\bar\tau}^s \int_{\max\{\varsigma,0\}}^{\min\{\varsigma+\bar\tau,s\}} e^{\epsilon t} dt \Phi_i^2(e_i(\varsigma)) r_i d\varsigma$$

$$\leq \epsilon \sum_{i=1}^n \int_{-\bar\tau}^s \left(\int_\varsigma^{\varsigma+\bar\tau} e^{\epsilon t} dt \right) \Phi_i^2(e_i(\varsigma)) r_i d\varsigma$$

$$\leq \epsilon \sum_{i=1}^n \int_{-\bar\tau}^s \bar\tau e^{\epsilon(\varsigma+\bar\tau)} \Phi_i^2(e_i(\varsigma)) r_i d\varsigma$$

$$\leq \epsilon \sum_{i=1}^n \int_{-\bar\tau}^s \bar\tau e^{\epsilon(\varsigma+\bar\tau)} e_i^2(\varsigma) \kappa_i^2 r_i d\varsigma$$

$$\leq \epsilon \bar\tau e^{\epsilon\bar\tau} ||L^2 R|| \int_{-\bar\tau}^s e^{\epsilon\varsigma} e^T(\varsigma) e(\varsigma) d\varsigma$$

$$\leq \epsilon \bar\tau e^{\epsilon\bar\tau} ||L^2 R|| \left(\int_{-\bar\tau}^0 e^{\epsilon\varsigma} e^T(\varsigma) e(\varsigma) d\varsigma \right.$$

$$\left. + \int_0^s e^{\epsilon\varsigma} e^T(\varsigma) e(\varsigma) d\varsigma \right). \tag{11}$$

From (9), (10) and (11),

$$e^{\epsilon s} V(s) - V(0)$$

$$\leq \frac{1}{2} \Big(\epsilon\alpha - \lambda\alpha + 2\epsilon||LQ|| + 2\epsilon\bar\tau e^{\epsilon\bar\tau}||L^2 R|| \Big)$$

$$\times \int_0^s e^{\epsilon t} e^T(t) e(t) dt + \epsilon\bar\tau e^{\epsilon\bar\tau}||L^2 R|| \int_{-\bar\tau}^0 e^{\epsilon t} e^T(t) e(t) dt$$

$$\leq \Big(\epsilon\bar\tau||L^2 R|| \int_{-\bar\tau}^0 e^{\epsilon t} dt \Big) ||\psi||^2 \equiv H_1 ||\psi||^2.$$

Thus,

$$V(t) \leq \Big(V(0) + H_1 ||\psi||^2 \Big) e^{-\epsilon t}, \quad \forall t > 0, \tag{12}$$

where

$$V(0) \leq \frac{1}{2} \Big(\beta + 2||QL|| + 2\bar\tau||L^2 R|| \Big) ||\psi||^2 \equiv H_2 ||\psi||^2.$$

It follows from (11) and (12) that

$$\frac{\alpha}{2} e^T(t) e(t) \leq V(t) \leq (H_1 + H_2) ||\psi||^2 e^{-\epsilon t}, \quad \forall t > 0.$$

Thus, we have

$$||e(t)|| \leq \sqrt{\frac{2}{\alpha}(H_1 + H_2)}||\psi||e^{-\frac{\varepsilon}{2}t}, \tag{13}$$

which implies the anti-synchronization error system (8) is globally exponentially stable. This completes the proof.

5   Illustrative Example

In order to show the effectiveness of the obtained results, an illustrative example is given as follows:

Example 1. Consider memristive system (4) in Example 1. As

$$\lambda = \min_{l=1,2}\{\lambda_{\min}\{D_l\}\} = 0.9.$$

Obviously, there exists a positive definite diagonal matrix $Q = \text{diag}\{0.5, 0.5\}$ such that

$$\daleth = -\hat{A},$$

to make

$$\Pi_l = -2\lambda Q L^{-1} - \daleth + 2||Q||||B||_2 I$$
$$= K + \begin{bmatrix} 2.5 & 10 \\ 0.1 & 2.5 \end{bmatrix} < 0,$$

then, the anti-synchronization error system (8) can achieve global exponential stability. To simulate the obtained result, let

$$K = \begin{bmatrix} -3.5 & -10 \\ -0.1 & -3.5 \end{bmatrix}.$$

Set the initial states of slave system (5) is $[3.5 \quad -0.7]$. The state and error trajectories of master system and slave system, are presented in Fig. 1, which illustrate the effectiveness of the obtained results.

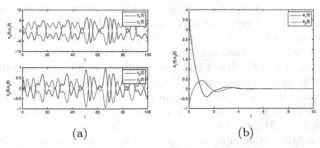

(a)   (b)

Fig. 1. State and error trajectories of master system (4) and slave system (5).

6   Conclusions

This paper investigated the global exponential anti-synchronization problem of coupled chaotic memristive neural networks with time-varying delays via the output function controller. A new scheme of memristive neural networks is designed, corresponding dynamics equation is set up, and take the PDC fuzzy strategy to analyze this system. A numerical example is presented to show the effectiveness of the obtained results.

Acknowledgement. This work was supported by the Natural Science Foundation of China under Grant 61125303, the Excellent Youth Foundation of Hubei Province of China under Grant 2010CDA081, National Priority Research Project NPRP 4-451-2-168, funded by Qatar National Research Fund.

References

1. Jo, S., Chang, T., Ebong, I., Bhadviya, B., Mazumder, P., Lu, W.: Nanoscale memristor device as synapse in neuromorphic systems. Nanotech. Lett. 10, 1297–1301 (2010)
2. Ananthanarayanan, R., Eser, S., Simon, H., Modha, D.: Proceedings of 2009 IEEE/ACM Conference High Performance Networking Computing, Portland, OR, November 2009
3. Smith, L.: Handbook of Nature-Inspired and Innovative Computing: Integrating Classical Models with Emerging Technologies, pp. 433–475. Springer, New York
4. Strukov, D., Snider, G., Stewart, D., Williams, R.: The missing memristor found. Nature 453, 80–83 (2008)
5. Chua, L.: Memristor-The missing circuit element. IEEE Trans. Circuits Theory 18, 507–519 (1971)
6. Sharifiy, M., Banadaki, Y.: General spice models for memristor and application to circuit simulation of memristor-based synapses and memory cells. J. Circuits Syst. Comput. 19, 407–424 (2010)
7. Choi, T., Shi, B., Boahen, K.: An on-off orientation selective address event representation image transceiver chip. IEEE Trans. Circuits Syst. I 51, 342–353 (2004)
8. Indiveri, G.: A neuromorphic VLSI device for implementing 2-D selective attention systems. IEEE Trans. Neural Networks 12, 1455–1463 (2001)
9. Liu, S., Douglas, R.: Temporal coding in a silicon network of integrate-and-fire neurons. IEEE Trans. Neural Networks 15, 1305–1314 (2004)
10. Li, C., Feng, G.: Delay-interval-dependent stability of recurrent neural networks with time-varying delay. Neurocomput. 72, 1179–1183 (2009)
11. Li, C., Feng, G., Liao, X.: Stabilization of nonlinear system via periodically intermittent control. IEEE Trans. Circuit Syst. II 54, 1019–1023 (2007)
12. Shen, Y., Wang, J.: An improved algebraic criterion for global exponential stability of recurrent neural networks with time-varying delays. IEEE Trans. Neural Networks 19, 528–531 (2008)
13. Song, Q.: Synchronization analysis in an array of asymmetric neural networks with time-varying delays and nonlinear coupling. Appl. Math. Comput. 216, 1605–1613 (2010)

14. Song, Q., Zhao, Z., Yang, J.: Passivity and passification for stochastic Takagi-Sugeno fuzzy systems with mixed time-varying delays. Neurocomput (2013). doi:10.1016/j.neurocom.2013.06.018

15. Cao, J., Chen, G., Li, P.: Global synchronization in an array of delayed neural networks with hybrid coupling. IEEE Trans. Syst. Man Cybern. B 38, 488–498 (2008)

16. Juang, C., Chen, T., Cheng, W.: Speedup of implementing fuzzy neural networks with high-dimensional inputs through parallel processing on graphic processing units. IEEE Trans. Fuzzy Syst. 19, 717–728 (2011)

17. Li, J., Kazemian, H., Afzal, M.: Neural network approaches for noisy language modeling. IEEE Trans. Neural Networks Learn. Syst. (2013). doi:10.1109/TNNLS.2013.2263557

18. Park, M., Kwon, O., Park, J., Lee, S., Cha, E.: Synchronization criteria for coupled neural networks with interval time-varying delays and leakage delay. Appl. Math. Comput. 218, 6762–6775 (2012)

19. Zhang, H., Ma, T., Huang, G., Wang, Z.: Robust global exponential synchronization of uncertain chaotic delayed neural networks via dualstage impulsive control. IEEE Trans. Syst. Man Cybern. B Cybern. 40, 831–844 (2010)

20. Dong, J., Wang, Y., Yang, G.: Control synthesis of continuous-time T-S fuzzy systems with local nonlinear models. IEEE Trans. Syst. Man Cybern. B Cybern. 39, 1245–1258 (2009)

21. Liu, X., Zhong, S.: T-S fuzzy model-based impulsive control of chaotic systems with exponential decay rate. Phys. Lett. A 370, 260–264 (2007)

22. Park, C., Cho, Y.: T-S model based indirect adaptive fuzzy control using online parameter estimation. IEEE Trans. Syst. Man Cybern. B Cybern. 34, 2293–2302 (2004)

23. Takagi, T., Sugeno, M.: Fuzzy identification of systems and its applications to modelling and control. IEEE Trans. Syst. Man Cybern. SMC-15, 116–132 (1985)

24. Zhao, W., Tan, Y.: Harmless delay for global exponential stability of Cohen-Grossberg neural networks. Math. Comput. Simul. 74, 47–57 (2007)

Computer Vision

Representative Video Action Discovery Using Interactive Non-negative Matrix Factorization

Hui Teng[1,2], Huaping Liu[2], Lianzhi Yu[1], and Fuchun Sun[2]

[1] School of Optical-Electrical and Computer Engineering,
University of Shanghai for Science and Technology, Shanghai
[2] Department of Computer Science and Technology, Tsinghua University, Beijing
State Key Lab. of Intelligent Technology and Systems, Beijing
hpliu@tsinghua.edu.cn

Abstract. In this paper, we develop an interactive Non-negative Matrix Factorization method for representative action video discovery. The original video is first evenly segmented into some short clips and the bag-of-words model is used to describe each clip. Then a temporal consistent Non-negative Matrix Factorization model is used for clustering and action segmentation. Since the clustering and segmentation results may not satisfy the user's intention, two extra human operations: MERGE and ADD are developed to permit user to improve the results. The newly developed interactive Non-negative Matrix Factorization method can therefore generate personalized results. Experimental results on the public Weizman dataset demonstrate that our approach is able to improve the action discovery and segmentation results.

Keywords: Interactive action summarization, Non-negative Matrix Factorization, video analysis.

1 Introduction

There has been a lot of interests in developing practical systems to automatically understand video data. Of the many related tasks, discovering representative actions from video clip is of considerable practical importance. Such algorithms could automatically extract representative actions within streaming or archival video and therefore significantly improve the efficiency of video understanding.

In Ref.[6], a Bayesian non-parametric model of sequential data is adopted to allow completely unsupervised activity discovery. The authors claim that this work need not predefine the relevant behaviors or even their numbers, as both of them are learned directly from data. However, such a method admits the following disadvantages: (1)Due to the complexity of non-parametric Beyasian method, its time burden is rather huge; (2) The number of behaviors, although need not to be determined by the user, is still sensitive to some parameters of the algorithm (especially, the Dirichlet prior parameter). That is to day, the task of determining the number of behavior does not diminish, but is replaced

© Springer International Publishing Switzerland 2015
X. Hu et al. (Eds.): ISNN 2015, LNCS 9377, pp. 205–212, 2015.
DOI: 10.1007/978-3-319-25393-0_23

by another task to determine a more uninterpreted parameter. (3) The inference algorithm may introduce randomness. This leads to inconsistent results from multiple runs when the human factor is incorporated in to the loop. Such a problem was pointed by Ref.[3]. In Ref.[1], a relevance feedback strategy is proposed to help action search and localization in video database. All of the above work do not consider how to use the human-machine interface to enhance the action discovery performance. Recently, Ref.[7] addresses this problem for image clustering by introducing some human operation, and Ref.[3] used interactive non-negative matrix method for document topic discovery. To the best of the knowledge of the authors, there is no related work to solve video action discovery using human operation. This motivates us to solve this problem. The main task of this work is to discover the action categories within a video sequence, and identify such actions in this video sequence. The main contributions are summarized as follows: (1)We develop an interactive non-negative matrix factorization method for representative video action discovery. (2)We design two human operations: ADD and MERGER to realize the relevance feedback and enhance the video summarization performance. (3)We develop a practical software system and perform extensive experimental validations for the proposed method.

The rest of this paper is organized as follows: Section 2 is about the video representation. In section 3 we give a detailed introduction about the proposed method and Section 4 presents the experimental results.

2   Video Representation

The first-of-all task to analyze a video is to transform it into some suitable structured form. In this work, we follow the popular Bag-of-Words framework which was successfully utilized many action analysis work. To this end, we use Spatio-Temporal Interest Points (STIPs) to detect interest points and obtain Histogram-Of-Gradients (HOG) and Histogram-Of-Optical flow (HOF) descriptors. The obtained default descriptors is of $d = 162$ dimensions. We evenly divide to original video into segments which length is T frames. The parameter T is specified by the users. It should ensure the action consistency within each segment. In this work, we select $T = 24$ frames,which means about one second. These segments, which are denoted as $\mathbf{P}_1, \mathbf{P}_2, \cdots, \mathbf{P}_N$, represent the basic units of the actions. The final action summary should include such segments. The value N is obtained by the ceil of the whole frame numbers divided by T.

To give a formal representation of the segments, we first cluster all of the descriptors in this video into K clusters.The parameter K is also a meta-parameter which is specified by the users. A larger K will give better accuracy, but will also slow down the summarization period. In this work we empirically set it as 128. The obtained K cluster centers are regarded as code-words. Then each descriptor is mapped to the nearest code-word and each segment can be represented as a K-dimensional BoW histogram[1]. We therefore can represent the whole video as $\{\mathbf{y}_1, \mathbf{y}_2, \cdots, \mathbf{y}_N\}$, where $\mathbf{y}_i$ is the K-dimensional BoW histogram for the i-th segment. After this period, each video can be represented as a matrix $\mathbf{Y} = [\mathbf{y}_1, \mathbf{y}_2, \cdots, \mathbf{y}_N] \in R^{K \times N}$.

3 Non-negative Matrix Factorization for Video Action Discovery

3.1 Basic Non-negative Matrix Factorization

Given the matrix $\mathbf{Y} \in R^{K \times N}$ which includes the low-level action information of the original video, where N is the number of examples in the video. We then face the problem of how to extract the representative action clips from the matrix $\mathbf{Y}$ and then project each column to the corresponding representative action clip, providing the action segmentation results. A representative method is the popular Non-negative Matrix Factorization (NMF) in Ref.[2], which solves the following optimization problem:

$$\min_{\mathbf{U},\mathbf{V}} \|\mathbf{Y} - \mathbf{U}\mathbf{V}\|_F^2 \qquad s.t.\ \mathbf{U} \geq 0,\ \mathbf{V} \geq 0 \tag{1}$$

where $\mathbf{U} \in R^{K \times r}$ and $\mathbf{V} \in R^{r \times N}$ are two non-negative matrices. The term-topic matrix $\mathbf{U}$ uncovers the latent topic structure of the actions and r is usually set by the users and chosen to be smaller than K or N.

Once the solutions of $\mathbf{U}$ and $\mathbf{V}$ are obtained, we can subsequently infer the topic presentations of segments, namely the topic-segment matrix $\mathbf{V}$ by projecting the segments into the latent topic space. Such a model was originally proposed in Ref.[3] and then was used in many fields such as document clustering and image clustering. However, in our work, since we deal with continuous video, the temporal consistence should be encouraged to reflect the continuity of action. Therefore the model is modified as

$$\min_{\mathbf{U},\mathbf{V}} \|\mathbf{Y} - \mathbf{U}\mathbf{V}\|_F^2 + \beta \sum_{i=1}^{N-1} \|\mathbf{V}_{i+1} - \mathbf{V}_i\|_F^2 \tag{2}$$
$$s.t.\ \mathbf{U} \geq 0,\ \mathbf{V} \geq 0$$

where β is a parameter to encourage the temporal consistency term, and $\mathbf{V}_i$ represents the i-th column of $\mathbf{V}$.

After obtaining the solutions $\mathbf{U}$ and $\mathbf{V}$, we can easily obtain the discovered representative actions and the temporal action segmentation results. The details are described as follows. For $\mathbf{U}$, each column $\mathbf{U}_i \in R^K$ corresponds a representative action clip. By searching $i^* = \underset{j \in [1,N]}{\operatorname{argmin}} \frac{\mathbf{U}_i^T \mathbf{y}_j}{\|\mathbf{U}_i\|_2 \cdot \|\mathbf{y}_j\|_2}$, we can use the the video clip $\mathbf{P}_{i^*}$ as the corresponding representative action clip. On the other hand, we use the column $\mathbf{V}_j \in R^r$ for $j = 1, 2, \cdots, N$ to determine the cluster assignment of the j-th video clip and therefore realize the action segmentation. Concretely speaking, we search the maximum element in the vector $\mathbf{V}_j$ and use the corresponding index as the clustering assignment results.

3.2 Interactive Non-negative Matrix Factorization

In Ref.[3], some interesting interaction operation, such as key-words operations are used for interactive topic discovery or refinement. Such operations cannot been exploited in the video scenarios. The main reason is that for document clustering, the dictionary atom is the conventional words (such as *dog*, *apple*, *play*, *eat*, and so on.) which have the semantic meanings, so we can use the keyword distribution of each topic to realize the visualization. However, it is impossible to construct such a dictionary for a video. In our case, the dictionary is learning using K-means clustering algorithm and therefore the words do not have any semantic meaning. As a result, the key-words based operation defined in Ref.[3] cannot be used. Due to the same reason, such a visualization manner is not suitable in our case. To this end, we developed two interaction operations: ADD and MERGE for the visualized video action discovery.

MERGE Operation. The merge operation tries to solve the problem that some similar video segments may be clustered into different topics. This is unavoidable due to at least two reasons: (1) The semantic gap between the human understanding and the adopted BoW model which is based on low-level feature descriptor. (2) The results are not consistent to the user's intention.

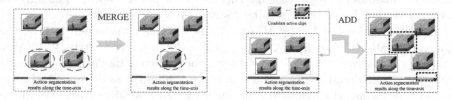

Fig. 1. MERGE operation   **Fig. 2.** ADD operation

To solve this problem, we permit the user to click the visualization action boxes and click the button *merge* to tell the computer that some segments should be merged into the same topic in the next iteration. This interaction also provides very important supervised information that we can exploit to enhance our model. In fact, the visualization of actions are shown as the video segments $\mathbf{P}_{t_1}, \mathbf{P}_{t_2}, \cdots, \mathbf{P}_{t_r}$. Without loss of generalization, we denote the selected merge segments as $\mathbf{P}_i$ and $\mathbf{P}_j$, then we add this pair into a set $\mathcal{M} = \mathcal{M} \cup \{(i,j)\}$, then we solve the following optimization problem in the next iteration:

$$\min_{\mathbf{U},\mathbf{V}} ||\mathbf{Y} - \mathbf{U}\mathbf{V}||_F^2 + \beta \sum_{i=1}^{N-1} ||\mathbf{V}_{i+1} - \mathbf{V}_i||_F^2 + \gamma \sum_{(i,j)\in\mathcal{M}} ||\mathbf{V}_i - \mathbf{V}_j||_F^2 \qquad (3)$$

$s.t. \ \mathbf{U} \geq 0, \ \mathbf{V} \geq 0.$

The main characteristic of this model is the third term which encourages the i-th and j-th segments to share the similar topic pattern, and γ is a trade-off parameter. Please note that the pair set $\mathcal{M}$ is set to empty for the first iteration. During iteration procure, once $\mathcal{M}$ is added with some pair elements, it always play roles in the subsequent iterations.

ADD Operation. Though the above model can successfully discover most of the representative actions from the video, it is still possible that some important action clips cannot be discovered automatically. To this end, a candidate list of new action clips should be presented to the user for performing the ADD operation. Such a list should be short and representative. Ideally, it should contain only the actions which are not included in the list of the discovered topical actions. That is to say, it should not be well reconstructed by the discovered representative actions. Based on the above discussion, we design a performance index to evaluate the novelty of each action clip. To this end, for each video segments, we define its confidence about the topic assignment. We regard $\bar{\mathbf{V}}_i$ as the L_1 normalized i-th column of $\mathbf{V}$, and then adopt its entropy function as $En(\mathbf{V}_i) = -\sum\limits_{j=1}^{r} \bar{\mathbf{V}}_i(j) \log \bar{\mathbf{V}}_i(j)$. Obviously, when there is only one element of $\mathbf{V}_i$ is nonzero and equal to one, then the entropy is zero and the confidence score is maximum. On the other hand, when all the elements of $\mathbf{V}_i$ are equal to $1/r$, then the entropy is equal to $\log_2 r$ and the confidence score is minimum. Therefore, it is very convenient to adopt the entropy to select the most uncertain video segments for the operation ADD. In this work, we sort the entropies (in descending order) of all video segments which are not visualized and not deleted in the former stages, and then select the top N_a segments for visualization in a specifically design region and the user can browse them in a short time and then select some ones to add in the next iterations. The number N_a should not be too large, otherwise the user will be strongly confused. In this paper, it is set to 5. That is to say, at each iteration stage, we provide 5 most uncertain video segments for the user for possible ADD operation.

Once some action of which representation is $\mathbf{y}_i$ is selected to be added, then we should increase the number of r by one in the next operation and make some adjustments. Concretely speaking, we augment the topic matrix as $\bar{\mathbf{U}} = [\mathbf{U} \ \mathbf{y}_i] \in R^{N \times (r+1)}$. The optimization problem then becomes:

$$\min_{\mathbf{V}} \|\mathbf{Y} - \bar{\mathbf{U}}\mathbf{V}\|_F^2 + \beta \sum_{i=1}^{N-1} \|\mathbf{V}_{i+1} - \mathbf{V}_i\|_F^2 + \gamma \sum_{(i,j)\in\mathcal{M}} \|\mathbf{V}_i - \mathbf{V}_j\|_F^2, s.t. \ \mathbf{V} \geq 0$$

(4)

Note that in the above model, $\bar{\mathbf{U}}$ is known and only $\mathbf{V}$ should be calculated.

3.3 Optimization Method

All of the model in (2), (3) and (4) can be efficiently solve by the regularized NMF method proposed in Ref.[4]. To this end, we should construct a nearest neighbor graph to encode the consistency information of the data points. Consider a graph with vertices where each vertex corresponds to a data point. Define the edge weight matrix $\mathbf{W} \in R^{N \times N}$ as follows:

$$\mathbf{W}_{ij} = \begin{cases} \beta, & \text{if } |i - j| = 1 \\ \gamma, & \text{if } \{i, j\} \in \mathcal{M} \text{ and } |i - j| \neq 1 \\ 0, & \text{otherwise.} \end{cases}$$

(5)

Define a diagonal matrix $\mathbf{D}$, whose entries are column sums of $\mathbf{W}$, i.e., $\mathbf{D}_{ii} = \sum_{j=1}^{N} \mathbf{W}_{ij}$. Then the reformulated optimization problem leads to the two new following update rules[4]:

$$\mathbf{U}_{ij} \leftarrow \mathbf{U}_{ij} \frac{(\mathbf{YV}^T)_{ij}}{(\mathbf{UVV}^T)_{ij}}, \mathbf{V}_{ij} \leftarrow \mathbf{V}_{ij} \frac{(\mathbf{U}^T\mathbf{V} + \mathbf{VW})_{ij}}{(\mathbf{U}^T\mathbf{UV} + \mathbf{VD})_{ij}} \qquad (6)$$

where the subscript ij represents the i, j-th element in the corresponding matrix. The detailed algorithm flow and analysis can be found in [4].

4 Performance Evaluation

4.1 Dataset

We use the well-known Weizman database[5] of 90 low-resolution video sequences showing 9 different people, each performing 10 natural actions such as run, walk, skip, jack, jump, pjump, side, wave2, wave1 and bend. To evaluate the performance of our interactive method, we have created a "stitched" version of the weizman dataset into uninterrupted sequences. Each sequence depicts a single person performing 10 actions for a total duration of approximately 700 frames.

How to evaluate our approach is still an open problem. Generally, if ground truth is available, many evaluation metrics are available for clustering, such as purity, and normalized mutual information. To this end, the ground truth for each video is established manually based on the exact actions in every single sequence.

4.2 Operation Process

Figure 3 illustrates the two operations. In some cases, similar actions may be extracted. Such case often occurs when the number r is set to a large value. The operation MERGE allows us to merge the similar actions selected by the user when he press the *merge* button, which is shown in Figure 3(a). On the other hand, we need to find as more actions in the whole video sequence as possible. Figure 3(b) demonstrates the process of this operation. Users select the new actions from the list of candidate actions. By pressing the *add* button, the selected actions are added into the clustering results. Note that when performing either ADD or MERGE operations, we modify the model to produce the expecting clustering performance according to users' operations and the action segmentation result is demonstrated along the time axis in different color bars.

4.3 Performance Evaluation on the Interactive Interaction

To evaluate the action segmentation results in each iteration, we adopt the purity and NMI indices which are popular in the community of clustering. Purity[8] is a simple and transparent evaluation measure. To compute purity, each cluster is assigned to the class which is most frequent in the cluster, and then the accuracy of this assignment is measured by counting the number of correctly assigned segments and dividing by N which is the total number of the whole video segments.

(a) MERGE Operation   (b) ADD Operation

Fig. 3. Demonstrations of MERGE and ADD operation

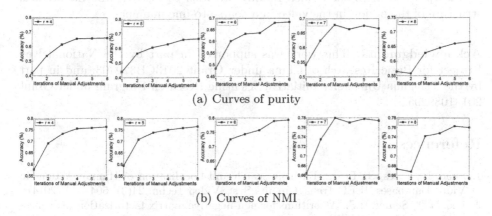

(a) Curves of purity

(b) Curves of NMI

Fig. 4. Clustering performance of purity and NMI.

Denote the ground-truch action clustering results as $\Omega = \{\omega_1, \omega_2, \ldots, \omega_g\}$ and $\mathbb{C} = \{c_1, c_2, \ldots, c_r\}$ as the practical clustering results, then the purity is defined as $Purity(\Omega, \mathbb{C}) = \frac{1}{N} \sum_k \max_j |\omega_k \cap c_j|$. Since high purity value is easy to achieve when the number of clusters is large and particularly, purity is 1 if each segment gets its own cluster. Thus we cannot use purity only to trade off the quality of the clustering against the number of clusters. Normalized mutual information(NMI)[8] is confident to make this tradeoff and can be information-theoretically interpreted $\mathbf{NMI}(\Omega, \mathbb{C}) = \frac{I(\Omega;\mathbb{C})}{[H(\Omega)+H(\mathbb{C})]/2}$, where I is the mutual information and $H(\cdot)$ represents the entropy.

Using these two evaluation metrics, we conduct experiments with Weizman dataset and there are 3 different users involved in this process. They make their adjustments to obtain the willing performance which is to find as more actions as possible during the whole process. By setting the different initial value of the number of clusters(r ranges from 4 to 8), we compute the average accuracy of the 9 video sequence. Figure 4 illustrates the clustering performance using our interactive method. We can see that the clustering performance can be significantly improved by adopting manual interactive adjustments. Note that when $r = 8$, the accuracy declined. Because when the number of cluster r is getting large, users have to compromise to the higher possibility of exploiting same actions so that they need to merge the very several actions, which results in the phenomenon that fewer number of actions leads to lower accuracy rate.

5   Conclusion and Future Work

This paper proposed an interactive method to detect representative actions within streaming or archival video. Incorporated with user's intention, expecting results have been obtained. However, there still exists a lot work to be further investigated. Firstly, we wish to extend the work on single video to video sets and discover more sensible behavior patterns for the end users; Secondly, we hope to develop more flexible interface and more high-level knowledge of the human can be incorporated in to the model. Finally, we wish to discover the hierarchical structure of the action in the video in a coarse-to-fine manner.

Acknowledgments. This work was supported in part by the National Key Project for Basic Research of China under Grant 2013CB329403; and in part by the Tsinghua University Initiative Scientific Research Program under Grant 20131089295.

References

1. Shao, L., Jones, S., Li, X.: Efficient Search and Localization of Human Actions in Video Databases. IEEE Trans. Circuits Syst. Video Techn. 24(3), 504–512 (2014)
2. Lee, D.D., Seung, H.S.: Algorithms for non-negative matrix factorization. Advances in Neural Information Processing Systems, 556–562 (2001)
3. Choo, J., Lee, C., Reddy, C.K., Park, H.: Utopian: User-driven topic modeling based on interactive nonnegative matrix factorization. IEEE Trans. on Visualization and Computer Graphics, 1992–2001 (2013)
4. Cai, D., He, X., Wu, X., Han, J.: Non-negative matrix factorization on manifold. In: Proc. of Eighth IEEE International Conference in Data Mining(ICDM), pp. 63–72 (2008)
5. Blank, M., Gorelick, L., Shechtman, E., Irani, M., Basri, R.: Actions as space-time shapes. In: Proc. of Tenth IEEE International Conference on Computer Vision (ICCV), pp. 1395–1402 (2005)
6. Hughes, M.C., Sudderth, E.B.: Nonparametric discovery of activity patterns from video collections. In: Computer Vision and Pattern Recognition Workshops (CVPRW), pp. 25–32 (2012)
7. Wang, M., Ji, D., Tian, Q., Hua, X.S.: Intelligent photo clustering with user interaction and distance metric learning. Pattern Recognition Letters, 462–470 (2012)
8. Evaluation of clustering. http://nlp.stanford.edu/IR-book/html/htmledition/evaluation-of-clustering-1.html

Image Retrieval Based on Texture Direction Feature and Online Feature Selection

Xiaohong Ma* and Xizheng Yu

School of Information and Communication Engineering,
Dalian University of Technology, Dalian, China
maxh@dlut.edu.cn

Abstract. In this paper, a new method for image texture representation is proposed, which represents image content using a 49 dimensional feature vector through calculating the variation of texture direction and the intensity of texture. In addition, the texture feature is grouped into a feature set with some other image texture representation methods, and then a new online feature selection method with a novel discrimination criterion is presented. We test the discriminating ability of every feature in the feature set utilizing the discrimination criterion, and select the optimal feature subset, which expresses image content in an even better fashion. The results of the computer simulation experiments show that the proposed feature extraction and feature selection method can represent image content effectively, and improve the retrieval precision visibly.

Keywords: Image retrieval, texture direction feature, online feature selection, discrimination criterion.

1 Introduction

With the development of computer technology, a mass of multimedia information grows out of Internet. We can get these datum on the Internet, but at the meanwhile, it becomes harder and harder for us to find useful information. In order to obtain datum that users are concerned about, content based image retrieval (CBIR) becomes a research focus in the field of computer vision. In tradition, image retrieval systems fulfill image indexing via keywords annotation, but it needs a good deal of manual operation, and keywords annotation depends much on people who label the images, there may be different understanding of the same image among different people. Compared with text-based image retrieval (TBIR), CBIR system extracts image visual features automatically.

For the past few years, researchers present many feature extraction methods. Liu *et al.* build micro-structure descriptor (MSD) [1] according the similarity of edge direction and statistic characteristic of color feature, so it blends color, texture, shape and spatial information together. Yang *et al.* describe image content with 4-5 kinds of prominent colors, and extracts dominant color descriptor (DCD) [2]. Balasubramani *et al.* extract edge histogram descriptor (EHD) [3]

---

* Corresponding author.

© Springer International Publishing Switzerland 2015
X. Hu et al. (Eds.): ISNN 2015, LNCS 9377, pp. 213–221, 2015.
DOI: 10.1007/978-3-319-25393-0_24

via calculating edge distribution of an image block with different edge operators. Young *et al.* calculate block difference of inverse probabilities (BDIP) and block variation of local correlation coefficients (BVLC) [4] on an image block, and then get the corresponding texture feature.

The feature selection technology basically narrows the semantic gap by selecting a feature subset. Feature selection methods are classified into Filter-based feature selection [5] and Wrapper-based feature selection [6]. The evaluation criterion of Filter-based feature selection is determined by properties of the data itself, so it is independent of learning algorithm. The frequently-used Filter-based feature selection algorithms are Relief algorithm [7] and Mitra algorithm [8]. Wrapper-based feature selection evaluates the performance of feature subset using learning algorithms, and then chooses the feature subset with higher precision rate.

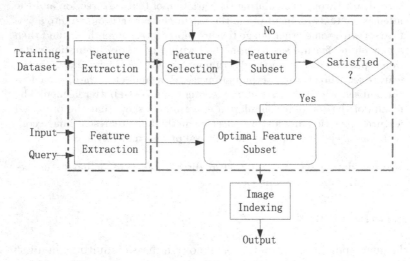

Fig. 1. Image Retrieval System Based on Feature Extraction and Selection.

In this paper, a new image retrieval method is proposed by means of combining feature extraction and selection. First, we compute the direction variation and intensity of pixel values in an image block, which is divided into different texture patterns. Two image blocks in neighborhood make up a pattern pair, we obtain the Texture Direction Descriptor by counting the number of the pattern pairs. Then, Texture Direction Descriptor constitutes an image feature set with other image features. Test the discriminating ability of every feature utilizing a discrimination criterion [8][9], and select the optimal feature subset with the best discriminating ability. The system chart of the proposed image retrieval system is shown in Fig. 1.

The outline of the paper is as follows. Section 2 presents the feature extraction approach in detail. Section 3 proposes the online feature selection method. Section 4 provides the results of experiments. Conclusions are given in Section 5 at the end of the paper.

2   Texture Direction Feature

In this paper, we propose a new texture feature extraction method. Methods for texture feature extraction generally obtain the texture image first, and then make a statistical analysis of the texture image. Different from previous methods, the proposed method calculates the variation of texture direction and intensity on original image directly.

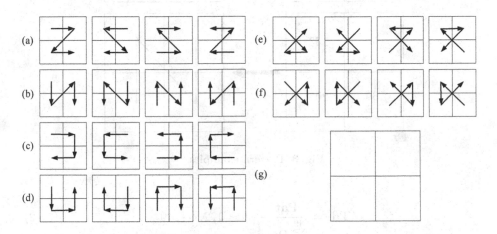

Fig. 2. Patterns of texture direction.

There are connections among pixels of an image, those with the most closely connections are the adjacent pixels, which compose the neighborhood. Image texture can be expressed by the variation of the pixels in neighborhood. Scan the pixels on a 2×2 image block according to the changing of gray value in ascending order, an image block is classified into one kind of patterns in Fig. 2.(a)-(f). If there are equivalent pixels on the image block, define it as non-direction pattern, as shown in Fig. 2.(g). Fig. 3. shows two examples of these patterns. Although the patterns discussed above contain the texture information of an image, we can't realize image retrieval yet. In order to extract the effective texture feature, take a 4×4 image block, which can be divided into four 2×2 subblock in further. A 2×2 subblock forms a neighborhood with the adjacent 2×2 subblock. As there are 7 kinds of patterns, there are 49 kinds of pattern pairs. Record the pattern pairs as **Pair**(j, k), define

$$\mathbf{Pat}_i = \mathbf{Pair}(j, k) \quad i = 1, 2, \cdots, 49;$$
$$j = 1, 2, \cdots, 7; \quad k = 1, 2, \cdots, 7 \tag{1}$$

Count the number of occurrences of pattern pairs, which is defined as the probabilities of emerged texture direction, namely

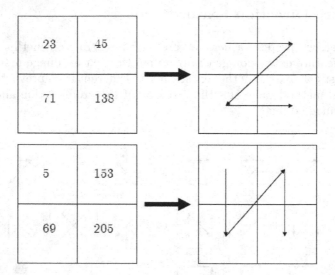

Fig. 3. Pattern examples.

$$\mathbf{Td}_i = \frac{\mathbf{Pat}_i}{\sum\limits_{i=1}^{49} \mathbf{Pat}_i} \quad i = 1, 2, \cdots, 49 \tag{2}$$

Td is the 49-dimension Texture direction descriptor (TDD), which is consisted by $\mathbf{Td}_i$, $i = 1, 2, \cdots, 49$.

TDD is provided with invariance of translation, rotation, zoom, but it just considers the variance of gray direction only, not the variance of gray intensity, so we represent texture intensity by the mean gray value of pattern pairs. When the i_{th} kind of pattern pair appears, record the two image blocks as $G_i(j)$, the texture intensity is defined by the gray intensity of different pattern pairs, namely

$$\mathbf{Ti}_i = \frac{1}{\mathbf{Pat}_i} \sum\limits_{j=1}^{\mathbf{Pat}_i} G_i(j) \quad i = 1, 2, \cdots, 49 \tag{3}$$

In Formula (3), $\mathbf{Pat}_i$ is the number of occurrences of the i_{th} pattern pair. $\mathbf{Ti}_i$, $i = 1, 2, \cdots, 49$ constitutes the 49-dimension texture intensity descriptor (TID) **Ti**.

3   Online Feature Selection

In this section, a method of feature selection with discrimination criterion (FSDC) is presented. The problem that must be tackled in online learning is that how to find out the more representative features. Compared with other machine learning problems, online learning in CBIR system should give out the results at a high rate of speed. In addition, the size of the training set must be small due to the curse of dimensionality. We propose a novel feature selection criterion, which is based on the similarity among different training samples. We just need a small scale of training samples, and the time consumed during training is low, so the proposed method is suitable for online learning in CBIR system.

Denote $F_x = [f_1(x), \cdots, f_k(x), \cdots, f_d(x)]^T$ as the d-dimensional feature vector of image x , the relevant image set is $D = \{x_i^D, i = 1, \cdots, p\}$, and the irrelevant image set is $I = \{x_j^I, j = 1, \cdots, q\}$, the label of relevant images and irrelevant images is $y(x_i^D) = 1$, $y(x_j^I) = -1$ respectively, D_k and I_k represents the projection of D and I along with the k_{th}Cdimensional feature, as given below.

$$\begin{cases} D_k = \{f_k(x_1^D), \cdots, f_k(x_i^D), \cdots, f_k(x_p^D)\} \\ I_k = \{f_k(x_1^I), \cdots, f_k(x_j^I), \cdots, f_k(x_q^I)\} \end{cases} \tag{4}$$

The relation between x and D, I is R_k, U_k respectively, as shown in Formula (5).

$$\begin{cases} R_k = \sum\limits_{i=1}^{p} (f_k(x) - f_k(x_i^D))^2, \quad k = 1, 2, \cdots, d \\ U_k = \sum\limits_{j=1}^{q} (f_k(x) - f_k(x_j^I))^2, \quad k = 1, 2, \cdots, d \end{cases} \tag{5}$$

The discriminating ability of each feature, otherwise known as the discrimination criterion

$$A_k = R_k/U_k, \quad k = 1, 2, \cdots, d \tag{6}$$

Realign A_k in ascending order as $\widetilde{A}_k$, and select the headmost features according to $\widetilde{A}_k$ of size B. Estimate the category of each sample, as given below.

$$\begin{cases} \widehat{y}(x_i) = 1 \, if \, \mathbf{DIF}_i \leq Thr \\ \widehat{y}(x_i) = -1 \, if \, \mathbf{DIF}_i > Thr \end{cases} \tag{7}$$

$\mathbf{DIF}_i$ and Thr is given in Formula (8) and (9).

$$\mathbf{DIF}_i = \sum_{k=1}^{B} (f_k(x) - f_k(x_i))^2, \quad i = 1, 2, \cdots, p + q \tag{8}$$

$$Thr = \sum_{k=1}^{B} (R_k + U_k)/(p + q) \tag{9}$$

Training error is

$$\phi = \frac{\sum\limits_{i=1}^{p+q} |y(x_i) - \widehat{y}(x_i)|}{(2 * (p + q))} \tag{10}$$

where $y(x_i)$ is the actual category of each training sample $\widehat{y}(x_i)$ is the estimated category of each training sample. The training error of feature subset $S_i, i =$

$1, 2, \cdots, B$ is ϕ^B on the training dataset, select the optimal feature subset by minimizing ϕ^B, the dimension of the optimal feature subset is

$$\widehat{B} = \arg\min_B \phi^B \qquad (11)$$

the possible value of $\widehat{B}$ is $1, 2, \cdots, d$.

4   Experimental Results

In this section, we evaluate the effectiveness of the proposed feature extraction and feature selection method. First, we describe the image databases. Thereafter, we present the results of an evaluation showing how similar images can be found, and how retrieval precision can be improved visibly.

4.1   Experimental Dataset

For this research, we conduct the experiments on two image databases. The first image database Wang (http://www.ist.psu.edu/docs/related/shtml) contains about 11000 images. The second image database Caltech101 (http://www.vision.caltech.edu/Image_Datasets/Caltech101/) contains 101 categories of images. The number of images in each category ranges from 33 to 800. By using our feature extraction and selection method, we can select the optimal feature subset that best discriminate among different classes of images, and search the images which are similar to the query.

4.2   Recall versus Precision

To evaluate the effectiveness of feature extraction and feature selection method, we compare the proposed method with MSD and BDIP&BVLC on dataset of images Wang and Caltech101. Fig. 4. illustrates two results on image database Wang, and Fig. 5.(a) is the comparison of recall and precision for the three methods, it can be perceived from Fig. 5(a), the proposed method outperforms the

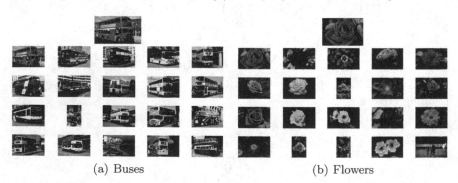

(a) Buses   (b) Flowers

Fig. 4. Examples of the results.

other two methods. Texture Direction Descriptor makes up a Dim dimensional feature set with MSD, DCD and EHD, selects a feature subset with Q dimension from the feature set, where $Q = 10, 20, 30, \cdots, Dim$. The feature subset with the highest precision rate is selected as the optimal feature subset. Experimental results show that precision rate reaches the highest when the dimension of feature subset is 80. Fig. 5.(b) is a curve of recall and precision for texture direction feature and proposed online feature selection method, it can be perceived from Fig. 5.(b) that retrieval efficiency is significantly improved after dealing with the proposed online feature selection method.

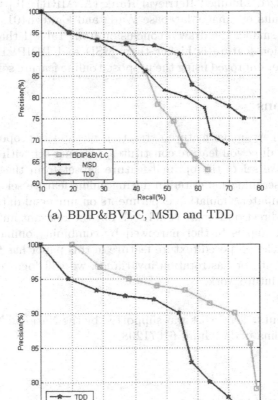

(a) BDIP&BVLC, MSD and TDD

(b) FSDC and TDD

Fig. 5. Comparison of recall versus precision.

Table 1. ARR and ANMRR

	ARR	ANMRR
BDIP&BVLC	0.8168	0.2132
MSD	0.8436	0.1963
TDD	0.8784	0.1739
FSDC	0.9250	0.1437

4.3 ARR and ANMRR

In order to verify the effectiveness of texture direction feature and online feature selection method in further, we adopt Average Retrieval Rate (ARR) [10] and Average Normalized Modified Retrieval Rank (ANMRR) [10] to evaluate the experimental results on image database Wang and Caltech101. The larger the ARR is, and the smaller the draw a conclusion from Table 1 that the efficiency of texture direction feature is higher than MSD and BDIP&BVLC, and the efficiency is further improved using the proposed online feature selection method.

5 Conclusions

In this paper, a new image feature extraction method is proposed. Then, the proposed texture direction feature constitutes a feature set with other low level visual features. We select the optimal feature subset from the feature set applying a novel discrimination criterion during online feature selection step. The results of the computer simulation experiments on universal databases indicate that the texture direction feature could seek out the relevant images effectively, and the semantic gap is further narrowed by combining online feature selection technology. The proposed texture feature in this paper has translation and scale invariance but not has rotation invariance, we will focus on the rotation invariance in the future work.

Acknowledgments. This work was supported by the National Natural Science Foundation of China under Grant 61071208.

References

1. Liu, G.H., Li, Z.Y., Zhang, L., Xu, Y.: Image retrieval based on micro structure descriptor. J. Pattern Recognition 44(9), 2123–2133 (2011)
2. Yang, N.C., Chang, W.H., Kuo, C.M.: A fast MPEG-7 dominant color extraction with new similarity measure for image retrieval. In: VCIR (2008)
3. Balasubramani, R., Kannan, D.V.: Efficient use of MPEG-7 color layout and edge histogram descriptors in CBIR systems. J. Global Journal of Computer Science and Technology 9(4), 157–163 (2009)
4. Young, D.C., Sang, Y.S., Nam, C.K.: Image retrieval using BDIP and BVLC moments. J. IEEE Transactions on Circuits and Systems for Video Technology 13(9), 951–957 (2003)

5. He, X.F., Ji, M., Zhang, C.Y.: A variance minimization criterion to feature selection using Laplacian Regularization. J. IEEE Transactions on Pattern Analysis and Machine Intelligence 33(10), 2013–2025 (2011)
6. Yang, J.B., Shen, K.Q., Ong, C.J.: Feature selection for MLP neural network: the use of random permutation of probabilistic outputs. J. IEEE Transactions on Neural Networks 20(12), 1911–1922 (2009)
7. Yew, S.O.: Relief-C: Efficient feature selection for clustering over noisy data. In: ICTAI (2011)
8. Mitra, P., Murthy, C.A., Pal, S.K.: Unsupervised feature selection using feature similarity. J. IEEE Transactions on Pattern Analysis and Machine Intelligence 24(3), 301–312 (2002)
9. Zhou, L.P., Wang, L., Shen, C.H.: Feature selection with redundancy-constrained class separability. J. IEEE Transactions on Neural Networks 21(5), 853–858 (2010)
10. Manjunath, B.S., Jens, R.O., Vasudevan, V.V.: Color and texture descriptors. J. IEEE Transactions on Circuits and Systems for Video Technology 11(6), 703–715 (2001)

Interlinked Convolutional Neural Networks
for Face Parsing

Yisu Zhou, Xiaolin Hu, and Bo Zhang

State Key Laboratory of Intelligent Technology and Systems
Tsinghua National Laboratory for Information Science and Technology (TNList)
Department of Computer Science and Technology
Tsinghua University, Beijing 100084, China

Abstract. Face parsing is a basic task in face image analysis. It amounts to labeling each pixel with appropriate facial parts such as eyes and nose. In the paper, we present a interlinked convolutional neural network (iCNN) for solving this problem in an end-to-end fashion. It consists of multiple convolutional neural networks (CNNs) taking input in different scales. A special interlinking layer is designed to allow the CNNs to exchange information, enabling them to integrate local and contextual information efficiently. The hallmark of iCNN is the extensive use of downsampling and upsampling in the interlinking layers, while traditional CNNs usually uses downsampling only. A two-stage pipeline is proposed for face parsing and both stages use iCNN. The first stage localizes facial parts in the size-reduced image and the second stage labels the pixels in the identified facial parts in the original image. On a benchmark dataset we have obtained better results than the state-of-the-art methods.

Keywords: Convolutional neural network, face parsing, deep learning, scene labeling.

1 Introduction

The task of image parsing (or scene labeling) is to label each pixel in an image to different classes, e.g., person, sky, street and so on [1]. This task is very challenging as it implies jointly solving detection, segmentation and recognition problems [1]. In recent years, many deep learning methods have been proposed for solving this problem including recursive neural network [2], multiscale convolutional neural network (CNN) [3] and recurrent CNN [4]. To label a pixel with an appropriate category, we must take into account the information of its surrounding pixels, because isolated pixels do not exhibit any category information. To make use of the context, deep learning models usually integrate multiscale information of the input. Farabet et al. [3] extract multi-scale features from image pyramid using CNN. Pinheiro et al. [4] solve the problem using recurrent CNN, where the coarser image is processed by a CNN first, then the CNN repeatedly takes its own output and the finer image as the joint input and proceeds. Socher

---

The original version of this chapter was revised: Contents in Table 1 have been corrected. The erratum to this chapter is available at https://doi.org/10.1007/978-3-319-25393-0_56

© Springer International Publishing Switzerland 2015
X. Hu et al. (Eds.): ISNN 2015, LNCS 9377, pp. 222–231, 2015.
DOI: 10.1007/978-3-319-25393-0_25

et al. [2] exploit structure of information using trees. They extract features from superpixels using CNN, combine nearby superpixels with same category recursively.

As a special case of image parsing, face parsing amounts to labeling each pixel with eye, nose, mouth and so on. It is a basic task in face image analysis. Compared with general image parsing, it is simpler since facial parts are regular and highly structured. Nevertheless, it is still challenging since facial parts are deformable. For this task, landmark extraction is a common practice. But most landmark points are not well-defined and it is difficult to encode uncertainty in landmarks like nose ridge [5]. Segmentation-based methods seem to be more promising [5][6].

In the paper, we present a deep learning method for face parsing. Inspired by the models for general image parsing [3][4], we use multiple CNNs for processing different scales of the image. To allow the CNNs exchange information, an interlinking layer is designed, which concatenates the feature maps of neighboring CNNs in the previous layer together after downsampling or upsampling. For this reason, the proposed model is called *interlinked CNN* or *iCNN* for short. The idea of interlinking multiple CNNs is partially inspired by [7] where multiple classifiers are interlinked.

Experiments on a pixel-by-pixel [5] labeling version of the Helen dataset [8] demonstrate the effectiveness of iCNN compared with existing models.

2   iCNN

The overall structure of the proposed iCNN is illustrated in Fig. 1. Roughly speaking, it consists of several traditional CNN in parallel, which accept input in different scales, respectively. These CNNs are labeled CNN-1, CNN-2, ... in the order of decreasing scale. The hallmark of the iCNN is that the parallel CNNs interact with each other. From left to right in Fig. 1, the iCNN consists of alternating convolutional layers and interlinking layers, as well as an output layer, which are described as follows.

2.1   Convolutional Layers

The convolutional layers are the same as in the traditional CNN, where local connections and weight sharing are used. For a weight kernel $w_{uvkq}^{(l)}$, the output of a unit at (i, j) in the l-th layer is

$$y_{ijq}^{(l)} = f\left(\sum_{k=1}^{C}\sum_{u=1}^{P_1}\sum_{v=1}^{P_2} w_{uvkq}^{(l)} y_{i+u,j+v,k}^{(l-1)} + b^{(l)}\right) \tag{1}$$

where P_1 and P_2 denote the size of the weight kernel in the feature map, C denotes the number of channels in the $(l-1)$-th layer, $b^{(l)}$ denotes the bias in the l-th layer, and $f(\cdot)$ is the activation function. Throughout the paper, tanh function is used as the activation function. If we use Q kernels $w_{uvkq}^{(l)}$, that is,

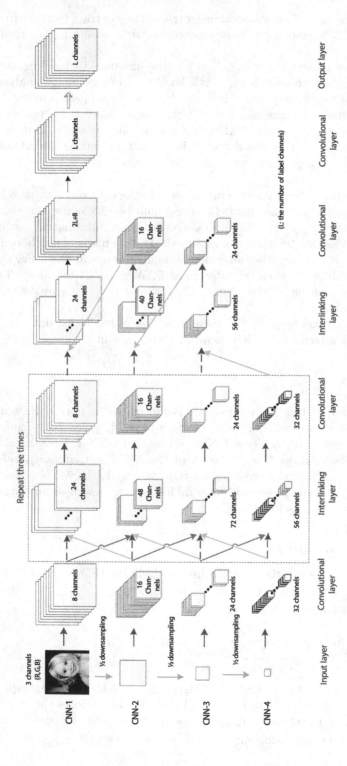

Fig. 1. The structure of the iCNN. Black solid arrows: convolution and nonlinear transformation as described in (1). Black dotted arrow: convolution. Black open arrow: softmax. Green arrows: downsampling (mean pooling). Black dashed arrows: pass the feature maps to the next layer. Blue arrows: downsampling (max pooling) the feature maps and pass them to the next layer. Yellow arrows: upsampling (nearest neighbor) the feature maps and pass them to the next layer. Best viewed in color.

$q = 1, \ldots, Q$, then a total number of Q feature maps (the q-th feature map consists of $y_{ijq}^{(l)}$ for all i, j) will be obtained in the l-th layer.

The operation in the bracket in (1) can be implemented by tensor convolution. The surrounding of feature maps in the $(l-1)$-th layer are padded with zeroes such that after convolution and activation the resulting feature maps in the l-th layer has the same size in the first two dimensions as the feature maps in the $(l-1)$-th layer.

2.2 Interlinking Layers

In conventional CNN [9][10], there are downsampling layers which perform local max pooling or average pooling. They can realize shift invariance, which is important for pattern recognition. Downsampling reduces the size of feature maps. This is not a problem for pattern recognition (instead it is preferred because it reduces the computational burden in subsequent layers), but becomes problematic for scene parsing if an end-to-end model is desired. The output of an end-to-end model should have the same size as the input image in the first two dimensions because we have to label every pixel. Considering this requirement, we do not perform downsampling in the first CNN (top row in Fig. 1). The other CNNs (other rows in Fig. 1) process the input in smaller scales, and we do not perform downsampling in their own previous feature maps, either (black dashed arrows in Fig. 1).

These parallel CNNs process different scales of the input, which contain different levels of fine to coarse information. To let each CNN utilize multi-scale information, a special layer is designed. Consider CNN-k. In this layer, the feature maps from its own previous layer and those from the previous layer of CNN-$(k-1)$ and CNN-$(k+1)$ are concatenated. But the three types of feature maps cannot be concatenated directly because they have different sizes in the first two dimensions: those from CNN-$(k-1)$ are larger than those from CNN-k and those from CNN-$(k+1)$ are smaller than those from CNN-k. Our strategy is to downsample those from CNN-$(k-1)$ and upsample those from CNN-$(k+1)$ such that they have the same size as those from CNN-k in the first two dimensions. Max pooling is used for downsampling and nearest neighbor interpolation is used for upsampling. By performing downsampling/upsampling and then concatenation, we have interlinked the parallel CNNs.

2.3 Output Integration

It has been seen that after either the convolutional layer or interlinking layer, the size of the feature maps of each CNN in the first two dimensions do not change. Only CNN-1's feature maps have the same size as the output tensor in the first two dimensions. To utilize the information of other CNNs, we perform the following steps for $k = 4, 3, 2$ in sequel:

1. upsample CNN-k's final feature maps to match the size of CNN-$(k-1)$'s feature maps in the first two dimensions,

2. concatenate these feature maps with those from CNN-$(k-1)$, and
3. perform convolution and nonlinear transformation using (1) to obtain a bunch of CNN-$(k-1)$'s final feature maps.

After these operations, an additional convolutional layer without nonlinear transformation is used in CNN-1 with L feature maps, where L denotes the number of different labels. See Fig. 1 for illustration.

2.4   Output Layer

Only CNN-1 has a softmax layer in the end, which output the labels of each pixel. The output is a 3D tensor with the first two dimensions corresponding to the input image and the third dimension corresponding to the labels. At each location of the pixel, the one-hot representation is used for labels, that is, there is only one element equal to one and others equal to zero along the third dimension.

2.5   Training

The cross-entropy function is used as the loss function. Same as other CNNs, any minimization technique can be used. Stochastic gradient descend is used in this project.

2.6   Parameter Setting

For this particular application, the input image has a size of either 64×64 or 80×80. There are two stages in the proposed face parsing pipeline where in the first stage the entire image is resized (downsampling) to 64×64 and in the second stage 64×64 and 80×80 patches are extracted in the original image to cover the eye/nose/eyebrow and the mouth, respectively. See the next section for details. For RGB images, the input has three channels. The input image is then downsampled to $1/2$, $1/4$ and $1/8$ size using a 2×2 mean pooling. In all convolutional layers and all CNNs, the size of the receptive field is set to 5×5 (the first two dimensions) except in the last convolutional layer of CNN-1 (the black dotted arrow) where 9×9 is used.

3   Face Parsing with iCNNs

Usually a face image for parsing is large, e.g., the images of Helen dataset [8] for this task are of the size 256×256 [5]. If we input such large images to the proposed iCNN, both training and testing are slow. To speed up the process we separate the face parsing procedure into two stages, and both stages use iCNN.

3.1   Stage 1: Facial Parts Localization

The goal of this stage is to localize the facial parts including the eyes, nose and so on with iCNN. Note that we do not label the Face Skin part in this project, since it has a large area, which is unsuitable for the proposed iCNN to process. The input image is preprocessed by subtracting the mean and dividing the norm. The input image as well as its label map is resized to 64×64 in the first two dimensions (both the input image and the output map are 3D tensors) by downsampling. The output tensor has 9 channels corresponding to the label maps of background, left eyebrow, left eye, right eyebrow, right eye, nose, upper lip, inner mouth and lower lip, respectively (Fig. 2). Except the first label map (background), each median axis of the label map is calculated, and scaled back to original image to obtain the estimation of the part location. For mouth related parts (upper lips, inner mouth, lower lips), a shared median axis is calculated. For the first five parts, 64×64 patches are extracted from the original input face image. For mouth-related parts, a 80×80 patch is extracted.

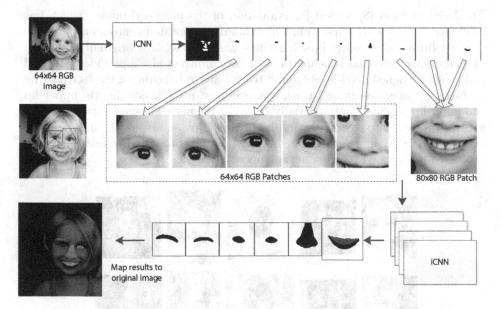

Fig. 2. The pipeline for face parsing. In the first stage, the entire image is resized to 64×64 with aspect ratio kept. It is input to an iCNN and obtain ten label maps where the first is the background and the others are facial components. The median point of each component except the two lips and in-mouth is calculated. Since the lips and in-mouth are processed together, these three parts are first merged together and then a joint median point is calculated. A 64×64 or 80×80 patch is extracted around the median point. In the second stage, with mirroring operation (right eye and eyebrows flipped), six small parts are processed by four iCNNs to get exact segmentation at the pixel level. Best viewed in color.

3.2 Stage 2: Fine Labeling

In the previous stage, we have extracted the five 64×64 patches and one 80×80 patch from the original image. Then we use four iCNNs to predict the labels of the pixels in each patch (Fig. 2). The four iCNNs are used for predicting eyebrows, eyes, nose, and mouth components, respectively. Note that one iCNN is used for predicting both left eyebrow and right eyebrow. Since the left eyebrow and right eyebrow are symmetric, during training the image patches of right eyebrows are flipped and combined with image patches of left eyebrows. Therefore this iCNN has only one label map in the output. In testing, the predicted label maps of right eyebrows are flipped back. Similarly, one iCNN is used for predicting both left eye and right eye. The iCNN for the nose has only one label map in the output and the iCNN for the mouth components has three label maps.

4 Experiments

4.1 Dataset

The Helen dataset [8] is used for evaluation of the proposed model, which has 2330 face images with dense sampled, manually-annotated contours around the eyes, eyebrows, nose, outer lips, inner lips and jawline. It is originally designed as a landmark detection benchmark database. Smith et al. [5] provides a resized and roughly aligned pixel-level ground truth data to benchmark the face parsing problem. It generates ground truth eye, eyebrow, nose, inside mouth, upper lip and lower lip segments automatically by using the manually-annotated contours as segment boundaries. Some examples of Helen are shown in Fig. 3, where the first line is the original database images with annotations, and second line is the processed pixel-based labeling for parsing.

Fig. 3. Example images (top) and corresponding labels (bottom) of the Helen dataset. Best viewed in color.

We use the same training, testing and validation partition as in [5]. The dense annotated data is separated into 3 parts: 2000 images for training, 230 images for validation, and 100 images for testing. The validation set is used to test whether model is converged.

4.2   Training and Testing

We train the iCNNs in stages 1 and 2 separately. For stage 1, the entire training images, as well as the corresponding ground truth label maps, are resized to 64×64 with aspect ratio kept. For stage 2, the training data are 64×64 or 80×80 patches extracted from the original 256×256 training images (see Section 3.1). The corresponding ground truth label maps are extracted from the original 256×256 ground truth label maps.

Stochastic gradient descent is used as the training algorithm. Since the number of images is small compared to number of parameters, to prevent overfitting and enhance our model, data argumentation is used. During stochastic gradient descent, a random $15°$ rotation, 0.9~1.1x scaling, and -10~10 pixels shifting in each direction are applied to each input every time when it enters the model.

In Stage 2, by visualizing the feature maps, we find that in the last convolutional layer of CNN-1 among the L feature maps there is a feature map, denotedy B, corresponding to the background part. We find that modulating this feature map by $\beta B + \beta_0$ can enhance the prediction accuracy. For each facial part, β and β_0 are obtained by maximizing the F-measure [5] on the validation set using the L-BFGS-B algorithm offered by SciPy, an open-source software.

For testing, each image undergoes stages 1 and 2 in sequel. Only the predicted labels in stage 2 are used for evaluation of the results.

All codes are written in Theano [11] and Pylearn2 [12].

4.3   Results

The evaluation metric is the F-measure used in [5]. From Table 1, it is seen that for most facial parts, iCNNs obtain the highest scores. Note that in our training data, the labels of Face Skin area are not used. As we can see in the table, this area is usually a high-score term for most methods, and omitting it will in no way enhance the overall performance of iCNNs. Even though, iCNNs achieves higher overall score than existing models. Some example labeling results are shown in Fig. 4 along with the results obtained in [5].

Table 1. Comparison with other models (F-Measure)

Model	Eye	Eyebrow	Nose	In mouth	Upper lip	Lower lip	Mouth (all)	Face Skin	Overall
[13]	0.533	n/a	n/a	0.425	0.472	0.455	0.687	n/a	n/a
[14]	0.679	0.598	0.890	0.600	0.579	0.579	0.769	n/a	0.733
[15]	0.770	0.640	0.843	0.601	0.650	0.618	0.742	0.886	0.738
[16]	0.743	0.681	0.889	0.545	0.568	0.599	0.789	n/a	0.746
[5]	**0.785**	0.722	**0.922**	0.713	0.651	0.700	0.857	0.882	0.804
iCNNs	0.778	**0.863**	0.920	**0.777**	**0.824**	**0.808**	**0.889**	n/a	**0.845**

Fig. 4. Labeling results on several example images obtained using the method in [5] (top) and the proposed method in this paper (middle). The bottom shows the ground truth labels. Best viewed in color.

5   Concluding Remarks

We propose an interlinked CNN (iCNN), where multiple CNNs process different levels of details of the input, respectively. Compared with traditional CNNs it features interlinked layers which not only allow the information flow from fine level to coarse level but also allow the information flow from coarse level to flow to the fine level. For face parsing, a two-stage pipeline is designed based on the proposed iCNN. In the first stage an iCNN is used for facial part localization, and in the second stage four iCNN are used for pixel labeling. The pipeline does not involve any feature extraction step and can predict labels from raw pixels. Experimental results have validated the effectiveness of the proposed method.

Though this paper focuses on face parsing, the proposed iCNN is not restricted to this particular application. It may be useful for other computer vision applications such as general image parsing and object detection.

Acknowledgement. The first author would like to thank Megvii Inc. for providing the computing facilities. This work was supported in part by the National Basic Research Program (973 Program) of China under Grant 2012CB316301 and Grant 2013CB329403, in part by the National Natural Science Foundation of China under Grant 61273023, Grant 91420201, and Grant 61332007, in part by the Natural Science Foundation of Beijing under Grant 4132046.

References

1. Tu, Z., Chen, X., Yuille, A.L., Zhu, S.C.: Image Parsing: Unifying Segmentation, Detection, and Recognition. International Journal of Computer Vision 63, 113–140 (2005)
2. Socher, R., Lin, C.C., Manning, C., Ng, A.Y.: Parsing natural scenes and natural languages with recursive neural networks. In: ICML, pp. 129–136 (2011)
3. Farabet, C., Couprie, C., Najman, L., LeCun, Y.: Learning Hierarchical Features for Scene Labeling. IEEE Transactions on Pattern Analysis and Machine Intelligence 35, 1915–1929 (2013)
4. Pinheiro, P., Collobert, R.: Recurrent convolutional neural networks for scene labeling. In: ICML, pp. 82–90 (2014)
5. Smith, B.M., Zhang, L., Brandt, J., Lin, Z., Yang, J.: Exemplar-based face parsing. In: CVPR, pp. 3484–3491 (2013)
6. Luo, P., Wang, X., Tang, X.: Hierarchical Face parsing via deep learning. In: CVPR, pp. 2480–2487 (2012)
7. Seyedhosseini, M., Sajjadi, M., Tasdizen, T.: Image segmentation with cascaded hierarchical models and logistic dsjunctive normal networks. In: ICCV, pp. 2168–2175 (2013)
8. Le, V., Brandt, J., Lin, Z., Bourdev, L., Huang, T.S.: Interactive facial feature localization. In: Fitzgibbon, A., Lazebnik, S., Perona, P., Sato, Y., Schmid, C. (eds.) ECCV 2012, Part III. LNCS, vol. 7574, pp. 679–692. Springer, Heidelberg (2012)
9. LeCun, Y., Bottou, L., Bengio, Y., Haffner, P.: Gradient Based Learning Applied to Document Recognition. Proceedings of the IEEE 86(11), 2278–2324 (1998)
10. Krizhevsky, A., Sutskever, I., Hinton, G.E.: Imagenet classification with deep convolutional neural networks. In: Advances in Neural Information Processing Systems (NIPS), pp. 1097–1105 (2012)
11. Bergstra, J., Breuleux, O., Bastien, F., Lamblin, P., Pascanu, R., Desjardins, G., Turian, J., Warde-Farley, D., Bengio, Y.: Theano: A CPU and GPU math expression compiler. In: Procesedings of the Python for Scientific Computing Conference (SciPy) (2010)
12. Goodfellow, I.J., Warde-Farley, D., Lamblin, P., Dumoulin, V., Mirza, M., Pascanu, R., Bergstra, J., Bastien, F., Bengio, Y.: Pylearn2: a Machine Learning Research Library. arXiv preprint arXiv:1308.4214 (2013)
13. Zhu, X., Ramanan, D.: Face detection, pose estimation and landmark localization in the wild. In: CVPR (2012)
14. Saragih, J.M., Lucey, S., Cohn, J.F.: Face Alignment throughsubspace constrained mean-shifts. In: CVPR (2009)
15. Liu, C., Yuen, J., Torralba, A.: Nonparametric Scene Parsing via Label Transfer. IEEE Transactions on Pattern Analysis and Machine Intelligence 33(12), 2368–2382 (2011)
16. Gu, L., Kanade, T.: A generative shape regularization model for robust face alignment. In: Forsyth, D., Torr, P., Zisserman, A. (eds.) ECCV 2008, Part I. LNCS, vol. 5302, pp. 413–426. Springer, Heidelberg (2008)

Image Tag Completion by Local Learning

Jingyan Wang[1,2,3], Yihua Zhou[4], Haoxiang Wang[5], Xiaohong Yang[6],
Feng Yang[6], and Austin Peterson[7]

[1] National Time Service Center, Chinese Academy of Sciences, Xi' an,
Shaanxi 710600, China
jingbinwang1@outlook.com
[2] Graduate University of Chinese Academy of Sciences, Beijing 100049, China
[3] Provincial Key Laboratory for Computer Information Processing Technology,
Soochow University Suzhou 215006, China
[4] Department of Mechanical Engineering and Mechanics, Lehigh University,
Bethlehem, PA 18015, US
[5] Department of Electrical and Computer Engineering, Cornell University, Ithaca,
NY 14850, USA
[6] College of Computer Science and Technology, Shandong University of Finance and
Economics, Jinan 250014, China
[7] Electrical and Computer Engineering Department,
The University of Texas at San Antonio, San Antonio, TX, 78249, USA
austin.peterson1@outlook.com

Abstract. The problem of tag completion is to learn the missing tags
of an image. In this paper, we propose to learn a tag scoring vector for
each image by local linear learning. A local linear function is used in
the neighborhood of each image to predict the tag scoring vectors of its
neighboring images. We construct a unified objective function for the
learning of both tag scoring vectors and local linear function parameters. In the objective, we impose the learned tag scoring vectors to be
consistent with the known associations to the tags of each image, and
also minimize the prediction error of each local linear function, while
reducing the complexity of each local function. The objective function
is optimized by an alternate optimization strategy and gradient descent
methods in an iterative algorithm. We compare the proposed algorithm
against different state-of-the-art tag completion methods, and the results
show its advantages.

Keywords: Image tagging, Tag completion, Local learning, Gradient
descent.

1 Introduction

Recently, social network has been a popular tool to share images. When a social network user uploads an image, the image is usually associated with a tag/
keyword which is used to describe the semantic content of this image. The tags
provided by the users are usually incomplete. Zhang et al. designed and implemented a fast motion detection mechanism for multimedia data on mobile

© Springer International Publishing Switzerland 2015
X. Hu et al. (Eds.): ISNN 2015, LNCS 9377, pp. 232–239, 2015.
DOI: 10.1007/978-3-319-25393-0_26

and embedded environment [25]. Recently, the problem image tag completion is proposed in the computer vision and machine learning communities to learn the missing tags of images [22,11,10,1,23]. This problem is defined as the problem of complete the missing elements of a tag vector of a given image automatically.

In this paper, we investigate the problem of image tag completion, and proposed a novel and effective algorithm for this problem based on local linear learning. We propose a novel and effective tag completion method. Instead of completing the missing tag association elements of each image, we introduce a tag scoring vector to indicate the scores of assigning the image to the tags in a given tag set. We propose to study the tag scoring vector learning problem in the neighborhood of each image. For each image in the neighborhood, we propose to learn a linear function to predict a tag scoring vector from a visual feature vector of its corresponding image feature. We propose to minimize the perdition error measure by the squared ℓ_2 norm distance over each neighborhood, and also minimize the squared ℓ_2 norm of the linear function parameters. Besides the local linear learning, we also proposed to regularize the learning of tag scoring vectors by the available tags of each image. We construct a unified objective function to learn both the tag scoring vectors and the local linear functions. We develop an iterative algorithm to optimize the proposed problem. In each iteration of this algorithm, we update the tag scoring vectors and the local linear function parameters alternately.

This rest parts of paper are organized as follows: in section 2, we introduced the proposed method. In section 3, we evaluate the proposed methods on some benchmark data sets. In section 4, the paper is concluded with future works.

2   Proposed Method

We assume that we have a data set of n images, and their visual feature vectors are $\mathbf{x}_i|_{i=1}^n$, where $\mathbf{x}_i \in \mathbb{R}^d$ is the d-dimensional feature vector of the i-th image. We also assume that we have a set of m unique tags, and a tag vector $\widehat{\mathbf{t}}_i = [\widehat{t}_{i1}, \cdots, \widehat{t}_{im}]^\top \in \{+1, -1\}^m$ for the i-th image $\mathbf{x}_i$, where $\widehat{t}_{ij} = +1$ if the j-th tag is assigned to the i-th image, and -1, otherwise. In real-world applications, the tag vector of an image $\mathbf{x}_i$ is usually incomplete, i.e., some elements of $\widehat{\mathbf{t}}_i$ are missing. We define a vector $\mathbf{v}_i = [v_{i1}, \cdots, v_{im}] \in \{1, 0\}^m$, where $v_{ij} = 1$ if $\widehat{t}_{ij}$ is available, and 0 if $\widehat{t}_{ij}$ is missing. We propose to learn a tag scoring vector $\mathbf{t}_i = [t_{i1}, \cdots, t_{im}] \in \mathbb{R}^m$, where t_{ij} is the score of assigning the j-th tag to the i-th image.

The set of the κ nearest neighbor of each image $\mathbf{x}_i$ is denoted as $\mathcal{N}_i$, and we assume that the tag scoring vector $\mathbf{t}_j$ of a image $\mathbf{x}_j \in \mathcal{N}_i$ can be predicted from its visual feature vector $\mathbf{x}_j$ using a local linear function $f_i(\mathbf{x}_j)$,

$$\mathbf{t}_j \leftarrow f_i(\mathbf{x}_j) = W_i \mathbf{x}_j, \ \forall \ j : \mathbf{x}_j \in \mathcal{N}_i, \tag{1}$$

where $W_i \in \mathbb{R}^{m \times d}$ is the parameter of the local linear function. To learn the tag scoring vector and the local function parameters, we propose the following minimization problem,

$$\min_{\mathbf{t}_i|_{i=1}^n, W_i|_{i=1}^n} \left\{ g(\mathbf{t}_i|_{i=1}^n, W_i|_{i=1}^n) = \sum_{i=1}^n \left(\sum_{j:\mathbf{x}_j \in \mathcal{N}_i} \|\mathbf{t}_j - W_i \mathbf{x}_j\|_2^2 + \alpha \|W_i\|_2^2 \right. \right.$$
$$\left. \left. + \beta(\mathbf{t}_i - \widehat{\mathbf{t}}_i)^\top diag(\mathbf{v}_i)(\mathbf{t}_i - \widehat{\mathbf{t}}_i) \right) \right\} \tag{2}$$

where α and β are tradeoff parameters. The objective function $g(\mathbf{t}_i|_{i=1}^n, W_i|_{i=1}^n)$ in (2) is a summarization of three terms over all the images in the data set. The first term, $\sum_{j:\mathbf{x}_j \in \mathcal{N}_i} \|\mathbf{t}_j - W_i \mathbf{x}_j\|_2^2$, is the prediction error term of the local linear predictor over the neighborhood of each image. The second, $\|W_i\|_2^2$, is to reduce the complexity of the local linear predictor. The last term, $(\mathbf{t}_i - \widehat{\mathbf{t}}_i)^\top diag(\mathbf{v}_i)(\mathbf{t}_i - \widehat{\mathbf{t}}_i)$, is a regularization term to regularize the learning of tag scoring vectors by the incomplete tag vectors, so that the available tags are respected. To optimize the minimization problem in (2), we propose to use the alternate optimization strategy [4,12] in an iterative algorithm.

- **Optimization of $\mathbf{t}_i|_{i=1}^n$** In each iteration, we optimize $\mathbf{t}_i|_{i=1}^n$ one by one, and the minimization of (2) with respect to $\mathbf{t}_i$ can be achieved with the following gradient descent update rule,

$$\mathbf{t}_i^{new} = \mathbf{t}_i^{old} - \eta \nabla_{\mathbf{t}_i} g(\mathbf{t}_j|_{j=1}^n, W_i|_{i=1}^n)|_{\mathbf{t}_i = \mathbf{t}_i^{old}}, \tag{3}$$

  where $\nabla_{\mathbf{t}_i} g(\mathbf{t}_j|_{j=1}^n, W_i|_{i=1}^n)$ is the sub-gradient function of $g(\mathbf{t}_j|_{j=1}^n, W_i|_{i=1}^n)$, with respect to $\mathbf{t}_i$,

$$\nabla_{\mathbf{t}_i} g(\mathbf{t}_j|_{j=1}^n, W_i|_{i=1}^n) = 2 \sum_{k:\mathbf{x}_i \in \mathcal{N}_k} (\mathbf{t}_i - W_k \mathbf{x}_i) + 2\beta diag(\mathbf{v}_i)(\mathbf{t}_i - \widehat{\mathbf{t}}_i), \tag{4}$$

  and η is the descent step.
- **Optimization of $W_i|_{i=1}^n$** In each iteration, we also optimized $W_i|_{i=1}^n$ one by one. When W_i is optimized, $W_j|_{j \neq i}$ are fixed. Gradient descent method is also employed to update W_i to minimize the objective in (2),

$$W_i^{new} = W_i^{old} - \eta \nabla_{W_i} g(\mathbf{t}_i|_{i=1}^n, W_i|_{i=1}^n)|_{W_i = W_i^{old}}, \tag{5}$$

  where $\nabla_{W_i} g(\mathbf{t}_i|_{i=1}^n, W_i|_{i=1}^n)$ is the sub-gradient function with respect to W_i,

$$\nabla_{W_i} g(\mathbf{t}_i|_{i=1}^n, W_i|_{i=1}^n) = 2 \sum_{j:\mathbf{x}_j \in \mathcal{N}_i} (\mathbf{t}_j - W_i \mathbf{x}_j) \mathbf{x}_j^\top + 2\beta W_i. \tag{6}$$

3   Experiments

3.1   Setup

In the experiments, we used two publicly accessed image-tag data sets, which are Corel5k data set [30,5,14] and IAPR TC12 data set [31,9,30]. In the Corel5k data set, there are 4,918 images, and 260 tages. We extract density feature, Harris shift feature, Harris Hue feature, RGB color feature, and HSV color feature as visual features for each image. Moreover, we remove 40% of the elements of the tag vectors to make the incomplete image tag vectors. In the IAPR TC12 data set, there are 19,062 images, and 291 tags. We also remove 40% elements of the tag elements to construct the incomplete tag vectors. To evaluate the tag completion performances, we used the recall-precision curve as performance measure. We also use mean average precision (MAP) as a single performance measure.

3.2   Results

We compared the proposed method to several state-of-the-art tag completion methods, including tag matrix completion (TMC) [22], linear sparse reconstructions (LSR) [10], tag completion by noisy matrix recovery (TCMR)[1], and tag completion via NMF (TC-NMF) [23]. The experimental result on two data sets are given in Fig. 1 and Fig. 2. From these figures, we can see that the proposed method LocTC performs best. Its recall-precision curve is closer to the top-right corner than any other methods, and its MAP is also higher than MAPs of other methods.

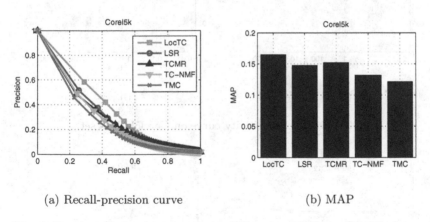

(a) Recall-precision curve   (b) MAP

Fig. 1. Results of comparison to state-of-the-art methods on Corel5k data set

In this section, we will study the sensitivity of the proposed algorithm to the two parameters, α and β. The curves of α and β on different data sets are given in Fig. 3 and Fig. 4. From these figures, we can see that the performances are stable to different valuse of both α and β.

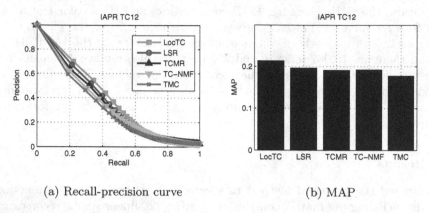

(a) Recall-precision curve   (b) MAP

Fig. 2. Results of comparison to state-of-the-art methods on IAPR TC12 data set

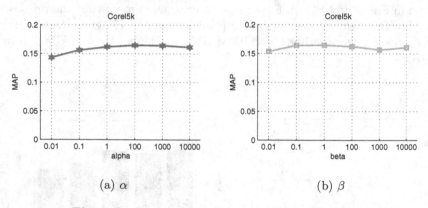

(a) α (b) β

Fig. 3. Parameter sensitivity curve on Corel5k data set.

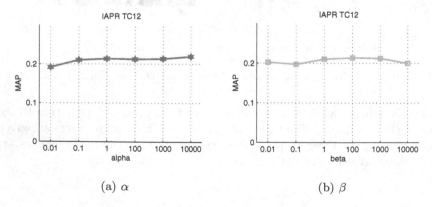

(a) α (b) β

Fig. 4. Parameter sensitivity curve on IAPR TC12 data set.

4   Conclusion and Future Works

In this paper, we study the problem of tag completion, and proposed a novel algorithm for this problem. We proposed to learn the tags of images in the neighborhood of each image. A local linear function is designed to predict the tag scoring vectors of images in each neighborhood, and the prediction function parameter is learned jointly with the tag scoring vectors. The proposed method is compared to state-of-the-art tag completion algorithms, and the results show that the proposed algorithm outperforms the compared methods. In the future, we will study how to incorporate these connections into our model and learn more effective tags. In this paper, we used one single local function for each neighborhood, and in the future, we will use more than than regularization to regularized the learning of tags [15,16], such as usage of wavelet functions to construct the local function [13]. Moreover, correntropy can also be considered as a alternative loss function to construct the local learning problem [17,24,6,34]. In the future, we also plan to extend the proposed algorithm for completion of tags of large scale image data set by using high performance computing technology [36,21,28,29,32,33,2,25,8,35,7,18,20,19], and completion of tags of gene/protein functions of bioinformatics problems [3,26,27,3].

Acknowledgements. This works was supported by an open research funding of the Provincial Key Laboratory for Computer Information Processing Technology, Soochow University, China, Grant No. KJS1324.

References

1. Feng, Z., Feng, S., Jin, R., Jain, A.K.: Image tag completion by noisy matrix recovery. In: Fleet, D., Pajdla, T., Schiele, B., Tuytelaars, T. (eds.) ECCV 2014, Part VII. LNCS, vol. 8695, pp. 424–438. Springer, Heidelberg (2014)
2. Gao, Y., Zhang, F., Bakos, J.D.: Sparse matrix-vector multiply on the keystone ii digital signal processor. In: 2014 IEEE High Performance Extreme Computing Conference (HPEC), pp. 1–6 (2014)
3. Hu, J., Zhang, F.: Improving protein localization prediction using amino acid group based physichemical encoding. In: Rajasekaran, S. (ed.) BICoB 2009. LNCS, vol. 5462, pp. 248–258. Springer, Heidelberg (2009)
4. Huang, S., Ma, Z., Wang, F.: A multi-objective design optimization strategy for vertical ground heat exchangers. Energy and Buildings 87, 233–242 (2015)
5. Huang, Y., Liu, Q., Zhang, S., Metaxas, D.: Image retrieval via probabilistic hypergraph ranking. In: Proceedings of the IEEE Computer Society Conference on Computer Vision and Pattern Recognition, pp. 3376–3383 (2010)
6. Li, L., Yang, J., Xu, Y., Qin, Z., Zhang, H.: Documents clustering based on max-correntropy nonnegative matrix factorization, pp. 850–855 (2015)
7. Li, T., Zhou, X., Brandstatter, K., Raicu, I.: Distributed key-value store on HPC and cloud systems. In: 2nd Greater Chicago Area System Research Workshop (GCASR) (2013)

8. Li, T., Zhou, X., Brandstatter, K., Zhao, D., Wang, K., Rajendran, A., Zhang, Z., Raicu, I.: Zht: A light-weight reliable persistent dynamic scalable zero-hop distributed hash table. In: 2013 IEEE 27th International Symposium on Parallel & Distributed Processing (IPDPS), pp. 775–787 (2013)

9. Li, Z., Liu, J., Xu, C., Lu, H.: Mlrank: Multi-correlation learning to rank for image annotation. Pattern Recognition 46(10), 2700–2710 (2013)

10. Lin, Z., Ding, G., Hu, M., Lin, Y., Sam Ge, S.: Image tag completion via dual-view linear sparse reconstructions. Computer Vision and Image Understanding 124, 42–60 (2014)

11. Lin, Z., Ding, G., Hu, M., Wang, J., Ye, X.: Image tag completion via image-specific and tag-specific linear sparse reconstructions. In: Proceedings of the IEEE Computer Society Conference on Computer Vision and Pattern Recognition, pp. 1618–1625 (2013)

12. Liu, L., Li, H., Xue, Y., Liu, W.: Reactive power compensation and optimization strategy for grid-interactive cascaded photovoltaic systems. IEEE Transactions on Power Electronics 30(1), 188–202 (2015)

13. Liu, Z., Abbas, A., Jing, B.Y., Gao, X.: Wavpeak: picking nmr peaks through wavelet-based smoothing and volume-based filtering. Bioinformatics 28(7), 914–920 (2012)

14. Wang, C., Yan, S., Zhang, L., Zhang, H.J.: Multi-label sparse coding for automatic image annotation, pp. 1643–1650 (2009)

15. Wang, J.J.Y., Bensmail, H., Gao, X.: Multiple graph regularized protein domain ranking. BMC Bioinformatics 13(1), 307 (2012)

16. Wang, J.J.Y., Bensmail, H., Gao, X.: Multiple graph regularized nonnegative matrix factorization. Pattern Recognition 46(10), 2840–2847 (2013)

17. Wang, J.J.Y., Wang, X., Gao, X.: Non-negative matrix factorization by maximizing correntropy for cancer clustering. BMC Bioinformatics 14(1), 107 (2013)

18. Wang, K., Kulkarni, A., Zhou, X., Lang, M., Raicu, I.: Using simulation to explore distributed key-value stores for exascale system services. In: 2nd Greater Chicago Area System Research Workshop (GCASR) (2013)

19. Wang, K., Zhou, X., Chen, H., Lang, M., Raicu, I.: Next generation job management systems for extreme-scale ensemble computing. In: Proceedings of the 23rd International Symposium on High-Performance Parallel and Distributed Computing, pp. 111–114 (2014)

20. Wang, K., Zhou, X., Li, T., Zhao, D., Lang, M., Raicu, I.: Optimizing load balancing and data-locality with data-aware scheduling. In: 2014 IEEE International Conference on Big Data (Big Data), pp. 119–128 (2014)

21. Wang, K., Zhou, X., Qiao, K., Lang, M., McClelland, B., Raicu, I.: Towards scalable distributed workload manager with monitoring-based weakly consistent resource stealing. In: Proceedings of the 24rd International Symposium on High-Performance Parallel and Distributed Computing, pp. 219–222 (2015)

22. Wu, L., Jin, R., Jain, A.: Tag completion for image retrieval. IEEE Transactions on Pattern Analysis and Machine Intelligence 35(3), 716–727 (2013)

23. Xia, Z., Feng, X., Peng, J., Wu, J., Fan, J.: A regularized optimization framework for tag completion and image retrieval. Neurocomputing (2014)

24. Xing, H.J., Ren, H.R.: Regularized correntropy criterion based feature extraction for novelty detection. Neurocomputing 133, 483–490 (2014)

25. Zhang, F., Gao, Y., Bakos, J.D.: Lucas-kanade optical flow estimation on the ti c66x digital signal processor. In: 2014 IEEE High Performance Extreme Computing Conference (HPEC), pp. 1–6 (2014)

26. Zhang, F., Hu, J.: Bayesian classifier for anchored protein sorting discovery. In: IEEE International Conference on Bioinformatics and Biomedicine, BIBM 2009, pp. 424–428 (2009)
27. Zhang, F., Hu, J.: Bioinformatics analysis of physicochemical properties of protein sorting signals (2010)
28. Zhang, F., Zhang, Y., Bakos, J.: Gpapriori: Gpu-accelerated frequent itemset mining. In: 2011 IEEE International Conference on Cluster Computing (CLUSTER), pp. 590–594 (2011)
29. Zhang, F., Zhang, Y., Bakos, J.D.: Accelerating frequent itemset mining on graphics processing units. The Journal of Supercomputing 66(1), 94–117 (2013)
30. Zhang, S., Huang, J., Li, H., Metaxas, D.: Automatic image annotation and retrieval using group sparsity. IEEE Transactions on Systems, Man, and Cybernetics, Part B: Cybernetics 42(3), 838–849 (2012)
31. Zhang, X., Liu, C.: Image annotation based on feature fusion and semantic similarity. Neurocomputing 149(PC), 1658–1671 (2015)
32. Zhang, Y., Zhang, F., Bakos, J.: Frequent itemset mining on large-scale shared memory machines. In: 2011 IEEE International Conference on Cluster Computing (CLUSTER), pp. 585–589 (2011)
33. Zhang, Y., Zhang, F., Jin, Z., Bakos, J.D.: An fpga-based accelerator for frequent itemset mining. ACM Transactions on Reconfigurable Technology and Systems (TRETS) 6(1), 2 (2013)
34. Zhang, Z., Chen, J.: Correntropy based data reconciliation and gross error detection and identification for nonlinear dynamic processes. Computers and Chemical Engineering 75, 120–134 (2015)
35. Zhao, D., Zhang, Z., Zhou, X., Li, T., Wang, K., Kimpe, D., Carns, P., Ross, R., Raicu, I.: Fusionfs: Toward supporting data-intensive scientific applications on extreme-scale high-performance computing systems. In: 2014 IEEE International Conference on Big Data (Big Data), pp. 61–70 (2014)
36. Zhou, X., Chen, H., Wang, K., Lang, M., Raicu, I.: Exploring distributed resource allocation techniques in the slurm job management system. Illinois Institute of Technology, Department of Computer Science, Technical Report (2013)

Haarlike Feature Revisited: Fast Human Detection Based on Multiple Channel Maps

Xin Zuo[1], Jifeng Shen[2], Hualong Yu[1], and Yuanyuan Dan[3]

[1] School of Computer Science and Engineering,
Jiangsu University of Science and Technology, Zhenjiang, Jiangsu, 212003, China
[2] School of Electronic and Informatics Engineering,
Jiangsu University, Zhenjiang, Jiangsu, 212013, China
[3] School of Environmental and Chemical Engineering,
Jiangsu University of Science and Technology, Zhenjiang, Jiangsu, 212003, China
shenjifeng@ujs.edu.cn

Abstract. Haarlike feature has achieved great success in detecting frontal human faces, but fewer attentions have been paid to the other objects such as pedestrian. The reason of the low detection rate for Haarlike feature is attributed to the usage in a naive way. In this paper, we have revisited Haarlike feature for object detection especially focus on pedestrians, but use it in a different way which is applied based on multiple channel maps instead of raw pixels and obtains a significant improvement. Furthermore, we have proposed an improved Haarlike feature that embeds statistical information from the training data which is based on the linear discriminative analysis criterion. The proposed feature works with the classical Gentle Boosting algorithm which is effective in training, and also running at real-time speed. Experiments based on INIRA dataset demonstrate that our proposed method is easy to implement and achieves the performance comparable to the state-of-the-arts.

Keywords: human detection, multiple channel maps, statistical information.

1 Introduction

Pedestrian detection has grasped much attention both from computer vision fields which is a nice delegate for generic object detection, and also has been considered as an important application in industrial field that can be applied to video surveillance, automated driver system and sports athlete evaluation. Pedestrian detection in unlimited environment is a very challenge task, for the sake of articulation, occlusion, illumination and view-changing, etc.

There're a large amount of papers published in the last decades, which focus on looking for light-weight discriminative feature or modeling its articulation as a deformable part model, etc. These attempts have lifted the accuracy from the 45% FPPI to 15% on the INRIA dataset with 0.1 FPPI in the last ten years. The running speed is also greatly improved with 20 fps on the 480x640 image which has far surpassed the classical methods. The most notable work is the HOG feature[1], which is almost

© Springer International Publishing Switzerland 2015
X. Hu et al. (Eds.): ISNN 2015, LNCS 9377, pp. 240–247, 2015.
DOI: 10.1007/978-3-319-25393-0_27

used in any object detection system. Another breakthrough is due to the proposal of integral channel feature[2], which is simple to compute, fast running in application and most importantly achieves the state-of-the-arts results.

There are many excellent reviews[3, 4] which nearly refers to all of the methods for pedestrian detection in the last decades. In this paper, we will only remind of some latest publication which is related to our proposed methods. Piotr [2], etc, have designed a detector which is based on integral channel features and trained with boosting algorithm. This detector can achieve a real-time running speed of 20 fps. Moreover, other researchers make use of the GPU and other cues in video to further improve the running speed with 100fps[5]. More recently, informed Haarlike feature[6] is proposed which generate a set of template from human parts instead of exhaustive sampling like traditional Haarlike features, and is very effective to calculate.

In this paper, we have revisited the Haarlike feature which is applied to multiple channel maps and work with the boosting algorithm that can greatly improve the accuracy of detector. Furthermore, due to the recent attempt of our earlier studies, we embed the statistical information from the training data into the Haarlike feature, which further increase the discriminative ability of the feature. Our method is faster to train which is less than 1 hour and also achieves real-time running speed of 20 fps for the 640x480 resolution image. Furthermore, the accuracy is comparable to the state-of-the-arts.

The rest paper is organized as follows: section 2 introduces the related work which our work based on. Section 3 describes our proposed SHF features in detail. Section 4 covers the experiment results and analysis and section 5 concludes this paper.

Fig. 1. Different channel of the input image

2   Related Works

2.1   Multiple Channel Maps

The ICF feature [11] models the feature C of image I as a channel generation function Ω, so the feature of image I can be represented as $C_i = \Omega_i(I)$, where

$\Omega = \{\Omega_i\}, i = 1, 2, ..., n$. C_i is the ith feature for image I, Ω_i is the ith channel generation function. The channel generation function can be linear, such as gray-level image of original image I or nonlinear, such as gradient image. Each channel represents a different feature space which derived from original image. Fig. 1 shows the common channels of the original image which reflect different aspects of input image. For example, the gradient image of different angle can reflect the direction which is similar to the Gabor filter, the canny image can represent the edge of human and different color space can reflect the color consistency in clothes, etc. The different channel is heterogeneous with each other, so extracting multiple channels is a procedure to obtain different information from a simple image.

2.2   Haarlike Feature (HF)

Haarlike feature is first proposed by C. Papageorgiou[7], who use them for object detection. It becomes popular when it combines with adaboost algorithm to realize fast face detection[8]. The HF is very similar to Garbor filter which can be used to detect lines with different orientation. The feature is calculated by sum up pixels in white area, and then substrate to the corresponding dark area. The formulation is shown in Eq. (1)

$$ f_{haar} = \sum_{i=1}^{n} w_i a_i = \boldsymbol{w}^T \boldsymbol{a} \tag{1} $$

Where w_i, a_i is the weight and sum of pixel values for each sub-region in these rectangles. n is the number of sub-regions.

The traditional way of using HF for object detection is to apply them simply to the raw pixel image. Although it is effective for rigid object such as faces, the performance deteriorates greatly when applies to the non-rigid object such as humans. In this paper, we have revisited Haarlike feature and applied them on the multiple channel maps and found that the performance is greatly improved. The exact analysis of this method will be postponed to the experimental section.

3   Statistical Haarlike Feature (SHF)

Due to the success of learning statistic information from the training data [9, 10], we also try to embed the distribution of positive and negative samples into the HF. If we look into the formula of Eq.(1), we can find that, the HF is similar to a image filter applied on the image with the size equals to the width and height of sub-region in each rectangle which corresponding to the HF. In this paper, we make use of the linear discriminative analysis to learn the weight which aims to get the optimal weight that have the largest intra-class invariance and smallest inter-class invariance. The formulation is shown in Eq.(2).

$$w_{opt} = \arg\max_{w} \frac{w^T S_b w}{w^T S_w w}$$

$$(2)$$

Where S_w and S_b is the within-class scatter matrix and between-class scatter matrix. The solution to the Eq. (2) can be obtained with the closed form $w^* = S_w^{-1}(m_1 - m_2)$. The m_1 and m_2 are the mean of the positive and negative samples in the area of the specific HF respectively. For example, in Fig. 2(b), the m_1 is a three dimensional vector with each element represents the mean value of the pixels in the left, middle and right area of this HF for all the positive training samples.

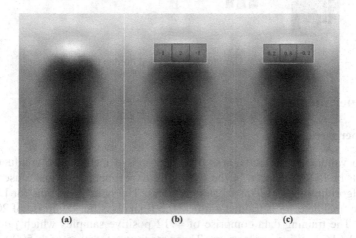

(a)   (b)   (c)

Fig. 2. The difference between HF and SHF on the second channel of LUV color space

The difference between HF and SHF is illustrated in Fig.2 (b-c). Fig.2 (a) shows the background image, which is the average of the second channel of LUV color space for the positive samples. The default weight of traditional HF is [1,-1, 1], but our learned weight equals [-0.2, 0.8,-0.2] on this specific HF. From Fig.2(c), we can get some intuition from this phenomenon that the brighter area of head is very discriminative, which need to be put on more weights.

The procedure of our proposed detector is shown in Fig.3. We can see that, it's very similar to the traditional boosting-based object detection framework, besides that, more information is embedded from the training data. The calculation of SHF feature can be implemented as an image filter, which can be very efficient to calculate. The procedure of our proposed method is described as follows. Firstly, the multiple channel maps are generated from the training set, which is similar to produce the integral channel feature (step 1, Fig.3). Secondly, each channel map is shrunk to one quarter of its original size which is similar to ACF[11](step 2, Fig.3). The second step is quite important, because it can reduce the side-effect from the image noise and misalignment. Thirdly, the SHF is calculated on the shrunk channel maps, which generates the final training data that fed into the Gentle boosting algorithm for feature selection. It's worth to mention that, the SHF learned for each HF in the feature pool is based on the Eq. (2) (step 3, Fig.3). The final detector is trained for four rounds, and in each round, hard negatives are mined to improve the performance of the detector.

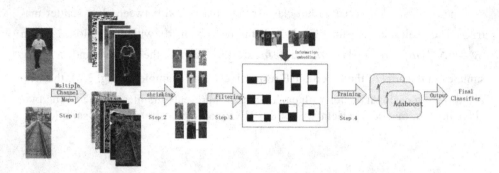

Fig. 3. Procedure of our proposed human detector

4   Experiments

4.1   Experiment Setting

In order to validate the effectiveness of our proposed method, we conduct a set of experiments on INRIA dataset. In the following experiments, we make use of gentle boosting algorithm with LUT as weak classifiers. The number of bins for LUT is set to 256. The final detector is trained 4 times, the final classifier comprise of 2048 weak classifiers. The training data comprise of 4912 positive samples which jittered from the original 614 positive annotations. These positives together with 5000 negatives which is initially random sampled, that fed to the boosting algorithm. For the bootstrap step, another 5000 hard negatives is mined which is prepared for the following round. The number of weak classifier in each round is 32, 128, 512 and 2048 respectively.

4.2   Comparison with HF and SHF

The first three feature selected by the boosting algorithm is shown in Fig. 4. The top three images are based on the HF, and the bottom ones belong to the SHF. The number in the top left corner of each image means the index of channels, the rectangle box in each window is the position of the feature. The real number in each sub-region of rectangle represents the weight for each box. For the HF, the real value in each sub-region is used to multiple the sum of values in this area, whereas, for the SHF, each pixel is multiple with the weight separately.

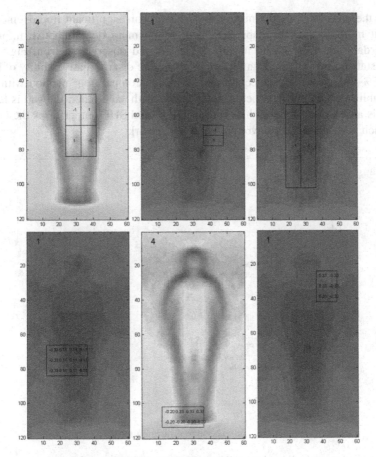

Fig. 4. Selected HF and SHF on different channel maps

We have noticed that, image gradient magnitude is a very discriminative channel, as both the learning algorithms choose it as its first feature; we also find that channels with gradient orientation of 120 degree is very discriminative, and the most discriminative area is focusing on the shoulder part, middle part and feet part of human. Another important channel is the L-part in the LUV color space, which mainly considers the coherence of color in some specific area.

4.3   Comparisons with State-of-the-Arts Algorithms

In this section, we have compared our proposed method with many other state-of-the-arts. The evaluation code is downloaded from piotr's toolbox which is available online. The state-of-the-arts methods include the ACF[11], ICF[2](ChnFtrs) and OICF[10] which is most close to our method. Other methods are also compared such as ConvNet[12], DPM[13](LatentSVM-V2), besides that, two baseline method(VJ[8] and HOG[1]) are also included. The comparison result is shown in Fig. 5, we can see that HF applied on the multiple channel maps instead of raw pixels can greatly decrease the miss rate, approximately to 47.87%. This significant improvement indicates that multiple channel maps can distill much more discriminative information from the data and that is why it works. The second observation from Fig.5 is that learning statistic information can further improve the discriminative ability of the HF, about 4.5% increase of the detection rate. We also compare our method with the recently published OICF[10], which is comparable with each other, the gap is less than 1%. It is also worth mention that, both of OICF and SHF outperforms the ConvNet [12], which is based on the convolution neural network.

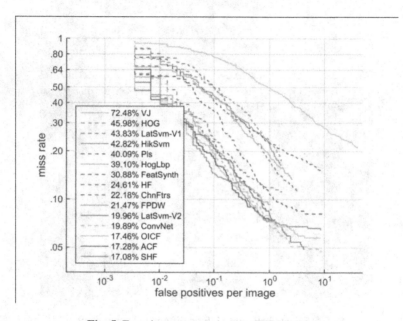

Fig. 5. Experiment comparison on INRIA dataset

4.4 Runtime Comparison

Our detector is implemented with Matlab 2014b and visual studio 2010 on HP workstation ZBook 17 (8 core CPU I7-4700MQ, 2.4GHZ, 32G). It takes less than 0.5 hours to train a four-stage detector and the running speed for a 640x480 image can reach 20 frames per second. The comparison of HF and our proposed SHF is shown in Tab.1. We can see that our proposed detector is slightly faster than the original HF. This is due to the higher discriminative ability which needs fewer amounts of features to exclude most of the negative patches.

Table 1.

Name	speed(fps) for 640x480 image
HF	18
Our proposed(SHF)	20

5   Conclusion

In this paper, we first have revisited the Haarlike feature which applied to the multiple channel maps instead of raw pixels of the image. Then we proposed an improved statistical Haarlike feature, which embeds the statistic information from the training data. Our proposed method is very effective to build a high performance human detector. Experiments based on INRIA dataset shows that our proposed method can get comparable performance with state-of-the-arts and also run at real-time speed for 640x480 images.

Acknowledgement. This project is supported by NSF of China (61305058, 61473086), Fundamental Research Funds for the Jiangsu University (13JDG093), Young Scientist Foundation of Jiangsu Province (BK20130471, BK20140566, BK20150470) and China Postdoctoral science Foundation (2014M561586).

References

1. Dala, N., Triggs, B.: Histograms of oriented gradients for human detection. In: Proceeding of IEEE Conference on Computer Vision and Pattern Recognition, vol. 1, pp. 886–893 (2005)
2. Dollar, P., Tu, Z., Perona, P., Belongie, S.: Integral channel features. In: British Machine Vision Conference (2009)
3. Dollár, P., Wojek, C., Schiele, B., Perona, P.: Pedestrian detection: an evaluation of the state of the art. IEEE Transactions on Pattern Analysis and Machine Intelligence 34(4), 743–761 (2012)
4. David, G.: Survey of pedestrian detection for advanced driver assistance systems. IEEE Transactions on Pattern Analysis and Machine Intelligence 32(7), 1239–1258 (2009)
5. Gool, L.V.: Pedestrian detection at 100 frames per second. In: IEEE Conference on Computer Vision and Pattern Recognition, pp. 2903–2910
6. Zhang, S., Bauckhage, C., Cremers, A.B.: Informed haar-like features improve pedestrian detection. In: IEEE Conference on Computer Vision and Pattern Recognition (2014)
7. Papageorgiou, C., Oren, M., Poggio, T.: A general framework for object detection. In: ICCV (1998)
8. Viola, P., Jones, M.: Rapid object detection using a boosted cascade of simple features. In: IEEE Conference on Computer Vision and Pattern Recognition, vol. 1, pp. 511–518 (2001)
9. Shen, J., Sun, C., Yang, W.: A novel distribution-based feature for rapid object detection. Neurocomputing 74(17), 2767–2779 (2011)
10. Shen, J., Zuo, X., Yang, W., Liu, G.: Real-time human detection based on optimized integrated channel features. In: Li, S., Liu, C., Wang, Y. (eds.) CCPR 2014, Part II. CCIS, vol. 484, pp. 286–295. Springer, Heidelberg (2014)
11. Dollár, P., Appel, R., Belongie, S., Perona, P.: Fast Pyramids for object detection. IEEE Transactions on Pattern Analysis and Machine Intelligence 36(8), 1532–1545 (2014)
12. Sermanet, P., Kavukcuoglu, K., Chintala, S., LeCun, Y.: Pedestrian detection with unsupervised multi-stage feature learning. In: IEEE Conference on Computer Vision and Pattern Recognition, pp. 3626–3633 (2013)
13. Pelzenszwalb, P., McAllester, D., Ramanan, D.: A discriminatively trained, multiscale, deformable part model. In: IEEE Conference on Computer Vision and Pattern Recognition (2008)

Wood Surface Quality Detection and Classification Using Gray Level and Texture Features

Deqing Wang[1,2,*], Zengwu Liu[2], and Fengyu Cong[1]

[1] Department of Biomedical Engineering,
Faculty of Electronic Information and Electrical Engineering,
Dalian University of Technology, Dalian, China
`deqing.wang@foxmail.com, cong@dlut.edu.cn`
[2] Dalian Scientific Test and Control Technology Institute, Dalian, China
`liuzengwu126@126.com`

Abstract. Computer vision methods can benefit wood processing industry. We propose a method to detect wood surface quality and classify wood samples into sound and defective classes. Gray level histogram statistical features and gray level co-occurrence matrix (GLCM) texture features are extracted from wood surface images and combined for classification. A half circle template is proposed to generate GLCM, avoiding calculating distances at each pixel every time and speeding up the algorithm greatly. The proposed approach uses more pixel information than traditional four-angle method, resulting in a significantly higher classification accuracy. Moreover the running time demonstrates our algorithm is efficient and suitable for real-time applications.

Keywords: Wood Surface Detection, Texture Image Classification, Gray Level Histogram Statistics, Gray Level Co-occurrence Matrix.

1 Introduction

In wood processing industry, detecting and classifying wood surface quality are very important processes. Traditionally these processes are carried out manually by human inspection, which are tiring and low efficient in long working time. Computer vision methods can bring dramatic changes to this with higher efficiency and less labor forces. Several works give comprehensive studies on this topic from algorithms to practical applications[1, 2].

Wood surface quality inspection is carried out according to surface defects such as knots, cracks, wood grains, tree rings, spots, dark areas, contrast changes, textures, and so on. The surface defects are loosely separated into two types: One is local textural irregularities and the other is global deviation of color and texture[3]. Wood surfaces with defects present typical texture features. There are a variety of techniques for discriminating textures, and they are generally divided into four categories: statistical approaches, geometrical/structural approaches, signal processing/filter based approaches, model based approaches[3, 4].

---

* Corresponding author.

© Springer International Publishing Switzerland 2015
X. Hu et al. (Eds.): ISNN 2015, LNCS 9377, pp. 248–257, 2015.
DOI: 10.1007/978-3-319-25393-0_28

From the feature extraction perspective, different approaches have been applied. Color histogram percentile features are extracted for knot detection[1]. The coefficients of a two-dimensional discrete wavelet transform are used to extract knots features through a cluster-based approach[5]. Gray level co-occurrence matrix (GLCM)[6] and Local binary patterns (LBP)[4] and their combination methods are used to detect wood knots[7, 8] or to classify different wood species based on wood grains[2]. In a way, most of the texture descriptors[9] and surface defect detection techniques[3, 10] can be applied to wood surface inspection. Gray level co-occurrence matrix is one of the most widely used methods in texture analysis. It has been widely applied to biomedical image analysis[11], synthetic aperture radar (SAR) image analysis[12], and industry detection[13]. Recently, many extensions to GLCM emerge[14, 15].

From the classification perspective, self-organizing map (SOM)[1, 7] , k-NN classifier[8] and neural network classifier[2, 5] are used in different wood surface inspection experiments. Support Vector Machine (SVM)[16] is also viable in this task.

Most of above researches are based on defective wood samples. As is different from these, in this paper, we detect wood surface in only one species without knowing the surface quality in advance. We don't consider detailed detects types, and just classify samples into two classes of sound and defective ones. Fig.1 shows some wood surface example images that need to be detected and classified.

We extract gray level histogram statistical features and gray level co-occurrence matrix texture features from the wood surface images simultaneously. The former is used to capture global gray level changes, and the latter is used to capture both global textures and local detects. The combination of these features can increase classification accuracy.

Conventionally, gray level co-occurrence matrix method is computed at four angles (0°, 45°, 90,° 135°)[6, 14], so four co-occurrence matrices need to be generated at each angle. Maria Petrou proposed a circle template method to compute GLCM[9]. In this method, only one matrix is generated, and the template covers more image details. In order to reduce repeat pixel pairs counting[9], we propose to use half circle template instead. This half template increase our algorithm running speed greatly.

After feature extraction, we use SVM and k-NN to test classification performance in a supervised manner.

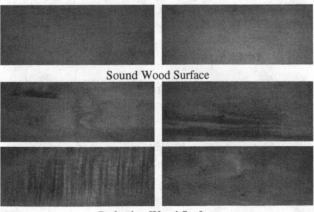

Sound Wood Surface

Defective Wood Surface

Fig. 1. Example of wood surface images

The remainder of this paper is organized as follows. In Sect. 2, we introduce gray level histogram statistical features and GLCM texture features. In Sect. 3, we describe our half circle template method to compute GLCM. In Sect. 4, we introduce our experimental setup with specific steps. In Sect. 5, we present results of classification accuracy and run time. Finally, we present our conclusions in Sect. 6.

2   Feature Extraction

We extract both gray level features using histogram statistics and texture features using gray level co-occurrence matrix.

2.1   Gray Level Histogram Features

We can calculate the image histogram features using the six classical statistical features[17]: mean $(\bar{n})$, variance (σ^2), skewness (b_s), kurtosis (b_k), histogram energy(b_E), histogram entropy(b_H).

2.2   Texture Features

Gray level co-occurrence matrix (GLCM) is a powerful descriptor to extract texture features, which represents the joint distribution of gray level pairs of neighboring pixels. The rotationally invariant co-occurrence matrix is constructed with all pairs of pixels at a fixed distance d from each other. It is written as[9]:

$$C_d(m,n) = \sum_i \sum_j \sum_{\hat{e}} \delta(m - g(i,j))\delta\left(n - g((i,j) + d\hat{e})\right) \tag{1}$$

where $\hat{e}$ is the unit vector pointing in a chosen direction, $g(i,j)$ is the gray value of pixel (i,j), $g((i,j) + d\hat{e})$ is the gray value of another pixel that is at distance d from pixel (i,j) and at the orientation indicated by unit vector $\hat{e}$, and $C_d(m,n)$ is the total number of pairs of pixels at distance d from each other identified in the image, such that the first one has gray value m and the second has gray value n $(m,n \in [0, L - 1])$. L is gray level number of original image, and we can reduce the separate gray levels number to G by

$$g(i,j)^{new} = \left[\frac{g(i,j)^{old}}{L-1} \times (G - 1)\right] \tag{2}$$

Then, normalized the co-occurrence matrix by dividing all its elements by the total number of pairs of pixels considered. The normalized matrix is a joint probability density function:

$$p(m,n) = \frac{C_d(m,n)}{\sum_{i=0}^{G-1}\sum_{j=0}^{G-1} C_d(i,j)} \tag{3}$$

Based on $p(m,n)$, we can compute the following statistical measures as texture features[9]:

- GLCM energy

$$G_E = \sum_{m=0}^{G-1} \sum_{n=0}^{G-1} p(m,n)^2 \qquad (4)$$

- GLCM entropy

$$G_H = -\sum_{m=0}^{G-1} \sum_{n=0}^{G-1} p(m,n) \log_2 p(m,n) \qquad (5)$$

- Contrast

$$G_{ct} = \frac{1}{(G-1)^2} \sum_{m=0}^{G-1} \sum_{n=0}^{G-1} (m-n)^2 p(m,n) \qquad (6)$$

- Correlation

$$G_{cn} = \frac{\sum_{m=0}^{G-1} \sum_{n=0}^{G-1} mnp(m,n) - \mu_x \mu_y}{\sigma_x \sigma_y} \qquad (7)$$

where

$$\mu_x = \sum_{m=0}^{G-1} m \sum_{n=0}^{G-1} p(m,n) \qquad (8)$$

$$\mu_y = \sum_{n=0}^{G-1} n \sum_{m=0}^{G-1} p(m,n) \qquad (9)$$

$$\sigma_x = \sum_{m=0}^{G-1} (m-\mu_x)^2 \sum_{n=0}^{G-1} p(m,n) \qquad (10)$$

$$\sigma_y = \sum_{n=0}^{G-1} (n-\mu_y)^2 \sum_{m=0}^{G-1} p(m,n) \qquad (11)$$

- Homogeneity

$$G_h = \sum_{m=0}^{G-1} \sum_{n=0}^{G-1} \frac{p(m,n)}{1+|m-n|} \qquad (12)$$

More than ten statistical measures can be computed from co-occurrence matrix[6], but we just used the common five features above in this paper.

3   Half Circle GLCM Template

Maria Petrou proposes a circle template method to compute GLCM[9]. The circle is defined as the locus of points at a fixed distance d from the center. Assume the center point to be at coordinate position $(0,0)$, the circle template is discrete points which have the most approximated distances of d. Fig 2 shows the templates of distances from 1 to 4.

Using the template we scan the image line by line from left to right. We can see that when the new template center moves to the old circle's right coordinate, the old center becomes the new circle's left coordinate[9]. Thus we count the same pair twice. In order to reduce the repeats, we propose to use half circle, as is shown in Fig. 3.

In half circle template, we calculate the coordinate codes in advance. The coordinates of distance from 1 to 4 are listed in Table 1.

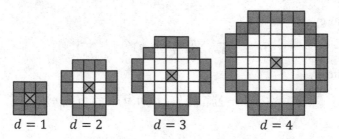

Fig. 2. Circle template

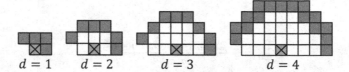

Fig. 3. Half circle template

Table 1. Half circle coordinate codes

Distance	$d=1$	$d=2$	$d=3$	$d=4$
Coordinate	(-1,1), (0,1), (1,1), (1,0)	(-2,1), (-1,2), (0,2), (1,2), (2,1), (2,0)	(-3,1),(-2,2), (-1,3),(0,3), (1,3),(2,2), (3,1),(3,0)	(-4,1),(-4,2), (-3,2),(-3,3), (-2,3),(-2,4), (-1,4),(0,4), (1,4),(2,4), (2,3),(3,3), (3,2),(4,2), (4,1),(4,0)

4   Experimental Setup

All experiments are carried out on computer of Intel Core i3-2120 3.30GHz CPU, 2GB RAM and 32-bit Windows XP SP3 system. All algorithms are implemented using C++ in Microsoft Visual Studio 2010.

Our database contains 414 wood samples[18], half of which is sound samples and another half part is defective. We select 100 samples in each class as the training samples, and the left 214 samples are used for testing. We conduct the experiment in the following steps:

- Step 1

We extract both gray level histogram features and GLCM features in the ith image to form a feature vector

$$v_i = [v_{i1}, v_{i2}, \cdots, v_{i10}, v_{i11}] = [\bar{n}, \sigma^2, b_s, b_k, b_E, b_H, G_E, G_H, G_{ct}, G_{cn}, G_h] \tag{13}$$

- Step 2

Normalized feature vector v_i. Let c_j represents all the jth components in feature vector v_i of training samples.

$$c_j = [v_{1j}, v_{2j}, \cdots, v_{nj}]^T \tag{14}$$

$j = 1, 2, \cdots 11$, and $n = 200$ for the training samples in our experiment. In order to eliminate noise in data, we let $\max_j$ represents the largest but three value in c_j, and $\min_j$ represents the smallest but three value in c_j. Reset all the values in c_j larger than $\max_j$ to $\max_j$, and all the values smaller than $\min_j$ to $\min_j$. The normalized value is

$$\hat{v}_{ij} = \frac{v_{ij} - \min_j}{\max_j - \min_j} \tag{15}$$

All the feature elements in training sample have been normalized to $[0,1]$. Normalize test sample feature vectors using (15) in the same way. The normalized feature vector is

$$\hat{v}_i = [\hat{v}_{i1}, \hat{v}_{i2}, \cdots, \hat{v}_{i10}, \hat{v}_{i11}] \tag{16}$$

- Step 3

Center the normalized feature vector. Let $\hat{c}_j$ represents all the jth components in normalized feature vector $\hat{v}_i$ of training samples.

$$\hat{c}_j = [\hat{v}_{1j}, \hat{v}_{2j}, \cdots, \hat{v}_{nj}]^T \tag{17}$$

Let $\hat{m}_j$ represents the mean of $\hat{c}_j$. $j = 1, 2, \cdots 11$, $n = 200$ for the training samples. We can compute the centered feature vector for both training samples and test samples.

$$\tilde{v}_i = [\tilde{v}_{i1}, \tilde{v}_{i2}, \cdots, \tilde{v}_{i10}, \tilde{v}_{i11}] = [\hat{v}_{i1} - \hat{m}_1, \hat{v}_{i2} - \hat{m}_2, \cdots, \hat{v}_{i10} - \hat{m}_{10}, \hat{v}_{i11} - \hat{m}_{11}] \tag{18}$$

- Step 4

After normalizing and centering the feature vectors, SVM and k-NN classifier are used for classification. SVM classifier is constructed using LIBSVM[16] with default nonlinear kernel of radial base function. We train the SVM model using training samples and then classify the testing samples using the trained model.

k-NN classifier is constructed using KD-tree searching method in ALGLIB[19]. $k = 10$ in the classifier and we use distance-weighted voting method to determine the class.

5   Experiment Result

5.1   Calculate GLCM

We display gray level co-occurrence matrix corresponding to wood surface images of Fig. 1. in Fig. 4. GLCM is computed using half circle template. We use 256 gray levels, so the matrices size is 256 by 256.

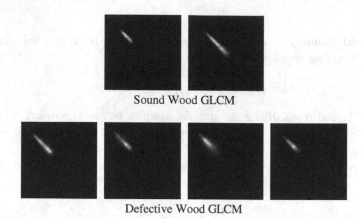

Sound Wood GLCM

Defective Wood GLCM

Fig. 4. GLCM image corresponding to Fig.1.

5.2   Classification Accuracy of Different Features

We test the classification performance using gray level (GL) histogram statistical features and GLCM textures features. The results are listed in Table 2 and Table 3.

In Table 2 and Table 3, both SVM and k-NN results show that classification accuracy using either of the two kinds of features separately are below 80%. While, using combination of gray level and GLCM features, the classification accuracy are increased greatly with a maximum value 87.38%.

5.3   Influence of GLCM Distance

Table 3 shows that GLCM distance affects classification accuracy. When distance changes from small to 8, the accuracy increases. When distance equals to 8, SVM gives a maximum accuracy. According to Table 3, distances range from 5 to 10 is the best for our wood surface detection and classification problem. We did not test larger distances, such as 50 or 100, because they are not practical.

5.4   Influence of GLCM Template

We test the classification accuracy using circle template in Table 4. Comparing Table 3 and Table 4, results show that the accuracy are nearly the same using half circle and circle template.

Many applications use four angles (0°, 45°, 90,° 135°) method to compute GLCM, thus four matrices have to be generated for all angles[6]. So there are 5×4=20 GLCM features in total. In order to make comparison with half circle method, we reduce four angles features dimension using principal component analysis (PCA) to 5 for pure GLCM test, and reduce to 11 for combination test. The results listed in Table 5 show that the four angles method dose not achieve a high classification accuracy as half circle method. This is because four angles just count four pixel pairs in generating GLCM, while half circle template take more pixel pairs which contain more image details.

From Table 6, we can see that in order to get a better classification using half circle template, our GLCM algorithm just costs about 90 milliseconds at distance 8. We haven't found works that give detailed GLCM run time. In work [2], author says their total algorithm using GLCM costs about 300 seconds, which is really a long time. Compared with this, our algorithm runs in a flash, and is suitable for real-time applications.

Table 2. Classification using only gray level features

Feature Number	Classifier	Classification accuracy
6	SVM	0.775701
Gray Level (GL)	k-NN	0.724299

Table 3. Classification accuracy of GLCM and combination method using half circle template

Feature number	Classifier	Classification accuracy at different GLCM distances			
		5	6	7	8
5	SVM	0.775701	0.785047	0.785047	0.775701
(GLCM)	k-NN	0.747664	0.761682	0.757009	0.752336
5+6	SVM	0.85514	0.850467	0.850467	**0.873832**
(GLCM+GL)	k-NN	**0.841121**	**0.836449**	0.831776	0.831776
Feature number	Classifier	Classification accuracy at different GLCM distances			
		9	10	11	12
5	SVM	0.78972	0.785047	0.785047	0.794393
(GLCM)	k-NN	0.771028	0.771028	0.771028	0.761682
5+6	SVM	**0.869159**	**0.869159**	0.85514	0.85514
(GLCM+GL)	k-NN	0.831776	0.831776	0.836449	0.831776

Table 4. Classification accuracy of GLCM and combination method using circle template

Feature number	Classifier	Classification accuracy at different GLCM distances			
		5	6	7	8
5	SVM	0.775701	0.785047	0.785047	0.775701
(GLCM)	k-NN	0.747664	0.761682	0.757009	0.752336
5+6	SVM	0.85514	0.850467	0.850467	0.873832
(GLCM+GL)	k-NN	0.841121	0.836449	0.831776	0.831776

Table 5. Classification accuracy of GLCM and combination method using four angles method

Feature number	Classifier	Classification accuracy at different GLCM distances			
		5	6	7	8
5, PCA	SVM	0.799065	0.771028	0.780374	0.780374
(GLCM)	k-NN	0.742991	0.757009	0.771028	0.775701
11, PCA	SVM	0.85514	0.85514	0.841121	0.85514
(GLCM+GL)	k-NN	0.799065	0.799065	0.808411	0.813084

Table 6. GLCM run time using different templates

	Distance	5	6	7	8
Half circle Template time (ms)	Matrix generation time	45.2126	66.0531	67.2126	80.715
	Feature computing time	9.33333	8.36232	9.19324	9.04831
	Total time	**54.5459**	**74.4155**	**76.4058**	**89.7633**
Circle Template time (ms)	Distance	5	6	7	8
	Matrix generation time	89.6667	127.961	131.309	157.802
	Feature computing time	7.97585	7.67633	7.51691	8.05314
	Total time	97.6425	135.638	138.826	165.855
Four angles time (ms)	Distance	5	6	7	8
	Matrix generation time	20.1014	20.5604	20.3478	20.2464
	Feature computing time	33.2126	33.5362	32.8696	32.9227
	Total time	53.314	54.0966	53.2174	53.1691

5.5   GLCM Run Time

In Table 6, we record the average GLCM algorithm run time with details including matrix generation time, feature computing time and the total GLCM time.

We can see that as the distance increases, the matrix generation time for half circle and circle increases, because the templates have more coordinates. But the feature computing time has little change, because no matter what the template is, matrix size is fixed at 256 by 256. In four angles case, neither of matrix generation time and feature computing time changes much, because both coordinate number and matrix size are fixed. As there are four matrices, the feature computing time is nearly four times of half circle and circle case. In a word, the matrix generation time is decided by the coordinate number in template, and the feature computing time is decided by matrix number and size.

6   Conclusion

In this paper, we propose a wood surface quality detection and two-class classification method. We extract gray level histogram statistics features and gray level co-occurrence matrix (GLCM) texture features respectively from wood images. Using combination of the two types of features, both global gray level and texture features and local defect features are captured, and classification accuracy is increased much higher than using them separately. We propose to use a half circle template to compute GLCM, which covers more image texture details than traditional four angles method and can also improve classification accuracy. The template contains many coordinates at a certain distance avoiding computing distances at each pixel every time, so the algorithm is speeded up dramatically. We analyze the influences of different features, different distances and different templates on classification accuracy and run time. All of our algorithms are implemented using C++. The results show that our method costs just several tens of milliseconds, and is quite suitable for real-time applications.

References

1. Kauppinen, H.: Development of a color machine vision method for wood surface inspection. Oulun yliopisto (1999)
2. Wang, K., Bai, X.: The pattern recognition methods of wood surface defects. Science Press, Beijing (2011)
3. Xie, X.: A review of recent advances in surface defect detection using texture analysis techniques. Electronic Letters on Computer Vision and Image Analysis 7(3), 1–22 (2008)
4. Pietikäinen, M., Hadid, A., Zhao, G., et al.: Computer vision using local binary patterns. Springer, London (2011)
5. Yu, G., Kamarthi, S.V.: A cluster-based wavelet feature extraction method and its application. Engineering Applications of Artificial Intelligence 23(2), 196–202 (2010)
6. Haralick, R.M., Shanmugam, K., Dinstein, I.H.: Textural features for image classification. IEEE Transactions on Systems, Man and Cybernetics. SMC-3(6), 610–621 (1973)
7. Silvén, O., Niskanen, M., Kauppinen, H.: Wood inspection with non-supervised clustering. Machine Vision and Applications 13(5-6), 275–285 (2003)
8. Mäenpää, T., Viertola, J., Pietikäinen, M.: Optimising Colour and Texture Features for Real-time Visual Inspection. Pattern Analysis & Applications 6(3), 169–175 (2003)
9. Petrou, M., Sevilla, P.G.: Image processing: dealing with texture. Wiley, Chichester (2006)
10. Karimi, M.H., Asemani, D.: Surface defect detection in tiling Industries using digital image processing methods: Analysis and evaluation. ISA Transactions 53(3), 834–844 (2014)
11. Torheim, T., Malinen, E., Kvaal, K., et al.: Classification of Dynamic Contrast Enhanced MR Images of Cervical Cancers Using Texture Analysis and Support Vector Machines. IEEE Transactions on Medical Imaging 33(8), 1648–1656 (2014)
12. Dumitru, C.O., Datcu, M.: Information content of very high resolution SAR images: study of feature extraction and imaging parameters. IEEE Transactions on Geoscience and Remote Sensing 51(8), 4591–4610 (2013)
13. Yang, S.-W., Lin, C.-S., Lin, S.-K., et al.: Automatic inspection system for defects of printed art tile based on texture feature analysis. Instrumentation Science & Technology 42(1), 59–71 (2014)
14. Su, H., Sheng, Y., Du, P., et al.: Hyperspectral image classification based on volumetric texture and dimensionality reduction. Frontiers of Earth Science 9(2), 225–236 (2015)
15. Roberti de Siqueira, F., Robson Schwartz, W., Pedrini, H.: Multi-scale gray level co-occurrence matrices for texture description. Neurocomputing 120, 336–345 (2013)
16. Chang, C.-C., Lin, C.-J.: LIBSVM: A library for support vector machines. ACM Transactions on Intelligent Systems and Technology 2(3), 27 (2011)
17. Wang, Q.: Digital Image Processing. Science Press, Beijing (2009)
18. Wood Board Processing Using Computer Vision. http://www.matlab.org.cn/wood/
19. Bochkanov, S., Bystritsky, V.: ALGLIB-a cross-platform numerical analysis and data processing library. ALGLIB Project. Novgorod, Russia (2011)

Aerial Scene Classification
with Convolutional Neural Networks

Sibo Jia, Huaping Liu, and Fuchun Sun

Department of Computer Science and Technology,
Tsinghua University, Beijing, China
State Key Lab. of Intelligent Technology and Systems, Beijing, China
hpliu@tsinghua.edu.cn

Abstract. A robust satellite image classification is the fundamental step
for aerial image understanding. However current methods with hand-
crafted features and conventional classifiers have limited performance. In
this paper we introduced convolutional neural network (CNN) method
into this problem. Two approaches, including using conventional classifier
with CNN features and direct classification with trained CNN models,
are investigated with experiments. Our method achieved 97.4% accuracy
on 5-fold cross-validation test of the UCMERCED LULC dataset, which
is 8% higher than state-of-the-art methods.

1   Introduction

The satellite image analysis has received great interest from both the academic
and industrial communities. However, the classification and understanding of
the aerial scenes admits many technical challenges such as the diversified classes
and obscure image details. To tackle these problems, many modern machine
learning methods have been developed to address the aerial scenes classification.
A detailed survey can be found in [5].

On the other hand, some deep learning methods, such as auto-encoder, con-
volutionary neural networks (CNN) and others, have been extensively studied in
image classification, speech recognition and machine learning[1,2,3]. All of the
successful applications show that stack generalization plays important roles in
the machine intelligence. However, to the best knowledge of the authors, the
deep learning method has never been used in the classification of aerial scenes.
This motivates us to perform experimental validations on the problem.

In this paper, we perform extensive experiments to show that a well-trained
CNN can get very surprisingly high recognition accuracy on public available
aerial scene dataset. Currently the best accuracy is about 90%, while our method
can achieve accuracy of 97%. The rest of this paper is organized as follows:
Section 2 gives a brief introduction about CNN. Section 3 presents the details
about the classification and Section 4 shows the experimental results.

2   Brief Introduction on CNN

Convolutional neural network (or CNN) is a widely used model for image and
video recognition, which features a feed-forward artificial neural network where

© Springer International Publishing Switzerland 2015
X. Hu et al. (Eds.): ISNN 2015, LNCS 9377, pp. 258–265, 2015.
DOI: 10.1007/978-3-319-25393-0_29

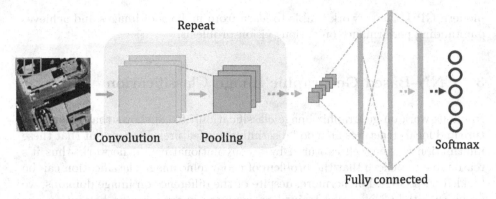

Fig. 1. An exemplary architecture of CNN

the individual neurons are tiled in such a way that they respond to overlapping regions in the visual field. Compared to other image classification algorithms, convolutional neural networks use relatively little pre-processing, as it can learn the filters that in traditional algorithms were hand-engineered. The lack of a dependence on prior-knowledge and the existence of difficult to design hand-engineered features is a major advantage for CNNs.

Figure 1 shows the typical architecture of a CNN network. It consists of multiple layers of small neurons which look at small portions of the input image, called receptive fields. The results of these collections are then tiled so that they overlap to obtain a better representation of the original image. Each neuron consists of a convolution operation with weights W^k and bias b_k and an activation operation $f(\cdot)$. Then the feature of the k-th neuron h^k is obtained by

$$h_{ij}^k = f((W^k * x)_{ij} + b_k)$$

where x is the output feature map of the previous layer. Between the convolutional layers exists local or global pooling layers, which combine the outputs of neuron clusters. When the convolutional and pooling layers are enough to fully cover the whole image region, they are connected to MLP (multilayer perceptron) layers and optionally softmax classification layers. The MLP layers produce a high dimensional vector which can be served as a compact feature of the image, while the softmax layer directly outputs the classification result of the input image. The network is optimized by backpropagation and stochastic gradient descent. It takes a 'mini-batch' of samples each time, compute the gradient $\nabla L(W)$, and obtain the update value V_{t+1} and updated weights W_{t+1} at iteration $t + 1$ given the previous weight update V_t and current weights W_t:

$$V_{t+1} = \mu V_t - \alpha \nabla L(W_t)$$
$$W_{t+1} = W_t + V_{t+1}$$

where the learning rate α is the weight of the negative gradient and momentum μ is the weight of the previous update[4]. Thanks to the computational power of modern GPU, the network is able to learn from millions of images and achieves outstanding performance on various vision problems.

3   CNN-Based Geographic Image Classification

Previous work on geographic image classification[7,5,6,8] shows that color, texture and local structures are good discriminative features. It turns out that these information can be well captured by a convolutional neural network, thus it's reasonable to believe that the problem of geographic image classification can be tackled with CNN. Furthermore, despite of the difference on image domains, we argue that the CNN model trained on common images can be helpful on our problem, since the size of a typical dataset for CNN training, e.g. ImageNet[9] is by far larger than the geographic image dataset we have at hand and the neural network will be able to learn enough discriminative features from common images which are also effective on geographic images.

We propose two approaches of geographic image classification using CNN. The first one is to use a off-the-shelf CNN model to extract high dimensional features of geographic images followed by a traditional classifier e.g. SVM. The other approach is to retrain a CNN model using geographic images based on a pretrained model, the process named 'finetuning', and use the new network directly for classification. We will not train a whole new model mainly because we lack the massive amount of training images. While the first method can be very easily applied as it doesn't need any training of neural networks, an adaptation of CNN models trained on common images to the target image domain will hopefully yield better performance. Thus both approaches are investigated in this work.

3.1   Classification Without CNN Retraining

Following the settings of other works, we constrained all the training and testing data to the LULC dataset[5], which contains 2100 land use images of 21 different classes. We used the CNN deep learning framework Caffe as our experiment platform[10], which provides an efficient implementation of deep learning and several off-the-shelf CNN models. The experiment is conducted as follows: high dimensional features of all the 2100 images in the Features of all the images in LULC dataset are extracted with a pretrained model, then part of the images are used to train a classifier while the rest serve as testing data. The training and testing split follows the form of a 5-fold cross validation.

There are three trained models provided by Caffe which we used for our classification problem: AlexNet[2], GoogLeNet[11] and CaffeNet which is an improved version of AlexNet. All three models are trained on the ImageNet dataset, generating features of which dimension ranges from 1024 to 4096. As for the classifiers, we tested SVM, KNN classifier and random forest. As the combinations of model and classifier are rather large, we conduct the experiment in two steps. First we try different classifiers on one of the trained models, then we use the best classifier setting to test other CNN models. Final result is reported as the average accuracy of cross validation on the best model/classifier combination.

3.2   Classification with CNN Retraining

In this experiment, we will train a CNN model using a trained model and images from the training set. We use CaffeNet as the model to finetune on, which is originally trained on the 1000-class ImageNet images. Instead of using the 4096-dimension features as we did in the previous experiment, this time we will use the softmax classification output. The only modification we make to the CaffeNet is to change the 1000-class softmax layer to a 21-class softmax layer corresponding to the LULC dataset, enabling the network to learn more discriminative features and a 21-class classifier for the LULC dataset. Before training begins, the parameters of every layer except the softmax layer are set to be identical as the trained CaffeNet model, while the softmax layer parameters are initiated randomly. Then the network is trained keeping the learning rate of previous layers smaller than that of the softmax layer, in order to learn the classifier and 'finetune' the convolution layers simultaneously.

We follow the same 5-fold cross validation setting as in the previous experiment. That means only 1680 images can be used to train the CNN model, which is far from enough for a typical deep learning scenario. Thus we extended the training set by flipping and rotating every image to form 7 new images, resulting in a training set 8 times the size to the original. This operation is reasonable for the LULC dataset because content of the photo taken from an aircraft is almost always invariant to flipping and rotation.

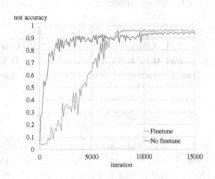

Fig. 2. Training with or without finetune

Due to the small training set, the network only took about 2 hours to convolve on a TITAN BLACK GPU. Testing error after the network convolves is lower than the error rate without retraining CNN. In order to confirm that the improvement is gained from the finetuned CNN instead of from the softmax classification layer alone, we ran the training process again, keeping everything the same except for fixing the parameters in the convolution layers, which is equivalently training a softmax classifier only. Curves for the training process are shown in Figure 2, which reveal that only training the classifier leads to a faster convolving speed but lower performance. This can be explained by the fact that fewer tunable parameters leads to less learning capacity. Through this experiment, the effect of finetuning the CNN network is also confirmed.

We trained 5 networks in total, each tested on the corresponding 20% testing set and collected the result afterwards. Typically CNN networks are not tested using cross validation, but we did so in order to make a fair comparison.

4   Experiment Results

In this section we report the results of the experiments on the LULC dataset. For every setting accuracies of the 5 cross validation test and average accuracy are reported. First we tested classification on the pre-trained ImageNet CNN features. Accuracy of different classifiers on the same CNN model CaffeNet is shown in Table 1. The best classifier, SVM achieved 94.3% overall accuracy. Fixing the classifier, we tested performance on different CNN models. Table 2 gives the result, showing that the accuracy of CaffeNet is slightly higher than other two models. The experiments show that the CNN network can produce discriminative features good enough to handle the geographic image classification problem, even if the network is not trained on this particular domain.

Table 1. Test result of different classifiers on CaffeNet

Setting	Cross validation accuracy	Overall
SVM	0.94 0.95 0.95 0.95 0.92	**0.943**
KNN Classifier	0.82 0.83 0.85 0.82 0.82	0.829
Random forest	0.89 0.90 0.91 0.88 0.88	0.895

Table 2. Test result of different models on SVM

Model	Cross validation accuracy	Overall
CaffeNet	0.94 0.95 0.95 0.95 0.92	**0.943**
AlexNet	0.93 0.94 0.9 0.95 0.92	0.940
GoogLeNet	0.91 0.93 0.95 0.93 0.91	0.923

Table 3. Test result of classification with new CNN models

Cross validation accuracy	Overall
1.00 0.95 0.96 0.96 0.97	**0.974**

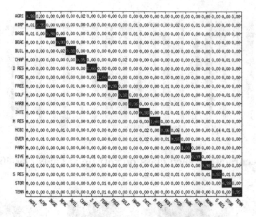

Fig. 3. Confusion matrix of 21 classes

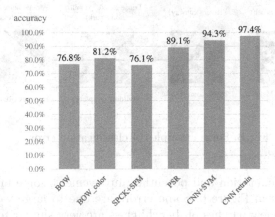

Fig. 4. Comparison with previously reported accuracies

For the finetuned network based on CaffeNet, the result is summarized in Table 3. Overall accuracy of the 5-fold cross validation is 97.4%, when trained on augmented images from part of the LULC dataset. The confusion matrix of the testset is shown in Figure 3. Figure 4 shows the comparison with previously reported accuracies[6]. Time consumption for classifying one image is ∼60ms on an Intel Xeon 2.8GHz CPU.

The statistics of the accuracy for every class is shown in Figure 5, calculated from all the tests of the cross validation. Compared with accuracies of other works, the CNN network is particularly good at capturing textures (*e.g.* chaparral) and structures (*e.g.* intersection), thanks to the learned filters and multi scale pooling.

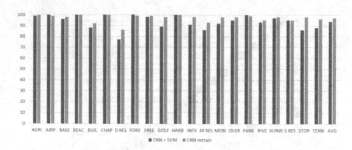

Fig. 5. Accuracy for each class on retrained CNN classification

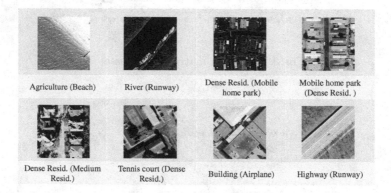

| Agriculture (Beach) | River (Runway) | Dense Resid. (Mobile home park) | Mobile home park (Dense Resid.) |
| Dense Resid. (Medium Resid.) | Tennis court (Dense Resid.) | Building (Airplane) | Highway (Runway) |

Fig. 6. Some examples of classification error

To explore the limitation and potential improvement, some misclassification samples are shown in Figure 6. Some errors are due to large variation of certain classes, *e.g.* a few patches of 'beach' class are very similar to 'agriculture', however other patches from different angle or scale can never be mistaken as 'agriculture'. This implies that for a practical geographic image classification system, it's necessary to consider neighboring patches to correctly classify hard patches occasionally occurred. One patch of 'tennis court' is classified as residential, as there are indeed many buildings around. This suggests that the current network still needs more training samples or training time to capture particular object like a tennis court. There are also classes containing complicated structures with subtle difference, like 'mobile home park', 'dense residential' and 'building', which might only be better distinguished if given much more training samples.

5   Conclusion

In this work we applied convolutional neural network to aerial image classification problem through two different approaches, and achieved the accuracy of

97.4%, much higher than previous state-of-the-art. Notice that all the training data we used was constrained within the LULC dataset. Analysis of the result showed that the performance may be further improved if given more training data. For future works we plan to extend the problem to aerial scene detection and understanding, and apply state-of-the-art methods of object detection based on CNN, hoping to achieve better performance.

Acknowledgments. This work was supported in part by the National Key Project for Basic Research of China under Grant 2013CB329403; and in part by the Tsinghua University Initiative Scientific Research Program under Grant 20131089295.

References

1. LeCun, Y., Bottou, L., Bengio, Y., Haffner, P.: Gradient-based learning applied to document recognition. In: Proceedings of the IEEE, pp. 2278–2324 (1998)
2. Krizhevsky, A., Ilya, S., Hinton, G.E.: ImageNet classification with deep convolutional neural networks. In: Advances in Neural Information Processing Systems, vol. 25, pp. 1097–1105 (2012)
3. Hannun, A.Y., Case, C., Casper, J., Catanzaro, B.C., Diamos, G., Elsen, E., Prenger, R., Satheesh, S., Sengupta, S., Coates, A., Ng, A.Y.: Deep speech: Scaling up end-to-end speech recognition. In: arXiv:1412.5567
4. Bottou, L.: Stochastic gradient descent tricks. In: Montavon, G., Orr, G.B., Müller, K.-R. (eds.) Neural Networks: Tricks of the Trade, 2nd edn., LNCS, vol. 7700, pp. 421–436. Springer, Heidelberg (2012)
5. Cheriyadat, A.M.: Unsupervised feature learning for aerial scene classification. IEEE Transactions on Geoscience and Remote Sensing, 439–451 (2014)
6. Chen, S., Tian, Y.: Pyramid of Spatial Relatons for Scene-Level Land Use Classification. IEEE Transactions on Geoscience and Remote Sensing, 1947–1957 (2015)
7. Lazebnik, S., Schmid, C., Ponce, J.: Beyond bags of features: Spatial pyramid matching for recognizing natural scene categories. In: 2006 IEEE Computer Society Conference on Computer Vision and Pattern Recognition, pp. 2169–2178 (2006)
8. Yang, Y., Newsam, S.: Bag-of-visual-words and spatial extensions for land-use classification. In: Proceedings of the 18th SIGSPATIAL International Conference on Advances in Geographic Information Systems, pp. 270–279 (2010)
9. Russakovsky, O., Deng, J., Su, H., Krause, J., Satheesh, S., Ma, S., Huang, Z., Karpathy, A., Khosla, A., Bernstein, M., Berg, A.C., Fei-Fei, L.: ImageNet large scale visual recognition challenge. In: arXiv:1409.0575 (2014)
10. Yangqing, J., Evan, S., Jeff, D., Sergey, K., Jonathan, L., Ross, G., Sergio, G., Trevor, D.: Caffe: Convolutional architecture for fast feature embedding. In: Proceedings of the ACM International Conference on Multimedia, pp. 675–678 (2014)
11. Szegedy, C., Liu, W., Jia, Y., Sermanet, P., Reed, S., Anguelov, D., Erhan, D., Vanhoucke, V., Rabinovich, A.: Going deeper with convolutions. In: arXiv:1409.4842

Signal Processing

A New Method for Image Quantization Based on Adaptive Region Related Heterogeneous PCNN

Yi Huang, Yide Ma, and Shouliang Li

School of Information Science and Engineering, Lanzhou University,
Lanzhou, China
ydma@lzu.edu.cn

Abstract. Based on the different strength of synaptic connections between actual neurons, this paper proposes a novel heterogeneous PCNN (HPCNN) algorithm to quantize images. HPCNN is constructed with traditional pulse coupled neural network (PCNN) models, which has different parameters corresponding to different image regions. It puts pixels of different gray levels to be classified broadly into two categories: the background regional ones and the object regional ones. Moreover, HPCNN also satisfies human visual characteristics (HVS). The parameters of HPCNN model are calculated automatically according to these categories and quantized results will be optimal and more suitable for human to observe. At the same time, the experimental results show the validity and efficiency of our proposed quantization method.

Keywords: PCNN, HPCNN, quantization, HVS.

1 Introduction

Pulse coupled neural networks (PCNN) model is a improved version of Echorn's cortical model described in in ref [1], which is inspired by mammalian primary visual cortex neurons. And in this paper, a new kind of heterogeneous pulse coupled neural networks model (HPCNN) is suggested, which is constructed with traditional pulse coupled neural network (PCNN) models having different parameters corresponding to different image regions.

Most reported researches in image processing field including PCNN have generally emphasized homogeneous architectures of artificial neural networks. However, the nervous system in real world exhibits great heterogeneity in both its constructing elements and its patterns of interconnection [2]. In most cases, PCNN model of image analysis is set globally with the same parameters. Actually, the parameter setting of neural networks should be different accordingly. This paper puts different pixels (namely different neurons) of gray levels to be classified broadly into two categories: the background regional ones and the object regional ones. The same PCNN parameter setting is applied to the same region. In other words, the same category of neurons are set up with the same PCNN parameters.

© Springer International Publishing Switzerland 2015
X. Hu et al. (Eds.): ISNN 2015, LNCS 9377, pp. 269–278, 2015.
DOI: 10.1007/978-3-319-25393-0_30

The concept of heterogeneous PCNN is firstly proposed in this work and it has focused on quantization. While most traditional quantization methods such as clustering algorithm [3], fuzzy algorithm [4], μ-law algorithm [5] and uniform algorithm only pay attention to the statistic property, but don't take the relationship between the quantization quality and human visual characteristics into account. In our previous work in ref [6], a novel quantization method with PCNN model was proposed, whose parameters were set globally. But the method only emphasized on homogeneous architecture of PCNN but not on heterogeneous one. Therefore, an improved HPCNN quantization algorithm is suggested here.

2   Heterogeneous PCNN Properties and Application

2.1   Heterogeneous Neural Network

Artificial neural networks are computational models inspired by animals' nervous systems. An artificial neuron is a mathematical model whose components are analogous to the components of actual neuron [7]. Generally, artificial neural networks can be described as follow:

Dataset D consists of N samples (x_p, y_p), where x_p is input, y_p is output and $p = 1, 2..., N$. The task of neural network learning is to draw a function f from the dataset D which satisfies formula $y_p = f(x_p)$ [8].

The neural network S is composed of M neurons s_1, s_2, ..., s_M. The weight of each neuron is w_i $(i = 1, 2..., M)$, where $w_i \geq 0$ and $\sum_{i=1}^{N} w_i = 1$. To input x_p, the i-th member of the network output is $f_i(x_p)$. So the output of neural network S with input x_p is as follow:

$$f(x_p) = \sum_{i=1}^{M} w_i f_i(x_p) \tag{1}$$

In practice, the element structure of each neuron and the synaptic connection strength between them are different. There are two different kinds of heterogeneous neural networks. One is defined by the difference in structure and the other in paraments. If the neuronal structure is different in actual neuron, the neural models do not share the same expression correspondingly, that is to say the function f_i is different. And if the synaptic connection strength is different, parameters of the same kind of neural model are not the same correspondingly. The later kind of heterogeneous neural network with different parameters is taken advantage in the following HPCNN.

2.2   Basic PCNN Model

PCNN is a kind of artificial neural network which does not need pre-training and learning compared with traditional network [9]. Echorn's cortical model is a bio-inspired neural network developed in light of synchronous dynamics of neuronal activity in cat visual cortex [10,11]. A simplified neuron model of the PCNN is showed in Fig. 1, which consists of three parts, the input module, the nonlinear

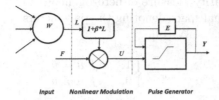

Fig. 1. A simplified neuron Model of the PCNN.

modulation module, and the pulse generating module. It can be described as the following equations [12,13]:

$$F_{i,j}[n] = I_{i,j} \tag{2}$$

$$L_{i,j}[n] = e^{-\alpha_L} L_{i,j}[n-1] + V_L \sum_{k,l} W_{i,j,k,l} Y_{k,l}[n-1] \tag{3}$$

$$U_{i,j}[n] = F_{i,j}[n](1 + \beta L_{i,j}[n]) \tag{4}$$

$$E_{i,j}[n] = e^{-\alpha_E} E_{i,j}[n-1] + V_E Y_{i,j}[n-1] \tag{5}$$

$$Y_{i,j}[n] = \begin{cases} 1, & U_{i,j}[n] > E_{i,j}[n] \\ 0, & \text{else} \end{cases} \tag{6}$$

where I is the input signal, F and L stand for feedback input and linking input of the neuron associated with neighborhood neurons which locate at (k,l) through synaptic weights $W_{i,j,k,l}$. Additionally, the feeding input F receives an input image I, which is normalized to gray intensity in advance. F and L are then modulated through linking strength β to yield internal activity U which is compared with the dynamic threshold E of the previous iteration to judge whether neuron fires ($Y_{ij[n]} = 1$) or not ($Y_{ij[n]} = 0$). The parameter α_E is the exponential decay coefficient of internal activity and V_E is the amplitude of E.

2.3   Heterogeneous PCNN and the Relation with HVS

In traditional image analysis, we take on PCNN models with the same parameters which can also be called homogeneous PCNN models. In practice, each neuron in the network corresponds to one pixel of an image and the parameters for each PCNN model should be different. It is obviously inaccurate to take on homogeneous PCNN models to perform quantization.

We selected a unified PCNN model, namely the function f_i is the same. Due to the synaptic connection strength between neurons is not all the same clearly, the parameters of each PCNN model are different. However, these five parameters (V_L, α_L, W, β and α_E) are considered the same in traditional applications. In this paper, we distinguish the differences of parameters based on different gray levels and employ HPCNN algorithm with different parameters to perform quantization.

We define the similarity measure of heterogeneous PCNN models u and v as $S_{u,v}$. Let N be a nominal space denoting the parameters (V_L, α_L, W, β and α_E) of PCNN model and $u, v \in N$ [14].

$$S_{u,v} = \begin{cases} 1, & if \ u = v \\ 0, & if \ u \neq v \end{cases} \tag{7}$$

If $S_{u,v}$ equals 0, the two PCNN models are heterogeneous, otherwise they are homogeneous.

We put different gray levels of pixels to be classified broadly into two categories: the background region and the object region. The same PCNN parameter setting is applied to the same region. It can be seen that the neurons in the same region are homogeneous while neurons in different regions are heterogeneous.

Human visual system (HVS) is an unique optical imaging system [15], whose characteristics can be theoretically described by the Web-Fechner-Law: Given different luminance value, the rate (ΔI) of minimum brightness increment ΔS_{min} that human eyes can percept and ambient brightness S is a constant:

$$\Delta I = \Delta S / S \tag{8}$$

Equation (8) can be changed to :

$$I = K \ln S + r \tag{9}$$

From the equations (2)-(6), we can find the firing moment of neurons as follow [16]:

$$n(m) = 1 + \frac{1}{\alpha_E} ln \frac{V_E}{cS_{ij}} + m \frac{1}{\alpha_E} ln \frac{cS_{ij} + V_E}{c'S_{ij}}, m = 0, 1, ..., N \tag{10}$$

Time matrix is defined to be a matrix recording the first time when each neuron fires. Then it equals the first two parts of (10) if the m equals 0 and can be expressed as

$$T_{ij}[n] = 1 + \frac{1}{\alpha_E} ln \frac{V_E}{cS_{ij}} \tag{11}$$

After transformation by substituting (11) into (9)

$$I_{ij} = K \ln S_{ij} + r = K(\alpha_E - \alpha_E T_{ij} + ln(V_E/c)) + r \tag{12}$$

A contrastive analysis of (9) and (12) can educe that the relation of time matrix T_{ij} and input stimulus S_{ij} is in accordance with that of the subjective brightness I and the objective brightness S. If we take time matrix as subjective response of human visual system and take input stimulus as objective brightness, we can see they all present logarithm relevant. HPCNN is consisted of PCNN models with different parameter setting in different regions. Therefore, this ensures HPCNN is also endowed with human visual characteristics.

3   Adaptive Region Related Heterogeneous Image Quantization Method

3.1   Automatic Parameter Setting

It's well known that the parameter setting of PCNN is complicated due to its dependency on the different gray layers, the large number of PCNN parameters and the interactions among them. The static property parameters V_L, α_L and β are set here referring to [17,18]. While dynamic property parameters α_E and V_E will be automatically set as the following steps.

The parameter α_E is the exponential decay coefficient of internal activity which yields a direct effect on quantization layers. α_E must be small enough to ensure dynamic threshold attenuates gradually. In this way, adjacent grayscales can be distinguished by different firing moment. In most cases, it is supposed that $E[0] = I_{max}$, then

$$
\begin{aligned}
E[1] &= e^{-\alpha_E} E[0] &= I_{max} * e^{-\alpha_E} \\
E[2] &= e^{-\alpha_E} E[1] &= I_{max} * e^{-2\alpha_E} \\
&\cdots &
\end{aligned}
$$

$$
E[k] = I_{max} * e^{-k\alpha_E} \tag{13}
$$

In order to make sure the minimum grey value of quantization layer k will be fired, let

$$
E[k] = I_{max} * e^{-k\alpha_E} = I_{min} \tag{14}
$$

and then

$$
\alpha_E = \frac{lnI_{max} - ln(I_{min} + 0.05)}{k} \tag{15}
$$

Where lnI_{min} is changed to $ln(I_{min} + 0.05)$ because that I_{min} should not be zero. k denotes the quantization layer and can be set on demand in quantization.

The parameter V_E is the amplitude of E. In line with firing properties of PCNN model, the pixel will be fired if its internal activity $U > E$ and the output of Y is assigned to 1 while the pixel's dynamic threshold E is set to V_E so that it will never be fired again. Theoretically, the value of V_E should be infinitely great. In practice, the value 200 is great enough to ensure that the neurons would be fired only once and it's set as

$$
V_E = 200 \tag{16}
$$

Taking these, automatic parameter setting method is realized.

3.2   Quantization Algorithm

The flow diagram of the adaptive region related HPCNN quantization algorithm is showed in Fig. 2(a), which can be described as follows:

Step 1: Subdivide the input image into background region B and object region O with PCNN segmentation.

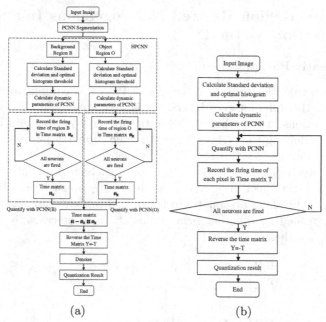

(a)   (b)

Fig. 2. (a) Workflow of HPCNN algorithm. (b) Workflow of PCNN algorithm

Step 2: Initialize parameters of HPCNN by getting standard deviation σ and optimal histogram threshold S' in different regions, setting the iteration number (namely the layers of quantization) and then calculating the dynamic property parameters.

Step 3: The neurons of HPCNN will be fired in region B and region O according to synchronous dynamics of neuronal activity, respectively. And then the output of function Y is recorded until all the neurons have been fired in both regions.

Step 4: Judge whether all the neurons in region B and region O have been fired and then output the time matrix T_B and T_O. And the time matrix T can be calculated

$$T = T_B + T_O \tag{17}$$

According to (12), the quantized image (Y_{out}) can be obtained by reversing the time matrix as

$$Y_{out} = -T \tag{18}$$

Step 5: Denoise image by mean filter to refrain pulse noises [19].

For comparison, the flow diagram of traditional PCNN algorithm [6] is shown in Fig. 2(b).

In *Step 1*, we use cross-entropy and Shannon entropy as criterion to perform segmentation [20,21]. The parameters of PCNN segmentation are set referring to [17]. Segmentation results of Cameraman (C) and Lena (L) show as Fig. 3. As the HPCNN algorithm depends on the result of the segmentation, this step is very critical. And PCNN has been proved to have stable performance due to its synchronous-pulsed feature [22].

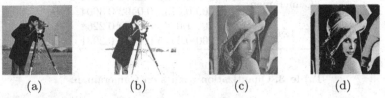

| (a) | (b) | (c) | (d) |

Fig. 3. Segmentation contrast. (a) Cameraman. (b) Cameraman after segmentation. (c) Lena. (d) Lena after segmentation.

Paraments of PCNN algorithm and HPCNN algorithm are listed in Table 1 and Table 2, respectively.

4 Experiment and Analysis

4.1 Experiment

To test and verify the effect of the proposed algorithm, a group of experiments are carried out rigorously on natural gray images from the standard image library. Two other quantization algorithms, μ-Law algorithm [5] and traditional PCNN algorithm [6], are used for comparison of the performances.

Mean Squared Error (MSE), Peak Signal-to-Noise Ratio (PSNR), Entropy and Compression Ratio (CR) in (19) are employed as evaluation indexes to evaluate the proposed algorithm quantitatively, and the results are displayed in Table 3 and Table 4 for Cameraman and Lena, where μ, P and H denote μ-Law, PCNN and HPCNN quantization algorithm, respectively. In addition, the processed images of Cameraman are shown in Fig. 4. The convergence time of PCNN is about 16.54s and that of HCPNN is about 15.06s on matlab7.11.0 and the hardware system platform is Inter(R) Core(TM) i3 CPU M330 @ 2.13GHz4G RAM with Windows 7.

$$CR = \frac{D(Y)}{D(I)} \tag{19}$$

where $I(\cdot)$ denotes the original image, $Y(\cdot)$ denotes the quantized image and function $D(\cdot)$ denotes the image bits.

Table 1. Parameters of PCNN.

Input Image	V_L	V_E	α_L	α_E	β
Cameraman	1	200	0.6118	0.0558	0.3125
Lena	1	200	0.7310	0.1754	0.0330

Table 2. Parameters of HPCNN.

Input Image		V_L	V_E	α_L	α_E	β
Cameraman	O	1	200	1.5439	0.0372	0.8333
	B	1	200	0.5741	0.0462	0.3604
Lena	O	1	200	1.0768	0.0399	0.2268
	B	1	200	0.6475	0.046	0.2931

Table 3. Quantization results on Cameraman.

Layer	16	32	64	96	128	200
CR(μ)	27.33%	33.30%	39.40%	42.40%	44.80%	46.90%
CR(P)	48.11%	56.61%	69.62%	77.97%	83.06%	87.51%
CR(H)	29.36%	37.66%	53.38%	57.95%	64.87%	73.72%
Entropy(μ)	1.6313	1.9896	2.4698	2.721	2.923	3.1657
Entropy(P)	2.7587	3.7149	4.6541	5.2247	5.6671	6.2978
Entropy(H)	2.0051	2.1375	3.1649	3.7770	4.1831	4.8103
MSE(μ)	0.1184	0.1149	0.1123	0.1114	0.1111	0.1106
MSE(P)	0.0807	0.0679	0.0641	0.0628	0.0612	0.0603
MSE(H)	0.0954	0.0692	0.0594	0.0443	0.0513	0.0521
PSNR(μ)	57.3981	57.5272	57.6268	57.6615	57.675	57.6937
PSNR(P)	59.0639	59.8136	60.0591	60.1517	60.2638	60.3205
PSNR(H)	58.3357	59.7280	60.3903	61.6676	61.0320	60.9591

Table 4. Quantization results on Lena.

Layer	16	32	64	96	128	200
CR(μ)	25.27%	35.85%	43.63%	45.46%	46.41%	47.08%
CR(P)	29.38%	47.80%	68.02%	78.41%	85.39%	93.35%
CR(H)	19.08%	29.34%	50.29%	59.47%	68.78%	80.63%
Entropy(μ)	1.4022	1.9658	2.4643	2.7155	2.9156	3.1751
Entropy(P)	2.0603	2.9962	3.9463	4.5248	4.9335	5.5797
Entropy(H)	1.3231	2.1159	3.0667	3.6158	4.0194	4.6471
MSE(μ)	0.5962	0.5711	0.5612	0.557	0.552	0.5508
MSE(P)	0.0279	0.0359	0.0085	0.011	0.0126	0.0128
MSE(H)	0.0154	0.0222	0.0076	0.0093	0.0105	0.011
PSNR(μ)	50.3768	50.5633	50.64	50.6722	50.711	50.72
PSNR(P)	69.7192	68.6191	74.887	73.7413	73.1567	73.096
PSNR(H)	72.2852	70.7014	75.3522	74.4903	73.9709	73.7395

4.2   Analysis and Discussion

In Table 3 and Table 4, MSE shows that the HPCNN algorithm processes with smaller deviation than traditional PCNN and μ-Law algorithm at most layers. PSNR shows the proposed algorithm bears best robustness from quantization layer 64 to layer 200. The entropy represents the average bits number of grayscale layers. It is obvious that HPCNN algorithm extracts less information because of

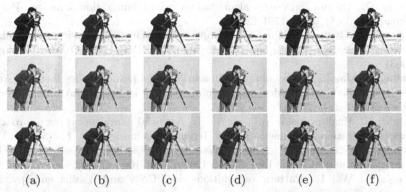

Fig. 4. Figures (a)∼(f) are different quantization layers from 16 to 200 quantized by μ-Law, PCNN and HPCNN algorithm from top to bottom, respectively.

the better compression performances than PCNN algorithm. Moreover, CR is an important index to measure whether the algorithm achieves better compression performance and the results show it does when compared with traditional PCNN algorithm. However, the quantized images of μ-Law algorithm as showed in Fig. 4 tend to be brighter and it brings down the image quality subjectively. While the traditional PCNN algorithm and HPCNN algorithm both satisfy human visual characteristics and the quantized images are more suitable for human to observe.

5 Conclusion

An adaptive image quantization method with HPCNN is proposed in this paper. HPCNN which satisfies human visual characteristics is adopted to quantize images. According to the experiment analysis and discussion above, we can intuitively see that the HPCNN algorithm is very suitable for performing quantization. In addition, our experimental results show that the proposed algorithm is qualified with smaller deviation and better robustness with higher compression ratio. As each neuron is corresponding to a pixel and they are heterogeneous to others with different gray levels, it is not fine enough to classify the neurons into just two categories for natural scence image. So it need to be classified finer. These existing problems in the method still need to be resolved in the next exploration.

References

1. Johnson, J.L., Ritter, D.: Observation of periodic waves in a pulse-coupled neural network. Opt. Lett. 18, 1253–1255 (1993)
2. Beer, R.D., Chiel, H.J., Sterling, L.S.: Heterogeneous neural networks for adaptive behavior in dynamic environments. Advances in Neural Information Processing Systems (1989)
3. Scheunders, P.: A comparison of clustering algorithms applied to color image quantization. Pattern Recognition Letters 18(11), 1379–1384 (1997)

4. Zdemir, D., Akarun, L.: A fuzzy algorithm for color quantization of images. Pattern Recognition 35(8), 1785–1791 (2002)
5. Kajitani, I., Otsu, N., Higuchi, T.: Improvements in myoelectric pattern classification rate with μ-LAW quantization. In: Proc. XVII IMEKO World Congress (2003)
6. Huang, Y., Ma, J., Du, S., Ma, Y.: Human Visual Characteristics Inspired Adaptive Image Quantization Method. Sampling Theory in Signal and Image Processing 13(2), 111–124 (2014)
7. Kumar, M., Raghuwanshi, N.S., Singh, R.: Artificial neural networks approach in evapotranspiration modeling: a review. Irrigation Science 29(1), 11–25 (2011)
8. Hansen, L.K., Salamon, P.: Neural network ensembles. IEEE transactions on Pattern Analysis and Machine Intelligence 12(10), 993–1001 (1990)
9. Zhang, Y., Wu, L.: Pattern recognition via PCNN and Tsallis entropy. Sensors 8(11), 7518–7529 (2008)
10. Eckhorn, R., Reitboeck, H.J., Arndt, M., Dicke, P.: Feature linking via stimulus-evoked oscillations: experimental results from cat visual cortex and functional implications from a network model. In: International Joint Conference on Neural Networks, pp. 723–730 (1989)
11. Eckhorn, R., Reitboeck, H.J., Arndt, M., Dicke, P.: Feature linking via synchronization among distributed assemblies: Simulations of results from cat visual cortex. Neural Computation 2(3), 293–307 (1990)
12. Lindblad, T., Kinser, J.M., Lindblad, T., Kinser, J.M.: Image Processing Using Pulse-Coupled Neural Networks, 2nd edn. Springer, NewYork (2005)
13. Wang, Z., Ma, Y., Cheng, F., Yang, L.: Review of pulse-coupled neural networks. Image and Vision Computing 28(1), 5–13 (2010)
14. Belanche Munoz, L., Valdes Ramos, J.J., Alquezar Mancho, R.: Heterogeneous neural networks: Theory and applications. Diss. PhD Thesis, Department of Languages and Informatic Systems, Polytechnic University of Catalonia, Barcelona, Spain (2000)
15. Gerhard, H.E., Wichmann, F.A., Bethge, M.: How sensitive is the human visual system to the local statistics of natural images? PLoS Computational Biology 9(1), e1002873 (2013)
16. Ma, Y., Li, L., Zhan, K., Wang, Z.: Pulse-coupled neural networks for digital image processing, pp. 107–112. Science Press, Beijing (2008)
17. Chen, Y., Park, S.K., Ma, Y., Ala, R.: A new automatic parameter setting method of a simplified PCNN for image segmentation. IEEE Transactions on Neural Networks 22(6), 880–892 (2011)
18. Chen, Y., Ma, Y., Kim, D.H., Park, S.K.: Region-Based Object Recognition by Color Segmentation Using a Simplified PCNN (2014)
19. Zhang, X., Yin, Z., Xiong, Y.: Adaptive switching mean filter for impulse noise removal. In: Congress on Image and Signal Processing, CISP 2008, vol. 3. IEEE (2008)
20. Yi-de, M., Qing, L., Zhi-Bai, Q.: Automated image segmentation using improved PCNN model based on cross-entropy. In: Proceedings of 2004 International Symposium on Intelligent Multimedia, Video and Speech Processing, pp. 743–746. IEEE (2004)
21. Pal, N.R.: On minimum cross-entropy thresholding. Pattern Recognition 29(4), 575–580 (1996)
22. Zhuo, W.: A PCNN-based method for vehicle license localization. In: 2011 IEEE International Conference on Computer Science and Automation Engineering (CSAE), vol. 3. IEEE (2011)

Noisy Image Fusion Based on a Neural Network with Linearly Constrained Least Square Optimization

Xiaojuan Liu, Lidan Wang *, and Shukai Duan

School of Electronics and Information Engineering, Southwest University,
Beibei. 400715 Chongqing, P.R. China
414823961@qq.com
{ldwang,,duansk}@swu.edu.cn

Abstract. Image fusion algorithm is a key technology to eliminate noise
through combining each image with different weight. Recently, convergence and convergence speed are two exiting problems which attract
more and more attention. In this paper, we originally propose a image fusion algorithm based on neural network. Firstly, the linearly constrained
least square(LCLS) model which can deal with image fusion problem is
introduced. In addition, in order to handle LCLS model, we adopt the
penalty function technique to construct a neural network. The proposed
algorithm has a simpler structure and faster convergence speed. Lastly,
simulation results show this fusion algorithm which has great ability to
remove different noise.

Keywords: LCLS, Neural network, Image fusion algorithm, Penalty
function.

1 Introduction

Image fusion is a process by combining more than one source images from different modalities or instruments into a single image with more information. The
successful fusion is of great importance in a lot of applications, such as biometric,
multi-media signal and remote sensing and so on. In the pixel level fusion, some
generic requirement can be imposed on the fusion results: Irrelevant features and
noise should be suppressed to a maximum extent in the fused image; All relevant

* The work was supported by Program for New Century Excellent Talents in University (Grant nos.[2013]47), National Natural Science Foundation of China (Grant
nos. 61372139, 61101233, and 60972155), Spring Sunshine Plan Research Project of
Ministry of Education of China (Grant no. z2011148), Technology Foundation for
Selected Overseas Chinese Scholars, Ministry of Personnel in China (Grant no. 2012-
186), University Excellent Talents Supporting Foundations in of Chongqing (Grant
no. 2011-65), University Key Teacher Supporting Foundations of Chongqing (Grant
no. 2011-65), Fundamental Research Funds for the Central Universities (Grant nos.
XDJK2014A009 and XDJK2013B011).

© Springer International Publishing Switzerland 2015
X. Hu et al. (Eds.): ISNN 2015, LNCS 9377, pp. 279–286, 2015.
DOI: 10.1007/978-3-319-25393-0_31

information contained in the source images should be preserved as much as possible; The fusion process should not introduce any artifacts or inconsistencies, which can distract or mislead the observer, or any subsequent image processing steps.

For the sake of overcoming such iteration, difficulty and the regularization methods have been proposed and widely researched [1] and [2]. The regularization method can be fast implemented by the efficient use of FFT and effectively implemented by real-time image processing based on neural networks [3]−[4]. On the other hand, choosing a proper regularization parameter is an important problem of the regularization method for a good image estimate. Several popular methods for choosing the optimal regularization parameter have been developed in [4]−[6], and [7]. Although the optimal regularization solution can lead to a perfect image estimate, only approximate optimal regularization parameters are available in practice since the noise variance needs compute. Because the regularization solution is very sensitive to the chosen regularization parameter, a good image estimate is still not guaranteed. Also, in a number of applications, the contaminating noise is usually non-Gaussian such as the uniform, the Laplacian, or a combination of them. For example, in blind-image identification, image models' parameters are always unknown [8] and [9]. For such a consideration, a robust entropic method was developed by the optimal selection of the regularization parameter [10]. An iterative method based on the high-order statistics was presented by [11], where two key parameters including the optimal regularization parameter need to be estimated.

In this paper, we introduce a new method which can deal with image fusion called the linearly constrained least squares (LCLS). We use recurrent neural network to implement the LCLS solution. Although some neural network methods have been proposed for data fusion in the literature [12], [14], these neural network approaches are difficult to obtain a good fusion solution and their implementations are also quite complex. The proposed algorithm has a simpler structure and faster convergence speed. Lastly, simulation results show this fusion algorithm which has great ability to remove different noise.

This paper is organized as follows. In Section 2, the LCLS method for image fusion is introduced. In Section 3, a neural network algorithm based on penalty function technique for the LCLS solutions is developed. In Section 4, the convergence property of the proposed neural network algorithm is established. Simulation is displayed in Section 5.

2   Problem Formulation and Model Description

In this section, we introduce image fusion model based on LCLS method and briefly describe its properties. Consider a multisensor system with K $(K \geq 2)$ sensors. Let the kth sensor measurement be expressed as[13]

$$\mathbf{X}_k(t) = a_k s(t) + n_k(t), (k = 1, ..., K; t = 1, ..., N)$$

where a_k is a scaling coefficient, N is the number of sensor measurements, $s(t)$ denotes the image signal, and n_k represents the additive Gaussian noise at the

kth sensor with zero mean. Moreover, $s(t)$ and n_k are mutually independent random processes. Using matrix and vector notations, the above representation can be written as

$$\mathbf{x}(t) = \mathbf{a}s(t) + \mathbf{n}(t)$$

where $\mathbf{a} = [a_1, ..., a_K]^T$, $\mathbf{x}(t) = [x_1(t), ..., x_K(t)]^T$, and $\mathbf{n}(t) = [n_1(t), ..., n_K(t)]^T$. The main goal of data fusion is to find an optimal fusion operator $\mathbf{w} \in \Re^K$ so that the uncertainty of the fused information is minimized. According to [13], LCLS method is formulated as the following constrained optimization problem:

$$min \quad f(\mathbf{w}) = \mathbf{w}^T R \mathbf{w}$$
$$s.t. \quad \mathbf{a}^T \mathbf{w} = 1. \tag{1}$$

Where $R = \frac{1}{N} \sum\limits_{t=1}^{N} x(t)x(t)^T$.

Actually, the mod $|w|$ is depending on system's output power, so model (1) is changed :

$$min \quad f(\mathbf{w}) = \mathbf{w}^T R \mathbf{w}$$
$$s.t. \quad \mathbf{a}^T \mathbf{w} = 1,$$
$$v \leq \mathbf{w} \leq u. \tag{2}$$

v and u are maximum output power and minimum output power respectfully.

Lemma 1. *[17] x^* is a solution to $VI(U, \Omega)$ if and only if x^* satisfied*

$$P_\Omega(x^* - \alpha U(x^*)) = x^*. \tag{3}$$

In equation (3) α is any positive constant, U is the gradient of the objective function in non-constraint programming and $P_\Omega : \Re^n \to \Omega$ is a projection operator which enforces vector ξ in feasible region Ω and defined by

$$P_\Omega(\xi) = \arg \min\limits_{\eta \in \Omega} \|\xi - \eta\|. \tag{4}$$

3   Neural Fusion Algorithm

Lemma 2. *If objective function and constraints of model (1) are continuously differentiable in feasible region, and for any $c > 0$, $\boldsymbol{w}^T R \boldsymbol{w} + c\|\boldsymbol{a}^T \boldsymbol{w} - 1\|^2$ exists a local minimum point w^*, then:*

$$\inf\{\boldsymbol{w}^T R \boldsymbol{w} | \boldsymbol{w} \in \Omega \cap B(\boldsymbol{w}^*, \delta)\} =$$
$$\lim\limits_{c \to \infty} \inf\{\boldsymbol{w}^T R \boldsymbol{w} + c\|\boldsymbol{a}^T \boldsymbol{w} - 1\|^2 | \boldsymbol{w} \in B(\boldsymbol{w}^*, \delta)\}. \tag{5}$$

We define $B(\mathbf{w}^*, \delta) = \{\mathbf{w} \in \Re^n \mid \; \| \mathbf{w}^* - \mathbf{w} \| < \delta\}$, $\delta > 0$. That is to say, if coefficient c is large enough, like result in [18], we can easily know the solution of $\mathbf{w}^T R\mathbf{w} + c\|\mathbf{a}^T\mathbf{w} - 1\|^2$ is a local solution for (2).

According to Lemma 1 and Lemma 2, we construct a recurrent neural network based on penalty technology as follows:

$$\dot{\mathbf{w}} = P_\Omega(\mathbf{w} - \bigtriangledown(\mathbf{w}^T R\mathbf{w} + c\|\mathbf{a}^T\mathbf{w} - 1\|^2)) - \mathbf{w}. \tag{6}$$

Theorem 1. *[12] If model (2) is a continuous convex optimization problem, the equilibria of equation (6) must be one of the KKT points and it is always stable in equilibria set.*

Remark 1. That is to say, if (2) is a convex optimization problem, and penalty factor c is enough large, (6) will find the solution of (2).

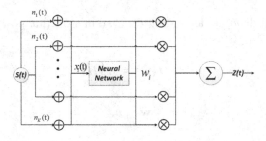

Fig. 1. A block diagram of the neural data fusion algorithm

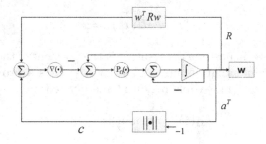

Fig. 2. Architecture of the continuous-time neural network in (6)

Figure 1 shows that how neural network (6) is embedded in a image fusion system. In order to embody architecture of (6) clearly, Figure 2 describes it in detail.

4   Simulation Results

In this section, we apply the proposed neural fusion algorithm to image fusion. Our platform is Matlab 2012b.

Firstly, we need to chose a colorful picture as original image, we need to produce noisy images by adding K different noises on the clear image, just like Figure 3. As everyone knows, each colorful image has three channel, namely, red channel, green channel and blue channel. And then, we add Gauss white noise with standard deviation 0.8 on every channel. Figure 7 shows the image after adding noise. Similarly, we add salt and pepper noise on original image just like Figure 9 shows. In order to verify the algorithm's performance, we use (6) to deal with image fusion under different number of sensors. In simulation, v sets to 0, u sets to 1.

Fig. 3. Clock full-color image

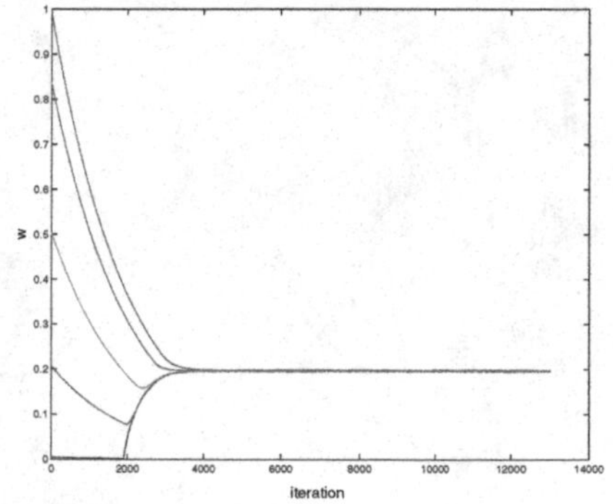

Fig. 4. The convergence process of red channel under 5 sensors

According to the results, we can easily conclude that more sensors can reap better fusion result. Compared with Figure 7, Figure 8 shows superior fusion result, meanwhile, Figure 9 and Figure 10 also describe the same conclusion. Figure 3 to Figure 5 state algorithm can convergent to optimal solution on every channel respectively.

Just as Figure 9 shows, salt pepper noise with standard deviation 0.8 is also added to original image. Simulation result shows great performance of algorithm.

In order to evaluate quality of fusion image, we introduce principle of mean error which is defined as follows:

$$E = \frac{1}{MN} \sum_{i=1}^{M} \sum_{j=1}^{N} [R(i,j) - F(i,j)]^2 \tag{7}$$

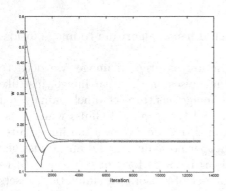

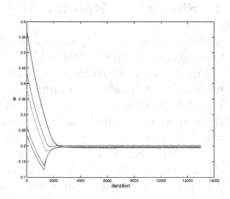

Fig. 5. The convergence process of green channel under 5 sensors

Fig. 6. The convergence process of blue channel under 5 sensors

Fig. 7. Clock image with Gaussian noise

Fig. 8. The fusion image using 100 sensors with Gaussian noise

Fig. 9. Clock image with salt pepper noise

Fig. 10. The fusion image using 100 sensors with salt pepper noise

Where M and N are row pixels number and column pixel's number. R(i,j) is original image pixel and F(i,j) is fusion image pixel. Obviously, the higher E the lower fusion quality.

Table 1. Comparison of different channels' mean error with Gaussian noise

precision: 10^3	5 sen.	10 sen.	50 sen.	100 sen.
Red channel	4.586	3.095	1.766	1.104
Green channel	5.104	3.578	2.074	1.266
Blue channel	5.606	4.162	2.536	2.266

Table 2. Comparison of different channels' mean error with salt pepper noise

precision: 10^3	5 sen.	10 sen.	50 sen.	100 sen.
Red channel	6.461	4.961	3.445	2.787
Green channel	7.543	5.957	4.575	3.745
Blue channel	8.695	7.071	5.841	5.307

Obviously, from Table 1 and Table 2 we can see that fusion quality is revelent to sensors number, more sensors correspond to clearer fusion image in face of different noise.

5   Conclusion

This paper proposes an image fusion algorithm based on neural network. Due to the penalty function technique and the use of differential equation, this algorithm has properties of high efficiency, simple structure, strong operability. In the aspect of theory, the convergence of the algorithm is guaranteed.A new fusion algorithm can effectively reduce the image noise. The simulation results verify the effectiveness of the algorithm.

References

1. Mammone, J.R., Podilchuk, C.I.: A general iterative method of image recovery. In: Visual Communications and Image Processing II (1987)
2. Katsaggelos, A.K., et al.: A regularized iterative image restoration algorithm. IEEE Transactions on Signal Processing 39, 914–929 (1991)
3. Sun, Y.: Hopfield neural network based algorithms for image restoration and reconstruction. I. Algorithms and simulations. IEEE Transactions on Signal Processing 48, 2105–2118 (2000)
4. Thompson, A.M., et al.: A Study of Methods of Choosing the Smoothing Parameter in Image Restoration by Regularization. Ranaon on Arn Analy and Mahn Nllgn 13, 326–339 (1991)

5. Galatsanos, N.P., Katsaggelos, A.K.: Methods for choosing the regularization parameter and estimating the noise variance in image restoration and their relation. IEEE Transactions on Image Processing 1, 322–336 (1992)
6. Kang, M.G., Katsaggelos, A.K.: General choice of the regularization functional in regularized image restoration. IEEE Transactions on Image Processing 4, 594–602 (1995)
7. Tekalp, A., et al.: Identification of image and blur parameters for the restoration of non- causal blurs. IEEE Transactions on Speech Signal Process 34, 963–972 (1986)
8. Hou, Z., et al.: Image denoising using robust regression. IEEE Signal Processing Letters 11, 243–246 (2004)
9. Zervakis, M.E., et al.: A Class Of Robust Entropic Functionals For Image Restoration. IEEE Transactions on Image Processing 4, 752–773 (1995)
10. Reeves, S.J., Mersereau, R.M.: Blur identification by the method of generalized cross-validation. IEEE Transactions on Image Processing 1, 301–311 (1992)
11. Hong, M.-C., et al.: Iterative Regularized Least-Mean Mixed-Norm Image Restoration. Opt. Eng. 41, 2515–2524 (2002)
12. Yan, Z., et al.: A collective neurodynamic optimization approach to bound-constrained nonconvex optimization. Neural Networks 55, 20–29 (2014)
13. Xia, Y., Leung, H., Bossé, E.: Neural data fusion algorithms based on a linearly constrained least square method. IEEE Transactions on Neural Networks 13, 320–329 (2002)
14. Xia, Y., Leung, H.: Performance analysis of statistical optimal data fusion algorithms. Information Sciences 277, 808–824 (2014)
15. Xia, Y., Mohamed, S.K.: Novel cooperative neural fusion algorithms for image restoration and image fusion. IEEE Transactions on Image Processing 16, 367–381 (2007)
16. Xia, Y., Leung, H.: A Fast Learning Algorithm for Blind Data Fusion Using a Novel L_2 -Norm Estimation. IEEE Sensors Journal 3, 666–672 (2014)
17. Xia, Y., Wang, J.: A general methodology for designing globally convergent optimization neural networks. IEEE Transactions on Neural Networks 9, 1331–1343 (1998)
18. Luenberger, D.G., Ye, Y.: Linear and nonlinear programming. Springer Science and Business Media 116 (2008)

A Singing Voice/Music Separation Method Based on Non-negative Tensor Factorization and Repeat Pattern Extraction

Yong Zhang and Xiaohong Ma*

School of Information and Communication Engineering,
Dalian University of Technology, Dalian, China
zhangyong@mail.dlut.edu.cn,
maxh@dlut.edu.cn

Abstract. In this paper, a novel singing voice/music separation method
is proposed based on the non-negative tensor factorization (NTF) and repeat pattern extraction technique (REPET) to separate the mixture into
an audio signal and a background music. Our system consists of three
stages. Firstly, we use the NTF to decompose the mixture into different
components, and similarity detection is applied to distinguish the components from each other, in order to classify the components into two classes
as the voice including voice/periodic music and the block music/voice;
next we utilize the REPET to extract the background music one step
further for the two classes, and the final background music is estimated
by adding the two backgrounds together, the left is added together as
the singing voice; finally the music spectrum and the voice spectrum are
filtered by harmonic filter and percussive filter respectively. To improve
the performance further, wiener filter is used to separate the voice and
music. Our method can improve the separation performance compared
with the other state-of-the-art methods on the MIR-1K dataset.

Keywords: NTF, REPET, Source Separation, Median Filter, Unsupervised Signal Processing.

1 Introduction

Single channel blind source separation (SCBSS) has been developed in recent
years. Tengtrairat *et al.* used the AR model to construct a pseudo-stereo hybrid model to achieve the SCBSS, which could separate multiple sources, but
the sources must have self-similarity [1]. Diamantaras *et al.* [2–5] addressed the
problem of binary source separation, which used the Taylor expansion and the
biased sources to convert the nonlinear mixed to linear mixed. Song *et al.* converted the SCBSS problem to a layer by layer separation problem [6].

Singing voice/music separation has attracted more and more attention and
widely been used. Such as the automatic synchronization, identification between

---

* Corresponding author.

© Springer International Publishing Switzerland 2015
X. Hu et al. (Eds.): ISNN 2015, LNCS 9377, pp. 287–296, 2015.
DOI: 10.1007/978-3-319-25393-0_32

music and lyrics [7, 8], singer identification [9], Kara OK, audio remixing and so on. However, the existing methods can not separate the singing voice and background music well, there is so much interference between the background music and the singing voice.

In recent years, singing voice/music separation has made some progress. Rafii *et al.* used the beat spectrum or the similarity matrix to identify the periodic or the self-similarity in the music, followed by the median filter to extract the background music [10, 11]. Antoine *et al.* proposed a method based on a general kernel regression framework, which utilized the backfitting algorithm to combine some methods in the way of kernels to achieve the singing voice/music separation [12].

There are some other methods extracting the pitch contours in the mixture to model the singing voice. Li *et al.* [13]used the pitch detection algorithm to estimate the pitch contours, then constructed the time frequency(T-F) mask through the pitch information to extract the singing voice and separate it from the background music. Hsu *et al.* also applied the pitch to extract the singing voice, at the same time estimated the background music [14].

Durrieu *et al.* used a source-filter model to estimate the singing voice and the non-negative matrix factorization (NMF) to approximate the background music, by using a iteration algorithm to estimate the corresponding parameters [15]. Huang *et al.* proposed a called robust principal component analysis (RPCA) method, which utilized a convex optimization algorithm to decompose the mixed power spectrum matrix, regarding the background music as a low rank matrix and the singing voice power spectrum matrix as a sparse matrix [16].

In this paper, The non-negative tensor factorization (NTF) is applied to the singing voice/music separation, which decomposes the mixed signals into multiple components. Then we use the similarity detection to distinguish the voice components and the music components, followed by the repeat pattern extraction technique (REPET) to extract the background music further. After that the extracted backgrounds are added together as a new music estimation, while the left is summed up as a new singing voice estimation. Finally the post-processing techniques are used to improve the separation results, which we call a harmonic filter and percussive filter. The experiment shows a relatively good result among the comparable methods.

The rest of the article is organized as follows: In Section 2 we present the NTF and the similarity detection; next the REPET techniques and the post-processing techniques are introduced, which utilize the harmonic filter, percussive filter and wiener filter to enhance the source separation further more. In Section 3, we evaluate our method on the MIR-1K database, and compare the background music and the singing voice at the same time with the other three state-of-the-art methods. In Section 4 we conclude this article in the end.

2   The Proposed Method

The NTF can decompose the mixed spectrum into a series of basic spectrums. However, the voice sepctrum and the music one can not always be separated well, that is to say, they are composed of several basic components, if the voice components the music components can be distinguished from each other through some methods, then the singing voice and background music will be separated through simply adding the corresponding components up together. Unfortunately, there is no effective methods that can distinguish the voice components from the music one, and we don't know the number of basic components. The REPET can extract the similar frames along time because the music has a certain periodic tempo, and the singing voice can be extracted much less when the REPET is used on the components from the NTF, because the REPET uses the median filter to extract the background music with the assumption that the singing voice is sparse in the time-frequency domain, which can improve the separation results. Therefore, a method based on the REPET and the NTF is proposed to achieve the singing voice/music separation.

This section presents the proposed method based on the NTF and the REPET. The system diagram is depicted as Fig. 1. The NTF is applied to decompose the mixture into different components as the first step and followed by the REPET to extract the background music of every components. The median filter and the wiener filter are in the end to improve the separation results further.

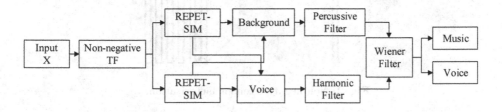

Fig. 1. System Diagram.

2.1 Non-negative Tensor Factorization

Audio signals construct the tensor through the time domain, frequency domain and the channels, actually it is the expansion of the signal amplitude spectrum on the channels. For a monaural signal in the mixed dual channel signals, we apply the short time fourier transform (STFT) to the mixture to get the time-frequency (T-F) spectrum $\mathbf{Spec} \in C^{F \times T \times I}$. And then we can obtain the amplitude spectrum $\mathbf{AmpSpec} \in R^{F \times T \times I}$ from the T-F spectrum $\mathbf{Spec}$ through modulo operation.

For a three order amplitude spectrum tensor $\mathbf{AmpSpec}$. F represents the identification ability of the frequency domain. T represents the time domain in the way of frames. I denotes the number of channels, $I = 1$ denotes the monaural, and $I = 2$ denotes the dual channels. In this article, we decompose the three order tensor into the sum of a series of weighted matrices, each weighted matrix represents a component spectrum $\mathbf{V}_i$, which is shown in Fig. 2, the formula in which is depicted as:

$$\mathbf{AmpSpec} \approx \sum_{i=1}^{N} \mathbf{Q}_i \mathbf{V}_i = \sum_{i=1}^{N} \mathbf{ComSpec}_i \tag{1}$$

$$\mathbf{V}_i = \mathbf{W}_{fi} \odot \mathbf{H}_{it} \tag{2}$$

where $\mathbf{Q}_i$ is the gain of the corresponding component matrix $\mathbf{V}_i$, $\mathbf{ComSpec}_i$ is the component amplitude spectrum, N is the component numbers. Here IS convergence criteria is applied to the NTF.

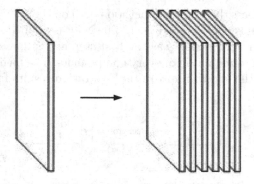

Fig. 2. Non-negative tensor factorization.

Each component spectrum corresponds to a signal in the time domain. When we apply the NTF to the amplitude spectrum, and the phase information is lost. Fortunately human ears are not sensitive to the phase information. So the sources are reconstructed through the wiener filter and the mixed T-F spectrum **Spec**. Finally we use the inverse short time fourier transform (ISTFT) to retrieve the signal in the time domain.

$$\mathbf{AmpSpec}_i = \frac{\mathbf{ComSpec}_i}{\sum\limits_{i=1}^{N} \mathbf{ComSpec}_i} \odot \mathbf{AmpSpec} \tag{3}$$

$$\mathbf{Spec}_i = \frac{\mathbf{AmpSpec}_i}{\mathbf{AmpSpec}} \odot \mathbf{Spec} = \frac{\mathbf{ComSpec}_i}{\sum\limits_{i=1}^{N} \mathbf{ComSpec}_i} \odot \mathbf{Spec} \tag{4}$$

where $\odot$ denotes the element-wise product. Division is the same. To be noticed, we use the phase information of the mixed signal to construct the source signal. It is not consistent to the phase of the practical signal, just a approximate estimation.

2.2 The Similarity Detection

The correlation coefficient is widely used to measure the correlation between two signals. Here the correlation coefficient is utilized to classify the component signals into two categories named as voice and music. Denote s_i as a time-domain component signal, $\mathbf{x}$ as a single channel mixture. Our evaluation criterion is defined as follow:

$$corr(\mathbf{s}, \mathbf{x}) = \frac{\sum\limits_{i=1}^{M} s_i x_i}{\sqrt{\sum\limits_{i=1}^{M} s_i^2 \sum\limits_{i=1}^{M} x_i^2}} \tag{5}$$

If the correlation coefficient is bigger than 0.5, we assume the component includes voice or periodic music. On the contrary, we treat it as a voice/music block. Through the simple similarity detection algorithm, the period music and the block music are divided. However, the block voice is divided into the music by mistake though it belongs to voice, so we need to apply REPET to both of the two categories.

2.3 The Repeat Pattern Extraction Technique

Singing voice and background music still disturb each other in the signals recovered from NTF. The main reason is that the background music does not have good periodicity. In other words, music signal is not a base signal, the same reason for the singing voice. So we utilize the REPET [12,13] to the two signals and post-processing to improve the separation results. Then the extracted background music is added up together as the final music estimation and the rest is summed up as the final voice estimation.

REPET-SIM is thus a generalization of REPET, an effective approach for separating the repeating background music from the non-repeating singing voice in a mixture, by identifying the repeating elements and smoothing of the non-repeating elements. In particular, REPET-SIM used a similarity matrix to identify the similar frames in the mixed spectrum, the following is the median filter to smooth out the non-repeating frames. The similarity matrix could identify the similar frames but not the fixed period.

2.4 Post-processing

There are always percussive elements in the music and harmonic elements in the singing voice. The percussive elements have self-similarity in the frequency domain, and the harmonic elements are self-similar in the time domain. Using these two prior knowledge, we use the median filter as a post-processing on the voice spectrum $\mathbf{AmpSpec}_{voice}$ named as harmonic filter and music spectrum $\mathbf{AmpSpec}_{music}$ named as percussive filter. To be more specific, it is the median filter [17]. On the voice spectrum across the time frames the median filter is utilized to enhance the harmonic elements and suppress the percussive elements;

percussive filter is applied on the music spectrum along a frame to enhance the percussive elements and suppress the harmonic elements.

Median filter is widely used in the image denoising, especially for removing the salt and pepper noise. Given an input vector $x(n)$, then $y(n)$ is the output of a median filter with length l, where l defines the order of the filter. Thus the median filter can be defined as:

$$y(n) = median\{x(n - k : n + k), k = floor(l/2)\} \tag{6}$$

For the special amplitude spectrum **AmpSpec**, denoting the i_{th} time frame as **AmpSpec**$_i$ and the h_{th} frequency slice as **AmpSpec**$_h$; for each time frame **AmpSpec**$_i$, **AmpSpec**$_{music}$ can be obtained from the percussive filter, for each frequency slice, using the harmonic filter we can get the **AmpSpec**$_{voice}$:

$$\mathbf{AmpSpec}_{music} = M\{\mathbf{AmpSpec}_i, l_{perc}\} \tag{7}$$

$$\mathbf{AmpSpec}_{voice} = M\{\mathbf{AmpSpec}_h, l_{harm}\} \tag{8}$$

where M represents the median filter. Apply the wiener filter to the filtered voice spectrum and the music spectrum as the follows:

$$\mathbf{AmpSpec}_{voice} = \frac{\mathbf{AmpSpec}_{voice}}{\mathbf{AmpSpec}_{voice} + \mathbf{AmpSpec}_{music}} \odot \mathbf{AmpSpec} \tag{9}$$

$$\mathbf{AmpSpec}_{music} = \frac{\mathbf{AmpSpec}_{music}}{\mathbf{AmpSpec}_{voice} + \mathbf{AmpSpec}_{music}} \odot \mathbf{AmpSpec} \tag{10}$$

The voice and the music spectrum soft mask are calculating through dividing the mixed spectrum respectively:

$$\mathbf{Mask}_{music} = \frac{\mathbf{AmpSpec}_{music}}{\mathbf{AmpSpec}} = \frac{\mathbf{AmpSpec}_{music}}{\mathbf{AmpSpec}_{voice} + \mathbf{AmpSpec}_{music}} \tag{11}$$

$$\mathbf{Mask}_{voice} = \frac{\mathbf{AmpSpec}_{voice}}{\mathbf{AmpSpec}} = \frac{\mathbf{AmpSpec}_{voice}}{\mathbf{AmpSpec}_{voice} + \mathbf{AmpSpec}_{music}} \tag{12}$$

Using the soft mask to multiply the mixed complex spectrum **Spec** will get the voice complex spectrum and the music complex spectrum. We use the ISTFT to reconstruct the singing voice and the background music in the end.

3   Evaluation

Experiments are measured on the MIR-1K dataset, including 1000 Chinese song clips, wav format 16kHz sampling rate. Background music and the singing voice are recorded in the left and right channel respectively. We create a set of 1000 mixtures by simply summing up the left and right channel into a monaural mixture for each song clip.

BSS Eval toolbox [20] is widely used to quantify the quality of the separation between a source and its estimation in BSS field. Source to interference ratio (SIR), sources to artifacts ratio (SAR), and signal to distortion ratio (SDR) are included [19]. We choose these metrics because they are widely used in the source separation evaluation.

In Durrieu's analysis, the window and the step size is 1024 and 512 samples respectively, and the number of iteration is set to 30, the same as [18]. As for the REPET-SIM, Hamming windows of 1024 samples and a step size of 512 samples are used. The minimal threshold between similar frames is set to 0, the minimal distance between consecutive frames to 0.1s, and the maximal number of repeating frames to 50 [11,18].

In our experiment, the window size of 512 samples is used, and the step size is 512 samples. The component's number is 2. The parameters of REPET-SIM are the same as the comparative method[11]. In the post-processing stage, we use a 17-order l_{harm} harmonic filter and a 21-order l_{perc} percussive filter[17]. Wiener filter is in the end which makes the sum of singing voice and music equal to the mixture.

Fig. 3 ~ 5 depicts the distribution of SDR, SIR and SAR [19] respectively. The SDR represents the ratio of two sources, the SIR indicts the interferences betweeen sources and the SAR implies the artifact's deviation from the algorithm itself. In Fig. 3 ~ 5, the left represents singing voice and the right denotes background music. The comparative methods include the Durrieu's, RPCA and REPET-SIM. Fig. 3 ~ 5 shows the average values of the 1000 clips, which is representative, the higher the better.

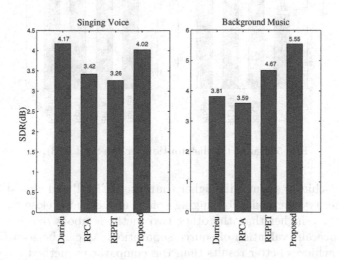

Fig. 3. Bar of the distribution for the SDR (dB).

Fig. 3 depicts the distribution of SDR, and it shows that our proposed method makes much better than the other comparative methods in background music.

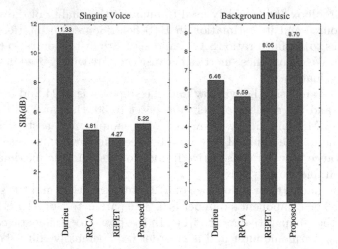

Fig. 4. Bar of the distribution for the SIR (dB).

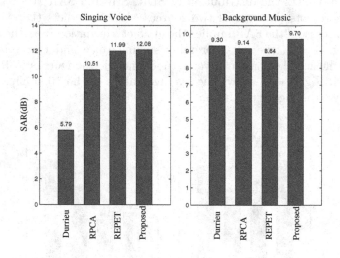

Fig. 5. Bar of the distribution for the SAR (dB).

Our method achieves about 1dB higher than the REPET and almost 2dB more than the other two methods. For singing voice our method is close to the results of Durrieu's but higher than the other two methods about 0.6 ~ 0.8dB. The SDR is the overall evaluation of source separation. As can be seen from SDR, our method achieves better results than the comparative methods.

Focusing on the SIR in Fig. 4, our proposed method achieves better results than the other comparative methods in music, about 0.7dB higher than the REPET and 2.0 ~ 3.0dB higher than the other two. However, our method is much lower than the Durrieu's in singing voice, but higher than the other two methods. This is related to the music/voice measures trade-off which can commonly be seen in singing voice/music separation. In other words, our proposed method is close to the comparative methods in SIR.

From Fig. 5, we can be see that our proposed method is better than the other state-of-the-art methods, an average increase about 0.4dB $\sim$ 0.5dB in the background music, and the singing voice is slightly higher than the REPET but much higher than the other two, especially the Durrieu's, which shows that our proposed method has fewer artifacts compared with the other methods in the estimations. In other words, our method improves the background music much more compared with the other methods, at the same time the singing voice achieves a little bit increase. On the whole, our proposed method realizes a relatively good result.

4 Conclusion

In this paper, we propose a new singing voice/music method based on the NTF and REPET. Our system consists of three stages. The NTF decomposes both the mixture and the sources into many components at the same time in the first stage, which can reduce the voice extracted by the REPET in the following stage; the harmonic filter, the percussive filter and the wiener filter can improve our performance further more in the end. The proposed method can improve the separation performance in the experiment on the MIR-1K dataset. Our later work will focus on the identification methods for the components.

Acknowledgments. This work was supported by the National Natural Science Foundation of China under Grant 61071208.

References

1. Tengtrairat, N., Gao, B., Woo, W.L.: Single-channel blind separation using Pseudo-stereo mixture and complex 2-D histogram. IEEE Transactions on Neural Networks and Learning Systems 24(11), 1722–1735 (2013)
2. Diamantaras, K.I., Papadimitriou, T.: Blind separation of three binary sources from one nonlinear mixture. Machine Learning for Signal Processing (2010)
3. Diamantaras, K.I., Papadimitriou, T., Vranou, G.: Blind separation of multiple binary sources from one nonlinear Mixture. In: IEEE International Conference on Acoustics, Speech and Signal Processing (2011)
4. Diamantaras, K.I., Papadimitriou, T.: Separating two binary sources from a single nonlinear mixture. In: IEEE International Conference on Acoustics Speech and Signal Processing (2010)
5. Diamantaras, K.I., Vranou, G., Papadimitriou, T.: Multi-Input Single-Output Nonlinear Blind Separation of Binary Sources. IEEE Transactions on Signal Processing 61(11), 2866–2873 (2013)
6. Song, J., Ma, X., Zhang, Y.: Binary source separation layer by layer for one sensor. In: IEEE International Conference on Intelligent Control and Information Processing (2014)

7. Wang, C.K., Lyu, R.Y., Chiang, Y.C.: An automatic singing transcription system with multilingual singing lyric recognizer and robust melody tracker. In: European Conference on Speech Communication and Technology (2003)

8. Fujihara, H., Goto, M., Ogata, J., Okuno, H.G.: Lyric Synchronizer: Automatic synchronization system between musical audio signals and lyrics. IEEE Journal of Selected Topics in Signal Processing 5(6), 1252–1261 (2011)

9. Zhang, T.: System and method for automatic singer identification. Research Disclosure (2003)

10. Rafii, Z., Pardo, B.: Repeating Pattern Extraction Technique (REPET): A simple method for music/voice separation. IEEE Transactions on Audio, Speech, and Language Processing 21(1), 71–82 (2013)

11. Rafii, Z., Pardo, B.: Music/Voice Separation Using the Similarity Matrix. In: IS-MIR (2012)

12. Liutkus, A., Fitzgerald, D., Rafii, Z.: Kernel additive models for source separation. IEEE Transactions on Signal Processing 21(21), 4298–4310 (2014)

13. Li, Y., Wang, D.L.: Separation of singing voice from music accompaniment for monaural recordings. IEEE Transactions on Audio, Speech, and Language Processing 15(4), 1475–1487 (2007)

14. Hsu, C.-L., Jang, J.-S.R.: On the improvement of singing voice separation for monaural recordings using the MIR-1K dataset. IEEE Transactions on Audio, Speech, and Language Processing 18(2), 310–319 (2010)

15. Durrieu, J., David, B., Richard, G.: A musically motivated mid-level representation for pitch estimation and musical audio source separation. IEEE Journal of Selected Topics in Signal Processing 5(6), 1180–1191 (2011)

16. Huang, P.-S., Chen, S.D., Smaragdis, P., Hasegawa-Johnson, M.: Singing-voice separation from monaural recordings using robust principal component analysis. In: IEEE International Conference on Acoustics, Speech and Signal Processing (2012)

17. Fitzgerald, D.: Harmonic/percussive separation using median filtering. In: 13th International Conference on Digital Audio Effects (2010)

18. Rafii, Z., Germain, F., Sun, D.L.: Combining Modeling of Singing Voice and Background Music For Automatic Separation of Musical Mixtures. In: ISMIR (2013)

19. Vincent, E., Gribonval, R., Févotte, C.: Performance measurement in blind audio source separation. IEEE Transactions on Audio, Speech, and Language Processing 14(4), 1462–1469 (2006)

20. BSS Eval toolbox, http://bass-db.gforge.inria.fr/bss_eval/

Automatic Extraction of Cervical Vertebrae from Ultrasonography with Fuzzy ART Clustering

Kwang Baek Kim[1,*], Doo Heon Song[2], Hyun Jun Park[3], and Sungshin Kim[4]

[1] Department of Computer Engineering, Silla University, Busan 617-736, Korea
gbkim@silla.ac.kr
[2] Department of Computer Games, Yong-In Songdam College, Yongin 449-040, Korea
dsong@ysc.ac.kr
[3] Department of Computer Engineering, Pusan National University, Busan 609-735, Korea
hyunjun@pusan.ac.kr
[4] School of Electrical and Computer Engineering, Pusan National University,
Busan 609-735, Korea
sskim@pusan.ac.kr

Abstract. Cervical vertebrae are important ramus communican that connect human body and the corpus. Muscles around cervical vertebrae such as deep cervical flexor and sternocleidomastoid muscle do key role to control chronicle neck pain thus monitoring such muscles near cervical vertebrae is important. In this paper, we propose a method to detect and analyze cervical vertebrae and related muscles automatically with fuzzy ART clustering from ultrasonography. The experiment verifies that our approach is consistent with human medical experts' decision to locate key measuring point for muscle analysis and successful in detecting cervical vertebrae accurately.

Keywords: Cervical Vertebrae, Fuzzy ART, Muscle Analysis, Ultrasonography, Deep Cervical Flexor.

1   Introduction

Neck pain is very common complaint affecting up to 70% of individuals at some point of their lives [1]. Clinical neck pain is associated with impairment of muscle performance and the functional impairments associated with neck pain and the cause-effect relationships between neck pain and motor control are well investigated [2]. Antevertebral deep cervical flexor (DCF) muscles such as longus coli and longus capitis do key role to stabilize cervical articulations and to preserve the lordotic curvature of the spine [3] and sternocleidomastoid muscle (SCM) is related with the rotation of the neck [4]. Strengthening of these muscles is important to treat the patients with neck pain provoked by various pathologies of cervical spine [5].

Using ultrasound image in muscle analysis is appropriate for its non-invasive, inexpensive, real time responses [6]. However, its limitations are often pointed out that sonographic images are dependent on the qualities of equipment and skills of expertise thus the diagnosis often misleads to subjective judgment [7]. Thus, we need an automatic

---

* Corresponding author.

© Springer International Publishing Switzerland 2015
X. Hu et al. (Eds.): ISNN 2015, LNCS 9377, pp. 297–304, 2015.
DOI: 10.1007/978-3-319-25393-0_33

image segmentation and identification tool for anatomical landmarks that can eliminate such subjectivity in the image analysis [8].

Unfortunately, there is almost no directly related research for such an automatic neck pain related muscle extractor/ analyzer by computer vision yet. A recent study tried to give an automatic segmentation of cervical vertebrae from X-rays [9] but not related to muscles of our interests. Our concern is to detect and extract muscles such as sternocleidomastoid and longus capitis/colli in conjunction with cervical vertebrae automatically from ultrasonography and measuring the thickness for further medical analysis [10, 11]. All three vision based approaches aim to locate measuring key point accurately to avoid manual subjective key point setting for muscle analysis.

Previously we applied simple average binarization and contour analysis algorithm to extract SCM and related objects [10] but for the rehabilitation purposes, locating cervical vertebrae and related DCF are more important and difficult due to low brightness contrast among objects. In our previous study [11], we applied fuzzy sigma binarization to overcome low brightness contrast by its adaptive thresholding characteristic. However, it does not consider the average brightness nor morphological characteristics of cervical vertebrae thus its performance is not stable especially when it forms the thickness measuring key points.

In this paper, we propose a fuzzy ART clustering [12] approach to find the key points accurately in detecting cervical vertebrae. With such a clustering approach and subsequent smearing technology to restore lost information, our software is more consistent with human medical experts' opinions in detecting cervical vertebrae and locating key points.

2   Overall Procedure for Automatic Cervical Vertebrae Analyzer

Obtained digital image follows DICOM (Digital imaging and Communications in medicine) standard format. In the main region of interest (ROI) part of the image as shown in Figure 1, there will be a blood vessel and two muscles are located above and below the vessel. The muscle above blood vessel is the sternocleidomastoid and the muscle below the blood vessel is the deep cervical flexors (DCF). Its lower part has irregular curve due to the border line of cervical vertebrae.

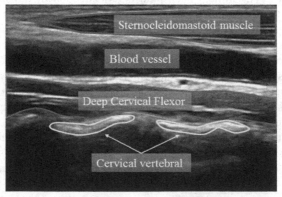

Fig. 1. ROI of ultrasound image

The cervical vertebrae area in the ultrasonography is shown as bright region under DCF since the area has high density. In order to detect cervical vertebrae, however, we need to remove other organ such as cartilage and subcutaneous fat area as noise that also have relatively high brightness. Thus, we need brightness enhancement procedure and subsequent noise removal/ image restoration process. The overall procedure for extracting cervical vertebrae can be summarized as Figure 2.

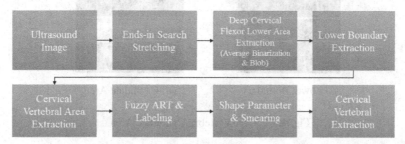

Fig. 2. Overall process for cervical vertebrae extraction

In this paper, we will only focus on explaining the role of fuzzy ART learning part.

3   Fuzzy ART Clustering for Cervical Vertebrae Detection

First, we need to extract DCF area since cervical vertebrae area is located under DCF and measuring DCF thickness is crucial in the muscle analysis. Figure 3 demonstrate the intermediate results of DCF extracting processes that should be preceded from Figure 2.

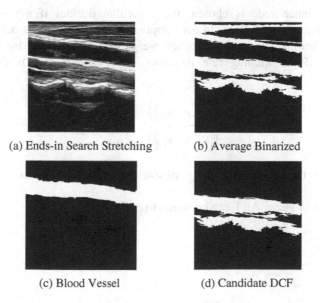

(a) Ends-in Search Stretching   (b) Average Binarized

(c) Blood Vessel   (d) Candidate DCF

Fig. 3. Extracting DCF

Then, from the lower boundary lines of DCF, we apply upper/under search to find a candidate region for cervical vertebra as shown in Figure 4.

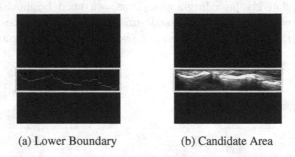

(a) Lower Boundary   (b) Candidate Area

Fig. 4. Cervical Vertebra Candidate Area

From the image shown as Fig. 4(b), we apply Fuzzy ART algorithm to make clusters to form cervical vertebrae. After fuzzy clustering, we apply quantization and labeling process to extract target objects. Fuzzy ART [12] is a powerful self-organizing pattern clustering algorithm but it needs to be modified and tuned with respect to the area of applications.

In our adoption of Fuzzy ART, we tuned the original algorithm as following;

Original Fuzzy ART may have maximum output vector less than 1 due to the disagreement between input vector and weight vector and that can cause inaccurate similarity test. Thus we set the weight of output vector to make the maximum value as 1.

When the winner node is chosen, the algorithm decides if a new cluster is initialized with similarity testing. In our adoption, if the similarity is acknowledged, we control the learning rate and connection weights dynamically so that the learning rate decreases if the winning rate increases. The learning rate is computed as following;

$$\beta = \frac{\left(\left\|x_i^p \wedge w_i^j\right\|\right)}{\left(\left\|x_i^p \vee w_i^j\right\|\right)} \times \frac{1}{f_{j*}} \tag{1}$$

where β denotes the learning rate and f_{j*} denotes the number of updates to the winner node.

Our adoption of Fuzzy ART can be summarized as Figure 5.

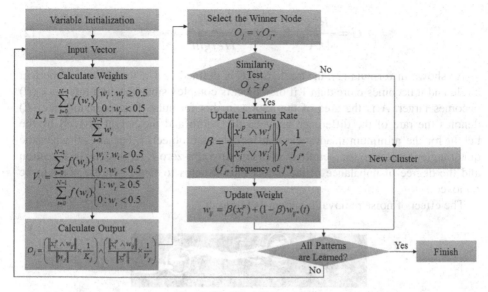

Fig. 5. Modified Fuzzy ART

Then the result of Fuzzy ART clustering is shown as Figure 6.

Fig. 6. The Effect of Fuzzy ART

The next step is to extract meaningful objects by local labeling procedure. Since the cervical vertebrae area is relatively bright, we apply the labeling to only highly bright clusters. The result of this labeling procedure is shown as Figure 7.

Fig. 7. After Labeling Procedure

In Figure 7, the extracted objects contain noises such as cartilage and subcutaneous fat. In order to remove noise objects, we test the circle rate and extension rate of objects in consideration. Formula (2) explains such shape parameters. The shape of cervical vertebrae is long curve with relatively consistent form while other noises have irregular shapes thus such morphological characteristics can discriminate target objects from noises with measurements explained in formula (2).

$$C = \frac{P^2}{4\pi A}, \quad E = \frac{|Width - Height|}{Height} \tag{2}$$

As shown in formula (2), the circle rate C is defined as 1 if the object is a perfect circle and it becomes more than 1 if the object is complex since the circumference (P) becomes larger. A is the area of the region in this formula. The extension rate (E) denotes the rate of the difference between the width and the height divided by the height for the minimum quadrangle that includes the object. E becomes 0 when the quadrangle is a square (or a circle of course). Any non-zero E represents the direction and the degree of unbalance of the object thus it helps to figure out the noise to be removed.

The effect of noise removal is shown as Figure 8.

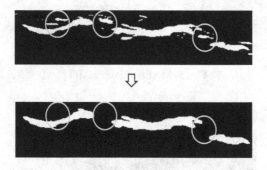

Fig. 8. Noise Removed

The resultant image may have lost information during the noise removal process. The horizontal smearing process is applied to compensate such information loss to obtain the final output image for extracting cervical vertebrae.

4   Experiment and Conclusion

The proposed method is implemented with C++ under Microsoft Visual Studio 2010 on the IBM-compatible PC with Intel(R) Core(TM) i7-2600 CPU @ 3.40GHz and 4GB RAM. The experiment uses fifty two 800 x 600 size DICOM format images.

The accuracy of the method or the utility of the automatic vision based cervical vertebrae analyzer can be measured by the agreement rate of locating thickness measuring key points with human expert – physical therapists. In order to avoid human subjectivity, our ground truths of measuring points are obtained by two physical therapists' agreements. As described in Table 1, the proposed system showed 92.3% agreement rate with multiple human experts' agreements.

Table 1. Muscle Thickness Measuring Key Points Extraction

	Proposed Method
Key Points	48/52 (92.3 %)

In our proposed method, key point is set to be the lowest point of the first cervical vertebrae object and the range of measurement is within 1 cm left and right of that key point. The thickness of the DCF muscle is then computed as the average length of vertical lines within that measuring range.

In our previous attempt [11], we used fuzzy sigma binarization instead of fuzzy ART of the proposed method. Also, the proposed method utilizes morphological information of cervical vertebrae in noise removal process. Figure 9 shows the difference of the proposed method and previous attempt [11].

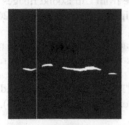

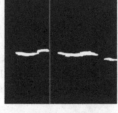

(a) Extracting Cervical Vertebrae [11]   (b) Extracting Cervical Vertebrae-Proposed

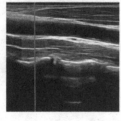

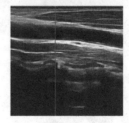

(a) Key Points [11]   (b) Key Point (Proposed)

Fig. 9. Performance Comparison

References

1. Fejer, R., Kyvik, K.O., Hartvigsen, J.: The prevalence of neck pain in the world population: a systematic critical review of the literature. European Spine Journal 15(6), 834–848 (2006)
2. Falla, D., Farina, D.: Neural and muscular factors associated with motor impairment in neck pain. Current Rheumatology Reports 9(6), 497–502 (2007)
3. Mayoux-Benhamou, M.A., Revel, M., Vallee, C., Roudier, R., Barbet, J.P., Bargy, F.: Longus colli has a postural function on cervical curvature. Surgical and Radiologic Anatomy 16(4), 367–371 (1994)

4. Middleditch, A., Oliver, J.: Functional anatomy of the spine, 2nd edn., ch. 3. 104-5. Butterworth-Heinemann (2002)
5. Ylinen, J., Kautiainen, H., Wirén, K., Häkkinen, A.: Stretching exercises vs manual therapy in treatment of chronic neck pain: a randomized, controlled cross-over trial. Journal of Rehabilitation Medicine 39(2), 126–132 (2007)
6. Cardinal, É., Bureau, N.J., Aubin, B., Chhem, R.K.: Role of ultrasound in musculoskeletal infections. Radiologic Clinics of North America 39(2), 191–201 (2001)
7. Mason, R.J., Broaddus, V.C., Martin, T., Gotway, M.B., King Jr, T.E., Schraufnagel, D., Nadel, J.A.: Murray and Nadel's Textbook of respiratory medicine, 5th edn., ch. 20. 455. Saunders (2010)
8. Enikov, E.T., Anton, R.: Image Segmentation and Analysis of Flexion-Extension Radiographs of Cervical Spines. Journal of Medical Engineering 2014 (2014)
9. Xu, X., Hao, H.W., Yin, X.C., Liu, N., Shafin, S.H.: Automatic segmentation of cervical vertebrae in X-ray images. In: The 2012 International Joint Conference on Neural Networks (IJCNN), pp. 1–8. IEEE (2012)
10. Kim, K.B., Yu, D.H., Hong, Y.S.: Extraction of Muscle from Ultrasound Images of Cervical Regions. Information-An International Interdisciplinary Journal 16(4), 2669–2678 (2013)
11. Kim, K.B., Park, H.J., Song, D.H., Han, S.S.: Extraction of Sternocleidomastoid and Longus Capitis/Colli Muscle Using Cervical Vertebrae Ultrasound Images. Current Medical Imaging Reviews 10(2), 95–104 (2014)
12. Carpenter, G.A., Grossberg, S., Rosen, D.B.: Fuzzy ART: Fast stable learning and categorization of analog patterns by an adaptive resonance system. Neural Networks 4(6), 759–771 (1991)

Fast Basis Searching Method of Adaptive Fourier Decomposition Based on Nelder-Mead Algorithm for ECG Signals

Ze Wang, Limin Yang, Chi Man Wong, and Feng Wan

Department of Electrical and Computer Engineering
Faculty of Science and Technology
University of Macau
fwan@umac.mo

Abstract. The adaptive Fourier decomposition (AFD) is a greedy iterative signal decomposition algorithm in the viewpoint of energy. Instead of using a fixed basis for decomposition, AFD uses an adaptive basis to achieve efficient energy extraction. In the conventional searching method, a new basis is searched from a large dictionary at every decomposition level. This usually results in a slow searching speed. To improve the efficiency, a fast searching method based on Nelder-Mead algorithm is proposed in this paper. The AFD with the proposed searching method is applied for electrocardiography (ECG) signals in which the selection ranges of four key parameters in the proposed searching method are determined based on simulation results of an artificial ECG signal. The simulation results of real ECG data shows that the computational time of the AFD based on the proposed searching method is just half of that based on the conventional searching method with similar reconstruction error.

Keywords: adaptive Fourier decomposition (AFD), Nelder-Mead algorithm, electrocardiography (ECG) signal.

1 Introduction

The adaptive Fourier decomposition (AFD) is a novel signal decomposition algorithm introduced by Qian et al [1,2,3]. It is based on the sequential extraction of energy starting from the high-energy mode to the low-energy mode [4]. The AFD can be considered as one kind of matching pursuit algorithm that is a type of sparse approximation based on a dictionary. For the classical Fourier decomposition, the simple sinusoidal basis function is applied as dictionary. Although the computational complexity of the classical Fourier decomposition is low, its energy extraction is not efficient since its basis cannot match signals adaptively [5]. The AFD overcomes such drawback by using the rational orthogonal system, $\{B_n\}_{n=1}^{\infty}$, as its basis where

$$B_n(e^{jt}) = \frac{\sqrt{1 - |a_n|^2}}{1 - \overline{a}_n e^{jt}} \prod_{k=1}^{n-1} \frac{e^{jt} - a_k}{1 - \overline{a}_k e^{jt}}, \tag{1}$$

© Springer International Publishing Switzerland 2015
X. Hu et al. (Eds.): ISNN 2015, LNCS 9377, pp. 305–314, 2015.
DOI: 10.1007/978-3-319-25393-0_34

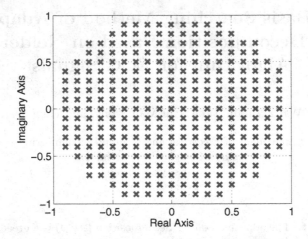

Fig. 1. Dictionary of a_n in the conventional searching method

$a_n \in \mathbb{D}$ $(n = 1, 2, \cdots)$, $\mathbb{D} = \{z \in \mathbb{C} : |z| < 1\}$, and $\mathbb{C}$ is the complex plane [1]. B_n is determined by the sequence of a_n [6]. In every decomposition level, a new value of a_n is searched in $\mathbb{D}$ to make sure that the decomposition achieves a high converging rate [1]. The crucial rule of this searching process is the maximal projection principle asserting that, for any $G_n \in H^2$, there exists $a_n \in \mathbb{D}$ such that

$$a_n = \arg \max \left\{ \left| \langle G_n, e_{\{a_n\}} \rangle \right|^2, : a_n \in \mathbb{D} \right\} \tag{2}$$

where G_n is the reduced remainder, and $e_{\{a_n\}}$ is the evaluator at a_n [1]. By defining

$$D(a_n) = \left| \langle G_n, e_{\{a_n\}} \rangle \right|^2, \tag{3}$$

the basis selecting problem in every decomposition level of AFD can be formulated as follows:

$$\text{maximize } D(a_n) = \left| \langle G_n, e_{\{a_n\}} \rangle \right|^2, \tag{4}$$
$$\text{subject to } |a_n| \leq 1.$$

Since the formula $D(a_n)$ is highly nonlinear and very complex, how to solve such optimization problem efficiently becomes the key problem of the AFD. Until now, only one simple searching method is implemented [7,8]. It is based on an discrete dictionary containing finite possible points of a_n as shown in Fig. 1. In every decomposition level, all points of a_n in this searching dictionary are verified one by one in order to find a_n that satisfies (4) [7]. In other words, for the conventional searching method, the higher density of the searching dictionary is, the more accurate the solution of (4) is but the longer the computational time is. Therefore, the overall performance of the AFD involves such trade-off between the accuracy and the computational time.

    In order to increase the efficiency of the basis searching process in the AFD, a fast basis searching process of AFD is presented in this paper. It is based on the Nelder-Mead algorithm that is one of the best known numerical algorithm for optimization problems [9]. Since it does not require derivative information, it is very suitable for solving (4) [10]. However, the Nelder-Mead algorithm is only able to find the local minimum point which is close to the given initial points. Therefore, the contrast function $D(a_n)$ is reformed to change the original maximization problem to the minimization problem. Moreover, in order to make sure the solution obtained from the Nelder-Mead algorithm is close to the global minimum point, suitable initial points of the Nelder-Mead algorithm are searched from a dictionary containing random available a_n points.

In the following parts, the Nelder-Mead algorithm is introduced first. Then, the fast basis searching method based on the Nelder-Mead algorithm is proposed. To verify the proposed searching method, the AFD with the proposed searching method is applied to artificial and real ECG signals. Selection ranges of four key parameters in the proposed searching method are determined based on simulation results of an artificial ECG signal. Based on the determined selection ranges, simulation results of real ECG data from the MIT-BIH Arrhythmia Database [14,15] shows that the computational time of the AFD based on the proposed searching method is only half of that based on the conventional searching method with similar reconstruction error.

2 Nelder-Mead Algorithm

The Nelder-Mead algorithm is one of the best known simplex method for finding local minimum of a function devised by Nelder and Mead [11]. For two variables, this method is a pattern search that compares function values at three vertices of a triangle [10]. In every searching step, based on the initial triangle, the worst vertex where the function value is largest is replaced with a new vertex [10]. Then, the new triangle is formed as the initial triangle for the next searching step.

To illustrate the Nelder-Mead algorithm, suppose that $f(x, y)$ is a function of two parameters x and y that is to be minimized. The Nelder-Mead solve this optimization problem based on following steps.

In the first step, the initial triangle BGW is established based on the given three points: $\mathbf{V}_k = (x_k, y_k)$, $k = 1, 2, 3$. The function value of $f(x, y)$ is evaluated at each of three points: $z_k = f(x_k, y_k)$ for $k = 1, 2, 3$. Suppose that $z_1 \leq z_2 \leq z_3$. Then, $\mathbf{B} = (x_1, y_1)$ denotes the best vertex, $\mathbf{G} = (x_2, y_2)$ denotes the good vertex, and $\mathbf{W} = (x_3, y_3)$ denotes the worst vertex [10].

In the second step, since the construction process needs the midpoint of the line segment joining $\mathbf{B}$ and $\mathbf{G}$, this midpoint needs to be evaluated by

$$\mathbf{M} = \frac{\mathbf{B} + \mathbf{G}}{2}. \tag{5}$$

In the final step, the worst vertex is updated. The decision logic is shown below. A reflected point $\mathbf{R}$ defined as (6) is needed to be tested. If the function value at $\mathbf{R}$ is smaller than the function value at $\mathbf{W}$, then the correct direction should be moved toward the minimum [10]. In addition, the extended point $\mathbf{E}$ defined as (7) should be tested [10]. If the function value at $\mathbf{E}$ is less than the function value at $\mathbf{R}$, then the better vertex than $\mathbf{R}$ is found [10].

$$\mathbf{R} = 2\mathbf{M} - \mathbf{W} \tag{6}$$

$$\mathbf{E} = 2\mathbf{R} - \mathbf{M} \tag{7}$$

If the function values at $\mathbf{R}$ and $\mathbf{W}$ are same, the contracted point $\mathbf{C}$ defined as (8) or (9) should be tested [10]. If the function value at $\mathbf{C}$ is not less than the value at $\mathbf{W}$, the points $\mathbf{G}$ and $\mathbf{W}$ should be shrunk toward $\mathbf{B}$. The point $\mathbf{G}$ is replaced by $\mathbf{M}$ [10]. The point $\mathbf{W}$ is replaced with $\mathbf{S}$ which is defined as (10) [10].

$$\mathbf{C}_1 = \frac{\mathbf{W} + \mathbf{M}}{2} \tag{8}$$

$$\mathbf{C}_2 = \frac{\mathbf{M} + \mathbf{R}}{2} \tag{9}$$

$$\mathbf{S} = \frac{\mathbf{B} + \mathbf{W}}{2} \tag{10}$$

In the Nelder-Mead algorithm, only function values are required to be evaluated in every step. The derivative of original function is not required. Therefore, the Nelder-Mead algorithm is easy to be implemented and suitable for the optimization problem of AFD shown in (4). However, it is only for finding local minimum of a function. How to adopt it to find the global maximum of (3) becomes the key problem.

3   Fast Basis Searching Method

The fast basis searching method is to find a_n which can achieve (4) efficiently by using the Nelder-Mead algorithm. Since the Nelder-Mead algorithm is for minimization problems, the original maximization problem shown in (4) should be represented as the minimization problem. From (3), it can be seen that the function value of $D(a_n)$ is always a non-negative number. Therefore, finding the global maximum of $D(a_n)$ is equivalent to finding the global minimum of $-D(a_n)$. Moreover, since a_n is determined by its magnitude ρ_{a_n} and phase α_{a_n}, the contrast function $D(a_n)$ can be represented as

$$Y(\rho_{a_n}, \alpha_{a_n}) = -\left| \left\langle G_n, e_{\{\rho_{a_n} e^{j\alpha_{a_n}}\}} \right\rangle \right|. \tag{11}$$

Based on the new contrast function $Y(\rho_{a_n}, \alpha_{a_n})$, the basis selecting problem can be reformulated as follows:

$$\text{minimize } Y(\rho_{a_n}, \alpha_{a_n}) = -\left| \left\langle G_n, e_{\{\rho_{a_n} e^{j\alpha_{a_n}}\}} \right\rangle \right| \tag{12}$$

$$\text{subject to } |\rho_{a_n}| \leq 1,$$

Algorithm 1. AFD based on proposed fast basis searching method

Input:

    $y(t)$: ECG signal;

    N_{decom}: Maximum decomposition level;

    N_{rand}: Total number of random points in the searching dictionary of initial points;

    $R_{\min}$: Minimum radius of outer circle;

    ρ_d and α_d: Distances of three initial points of Nelder-Mead algorithm.

1: Initialize $a_1 = 0$ and $N = 1$;

2: $G_1 \leftarrow y(t) + j\mathcal{H}\{y(t)\}$;

3: $e_{a_1} \leftarrow B_1 \leftarrow \frac{\sqrt{1-|a_1|^2}}{1-\overline{a}_1 e^{jt}}$;

4: Generate $\mathbf{u}$ containing N_{rand} random numbers based on the standard uniform distribution on $(0,1)$;

5: Generate searching dictionary $\mathbf{A}$ of initial points containing N_{rand} points in which $(\rho_{a_n k}, \alpha_{a_n k})$ follows (13);

6: **while** $N \leq N_{\text{decom}}$ **do**

7:     $G_{N+1} \leftarrow \left(G_N(t) - \langle G_N, e_{\{a_N\}}\rangle e_{\{a_N\}}\right)\frac{1-\overline{a}_N e^{jt}}{e^{jt}-a_N}$;

8:     Find the point that contains the maximum value of (3) in $\mathbf{A}$ denoting as (ρ_1, α_1);

9:     Apply Nelder-Mead method to find the point that contain minimum value of (11) by using (ρ_1, α_1), $(\rho_1 + \rho_d, \alpha_1)$ and $(\rho_1, \alpha_1 + \alpha_d)$ as initial points. Denote the obtained point as $(\rho_{a_n(\max)}, \alpha_{a_n(\max)})$;

10:     $a_{N+1} \leftarrow \rho_{a_n(\max)} e^{j\alpha_{a_n(\max)}}$;

11:     $B_{N+1} \leftarrow \frac{\sqrt{1-|a_{N+1}|^2}}{1-\overline{a}_{N+1}e^{jt}}\frac{e^{jt}-a_N}{\sqrt{1-|a_N|^2}}B_N$;

12:     $F_{N+1} \leftarrow \langle G_{N+1}, e_{\{a_{N+1}\}}\rangle B_{N+1}$;

13:     $N \leftarrow N + 1$;

14: **end while**

15: $y_r \leftarrow \text{Re}\left\{\sum_{n=1}^{N} F_n\right\}$;

Output:

    y_r: reconstructed signal;

    F_n: decomposition composition at n decomposition level where $n = 1, 2, 3, \cdots, N$.

---

which can be applied for the Nelder-Mead algorithm.

To make sure that the Nelder-Mead algorithm can obtain the global minimum instead of local minimum, initial points should be selected around the target global minimum point. In order to find such suitable points, in the proposed searching method, a searching dictionary containing random possible points is established. This process of searching initial points is similar to the conventional basis searching process of AFD. All points in the searching dictionary is verified one by one. The point which contains the minimum value of $Y(\rho_{a_n}, \alpha_{a_n})$ is applied to be the first initial point (ρ_1, α_1). Another two initial points can be selected around the first initial point such that $(\rho_1 + \rho_d, \alpha_1)$ and $(\rho_1, \alpha_1 + \alpha_d)$ where ρ_d and α_d are two small numbers. It should be noticed that the number of points in the searching dictionary is not required to be very large since the initial points are not required to be the accurate optimization solution. Moreover, according to [12], for ECG signals, the optimization solution points of a_n's normally are distributed in an outer circle. Therefore, the searching space of the initial points can be reduced to the outer circle $\{(\rho_{a_n}, \alpha_{a_n}) | R_{\min} \leq \rho_{a_n} < 1\}$ for ECG signals where $R_{\min}$ denotes the minimum radius of the outer circle. From the searching process, it can be seen that the searching dictionary is very important in order to find suitable initial points. All points in the outer circle should have equivalent probability to be selected into the searching dictionary. Suppose that the random points in the searching dictionary are defined as $\{(\rho_{a_n 1}, \alpha_{a_n 1}), (\rho_{a_n 2}, \alpha_{a_n 2}), \cdots, (\rho_{a_n N_{\text{rand}}}, \alpha_{a_n N_{\text{rand}}})\}$ where N_{rand} denotes the total number of random points, and $\mathbf{u}$ is an array of random numbers in $(0, 1)$ with uniformly distribution. To make sure that these random points are distributed in the outer circle uniformly, they should satisfy

$$(\rho_{a_n k}, \alpha_{a_n k}) = \left(2\pi u_k, \sqrt{(1 - R_{\min}^2) u_k} \right). \tag{13}$$

where $(\rho_{a_n k}, \alpha_{a_n k})$ and u_k denote the k-th random point in the searching dictionary and the k-th number in $\mathbf{u}$.

After obtaining three initial points, the Nelder-Mead algorithm is able to be applied for (12). The detailed process of the AFD based on the proposed fast basis searching method is shown in Algorithm 1 where $\mathcal{H}\{\cdot\}$ denotes the Hilbert transform which is used to transfer the original ECG signal to its analytical representation following the requirement of AFD [13].

4   Simulation Results and Discussions

In order to determine suitable selection ranges of parameters N_{rand}, $R_{\min}$, ρ_d and α_d shown in Algorithm 1, simulations of an artificial ECG signal are performed. Based on the obtained suitable selection ranges, the proposed searching method is evaluated by real ECG data from the MIT-BIH Arrhythmia Database [14,15] and compared with the conventional searching method. Two parameters are considered to evaluate the searching performances. One is the reconstruction error at 100 decomposition level which shows whether the decomposition components obtained by the searching method can express the original signal. It is defined as the ratio of the energy of reconstructed signal by using first 100 decomposition components to the energy of the original ECG signal. Another one is the computational time which indicates the efficiency of the searching method. The following simulations are completed in MATLAB 2014a. The system platform is Intel(R) Core(TM) i7-4770 CPU @ 3.40GHz with 12GB RAM.

4.1   Parameters Determination

There are total four parameters N_{rand}, $R_{\min}$, ρ_d and α_d that can affect searching results in the proposed fast searching method. N_{rand} and $R_{\min}$ are related to the process of generating random points in the searching dictionary of initial points. ρ_d and α_d are related to the performance of Nelder-Mead algorithm. In this part, an artificial ECG signal is applied for simulations. It is based on the nonlinear

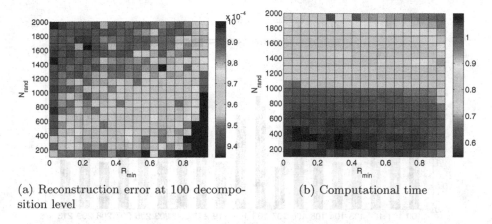

(a) Reconstruction error at 100 decomposition level

(b) Computational time

Fig. 2. Effects of different $R_{\min}$ and N_{rand}

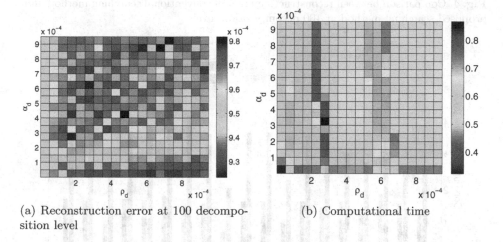

(a) Reconstruction error at 100 decomposition level

(b) Computational time

Fig. 3. Effects of different ρ_d and α_d

dynamic model proposed in [16]. The sampling frequency is 256Hz. The heart rate is 60 beats per minutes.

N_{rand} and $R_{\min}$ are determined first. To make sure the convergence of the Nelder-Mead algorithm, ρ_d and α_d are all set as small numbers such that $\rho_d = 0.0005$ and $\alpha_d = 0.004$. N_{rand} is evaluated from 100 to 2000. $R_{\min}$ is evaluated from 0 to 0.95. For every pair of N_{rand} and $R_{\min}$, 100 simulations are carried out to calculate mean results. Fig. 2 shows simulation results of different N_{rand}'s and $R_{\min}$'s. It can be seen that the reconstruction error at 100 decomposition level is all smaller than 0.001. In other words, no matter what values N_{rand} and $R_{\min}$ are, the decomposition components are all able to represent the original signal. Fig. 2(b) shows computational time of different N_{rand}'s and $R_{\min}$'s. It can be seen that the computational time increases when N_{rand} increases, which

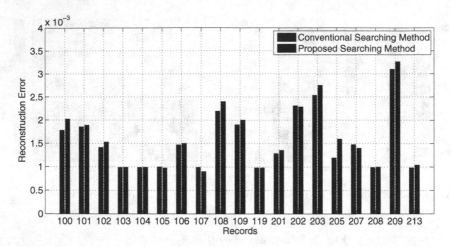

Fig. 4. Comparison between reconstruction error of conventional searching method and proposed searching method at 100 decomposition level

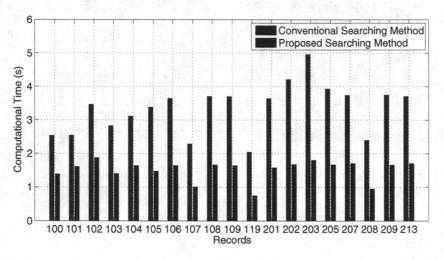

Fig. 5. Comparison between computational time of conventional searching method and proposed searching method

is due to that all points in the searching dictionary of initial points should be evaluated one by one in order to find suitable initial points. Combining these two simulation results, it is better to select N_{rand} from 500 to 1000 and select $R_{\min}$ from 0 to 0.4 to make sure that the computational time and reconstruction error are all low at the same time.

By setting N_{rand} and $R_{\min}$ as 600 and 0.1 separately, parameters ρ_d and α_d are evaluated. ρ_d and α_d are all evaluated from 1×10^{-6} to 1×10^{-3}. For every pair of ρ_d and α_d, 100 simulations are carried out to calculate mean results.

Fig. 3 shows simulation results of different α_d and ρ_d. From Fig. 3(a) and Fig. 3(b), it can be seen that reconstruction error and computational time are almost same for different values of ρ_d and α_d. However, there still are little differences. In order to obtain low reconstruction error and low computational time at the same time, it is better to select α_d from 1×10^{-4} to 5×10^{-4} and select ρ_d from 3×10^{-4} to 1×10^{-3}.

4.2 Comparison between Proposed Searching Method and Conventional Searching Method

After determining selection ranges of parameters N_{rand}, R_{min}, ρ_d and α_d, the decomposition performances of proposed searching method are evaluated by real ECG data from the MIT-BIH Arrhythmia Database and compared with conventional searching method. In order to make sure that the conventional searching method can obtain the solution that is close to the accurate optimization solution of (4), 7705 points are considered in its searching dictionary. For the proposed fast searching method, parameters N_{rand}, R_{min}, ρ_d and α_d are set as 800, 0.1, 5×10^{-4} and 3×10^{-4} separately. Fig. 4 shows the comparison of reconstruction error at 100 decomposition level for 19 different records. It can be seen that they are all small and almost same. Therefore, the decomposition components obtained from these two different searching methods are all able to represent the original signals. Fig. 5 shows the comparison of computational time. For all records, the computational time of the proposed searching method is almost half of the computational time of the conventional searching method. In other words, the proposed searching method is faster than the conventional searching method significantly.

5 Conclusion

A fast basis searching method of AFD based on the Nelder-Mead algorithm is proposed for ECG signals. In order to adopt the Nelder-Mead algorithm to the AFD, the original maximization problem for searching basis in the AFD is reformulated to the minimization problem. Moreover, a random searching dictionary is established for ECG signals to find suitable initial points of the Nelder-Mead algorithm. Based on simulation results of the artificial ECG signal, selection ranges of four key parameters N_{rand}, R_{min}, ρ_d and α_d in the proposed fast searching method are determined. The simulation results of real ECG data shows that the computational time of the AFD based on the proposed searching method is just half of that based on the conventional searching method with similar reconstruction error.

Acknowledgments. This work is supported in part by the Macau Science and Technology Development Fund under grants FDCT 036/2009/A and FDCT 055/2015/A2 and the University of Macau Research Committee under grants MYRG139(Y1-L2)-FST11-WF, MYRG079(Y1-L2)-FST12-VMI, MYRG069(Y1-L2)-FST13-WF, MYRG2014-00174-FST and MYRG2016-00240-FST.

References

1. Qian, T., Zhang, L.M., Li, Z.X.: Algorithm of Adaptive Fourier Decomposition. IEEE Trans. Signal Process. 59, 5899–5906 (2011)
2. Qian, T.: Adaptive Fourier Decompositions and Rational Approximations–Part I: Theory. Int. J. Wavelets Multiresolut Inf. Process. 12, 1461008 (2014)
3. Zhang, L., Hong, W., Mai, W., Qian, T.: Adaptive Fourier Decomposition and Rational Approximation–Part II: Software System Design and Development. Int. J. Wavelets Multiresolut Inf. Process. 12, 1461009 (2014)
4. Wang, Z., Wong, C.M., da Cruz, J.N., Wan, F., Mak, P.I., Mak, P.U., Vai, M.I.: Muscle and Electrode Motion Artifacts Reduction in ECG Using Adaptive Fourier Decomposition. In: 2014 IEEE International Conference on Systems, Man and Cybernetics, pp. 1456–1461. IEEE, San Diego (2014)
5. Qian, T., Wang, Y.B.: Adaptive Fourier Series–a Variation of Greedy Algorithm. Adv. Comput. Math. 34, 279–293 (2011)
6. Qian, T., Wang, Y.B.: Remarks on Adaptive Fourier Decomposition. Int. J. Wavelets Multiresolut Inf. Process. 11, 1350007 (2013)
7. Zhang, L., Hong, W., Mai, W., Qian, T.: Adaptive Fourier Decomposition and Rational Approximation–Part II: Software System Design And Development. Int. J. Wavelets Multiresolut Inf. Process. 12, 1461009 (2014)
8. AFDs Software, http://www.fst.umac.mo/en/staff/documents/fsttq/afd_form/index.html
9. Nocedal, J., Wright, S.J.: Numerical Optimization. Springer, New York (2006)
10. Klein, K., Neira, J.: Nelder-Mead Simplex Optimization Routine for Large-scale Problems: a Distributed Memory Implementation. Comput. Econ. 43, 447–461 (2014)
11. Nelder, J.A., Mead, R.: A Simplex Method for Function Minimization. Comput. J. 7, 308–313 (1965)
12. Ma, J., Zhang, T., Dong, M.: A Novel ECG Data Compression Method Using Adaptive Fourier Decomposition with Security Guarantee in e-Health Applications. IEEE J. Biomed. Health Inform. 19, 986–994 (2014)
13. Qian, T.: Adaptive Fourier Decompositions and Rational Approximations, Part I: Theory. Int. J. Wavelets Multiresolut. Inf. Process. 12, 1461008 (2014)
14. Moody, G.B., Mark, R.G.: The Impact of the MIT-BIH Arrhythmia Database. IEEE Eng. Med. Biol. Mag. 20, 45–50 (2001)
15. Goldberger, A.L., Amaral, L.A.N., Glass, L., Hausdorff, J.M., Ivanov, P.C., Mark, R.G., Mietus, J.E., Moody, G.B., Peng, C.K., Stanley, H.E.: PhysioBank, PhysioToolkit, and PhysioNet: Components of a New Research Resource for Complex Physiologic Signals. Circulation 101, e215–e220 (2000)
16. McSharry, P.E., Clifford, G.D., Tarassenko, L., Smith, L.A.: A Dynamical Model for Generating Synthetic Electrocardiogram Signals. IEEE Trans. Biomed. Eng. 50, 289–294 (2003)

Frequency Recognition Based on Wavelet-Independent Component Analysis for SSVEP-Based BCIs

Limin Yang, Ze Wang, Chi Man Wong, and Feng Wan

Department of Electrical and Computer Engineering
Faculty of Science and Technology
University of Macau
fwan@umac.mo

Abstract. Among the EEG-based BCIs, SSVEP-based BCIs have gained much attention due to the advantages of relatively high information transfer rate (ITR) and short calibration time. Although in SSVEP-based BCIs the frequency recognition methods using multiple channels EEG signals may provide better accuracy, using single channel would be preferable in a practical scenario since it can make the system simple and easy-to-use. To this goal, we propose a new single channel method based on wavelet-independent component analysis (WICA) in the SSVEP-based BCI, in which wavelet transform (WT) is applied to decompose a single channel signal into several wavelet components and then independent component analysis (ICA) is applied to separate the independent sources from the wavelet components. Experimental results show that most of the time the recognition accuracy of the proposed single channel method is higher than the conventional single channel method, power spectrum (PS) method.

Keywords: wavelet-independent component analysis (WICA), SSVEP, frequency recognition, BCI.

1   Introduction

A brain-computer interface (BCI) is a system that translates human intentions into command and control signals, providing a direct communication between a human or animal brain and an external device. Most modern BCIs rely on non-invasive scalp electroencephalogram (EEG) measurements which are easy to implement. In EEG-based BCI systems, three types of signals including event-related desynchronization/synchronization (ERD/ERS), P300 and steady-state visual evoked potential (SSVEP) are most commonly used [1-2].

SSVEP is a continuous brain response evoked over occipital scalp areas with the same frequency as rapidly repetitive visual stimulus [3]. According to this mechanism, SSVEP-based BCI systems use multiple visual stimuli that flicker at different frequencies simultaneously while the participants are required to focus on the stimulus they intend to select, which elicits the corresponding stimulation frequency in the EEG [4]. Due to its less calibration time and higher information transfer rate (ITR),

© Springer International Publishing Switzerland 2015
X. Hu et al. (Eds.): ISNN 2015, LNCS 9377, pp. 315–323, 2015.
DOI: 10.1007/978-3-319-25393-0_35

and small number of electrodes compared to other types of BCIs, SSVEP-based BCI has been increasingly studied in recent years [2]. A key step is to recognize the frequency of SSVEPs, for which a number of frequency recognition methods using either multiple channel or single channel EEG signals have been proposed in the literature. Multichannel detection methods such as canonical correlation analysis (CCA), independent component analysis (ICA) and minimum energy combination (MEC) make use of multiple channel EEG signals and channel covariance information which may achieve a high signal-to-noise ratio (SNR) [1,5,6]. However, using multiple channels may not only lead to less comfortable and inconvenient user experience but also increase the complexity of implementation in a real BCI system. In contrast, using single channel methods reduce the complexity of the implementation. One of the most widely-used single channel methods is the power spectrum (PS) method in which fast Fourier transform (FFT) is applied to estimate the power spectrum of the single channel EEG signals. Among all the frequency components in the evoked EEG spectrum that are the same as the stimulus frequencies, the one with maximum power determines the button that the user intends to select [4]. The PS method is simple, but unfortunately, it is sensitive to noise [4].

This paper proposes a new frequency recognition method for single channel based on wavelet-independent component analysis (WICA). In this method, wavelet transform (WT) is applied to decompose a single channel signal into several wavelet components and then independent component analysis (ICA) is applied to separate the independent sources from the wavelet components. The proposed frequency recognition method is verified by real SSVEP signals and compared with the PS method. From experimental results, the recognition accuracy of the proposed method is higher than that of the PS method most of the times.

2   Wavelet-Independent Component Analysis

The wavelet-independent component analysis (WICA) is a combined technique of the wavelet transform (WT) and the independent component analysis (ICA) for separating independent sources from the observation obtained from a single channel [7].

ICA is one kind of blind source separation (BSS) algorithms which aims to recover a set of unknown mutually independent source signals from their observed mixtures without knowledge of the mixing coefficients [8]. Suppose that there are N observed mixtures $\mathbf{x}(t) = [x_1(t), x_2(t), ..., x_N(t)]^T$ which can be modeled as

$$\mathbf{x}(t) = \mathbf{A}\mathbf{s}(t) \tag{1}$$

where source signals $\mathbf{s}(t) = [s_1(t), s_2(t), ..., s_M(t)]^T$ are assumed to be independent, and $\mathbf{A}$ is an $N \times M$ unknown full column-rank mixing matrix [8]. The BSS algorithm is to estimate the unknown $\mathbf{A}$ and $\mathbf{s}(t)$ from measurements $\mathbf{x}(t)$. The goal of ICA is to determine an $M \times N$ demixing matrix $\mathbf{W}$, with M output signals

$$\mathbf{y}(t) = \mathbf{W}\mathbf{x}(t) = \mathbf{W}\mathbf{A}\mathbf{s}(t) = \mathbf{P}\mathbf{D}\mathbf{s}(t) \tag{2}$$

where $\mathbf{P} \in \mathbf{R}^{M \times M}$ is a permutation matrix, and $\mathbf{D} \in \mathbf{R}^{M \times M}$ is a diagonal scaling matrix [8]. Then, the source signals are recovered up to scaling and permutation. It is well known that ICA cannot be applied for single channel data analysis directly because the number of observed mixtures should not be less than the number of sources [7]. For single channel data analysis, in [7] WT is proposed to generate the multiple components from the single channel data for ICA.

WT performs a type of time-frequency transforms, which overcomes the limitations in time and frequency resolution occurring with the classical Fourier transform and its variants [9]. WT is one of the leading techniques for processing non-stationary signals, and thus is well suited for EEG signals which are non-stationary [10]. For the classical discrete WT (DWT), the signal $f(t)$ is represented as

$$f(t) = \sum_k c_k^R \phi_R (t - 2^{-R} k) + \sum_{j=R}^{J-1} d_k^j \psi_j (t - 2^{-j} k) \tag{3}$$

where $\phi_j(t)$ and $\psi_j(t)$ present appropriate dilations and translations of the scaling functions $\phi(t)$ and the mother wavelet $\psi(t)$ which are defined as

$$\phi_j(t) = 2^{j/2} \phi(2^j t) \text{ and } \psi_j(t) = 2^{j/2} \psi(2^j t) \tag{4}$$

for integer j [9]. However, the coefficient sequences of the classical DWT are decimated at each decomposition level [9]. In other words, decomposition components contain different signal lengths of different decomposition levels. Such sets of signals are not suitable as inputs of ICA. Therefore, in the proposed SSVEP-BCI frequency recognition method, the stationary wavelet transform (SWT) is applied. In SWT, the filters at each decomposition level are modified by padding them out with zeros instead of decimating coefficient sequences in order to keep sequences at each new decomposition level have same length as the original sequences [9].

For WICA, by combining WT and ICA together, the limitation of ICA that it cannot be applied to single channel signal is overcome.

3   SSVEP Frequency Recognition Method Based on WICA

3.1   EEG Recordings

The EEG signals applied for examples were recorded from Oz channel (according to the international 10-20 system) and the ground was located at forehead while the reference was at the left mastoids. Circuit impedance was kept below 10 kΩ. The signals were amplified by a g.tec amplifier, g.USBamp (Guger Technologies, Graz, Austria) with a sampling rate at 600 Hz and filtered by a band-pass filter (0.5~30Hz) and a 50 Hz notch filter to avoid the baseline drift, high frequency noise and powerline interference.

3.2   Frequency Recognition Method Based on WICA

ICA is an effective technique used for EEG signals denoising and extraction in a number of studies [10, 11]. However, the ICA algorithm is not able to be applied directly to the frequency recognition of SSVEP-based BCIs when only a single channel is available. In other words, the input of ICA must be a matrix instead of a vector. To overcome this limitation, it is necessary to construct the input matrix of ICA using WT for decomposition. The combination of WT and ICA is called WICA technique.

Assume that $x(t)$ is the EEG signal recorded from a single channel, $f_{stimulus}$ is the real stimulus frequency of $x(t)$, and there are a group of reference signals $r_1(t)$, $r_2(t),\ldots,r_N(t)$ with N different stimulus frequencies $f_1, f_2,\ldots, f_N$.

$$\begin{cases} r_1(t) = \sin(2\pi f_1 t) \\ r_2(t) = \sin(2\pi f_2 t) \\ r_3(t) = \sin(2\pi f_3 t) \quad , t = \dfrac{1}{F_s}, \dfrac{2}{F_s}, \ldots, \dfrac{T}{F_s} \\ \ldots \\ r_N(t) = \sin(2\pi f_N t) \end{cases} \tag{5}$$

where T is the number of sampling points and F_s is the sampling rate.

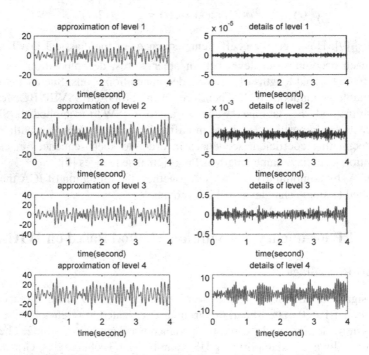

Fig. 1. Example of the SWT decomposition results (M=4)

The procedure of proposed method follows three main steps. First of all, SWT is applied to $x(t)$ for the wavelet decomposition specifically with mother wavelet *symlet* at decomposition level M in this study. Both the approximation and detail components

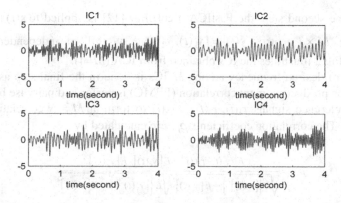

Fig. 2. Example of the output ICs from WICA (M=4)

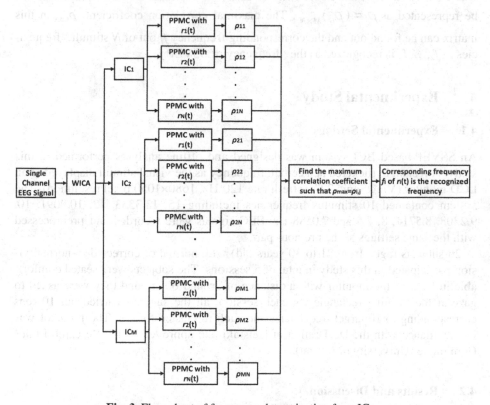

Fig. 3. Flow chart of frequency determination from ICs

in each level of SWT are generated so that M approximation components and M detail components can be obtained in total. Fig. 1 illustrates the wavelet decomposed components (M=4). The M decomposed approximation components are selected to form an $M \times T$ input matrix $\mathbf{x}(t) = [x_1(t), x_2(t),..., x_M(t)]^T$ of ICA while the detail components are eliminated.

Then in the second step, the FastICA algorithm [12] is applied to $\mathbf{x}(t)$ in order to find out the $M \times T$ matrix $\mathbf{y}(t) = [y_1(t), y_2(t), ..., y_M(t)]^T$ of independent components (ICs). Fig.2 illustrates the ICs obtained from ICA (M=4).

Selection of desired frequency among M ICs is done in the final step as shown in Fig.3. Pearson product-moment correlation (PPMC) is performed pairwise between M ICs and N reference signals $r_1(t)$, $r_2(t)$,...,$r_N(t)$ to form an $M \times N$ correlation coefficient matrix. The correlation coefficients $\rho_{i,j}$ are calculated by

$$\rho_{i,j} = \frac{E\left[y_i(t) \cdot r_j(t)\right] - E\left[y_i(t)\right] \cdot E\left[r_j(t)\right]}{\sqrt{E\left\{[y_i(t)]^2\right\} - E[y_i(t)]^2}\sqrt{E\left\{[r_j(t)]^2\right\} - E[r_j(t)]^2}} \tag{6}$$

where $i = 1, 2, ..., M$, and $j = 1, 2, ..., N$. Thus, the correlation coefficient matrix can be represented as $\rho = (\rho_{ij})_{M \times N}$. The maximal correlation coefficient ρ_{max} in this matrix can be found out and the corresponding frequency f_j out of N stimulus frequencies $f_1, f_2, ..., f_N$ is recognized as the frequency of SSVEP f_s.

4   Experimental Study

4.1   Experimental Settings

An SSVEP-based BCI system was designed and offline analyses performed in this study. A rectangle flickered at a certain frequency as the visual stimuli displayed in an LCD monitor (ViewSonic 22", refresh rate 120 Hz, 1680×1050 pixel resolution). The system contained 10 stimulus frequencies including 15, 13.3333, 12, 10.9091, 10, 9.2308, 8.5714, 8, 7.5 and 7.0588 Hz. EEG signals were recorded and preprocessed with the same settings as the previous part.

26 subjects (aged from 21 to 30 years old) with normal or corrected-to-normal vision participated in this study in total of 5 sessions. The subjects were seated comfortable in front of the monitor with a distance of around 60 cm and they were asked to gaze at the flashing rectangle. In each session, all the subjects carried out 10 runs corresponding to 10 target frequencies and each run lasted for 4 s. The protocol was in accordance with the Declaration of Helsinki and approved by the Research Ethics Committee (University of Macau).

4.2   Results and Discussion

The recorded EEG signals were analyzed using different time window lengths ranging from 1 to 4 s and the EEG data were preprocessed by band-pass filtering between 6-20 Hz. The recognition accuracy was calculated as the evaluation index of the performance of the proposed method and the recognition accuracies of the PS method were also presented for comparison. Let f_s be the frequency determined by the recognition algorithms and let $f_{stimulus}$ be the real stimulus frequency of a segment of signal.

In one segment, if f_s equals to $f_{stimulus}$, this segment is considered as the correct recognized segment. Then the recognition accuracy can be defined as

$$\text{Recognition Accuracy} = \frac{N_{correct}}{N_{total}} \times 100\% \quad (7)$$

where $N_{correct}$ is the number of correct recognized segments, and N_{total} is the number of total segments.

Table 1. Comparison of recognition accuracies between PS and WICA method with different window lengths

Subject	Window Length (L)							
	$L = 1$ s		$L = 2$ s		$L = 3$ s		$L = 4$ s	
	PS	WICA	PS	WICA	PS	WICA	PS	WICA
S1	67%	**70%**	82%	**86%**	83%	**90%**	82%	**94%**
S2	**60%**	54%	**81%**	81%	**81%**	81%	88%	**89%**
S3	29%	**37%**	41%	**52%**	49%	**55%**	48%	**60%**
S4	64%	**65%**	81%	**87%**	91%	**93%**	92%	**96%**
S5	67%	**69%**	77%	**85%**	79%	**89%**	80%	**91%**
S6	**23%**	21%	**29%**	29%	**34%**	30%	36%	**37%**
S7	**26%**	20%	36%	**39%**	41%	**49%**	36%	**55%**
S8	31%	**36%**	43%	**47%**	52%	**54%**	**58%**	51%
S9	56%	**60%**	77%	**80%**	81%	**82%**	80%	**88%**
S10	51%	**54%**	67%	**70%**	67%	**75%**	70%	**73%**
S11	25%	**29%**	17%	**28%**	17%	**44%**	30%	**43%**
S12	29%	**30%**	37%	**40%**	37%	**52%**	42%	**63%**
S13	27%	**28%**	31%	**42%**	33%	**47%**	26%	**51%**
S14	24%	**25%**	23%	**36%**	24%	**42%**	22%	**53%**
S15	26%	**27%**	29%	**41%**	29%	**44%**	30%	**53%**
S16	28%	**31%**	40%	**51%**	51%	**63%**	52%	**62%**
S17	19%	**22%**	17%	**27%**	20%	**36%**	20%	**40%**
S18	**48%**	46%	70%	**77%**	79%	**88%**	82%	**85%**
S19	**30%**	23%	**33%**	33%	43%	**49%**	36%	**46%**
S20	**41%**	38%	60%	**63%**	62%	**65%**	68%	**72%**
S21	**41%**	**41%**	**61%**	**61%**	68%	**70%**	72%	**80%**
S22	35%	**38%**	47%	**52%**	49%	**63%**	54%	**66%**
S23	35%	**36%**	48%	**56%**	51%	**68%**	58%	**68%**
S24	30%	**31%**	42%	**49%**	49%	**60%**	50%	**67%**
S25	**25%**	23%	34%	**39%**	39%	**41%**	**36%**	34%
S26	**19%**	18%	18%	**23%**	19%	**32%**	26%	**36%**
Mean	36.52%	**37.42%**	46.95%	**52.89%**	51.13%	**60.06%**	52.85%	**63.60%**

The accuracies of the frequency recognition using WICA and PS method with different time window lengths are shown in Table 1. The mean recognition accuracies of the PS method were found to be 36.52%, 46.95%, 51.13% and 52.85% in the time windows from 1 to 4 s respectively, while the mean accuracies of proposed WICA method were 37.42%, 52.89%, 60.06% and 63.60% in the time windows from 1 to 4 s respectively. When the lengths of time windows were the same, the accuracies of the proposed method were higher than that of the PS method most of the times. And the mean accuracy of the proposed method was on average 6.7% higher than that of the PS method in the same time windows. This indicates that in terms of the recognition accuracy, the proposed method has a better performance than PS method does. Although the accuracies of both methods increased as the lengths of time windows increased, the difference of the average accuracies between these two methods increased from 1% to 11%. This indicates that the improvement of recognition accuracy of the proposed method is more notable in longer time windows.

The proposed frequency recognition method can perform better than the traditional PS method mainly due to the noisy reduction of the WICA [11]. In the WICA, only approximation components of SWT are applied as the inputs of ICA, while the detail components that mainly contain high frequency noise are eliminated. Moreover, after the ICA, the IC containing maximal correlation coefficient is selected, which removes ICs that mainly contain noise. In order to compare the SNRs of the EEG signals after applying the proposed method and the PS method, the SNR of SSVEP signals is defined as

$$SNR = \frac{n \cdot X(f_{stimulus})}{\sum_{f=f_1}^{f_2} X(f)} \tag{8}$$

where $X(f)$ is the amplitude spectrum ranged from f_1=6 Hz to f_2=20 Hz calculated by FFT, and n is the number of points of $X(f)$ from f_1 to f_2.

Based on real EEG signal analyzing results, the averaged SNRs of signals after applying the proposed method are 3.748±1.335 while the averaged SNRs of signals after applying the PS method are only 3.078±1.227. It can be seen that applying the proposed method can achieve higher SNRs which means that the proposed method is less sensitive to noise than the PS method. As a consequence, it is reasonable that the proposed method can obtain higher recognition accuracy than the traditional PS method.

However, it is worth noting that in a few cases, the recognition accuracy of the proposed method is worse than that of the PS method. In the future work, the reason should be further explored.

5   Conclusion

In this study, a single channel frequency recognition method based on WICA is proposed for SSVEP-based BCIs. The proposed method utilizes WT to decompose the single channel EEG signals into several wavelet components and approximation components are selected to form the input matrix of ICA. Then ICA is applied to separate the ICs from the selected wavelet components. PPMC is performed between ICs and reference signals to build a correlation coefficients matrix. The maximal correlation coefficient can be found out and the corresponding frequency of reference signal is recognized as the frequency of SSVEP consequently. Experiment results on 26 subjects demonstrated that the proposed method based on WICA can achieve higher recognition accuracy than the PS method most of the times.

Future study may focus on the comparison between the proposed method and typical multichannel methods.

Acknowledgements. This work is supported in part by the Macau Science and Technology Development Fund under grant FDCT 036/2009/A and FDCT 055/2015/A2 and the University of Macau Research Committee under grants MYRG139(Y1-L2)-FST11-WF,   MYRG079(Y1-L2)-FST12-VMI,   MYRG069(Y1-L2)-FST13-WF, MYRG2014-00174-FST and MYRG2016-00240-FST.

References

1. Bin, G., Gao, X., Yan, Z., Hong, B., Gao, S.: An Online Multi-Channel SSVEP-Based Brain–Computer Interface Using a Canonical Correlation Analysis Method. J. Neural Eng. 6, 046002 (2009)
2. Guger, C., Allison, B.Z., Großwindhager, B., Prückl, R., Hintermüller, C., Kapeller, C., Bruckner, M., Krausz, G., Edlinger, G.: How Many People Could Use an SSVEP BCI? Front Neurosci. 6 (2012)
3. Wu, Z., Yao, D.: Frequency Detection with Stability Coefficient for Steady-State Visual Evoked Potential (SSVEP)-Based BCIs. J. Neural Eng. 5, 36–43 (2008)
4. Wu, Z.: SSVEP Extraction Based on the Similarity of Background EEG. PloS One 9, e93884 (2014)
5. Lopez, M., Pelayo, A., Madrid, F., Prieto, E., Statistical Characterization, A.: of Steady-State Visual Evoked Potentials and Their Use in Brain–Computer Interfaces. Neural Processing Lett. 29, 179–187 (2009)
6. Nan, W., Wong, C.M., Wang, B., Wan, F., Mak, P.U., Mak, P.I., Vai, M.I.: A Comparison of Minimum Energy Combination and Canonical Correlation Analysis for SSVEP Detection. In: 5th International IEEE/EMBS Conference on Neural Engineering, pp. 469–472. IEEE Press, Cancun (2011)
7. Lin, J., Zhang, A.: Fault Feature Separation Using Wavelet-ICA Filter. NDT&E Int. 38, 421–427 (2005)
8. Comon, P.: Independent Component Analysis: a New Concept? Signal Processing 36, 287–314 (1994)
9. Nason, G.P., Silverman, B.W.: The Stationary Wavelet Transform and Some Statistical Applications. Wavelets and Statistics 103, 281–299 (1995)
10. Kirkove, M., Francois, C., Verly, J.: Comparative Evaluation of Existing and New Methods for Correcting Ocular Artifacts in Electroencephalographic Recordings. Signal Processing 98, 102–120 (2014)
11. Sheoran, M., Kumar, S., Kumar, A.: Wavelet-ICA Based Denoising of Electroencephalogram Signal. Int. J. Inform. Comput. Technol. 4, 1205–1210 (2014)
12. Hyvarinen, A., Karhunen, J., Oja, E.: A fast fixed-point algorithm for independent component analysis. Neural Comput. 9, 1483–1492 (1997)

Machine Learning

An MCMC Based EM Algorithm
for Mixtures of Gaussian Processes

Di Wu, Ziyi Chen, and Jinwen Ma

Department of Information Science, School of Mathematical Sciences and LMAM,
Peking University, Beijing, 100871, China
jwma@math.pku.edu.cn

Abstract. The mixture of Gaussian processes (MGP) is a powerful statistical
learning model for regression and prediction and the EM algorithm is an effective method for its parameter learning or estimation. However, the feasible
EM algorithms for MGPs are certain approximations of the real EM algorithm
since Q-function cannot be computed efficiently in this situation. To overcome
this problem, we propose an MCMC based EM algorithm for MGPs where Q-
function is alternatively estimated on a set of simulated samples via the Markov
Chain Monte Carlo (MCMC) method. It is demonstrated by the experiments on
both the synthetic and real-world datasets that our proposed MCMC based EM
algorithm is more effective than the other three EM algorithms for MGPs.

Keywords: Mixture of Gaussian processes, EM algorithm, Classification,
Multimodality, Prediction.

1 Introduction

Gaussian Process (GP) is a powerful statistical learning model in machine learning and
pattern recognition [1]. However, a single GP model could not deal with the multimodality dataset. Then, Tresp [2] suggested a Mixture of Gaussian Processes (MGP) to
model a general multimodality dataset. In fact, it takes a similar architecture of mixture
of experts (ME). In the MGP model, there are a number of GPs being mixed together.

For the parameter learning of MGP, there are generally two kinds of approaches: (a).
The EM algorithms [2-8]; (b). The MCMC methods [9-12]. Obviously, EM algorithm is
a widely used method to deal with the multimodality dataset, but the time complexity of
the EM algorithm for MGPs is of exponential order. Actually, it is that the computation
of Q-function must be over all the possible values of the latent variables and is therefore
of exponential order. Then, certain simplifications or approximations of EM algorithm
were established, such as the hard-cut EM algorithm [3], the LOOCV EM algorithm [4]
and the variational EM algorithm [5-8]. On the other hand, the MCMC methods need to
estimate all the latent variables as well as the parameters on different sets of simulated
samples via the MCMC methods. In fact, they are very time consuming, especially for
the high dimensional and complicated cases.

In this paper, we propose an MCMC based EM algorithm for MGPs in which Q-
function is estimated on a set of simulated samples of latent variables via the MCMC
method. In this way, the MCMC sampling is utilized to simplify the computation of

© Springer International Publishing Switzerland 2015
X. Hu et al. (Eds.): ISNN 2015, LNCS 9377, pp. 327–334, 2015.
DOI: 10.1007/978-3-319-25393-0_36

Q-function and thus the EM algorithm becomes feasible and efficient for prediction and classification. It is demonstrated by the experiments on three synthetic and real datasets that our proposed MCMC based EM algorithm is feasible and effective and even outperform three other current EM algorithms on prediction.

The remainder of this paper is organized as follows. GP model and MGP model are reviewed in Section 2. In Section 3, we propose the MCMC based EM algorithm for MGPs. The experimental results are contained in Section 4. Finally, we make a brief conclusion in Section 5.

2 The GP and MGP Models

In this section, we revisit the GP and MGP models and make a review on related works.

2.1 The GP Model

$y(\mathbf{x}) \in R$ is a stochastic process, where $\mathbf{x} \in R^P$. For arbitrary $N \in Z^+$ and arbitrary dataset $\mathbf{X} = \{\mathbf{x}_n | n = 1 \sim N\}$, the definition of the Gaussian process is given as follows. If $[y(\mathbf{x}_1), ..., y(\mathbf{x}_N)]^T$ is subject to a N-dimensional Gaussian distribution $N\{[m(\mathbf{x}_1), ..., m(\mathbf{x}_N)]^T, \mathbf{C}(\mathbf{X}, \mathbf{X}; \theta)\}$, then $y(\mathbf{x})$ is said to follow a Gaussian process, where $m(\mathbf{x})$ is a mean function and $\mathbf{C}(\mathbf{X}, \mathbf{X}; \theta) = \left[C(\mathbf{x}_n, \mathbf{x}_j; \theta) \big| n, j = 1 \sim N \right]$ represents a $N \times N$ kernel matrix in which $C(\mathbf{x}, \tilde{\mathbf{x}}; \theta)$ is a kernel function [1]. The GP model is written as

$$y(\mathbf{x}) \sim GP[m(\mathbf{x}), C(\mathbf{x}, \tilde{\mathbf{x}}; \theta)]. \tag{1}$$

Here, we assume the mean function $m(\mathbf{x}) \equiv 0$ for simplify and utilize the kernel function

$$C(\mathbf{x}, \tilde{\mathbf{x}}; \theta) = \theta_{P+1} \exp\left[-\frac{1}{2} \sum_{p=1}^{P} \theta_p \left(x_p - \tilde{x}_p\right)^2\right] + \theta_{P+2}\delta(\mathbf{x} - \tilde{\mathbf{x}}),$$

where $\delta(\mathbf{x})$ is the Kronecker delta function, $\theta = [\theta_1, ..., \theta_{P+2}]$, $\mathbf{x} = [x_1, ..., x_P]^T$ and $\tilde{\mathbf{x}} = [\tilde{x}_1, ..., \tilde{x}_P]^T$.

2.2 The MGP Model

To deal with the multimodality dataset, Tresp [2] extended the GP model to an MGP model where a number of components are involved and each component fulfills a GP model.

The sample set is denoted as $D = \{(\mathbf{x}_n, y_n) | n = 1 \sim N\}$. Suppose that there are K components and the stochastic process of the k-th component, $y_k(\mathbf{x})$, is subject to a GP model defined in Eq. (1):

$$y_k(\mathbf{x}) \sim GP[0, C(\mathbf{x}, \tilde{\mathbf{x}}; \theta_k)], \tag{2}$$

where $k = 1 \sim K$.

We give the details of the MGP model, which consists of three distributions.

a) The association among components is introduced by an unknown indicator variable z_{nk}, i.e., if the n-th sample belongs to the k-th component, $z_{nk} = 1$; otherwise, $z_{nk} = 0$. z_{nk} is subject to

$$P(z_{nk} = 1) = \pi_k,$$

where $\sum_{k=1}^{K} \pi_k = 1$.

b) Given the indicator variables, the input variable $\mathbf{x}_n$ is subject to a Gaussian distribution

$$\mathbf{x}_n \big| (z_{nk} = 1) \sim N(\mu_k, \Sigma_k).$$

c) $y_k(\mathbf{x})$ is given by Eq. (2).

3   The MCMC Based EM Algorithm for MGPs

In this section, we establish an MCMC based EM algorithm for MGPs and present a prediction method for the MGP model.

3.1   Algorithm Design

In the MCMC based EM algorithm, the indicator variable z_{nk} are treated as latent variable and $\tilde{Z}^{(r)} = \left\{ z_{nk}^{(ri)} \big| n = 1 \sim N, k = 1 \sim K, i = 1 \sim I^{(r)} \right\}$ is a set of simulated samples of indicator variables, where r is the current number of iterations in EM algorithm and $I^{(r)}$ is the number of simulated samples in the r-th iteration. Denote the parameter set $\Theta = \left\{ \pi_k, \mu_k, \Sigma_k, \theta_k \big| k = 1 \sim K \right\}$. Then the detail of MCMC based EM algorithm is given by the following four steps.

a) Calculate the indicator variables by k-means cluster and initialize the parameters $\Theta^{(0)}$ by Maximum Likelihood Estimate. Set $r = 1$.

b) E-step: draw the simulated samples of indicator variables $\tilde{Z}^{(r)}$ by Gibbs sampling.

c) M-step: calculate $\Theta^{(r)}$ by maximizing $\hat{Q}$-function.

d) If the relative variation of $\hat{Q}$-function is less than the threshold, stop; otherwise, set $r \leftarrow r + 1$ and return to b).

Details of E-step: we used Gibbs sampling method to draw the simulated samples. Denote an indicator variable set $Z_{-n} = \left\{ z_{jk} \big| j \neq n; k = 1 \sim K \right\}$. Then the posterior probability of the n-th indicator variable utilized in Gibbs sampling is

$$P(z_{nk} = 1 | Z_{-n}, D, \Theta) = \frac{P(z_{nk} = 1; Z_{-n}, D | \Theta)}{\sum_{j=1}^{K} P\left(z_{nj} = 1; Z_{-n}, D \big| \Theta \right)},$$

where $P(z_{nk} = 1; Z_{-n}, D | \Theta) \propto \pi_k p(\mathbf{x}_n | \mu_k, \Sigma_k) \, p\left(\mathbf{y}_{+n,k} \big| \mathbf{X}_{+n,k}, \theta_k \right) \big/ p\left(\mathbf{y}_{-n,k} \big| \mathbf{X}_{-n,k}, \theta_k \right)$, $\mathbf{y}_{-n,k} = \left[y_j \big| z_{jk} = 1 \text{ and } j \neq n \right]$ and $\mathbf{y}_{+n,k} = \left[y_j \big| z_{jk} = 1 \text{ or } j = n \right]$ are vectors of y_j, and $\mathbf{X}_{-n,k}$ and $\mathbf{X}_{+n,k}$ are the same meaning as $\mathbf{y}_{-n,k}$ and $\mathbf{y}_{+n,k}$, respectively.

Details of M-step: $\hat{Q}$-function, a Monte Carlo estimator of Q-function, is given by

$$\hat{Q}\left(\Theta, \hat{Z}^{(r)}\right) = \frac{1}{I^{(r)}} \sum_{i=1}^{I^{(r)}} \sum_{k=1}^{K} \left\{ \sum_{n=1}^{N} z_{nk}^{(ri)} \left[\ln \pi_k + \ln p\left(\mathbf{x}_n | \mu_k, \Sigma_k\right) \right] + \ln p\left(\mathbf{y}_k^{(ri)} | \mathbf{X}_k^{(ri)}, \theta_k\right) \right\},$$

where $\mathbf{y}_k = [y_n | z_{nk} = 1, n = 1 \sim N]$ is a vector of y_n and $\mathbf{X}_k$ is the same meaning as $\mathbf{y}_k$.

We calculate the $\pi_k^{(r)}$ by Lagrange's method of multipliers and compute $\mu_k^{(r)}$ and $\Sigma_k^{(r)}$ by taking the derivatives to be zero

$$\pi_k^{(r)} = \frac{N_k^{(r)}}{\sum_{k=1}^{K} N_k^{(r)}}, \quad \mu_k^{(r)} = \frac{1}{N_k^{(r)}} \sum_{i=1}^{I^{(r)}} \sum_{n=1}^{N} z_{nk}^{(ri)} \mathbf{x}_n,$$

$$\Sigma_k^{(r)} = \frac{1}{N_k^{(r)}} \sum_{i=1}^{I^{(r)}} \sum_{n=1}^{N} z_{nk}^{(ri)} \left(\mathbf{x}_n - \mu_k^{(r)}\right)\left(\mathbf{x}_n - \mu_k^{(r)}\right)^T,$$

where $N_k^{(r)} = \sum_{i=1}^{I^{(r)}} \sum_{n=1}^{N} z_{nk}^{(ri)}$. The parameter $\theta_k^{(r)}$ is computed by gradient method [4].

3.2 On Prediction

The purpose of prediction is to predict the output at a new input $\mathbf{x}^*$. After the parameters Θ and simulated samples of indicator variables $\hat{Z}$ are estimated by the MCMC based EM algorithm, we calculate

$$\alpha_k = p\left(z_k^* = 1 \big| \mathbf{x}^*, \hat{\Theta}\right) = \frac{\hat{\pi}_k p\left(\mathbf{x}^* | \hat{\mu}_k, \hat{\Sigma}_k\right)}{\sum_{k=1}^{K} \hat{\pi}_k p\left(\mathbf{x}^* | \hat{\mu}_k, \hat{\Sigma}_k\right)},$$

where z_k^* denotes the indicator variable of the point to be predicted.

If the point to be predicted belongs to the k-th component, the predictive output and its variance with the i-th simulated sample of indicator variables are given by

$$\hat{y}_k^{(i)} = \mathbf{C}\left(\mathbf{x}^*, \hat{\mathbf{X}}_k^{(i)}; \hat{\theta}_k\right) \mathbf{C}\left(\hat{\mathbf{X}}_k^{(i)}, \hat{\mathbf{X}}_k^{(i)}; \hat{\theta}_k\right)^{-1} \hat{\mathbf{y}}_k^{(i)},$$

$$\hat{\delta}_k^{(i)} = C\left(\mathbf{x}^*, \mathbf{x}^*; \hat{\theta}_k\right) - \mathbf{C}\left(\mathbf{x}^*, \hat{\mathbf{X}}_k^{(i)}; \hat{\theta}_k\right) \mathbf{C}\left(\hat{\mathbf{X}}_k^{(i)}, \hat{\mathbf{X}}_k^{(i)}; \hat{\theta}_k\right)^{-1} \mathbf{C}\left(\mathbf{x}^*, \hat{\mathbf{X}}_k^{(i)}; \hat{\theta}_k\right)^T,$$

where $\mathbf{C}\left(\mathbf{x}^*, \hat{\mathbf{X}}_k; \hat{\theta}_k\right) = \left[C\left(\mathbf{x}^*, \mathbf{x}_n; \hat{\theta}_k\right) \big| z_{nk} = 1, n = 1 \sim N \right]$ represents a row vector of $C\left(\mathbf{x}^*, \mathbf{x}_n; \hat{\theta}_k\right)$. Therefore, the overall predictive output and its variance are obtained.

$$\hat{y} = \frac{1}{\tilde{I}} \sum_{i=1}^{\tilde{I}} \sum_{k=1}^{K} \alpha_k \hat{y}_k^{(i)}, \quad \hat{\delta} = \frac{1}{\tilde{I}} \sum_{i=1}^{\tilde{I}} \sum_{k=1}^{K} \alpha_k \left[\hat{\delta}_k^{(i)} + \left(\hat{y}_k^{(i)}\right)^2 \right] - \hat{y}^2,$$

where $\tilde{I}$ is the number of simulated samples.

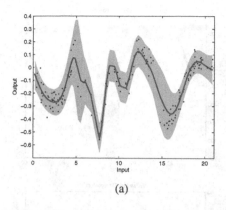
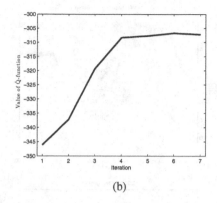

(a)   (b)

Fig. 1. (a) The plot of a simulated dataset: blue points are test samples, red line is prediction and shaded area is its 95% confidence band (b) The values of $\hat{Q}$-function with the iterations of the MCMC based EM algorithm

Table 1. The true value (TV), estimated value (EV) and relative error (RE) of the parameters for the simulated dataset

		π_k	μ_k	Σ_k	θ_k
$g=1$	TV	0.3333	2.5000	1.4414	[0.5000, 0.1000, 0.0025]
	EV	0.3281	2.5683	1.4980	[0.4676, 0.0578, 0.0026]
	RE	1.6%	2.7%	3.9%	[6.5%, 42.2%, 4.0%]
$g=2$	TV	0.3333	10.0000	5.7937	[1.0000, 0.2500, 0.0025]
	EV	0.3320	9.8433	5.7716	[1.1483, 0.2175, 0.0020]
	RE	0.4%	1.6%	0.4%	[14.8%, 13.0%, 20.0%]
$g=3$	TV	0.3333	18.0000	2.0598	[0.2000, 0.0500, 0.0025]
	EV	0.3398	17.8143	2.0349	[0.2437, 0.0303, 0.0028]
	RE	2.0%	1.0%	1.2%	[21.9%, 39.4%, 12.0%]

4   Experimental Results

In this section, we conduct the experiments of the MCMC based EM algorithm on a simulated dataset generated from an MGP model, a toy dataset given in [4] and a motor-cycle dataset given in [9] and also make certain comparison with the other competitive EM algorithms.

4.1   On the Simulated Dataset

We conduct an experiment on the simulated dataset generated by an MGP model of 1-dimensional input without sigular points. There are 192 training samples and 192 test samples of three components, i.e., $K = 3$. In fig.1(a), the test samples and prediction results are illustrated and we show the values of $\hat{Q}$-function with the iteration of our EM algorithm in Fig.1(b). It can be seen that the MCMC based EM algorithm is effective since the value of $\hat{Q}$-function always increases with the iteration. The indicator variables is calculated by maximum posterior probability criterion and the average classification

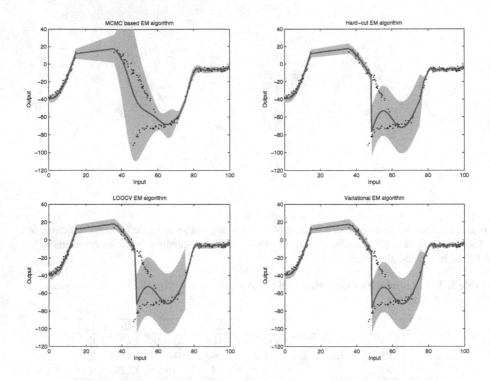

Fig. 2. The plot of a toy dataset: blue points are test samples, red line is prediction and shaded area is its 95% confidence band

accuracy rate is 98.9%. The estimations of the parameters are shown in Table 1 and we could observe that all relative errors are small expect the θ_k. The MGP model is effective on prediction with imprecise θ_k and the root mean square error (RMSE) of the prediction on test samples is 0.0653.

4.2 On Toy and Motorcycle Datasets

We compare MCMC based EM algorithm with hard-cut EM algorithm [3], LOOCV EM algorithm [4] and variational EM algorithm [4] on a toy dataset and a motorcycle dataset, respectively.

Table 2. The RMSEs of the predictions with the four algorithms on the two datasets

	Toy dataset	Motorcycle dataset
MCMC based EM algorithm	13.94	24.20
Hard-cut EM algorithm	16.81	25.31
LOOCV EM algorithm	16.94	25.85
Variational EM algorithm	16.91	29.69

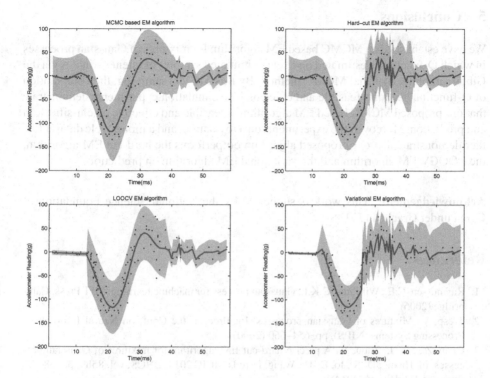

Fig. 3. The plot of motorcycle dataset: blue points are 133 samples, red line is prediction and shaded area is its 95% confidence band

The toy dataset [4] is a simulated dataset generated from four functions, i.e.,

$$
\begin{aligned}
f_1(x) &= 0.25x^2 - 40 + \sqrt{7}\varepsilon, \quad x \sim U(0, 15), \\
f_2(x) &= -0.0625(x - 18)^2 + 0.5x + 20 + \sqrt{7}\varepsilon, \quad x \sim U(35, 60), \\
f_3(x) &= 0.008(x - 60)^3 - 70 + \sqrt{4}\varepsilon, \quad x \sim U(45, 80), \\
f_4(x) &= -\sin(x) - 6 + \sqrt{2}\varepsilon, \quad x \sim U(80, 100),
\end{aligned}
\tag{3}
$$

where $\varepsilon \sim N(0, 1)$. We generate 50 training samples and 50 test samples from each function in Eq. (3) and utilize an MGP model with four components to learn this model. The motorcycle dataset [7] is a real world dataset, where 133 points of three components are observed by accelerometer readings. The predictions of four algorithms on the two datasets are shown in Fig. 2 and Fig. 3, respectively. The 200 blue points in Fig.2 are all test points and the blue points in Fig.3 are 100 training points and 33 test points. The predictions of MCMC based EM algorithm are better than those of the other three algorithms. The RMSEs of prediction are shown in Table 2 and the RMSEs of MCMC based EM algorithm are smaller than those of the other three algorithms on the two datasets.

5   Conclusions

We have established an MCMC based EM algorithm for mixtures of Gaussian processes in which Q-function is estimated on a set of simulated samples of latent variables via the Gibbs sampling, a special MCMC method. By this MCMC sampling, the computation of Q-function becomes feasible and efficient. The simulation experiments demonstrate that our proposed MCMC based EM algorithm is feasible and effective on classification and prediction. Moreover, the experiments on a toy dataset and a motorcycle dataset further demonstrate that our proposed algorithm outperforms the hard-cut EM algorithm, the LOOCV EM algorithm and the variational EM algorithm on prediction.

Acknowledgement. This work is supported by the National Science Foundation of China under Grant 61171138.

References

1. Rasmussen, C.E., Williams, C.K.I.: Gaussian process for machine learning. MIT Press, Cambridge (2006)
2. Tresp, V.: Mixtures of Gaussian processes. In: Proc. of the Conf. on Neural Information Processing Systems (NIPS), pp. 654–660 (2000)
3. Chen, Z., Ma, J., Zhou, Y.: A precise hard-cut EM algorithm for mixtures of Gaussian processes. In: Huang, D.-S., Jo, K.-H., Wang, L. (eds.) ICIC 2014. LNCS, vol. 8589, pp. 68–75. Springer, Heidelberg (2014)
4. Yang, Y., Ma, J.: An efficient EM approach to parameter learning of the mixture of Gaussian processes. In: Liu, D., Zhang, H., Polycarpou, M., Alippi, C., He, H. (eds.) ISNN 2011, Part II. LNCS, vol. 6676, pp. 165–174. Springer, Heidelberg (2011)
5. Nguyen, T., Bonilla, E.: Fast allocation of Gaussian process experts. In: Proceedings of the 31st International Conference on Machine Learning (ICML), pp. 145–153 (2014)
6. Sun, S., Xu, X.: Variational inference for infinite mixtures of Gaussian processes with applications to traffic flow prediction. IEEE Transactions on Intelligent Transportation Systems 12(2), 466–475 (2011)
7. Yuan, C., Neubauer, C.: Variational mixture of Gaussian process experts. In: Advances in Neural Information Processing Systems, vol. 21, pp. 1897–1904 (2008)
8. Lazaro-Gredilla, M., Vaerenbergh, S.V., Lawrence, N.D.: Overlapping mixtures of Gaussian processes for the data association problem. Pattern Recognition 45, 1386–1395 (2012)
9. Rasmussen, C.E., Ghahramani, Z.: Infinite mixture of Gaussian process experts. In: Advances in Neural Information Processing Systems, vol. 2, pp. 881–888 (2002)
10. Tayal, A., Poupart, P., Li, Y.: Hierarchical double Dirichlet process mixture of Gaussian processes. Association for the Advancement of Artificial Intelligence (2012)
11. Meeds, E., Osindero, S.: An alternative infinite mixture of Gaussian process experts. In: Advances in Neural Information Processing Systems, vol. 18, pp. 883–890 (2006)
12. Sun, S.: Infinite mixtures of multivariate Gaussian processes. In: Proceedings of the International Conference on Machine Learning and Cybernetics, pp. 1–6 (2013)

Automatic Model Selection of the Mixtures of Gaussian Processes for Regression

Zhe Qiang and Jinwen Ma

Department of Information Science, School of Mathematical Sciences and LMAM
Peking University, Beijing, 100871, China
jwma@math.pku.edu.cn

Abstract. For the learning of mixtures of Gaussian processes, model selection is an important but difficult problem. In this paper, we develop an automatic model selection algorithm for mixtures of Gaussian processes in the light of the reversible jump Markov chain Monte Carlo framework for Gaussian mixtures. In this way, the component number and the parameters are updated according the five types of random moves and model selection can be made automatically. The key idea is that the moves of component splitting or merging preserve the zeroth, first and second moments of the components so that the covariance parameters of the new components can be related to the origin ones. It is demonstrated by the simulation experiments that this automatic model selection algorithm is feasible and effective.

Keywords: Mixtures of Gaussian processes, Reversible jump MCMC, Model selection, Regression, Split and merge moves.

1 Introduction

As a powerful statistical learning tool, Gaussian Process (GP) is widely used in machine learning and pattern recognition [1]. Since a single GP model cannot deal with the multimodality dataset, the Mixture of Gaussian Processes (MGP) [2] have been developed to model a general multimodality dataset. Obviously, MGP can also overcome a major disadvantage of the Gaussian process modeling that calculating the inversion of an $N \times N$ covariance matrix requires the time complexity of $O(N^3)$ for a training dataset with N points. Structurally, Shi et al. [3],[4] considered MGP as a hierarchical model and fit the data in two levels. Moreover, they proposed the hybrid Markove Chain Monte-Carlo algorithm to train the covariance parameters and then to utilize BIC to determine the number K of GP components in the mixture. However, BIC is not so effective for the model selection on the mixtures of Gaussian processes.

On the other hand, Green [5] proposed the reversible jump Markov chain Monte Carlo (RJMCMC) framework to determine the dimension of parameters through the Markov chain Monte Carlo simulation. Later on, according to this theory, Richardson and Green [6] developed a RJMCMC approach to decide the number of actual components in the Gaussian mixture. Although this approach

© Springer International Publishing Switzerland 2015
X. Hu et al. (Eds.): ISNN 2015, LNCS 9377, pp. 335–344, 2015.
DOI: 10.1007/978-3-319-25393-0_37

is effective for Gaussian mixtures, its idea can also used to design the learning algorithm for MGPs and solve the model selection problem in this case. However, the structure of the likelihood function of MGP is quite different from that of Gaussian mixture so that the split and merge moves in the algorithm cannot be implemented directly.

In this paper, we develop an automatic model selection algorithm for MGPs in the light of the RJMCMC framework by making the split and merge moves feasible and effective. As we find, the difficulty in the split moves focuses on that the overall dispersion should keep relatively constant when a covariance matrix is split into two matrices. In order to solve this difficulty, we can keep the first two moments of the components to be same during a split or merge move. By mathematics analysis, we find that once the sampling interval is set small enough, these moments remains almost constant. In this way, our automatic model selection algorithm can be constructed effectively. It is demonstrated by simulation experiments that this automatic model selection algorithm is feasible and effective in the same way as the RJMCMC algorithm for Gaussian mixtures.

The rest of the paper is organised as follows. In Section 2, we revisit the mixture of Gaussian processes and give the Bayesian forms of the priors of the parameters. Section 3 presents the automatic model selection algorithm with five move types. Simulation experiments are conducted in Section 4. We conclude briefly in Section 5.

2   The Hierarchical Mixtures of Gaussian Processes

2.1   The Basic Model

For clarity, we consider MGP as the hierarchical mixture of Gaussian processes described Shi et al. [3]. Clearly, it tries to model a large dataset with groups of repeated measurements. Actually, the lower-level model fits the measurements in the same group, while the higher-level model tries to describe the heterogeneity among different groups. For example, the dataset used in [4] is a set of repeated standing-up trajectories corresponding to different paraplegic patients. For this case, the lower-level model fitted those standing-up trajectories of paraplegic patients in the same type, while the higher-level model described the heterogeneity among different types of paraplegic patients.

Mathematically, in the higher-level model, we let $\{(x_{mn}, y_{mn}), m = 1, \cdots, M\}$ be M curves belong to a number K of Gaussian processes, which can represent all the points on the training curves. The mixture model can be given by

$$\boldsymbol{y}_m \sim \sum_{k=1}^{K} \pi_k GP(\boldsymbol{\theta}_k). \tag{1}$$

In the lower-level model, each curve is assumed to be a function $f_m(\cdot)$ plus a white noise, that is,

$$y_{mn} = f_m(x_{mn}) + \epsilon_{mn},$$

where $\epsilon_{mn} \sim \mathcal{N}(0, \sigma_k^2)$. The m-th curve follows a Gaussian process if any finite subsequence or subset of the curve, say $\mathcal{D}_m = \{\boldsymbol{X}_m, \boldsymbol{Y}_m\} = \{(x_{mn}, y_{mn}), n = $

$1, \cdots, N_m\}$, follows a Gaussian distribution if $\mathbf{Y}_m \sim \mathcal{N}(0, \mathbf{\Sigma}(\mathbf{X}_m, \mathbf{X}_m; \theta_k))$, where

$$\mathbf{\Sigma}(x_{mi}, x_{mj}; \theta_k) = C(x_{mi}, x_{mj}; \theta_k) + \sigma_k^2 \delta_{ij},$$

where

$$C(x_{mi}, x_{mj}; \theta_k) = v_k \exp\left(-\frac{1}{2} w_k (x_{mi} - x_{mj})^2\right).$$

So, all the covariance parameters are $\theta_k = (w_k, v_k, \sigma_k^2)$.

For a test input x^*, if we assume it is on the m-th curve, and the m-th curve is belong to the k-th component, i.e., Gaussian process, the mean and variance of its prediction can be obtained as follows:

$$E[f_m(x^*)|\mathcal{D}_m] = \sigma^T(x^*)\Sigma^{-1}(\mathbf{X}_m, \mathbf{X}_m; \theta_k)\mathbf{Y}_m;$$
$$Var[f_m(x^*|\mathcal{D}_m)] = C(x^*, x^*) - \sigma^T(x^*)\Sigma^{-1}(\mathbf{X}_m, \mathbf{X}_m; \theta_k)\sigma(x^*),$$

where $\sigma(x^*) = (C(x^*, x_{m,1}), \cdots, C(x^*, x_{m,N_m}))^T$.

For the parameter learning of this hierarchical mixture model of Gaussian process, we introduce the latent variables z_m as follows:

$$f_m(\mathbf{X}_m)|z_m = k \sim GP(\theta_k),$$

and assume the mixing form of Eq.(1).

2.2   The Priors and its Bayesian Form

In the hierarchical mixtures of Gaussian processes, we adopt the forms of the priors as given in [4], that is,

$$w_k \sim I\Gamma(\frac{1}{2}, \frac{1}{2}), v_k \sim \mathcal{LN}(-1, 1^2), \sigma_k^2 \sim \mathcal{LN}(-3, 3^2), k = 1, \cdots, K$$

where $I\Gamma$ represents the inverse gamma distribution, and

$$(\pi_1, \cdots, \pi_K) \sim Dir(1, \cdots, 1).$$

In the Bayesian analysis, the priors are as important as the posteriors and likelihood function. Supposing that $\Theta = (\theta_1, \ldots, \theta_K)$, $\pi = (\pi_1, \ldots, \pi_K)$, $\mathcal{D}$ is the set of training data, we then have

(i) The posterior of the parameters:
$$p(\Theta, \pi|\mathcal{D}) \propto p(\Theta, \pi)L(\mathcal{D}|\Theta, \pi);$$

(ii) The likelihood of the mixture model:
$$L(\mathcal{D}|\Theta, \pi) = \prod_{m=1}^{M} \sum_{k=1}^{K} \pi_k \mathcal{N}(\mathbf{y}_m|0, \Sigma_k(\mathbf{x}_m));$$

(iii) The prior distribution:
$$p(\Theta, \pi) = p(\pi) \prod_{k=1}^{K} p(\theta_k).$$

In the next section, we will sample parameters from $p(\Theta, \pi|\mathcal{D})$.

3   The Markov Chain Monte Carlo Algorithm for the Hierarchical Mixtures of Gaussian Processes

In this section, we construct the Markov Chain Monte Carlo (MCMC) algorithm for the hierarchical mixtures of Gaussian processes in a similar way as the MCMC algorithm for Gaussian mixtures in [6]. Actually, the mathematical framework of our MCMC algorithm keeps the same, including a number of moves of the components or their parameters with time, but the components become Gaussian processes.

3.1   The Move Types

In [6], six types of moves were used. But here, since the adopted priors have no hyperparameters, there is no need for updating the hyperparameters. So, remaining five types of moves can be given as follows:

(a) $\boldsymbol{\pi} = (\pi_1, \cdots, \pi_K)$;
(b) $\boldsymbol{\Theta} = (\boldsymbol{\theta}_1, \cdots, \boldsymbol{\theta}_K)$, where $\boldsymbol{\theta}_k = (w_k, v_k, \sigma_k^2)$;
(c) $\boldsymbol{z} = (z_1, \cdots, z_M)$;
(d) Split or merge;
(e) Remove empty components.

For clarity, we refer to one process of implementing these five moves completely as a sweep, being set as the basic step of the MCMC algorithm.

3.2   The Moves with the Component Number Fixed

The moves with fixed component number include the first three steps(a)(b)(c) in section 3.1. For this part, we adopt the algorithm in [4]:

For $\boldsymbol{\pi}$ and $\boldsymbol{z}$, we use Gibbs sampling:

(i) sample z_m from $(z_m = k | \mathcal{D}, \boldsymbol{\Theta}, \boldsymbol{\pi}) \propto \pi_k p(\boldsymbol{y}_m | \boldsymbol{\theta}_k), m = 1, \cdots, M, k = 1, \cdots, K$

(ii) sample $\boldsymbol{\pi}$ from $(\pi_1, \cdots, \pi_K) | \boldsymbol{z} \sim Dir(1 + c_1, \cdots, 1 + c_K)$

Here, c_k represents the element number in the set $\{m = 1, \cdots, M | z_m = k\}$ and $Dir(\cdot)$ represents Dirichlet distribution.

And for $\boldsymbol{\Theta}$, we sample it from its posterior:

$$p(\boldsymbol{\Theta} | \mathcal{D}, \boldsymbol{z}) \propto \prod_{k=1}^{K} p(\boldsymbol{\theta}_k | \mathcal{D}_m, \boldsymbol{z})$$

Then, the posterior of $\boldsymbol{\theta}_k, k = 1, \cdots, K$ are independent with each other and we can sample each $\boldsymbol{\theta}_k$ separately. Here we adopt Hybrid Monte Carlo or Hamiltonian Monte Carlo to sample $\boldsymbol{\theta}_k$. Actually, this sampling method used to be adopted to simulate a physical system where a puck moves up and down along with a smooth surface and the total energy remains constant. However, in our application, we restate the potential energy as $\mathcal{E}(\boldsymbol{\theta}_k) = -\log p(\boldsymbol{\theta}_k | \mathcal{D}_m, \boldsymbol{z})$ and the kinetic energy as: $\mathcal{K}(\phi_k) = \frac{1}{2}\sum \phi_{k,i}^2$, $\phi_{k,i} \sim \mathcal{N}(0,1)$, $\phi_k = (\phi_{k,1}, \phi_{k,2}, \phi_{k,3})$. Then the total energy of the hamiltonian system for θ_k is $\mathcal{H}(\theta_k, \phi_k) = \mathcal{E}(\theta_k) + \mathcal{K}(\phi_k)$.

For the convenience of calculation, we split $\mathcal{H}(\theta_k, \phi_k)$ as follows:

$$\mathcal{H}(q,p) = \underbrace{-\frac{1}{2}\sum_{m\in\{z_m=k\}}\log p(\boldsymbol{y}_m|\theta_k)}_{\frac{U_0}{2}} + \underbrace{[-\log p(\theta_k) + \mathcal{K}(\phi_k)]}_{U_1} \underbrace{-\frac{1}{2}\sum_{m\in\{z_m=k\}}\log p(\boldsymbol{y}_m|\theta_k)}_{\frac{U_0}{2}}$$

thus the sample update step is:

(i) From the current state (θ_k, ϕ_k), we use a leapfrog step with step size ϵ to calculate the new state (θ_k^*, ϕ_k^*):

$$\phi_k^* = \phi_k - \frac{\epsilon}{2}\frac{\partial U_0}{\partial\theta_k}(\theta_k)$$

$$\theta_k^* = \theta_k + \epsilon\phi_k^*$$

$$\phi_k^* = \phi_k^* - \frac{\epsilon}{2}\frac{\partial U_1}{\partial\theta_k}(\theta_k^*, \phi_k^*)$$

$$\phi_k^* = \phi_k^* - \frac{\epsilon}{2}\frac{\partial U_0}{\partial\theta_k}(\theta_k^*)$$

(ii) Then new current state is:

$$(\theta_k^*, \phi_k^*) = \begin{cases} (\theta_k^*, \phi_k^*) & prob = \min(1, \exp(-\mathcal{H}(\theta_k^*, \phi_k^*) + \mathcal{H}(\theta_k, \phi_k))) \\ (\theta_k, -\phi_k) & otherwise \end{cases}$$

(iii) Finally, generate $\nu_{k,i} \sim \mathcal{N}(0,1)$, and update ϕ_k as: $\phi_k^* = \alpha\phi_k^* + \sqrt{1-\alpha^2}\nu_k$.

In addition, according to the advise in [7], we set $\epsilon = 0.5N_m^{-1/2}$, $\alpha = 0.95$.

For $k = 1, \cdots, K$, by repeating (i)(ii)(iii) $n = 20$ times, we finish one sweep of step(b). Note that $w_k, v_k, \sigma_k^2 > 0$, through the analysis of the method of handling constraints in [8], we reject the current state when $w_k, v_k, \sigma_k^2 < 0$

3.3 The Moves with the Component Number Changed

The moves with the component number changed include the last two steps(d)(e). For move(e), all the components whose $\pi_k < 1\%$ are considered to be empty components and we delete them directly.

The Detailed Balance Framework. For move(d), we construct a detailed balance framework for the move with the component number changed so that the covariance parameters of the new components can be related to the original ones. The key to success is that the first two moments remain almost constant. For convenience we shall denote the splitted component as k^*th component and the two new components as k_1, k_2.

(I) Actually, for the zeroth moment of $\boldsymbol{y}_m$, by simple mathematics calculation we find it is $\sum_{k=1}^K \pi_k$, thus keeping the zeroth moment constant means $\pi_{k^*} = \pi_{k_1} + \pi_{k_2}$.

(II) In the case of the first moment, it is $\boldsymbol{0}$ since we assume that the Gaussian process is zero mean. Thus its first moment always keep constant.

(III) However, the second moment is somewhat complicated and it is $\sum_{k=1}^{K} \pi_k \Sigma_k$. Then keeping the second moment constant means $\pi_{k^*} \Sigma_{k^*} = \pi_{k_1} \Sigma_{k_1} + \pi_{k_2} \Sigma_{k_2}$.

(i) For the diagonal entry, let $i = j$, then we get a balance formula:
$$\pi_{k^*}(v_{k^*} + \sigma_{k^*}^2) = \pi_{k_1}(v_{k_1} + \sigma_{k_1}^2) + \pi_{k_2}(v_{k_2} + \sigma_{k_2}^2)$$

(ii) For the off diagonal entry, let $i \neq j$, we have:
$$\pi_{k^*}(v_{k^*} \exp\left(-\frac{w_{k^*}(x_i - x_j)^2}{2}\right)) = \pi_{k_1}(v_{k_1} \exp\left(-\frac{w_{k_1}(x_i - x_j)^2}{2}\right))$$
$$+ \pi_{k_2}(v_{k_2} \exp\left(-\frac{w_{k_2}(x_i - x_j)^2}{2}\right)) \quad (2)$$

For convenience, denote Eq.(2) as a function of $(x_i - x_j)^2$:
$$f(x) := \pi_{k^*}(v_{k^*} \exp\left(-\frac{w_{k^*}}{2}x\right)) - \pi_{k_1}(v_{k_1} \exp\left(-\frac{w_{k_1}}{2}x\right))$$
$$- \pi_{k_2}(v_{k_2} \exp\left(-\frac{w_{k_2}}{2}x\right)) \equiv 0 \quad (3)$$

In fact, $f(x)$ is exponential decreasing about x and x is in $\{\epsilon^2, 2^2\epsilon^2, \cdots, N_m^2\epsilon^2\}$, thus we can conclude that when x gets the minimum ϵ^2, $f(x)$ reaches the maximum. Therefore, in order to keep $f(x) \approx 0$ we just need keep $f(\epsilon^2) \approx 0$. Do Taylor's expansion of $f(x)$ at $x = 0$ and denote the n-th term as a_n for convenience. What surprise us is that when $x = \epsilon^2$ and if ϵ is small enough, a_n will be monotonic decreasing, then a_0 will be the largest term and a_1 will be the second largest term. Therefore just let a_0, a_1 be zero, the second moment will keep almost constant.

Then, we have got our detailed balance framework:
$$\pi_{k^*} = \pi_{k_1} + \pi_{k_2} \tag{4a}$$
$$\pi_{k^*}\sigma_{k^*}^2 = \pi_{k_1}\sigma_{k_1}^2 + \pi_{k_2}\sigma_{k_2}^2 \tag{4b}$$
$$\pi_{k^*}v_{k^*} = \pi_{k_1}v_{k_1} + \pi_{k_2}v_{k_2} \tag{4c}$$
$$\pi_{k^*}v_{k^*}w_{k^*} = \pi_{k_1}v_{k_1}w_{k_1} + \pi_{k_2}v_{k_2}w_{k_2} \tag{4d}$$

The Merge and Split Formula. For merge move, we can use Eq.(4) to merge two components, so Eq.(4) is also our merge formula. For split move, reversible jump in [5] is needed. According to the theory above, 4 dimensions random variable $\boldsymbol{u} = (u_1, u_2, u_3, u_4)$ need to be generated to decide these new parameters: $u_i \sim \boldsymbol{Beta}(2,2), i = 1, \cdots, 4$. By combining these random parameters, the balance formula Eq.(4) and the reversible jump theory, we can write the split formula as:
$$\pi_{k_1} = u_1\pi_{k^*}, \quad \pi_{k_2} = (1 - u_1)\pi_{k^*}, u_1 \in (0, 1) \tag{5a}$$
$$\sigma_{k_1}^2 = u_2\sigma_{k^*}^2 \frac{\pi_{k^*}}{\pi_{k_1}}, \quad \sigma_{k_2}^2 = (1 - u_2)\sigma_{k^*}^2 \frac{\pi_{k^*}}{\pi_{k_2}}, u_2 \in (0, 1) \tag{5b}$$
$$v_{k_1} = u_3v_{k^*} \frac{\pi_{k^*}}{\pi_{k_1}}, \quad v_{k_2} = (1 - u_3)v_0^{k^*} \frac{\pi_{k^*}}{\pi_{k_2}}, u_3 \in (0, 1) \tag{5c}$$
$$w_{k_1} = \frac{1 - u_4}{u_3}w_{k^*}, \quad w_{k_2} = \frac{u_4}{1 - u_3}w_{k^*}, u_4 \in (0, 1) \tag{5d}$$

It is easy to validate that split formula Eq.(5) matches with merge formula Eq.(4).

The split and merge move is a Markov birth-death chain, and we set the split probability and merge probability as $b_k, d_k = 1 - b_k$ respectively, depending on k: $d_1 = 0, b_{k_{max}} = 0, b_k = d_k = 0.5, \forall k = 2, \cdots, k_{max} - 1$, where k_{max} is the maximum component number that we set according to individual cases. Then based on the acceptance probability formula of reversible jump move in [5] the acceptance ratio for a split move is $\min(1, A)$ and a merge move is $\min(1, A^{-1})$ where

$$A = \prod_{m=1}^{M} \frac{l(\boldsymbol{Y}_m|\boldsymbol{\theta}_{k+1})}{l(\boldsymbol{Y}_m|\boldsymbol{\theta}_k)} \times \frac{d_{k+1}}{b_k} \times \underbrace{\frac{p(\boldsymbol{\theta}_{k+1})}{p(\boldsymbol{\theta}_k)}}_{T_1} \times \underbrace{\frac{1}{Beta(u|\boldsymbol{\theta}_{k+1},\boldsymbol{\theta}_k)}}_{T_2} \times \underbrace{\left|\frac{\partial \boldsymbol{\theta}_{k+1}}{\partial(\boldsymbol{\theta}_k, \boldsymbol{u})}\right|}_{T_3}$$

Here, k^*, k_1, k_2 are chosen randomly from the K components.

4 Experimental Results

Our experiments are conducted on a set of simulated data generated from a mixture of three Gaussian processes with $\theta_1 = (1.0, 0.2, 0.0025), \theta_2 = (0.5, 1.0, 0.001)$, $\theta_3 = (10, 0.2, 0.0005)$, respectively, each with 3 curves. The data points are with $t = -4 : 0.08 : 4$, being equally spaced. We have two kinds of prediction, type I prediction: choosing half data randomly from each of the 9 curves as training data, the rest as test data; type II prediction: generating a new curve from the first GP, choosing half data randomly on this curve as known data, using this half and training parameters from the type I prediction to simulate the other unknown half data. In the later two subsections, the two type of prediction is used to verify the component number fixed algorithm and the component number varied algorithm for regression and model selection. For both of the two algorithms, we run them for 20000 sweep. Here we discard the first 10000 iterations and select one sample from each 200 iterations, in order to have approximately independent draws.

4.1 The Component Number Fixed

First, we fix $k = 3$, and use only the moves with the component number fixed for training. The log-likelihood tends to stabilize after about 1700 iteration. Table (1) presents the predictions. We will compare it with the component number varied result in the next subsection.

4.2 The Component Number Varied

Then we use our component number varied method to train and predict the same curves as before. Fig.(1) is $p(k < j|\mathcal{D}), k = 1, \cdots, 6$, from which we can conclude

Table 1. RMSE and correlation coefficient(r) between true and predicted responses

Training data:half data on the first 9 curves		
model: fixed component number of GP mixture regression model		
Test data	RMSE	r
the first GP	0.2329	0.9485
the second GP	0.3612	0.8520
the third GP	0.2266	0.9024
the 10th curve	0.2037	0.7566
Training data:half data on the first 9 curves		
model: varied component number of GP mixture regression model		
Test data	RMSE	r
the first GP	0.0602	0.9900
the second GP	0.0253	0.9982
the third GP	0.0645	0.9867
the 10th curve	0.0493	0.9820

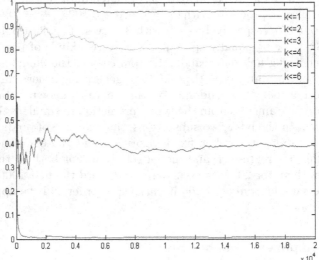

Fig. 1. $p(k < j | \mathcal{D}), j = 1, \cdots, 6$ for 40000 iterations

Table 2. posterior of k for Gaussian process mixture model

curve number	$p(k|\mathcal{D})$	proportion (%) of moves accepted	
		split	merge
9	p(1)=0.0001 p(2) = 0.0563 p(3)=0.3811	14	2
	p(4)=0.3677 p(5) = 0.1558 $\sum_{k>6} p(k) = 0.0391$		

that the algorithm has converged after about 10000 iteration. Additionally, we also present the posterior of k for mixtures of Gaussian processes in Table (2). From this table, we can conclude this model favors $3 - 4$ components. In this

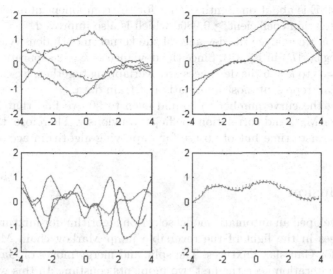

Fig. 2. predict curve with the component number varied method on 9 curves, the solid line represent the real value, and the dash line represents the predict line with 95% confidence interval

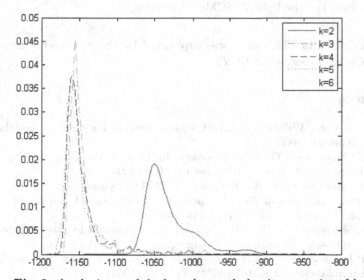

Fig. 3. the deviance of the kernel smooth density posterior of k

part, we initially add the birth and death moves in [6] to the algorithm in order to increase the opportunities for split and merge thus to speed up convergence of Markov chain, but finally it doesn't work, and we delete this type move.

The final predict results are presented in Fig.(2). We can see that predict curves almost overlap with the true curves. For type II prediction, the 95%

confidence interval is quite small. RMSE and correlation coefficient are in Table (1). The RMSE is about one tenth of that for the component number fixed case and the correlation coefficient ≥ 0.9800 which is also improved.

In addition, we analyze the deviance of the kernel smooth density of posterior of $k --2 * \log p(k|\mathcal{D})$ in Fig.(3). Since the deviance of $k = 2$ separate with $k = 3$ and from $k = 3$ to $k = 6$ the deviances are overlapping together, we can conclude that $k = 3$ can represent most information of train data.

Increasing the curve number to 30 and even to 90, we find that it does not improve the RMSE and correlation coefficient obviously. Therefore, this become a burden to waste time but of no use for improving algorithm accuracy and is unnecessary.

5   Conclusion

We have developed an automatic model selection algorithm for mixtures of Gaussian processes in the light of the reversible jump Markov chain Monte Carlo framework for Gaussian mixtures. The split and merge moves of Gaussian processes in the iteration keep the first two moments constant. In this way, the automatic model selection algorithm makes it possible to do the Bayesian analysis of both parameter estimation and model selection for the mixtures of Gaussian processes. Moreover, it is demonstrated that this developed algorithm is feasible and outperform the the hybrid MCMC algorithm.

Acknowledgments. This work was supported by the National Science Foundation of China under Grant 61171138.

References

1. Rasmussen, C.E., Williams, C.K.I.: Gaussian process for machine learning. MIT Press, Cambridge (2006)
2. Tresp, V.: Mixtures of Gaussian processes. In: Proc. of the Conf. on Neural Information Processing Systems (NIPS), pp. 654–660 (2000)
3. Shi, J.Q., Murray-Smith, R., Titterington, D.M.: Bayesian regression and classification using mixtures of Gaussian processes. International Journal of Adaptive Control and Signal Processing 17(2), 149–161 (2003)
4. Shi, J.Q., Murray-Smith, R., Titterington, D.M.: Hierarchical Gaussian process mixtures for regression. Statistics and Computing 15(1), 31–41 (2005)
5. Green, P.J.: Reversible jump markov chain monte carlo computation and bayesian model determination. Biometrika 82(4), 711–732 (1995)
6. Richardson, S., Green, P.J.: On Bayesian Analysis of Mixtures with an Unknown Number of Components (with discussion). Journal of the Royal Statistical Society: Series B (Statistical Methodology) 59(4), 731–792 (1997)
7. Rasmussen, C.E.: Evaluation of Gaussian processes and other methods for non-linear regression. Diss. University of Toronto (1996)
8. Neal, R.M.: MCMC using Hamiltonian dynamics. In: Handbook of Markov Chain Monte Carlo (2011)

An Effective Model Selection Criterion for Mixtures of Gaussian Processes

Longbo Zhao, Ziyi Chen, and Jinwen Ma[*]

Department of Information Science, School of Mathematical Sciences and LMAM,
Peking University, Beijing, 100871, China
jwma@math.pku.edu.cn

Abstract. The Mixture of Gaussian Processes (MGP) is a powerful statistical learning framework in machine learning. For the learning of MGP on a given dataset, it is necessary to solve the model selection problem, i.e., to determine the number C of actual GP components in the mixture. However, the current learning algorithms for MGPs cannot solve this problem effectively. In this paper, we propose an effective model selection criterion, called the Synchronously Balancing or SB criterion for MGPs. It is demonstrated by the experimental results that this SB criterion is feasible and even outperforms two classical criterions: AIC and BIC, for model selection on MGPs. Moreover, it is found that there exists a feasible interval of the penalty coefficient for correct model selection.

Keywords: Mixture of Gaussian processes, Model selection, EM algorithm, Parameter learning, Likelihood.

1 Introduction

The Gaussian Process (GP) model is a powerful tool for machine learning. However, it has two limitations. Firstly, it can only fit a single modality dataset. Secondly, for the GP model, the learning algorithm has a large computational complexity $O(N^3)$[1], where N is the number of training samples. In order to solve these issues, Tresp [2] proposed the mixture of Gaussian processes (MGP) in 2000. From then on, various MGP models have been proposed and can be classified into two main forms: the conditional models [2-5] and the generative models [1, 6]. Here, we adopt the generative model since it can infer missing inputs from outputs [7]. In fact, with different number of GP components, the MGP model may lead to quite different experimental results for regression and classification. So, it is critical to know the true number of GP components in the mixture or dataset and thus to get the reasonable result. That is, we must determine the number C of GP components in the mixture for the parameter learning, which is referred to as the model selection problem for the learning of the mixture.

---

[*] Corresponding author.

© Springer International Publishing Switzerland 2015
X. Hu et al. (Eds.): ISNN 2015, LNCS 9377, pp. 345–354, 2015.
DOI: 10.1007/978-3-319-25393-0_38

For model selection, there are some classical criterions like AIC [8], BIC [9], etc., which have been demonstrated effectively for Gaussian Mixtures. However, for MGPs, these criterions do not fit well. In order to solve this model selection problem, we try to improve AIC, BIC criterion and propose a new and effective model selection criterion for model selection on MGPs, called the Synchronously Balancing or SB criterion.

For parameter learning, EM algorithm is an effective way for finite mixtures [10]. However, for the MGP model, the approximations in the implementation of E-step or M-step must be made since it cannot be computed efficiently yet. Among these approximation versions of the EM algorithm for MGPs, we adopt the recently proposed hard-cut EM algorithm [11]. However, the EM algorithm has the local maxima problem. To solve this problem, we further implement the SMEM algorithm [12] after the convergence of the hard-cut EM algorithm.

The rest of the paper is organized as follows. Section 2 introduces the GP and MGP models. Section 3 presents the SB criterion and gives the model selection framework. In Section 4, we test the SB criterion on three synthetic datasets and compare it with AIC, BIC. Moreover, we apply our SB criterion on an artificial toy dataset to select the number of actual components. Finally, we make a brief conclusion in Section 5.

2   The GP and MGP Models

2.1   The GP Model

Given a dataset consisting of N samples $D = \{X, Y\} = \{(x_i, y_i): i = 1,2, \cdots, N\}$, where x_i is a Q-dimensional input vector, and y_i is an output, a GP model is mathematically defined as follows:

$$Y \sim N\big(m(X), K(X, X)\big) \tag{1}$$

where

$$m(X) = [m(x_1), m(x_2), \cdots, m(x_N)]^T \tag{2}$$

$$K(X, X) = \big[K(x_i, x_j)\big]_{N \times N} \tag{3}$$

denote the mean vector and covariance matrix, respectively. As in most cases, we can set $m(X) = 0$, and adopt the squared exponential (SE) covariance function [13]:

$$K(x_i, x_j | \theta) = f^2 exp\left(-\frac{l^2}{2}\|x_i - x_j\|^2\right) + \sigma^2 I(i = j) \tag{4}$$

where $\theta = \{f, l, \sigma\}$ denotes the parameters of the GP model. Therefore, the log-likelihood function of the outputs can be derived as follows:

$$\log p(Y|X, \theta) = \log\big[N\big(Y|0, K(X, X|\theta)\big)\big] \tag{5}$$

and we can obtain the estimation of these parameters via maximum likelihood estimation (MLE), that is

$$\hat{\theta} = argmax_\theta log[N(Y|0, K(X, X|\theta))] \tag{6}$$

2.2   The MGP Model

An MGP model is comprised of multiple Gaussian Process components, and in each component, the corresponding outputs are subject to a certain Gaussian Process. These Gaussian Processes have different parameters and are independent.

For our generative MGP model, the samples are partitioned into the GP components with the following probability

$$p(z_i = c) = \pi_c; \quad c = 1, 2, \cdots, C \; i.i.d \; for \; i = 1, 2, \cdots, N \tag{7}$$

where $z_i = c$ means that the i-th sample belongs to the c-th GP component.

Given the partition of the samples, each input x_i is subject to a Gaussian distribution, that is

$$p(x_i| z_i = c) \sim N(\mu_c, S_c); \quad c = 1, 2, \cdots, C \; i.i.d \; for \; i = 1, 2, \cdots, N \tag{8}$$

Denote $I_c = \{i|z_i = c\}$, $X_c = \{x_i|z_i = c\}$ and $Y_c = \{y_i|z_i = c\}$ as the indexes, inputs and outputs of the samples in the c-th GP component, respectively. Given X_c, the corresponding outputs Y_c is subject to the GP given by Eq.(2) with the parameters $\theta_c = \{f_c, l_c, \sigma_c\}$, and these GP components are independent.

In summary, Eqs. (1), (7), (8) completely define the generative MGP model. Based on the definition, the log-likelihood function is derived as follows:

$$log\, p(Y|X, \Theta, \Psi) = \sum_{c=1}^{C}\{\sum_{i \in I_c}[log\pi_i + logN(x_i|\mu_c, S_c)] + logN(y_c|0, K_c)\} \tag{9}$$

where $\Theta = \{\theta_c\}_{c=1}^{C}$ and $\Psi = \{\mu_c, S_c\}_{c=1}^{C}$ denote the whole set of parameters for outputs and inputs, respectively.

3   The SB Criterion and Model Selection Framework

3.1   The SB Criterion

The objective functions of the AIC and BIC criterion can both be expressed as follows:

$$F = \log likelihood - \delta * penalty \tag{10}$$

where $\delta > 0$ denotes the penalty coefficient, $\log likelihood$ and $penalty$ denote the log-likelihood function and penalty term, respectively. For AIC criterion, $\delta = 1$ and $penalty = KC$, where K denotes the number of parameters in each component, and C denotes the number of components. For BIC criterion, $\delta = 0.5$ and $penalty = KC\log N$, where N denotes the number of samples. These two criterions work effectively for Gaussian Mixture Model. However, for the MGP model, the change of *penalty* with C is too small in comparison with that of $\log likelihood$, so that the selected value of C tends to be large. In order to solve this problem, we try to improve these two criterions to make the changes of the log-likelihood and the penalty term synchronously balanced and construct the following effective criterion:

$$F = \log likelihood - \delta N \log C \qquad (11)$$

Compared with AIC and BIC, such a penalty has a much larger variation with C so that the log-likelihood and penalty are more balanced.

3.2   Model Selection with SB Criterion

Our proposed model selection framework combines the advantages of the SB criterion, the hard-cut EM algorithm [11], and the SMEM algorithm [12]. More specifically, for some values of C, we train the MGP model with hard-cut EM algorithm, update the estimated parameters via SMEM algorithm to avoid local maxima, and then select the best value of C according to the SB criterion.

Before establishing our framework for model selection, we first introduce the hard-cut EM algorithm as well as the SMEM algorithm used in this framework.

The Hard-Cut EM Algorithm and the SMEM Algorithm. The main idea of the hard-cut EM algorithm is to partition the samples into the corresponding GP components according to the maximum a posterior (MAP) criterion in E-step, that is

$$z_i = argmax_{1 \leq c \leq C} \pi_c N(x_i|\mu_c, S_c) N(y_i|0, l_c^2 + \sigma_c^2) \qquad (12)$$

With the known partition, the parameters of each GP component are estimated via MLE respectively in M-step, i.e.

$$\pi_c = \frac{1}{N}\sum_{i=1}^{N} I(z_i = c), \mu_c = \frac{\sum_{i=1}^{N} I(z_i=c)x_t}{\sum_{i=1}^{N} I(z_i=c)}, S_c = \frac{\sum_{i=1}^{N} I(z_i=c)(x_i-\mu_c)(x_i-\mu_c)^T}{\sum_{i=1}^{N} I(z_i=c)} \qquad (13)$$

and θ_c is learnt by Eq. (6).

In the SMEM algorithm, merge and split operations are implemented after the convergence of EM iterations in order to avoid local maxima.

For convenience, we denote

$$post_{ci} = p(z_i = c|x_i, y_i) = \frac{\pi_c N(x_i|\mu_c, S_c) N(y_i|0, l_c^2 + \sigma_c^2)}{\sum_{c=1}^{C} \pi_c N(x_i|\mu_c, S_c) N(y_i|0, l_c^2 + \sigma_c^2)} \qquad (14)$$

as the posterior probability of the t-th sample belonging to the c-th GP component obtained from EM iterations, and denote $post_c = (post_{c1}, post_{c2}, \cdots, post_{cN})$. When $post_u$ and $post_v$ are almost equal, we can merge the u-th and the v-th GP components into one component. So, we define the merge criterion as the similarity between $post_u$ and $post_v$:

$$O_{merge}(u, v) = \frac{post_u post_v^T}{\|post_u\|\|post_v\|}, u \neq v \qquad (15)$$

where $\|\cdot\|$ denotes the Euclidean vector norm, and we can merge the two GP components with the largest $O_{merge}(u, v)$.

After the merge operation, we attempt to split each GP component into two GP components, called the k_1-th and the k_2-th components, and estimate $post_{k_1}$ and $post_{k_2}$ by minimizing $O_{merge}(k_1, k_2)$. Then, we only accept the split of the k^*-th GP component which leads to the smallest minimum $O_{merge}(k_1, k_2)$.

The Model Selection Framework. Denote the set of candidate values of C as

$$S = \{C | l \leq C \leq L\} \tag{16}$$

For each element C from the set S, we learn the MGP model with C components via the hard-cut EM algorithm and SMEM algorithm in turn to get the maximum likelihood:

Step 1 Initialization: Set $s = 1, BestL_C = -Inf$ and initialize the parameters $\{\Theta^0, \Psi^0\}$ in the MGP model.

Step 2 Parameter Learning:

At phase s, we perform the hard-cut EM algorithm with the initial parameters $\{\Theta^{s-1}, \Psi^{s-1}\}$. After convergence, we obtain the estimated parameters $\{\widetilde{\Theta}^s, \widetilde{\Psi}^s\}$, and the corresponding log-likelihood function L_C. If $L_C > BestL_C$, then set $BestL_C = L_C$.

Then, implement the SMEM algorithm [12] with the initial parameters $\{\widetilde{\Theta}^s, \widetilde{\Psi}^s\}$, and we can obtain the updated parameters $\{\Theta^s, \Psi^s\}$, and the corresponding log-likelihood L_C. If $L_C > BestL_C$, then set $BestL_C = L_C$.

Step 3 Set $s = s + 1$, if $s = MaxTime$, terminate and output $BestL_C$; otherwise, return to Step2.

After the learning process above, we have obtained the maximum log-likelihood $BestL_C$, for each C from the candidate set S. Then according to the SB criterion, we obtain the appropriate number of GP components as follows:

$$C^* = argmax_{C \in S} \{BestL_C - \delta N \log C\} \tag{17}$$

4 Simulation Experiments

In order to test the effectiveness and accuracy of our proposed SB criterion for model selection, we generate three typical synthetic datasets from the MGP model, and then apply the SB criterion to these datasets with various values of penalty coefficient δ and compare the SB criterion with two classical model selection criterions, AIC and BIC, on the large synthetic dataset. Moreover, we carry out the same experiment on an artificial toy dataset. Finally, by summarizing these experimental results, we obtain an appropriate empirical interval for δ that leads to reliable model selection.

4.1   On Three Typical Synthetic Datasets of MGP

Three synthetic datasets are generated from the MGP models with different sizes. For the small synthetic dataset, there are 939 samples and 5 GP components, as shown in Fig.1. The medium synthetic dataset has 2400 samples and 8 GP components, as plotted in Fig.2. The large synthetic dataset has 10000 samples and 10 GP components, as plotted in Fig.3.

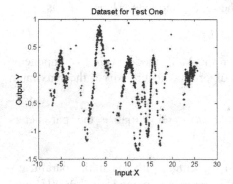

Fig. 1. The small synthetic dataset   **Fig. 2.** The medium synthetic dataset

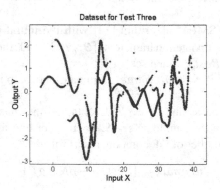

Fig. 3. The large synthetic dataset

Then, we apply the model selection framework above to these synthetic datasets with some $\delta \in (0,3)$. The candidate sets for the small, medium and large datasets are $S = \{2,3,\cdots,10\}$, $S = \{3,4,\cdots,13\}$ and $S = \{5,6,\cdots,15\}$, respectively. We repeat the experiment 18 times on the small dataset and 15 times on the medium and large datasets. For each value of δ, the number of experiments where the estimated value of C does not equal to the true value is shown in Figs.4-6 for the three datasets, respectively.

It can be seen from Figs. 4-6 that our proposed model selection framework selects the correct value of C with very high probability when the penalty coefficient δ lies in a suitable interval, whereas the error increases when δ gets away from this interval, since appropriate value of δ ensures the balance between the log-likelihood and the penalty. The suitable intervals are (1.0,1.8), (1.3,2.2) and (1.10,1.75) for the small, medium and large datasets, respectively. Particularly, with the best value of δ, our proposed model selection framework based on the SB criterion gives correct result for all the 15 times on both the medium and the large synthetic dataset, whereas the large dataset has heavy overlaps among the GP components that makes model selection even more difficult, which firmly demonstrates the strong ability of our proposed model selection framework.

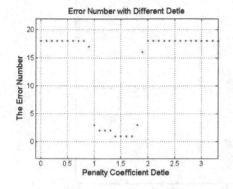

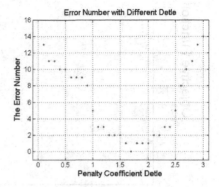

Fig. 4. Model selection result on the small synthetic dataset

Fig. 5. Model selection result on the medium synthetic dataset

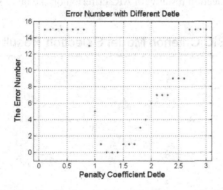

Fig. 6. Model selection result on the large synthetic dataset

4.2   Experimental Results with AIC and BIC Criterion

To compare our proposed SB criterion with two classical criterions, AIC and BIC, we also apply AIC and BIC criterions to the three synthetic datasets above. Figs.7 & 8 show the objective functions of AIC and BIC criterions against the value of C on the large synthetic dataset, respectively. From Figs.7 & 8, it can be seen that these two criterions prefer to select the maximum value of C from the candidate set, since the log-likelihood is dramatically increasing with C for MGP models whereas the penalty is relatively stable, so that the objective functions also increase with C. In contrast, due to the synchronous balance between the log-likelihood and the penalty, our proposed SB criterion significantly outperforms the AIC and BIC criterion on the synthetic dataset.

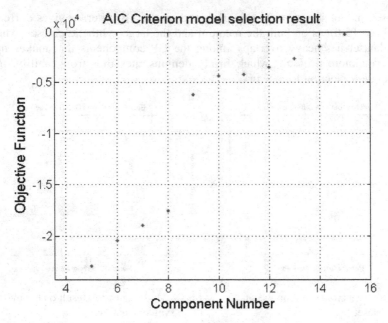

Fig. 7. The model selection result with AIC criterion on the large synthetic dataset

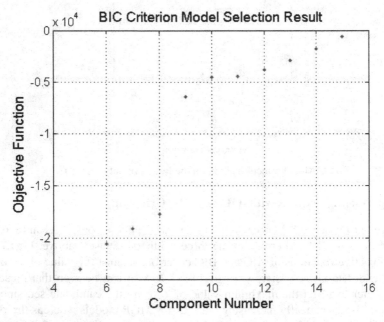

Fig. 8. The model selection result with BIC criterion on the large synthetic dataset

4.3   On an Artificial Toy Dataset

The artificial toy dataset is used to test some MGP models since it is highly multi-modal [6, 7, 11, 14]. The dataset consists of four groups, and each group is generated from a continuous function with different levels of Gaussian noise. In our experiment, we generate 200 samples for each group, as shown in Fig.9. Then, we apply the SB criterion and repeat the experiment 25 times with the candidate set $S = \{2, 3, \cdots, 10\}$. For some values of δ, the number of experiments where the estimated value of $C \neq 4$(the true number of components) is shown in Fig.10. It can be observed from Fig. 10. that the SB criterion makes mistakes only twice among the 25 times, which means it can select the true number of components with very high probability, when δ comes from $(1.25, 2.15)$, whereas the performance becomes poorer when δ gets too large or too small, as also shown in the experiments on the synthetic datasets above. Since the Toy dataset does not come from MGP models and is more similar to a real dataset, our proposed model selection framework also demonstrates potential applicability.

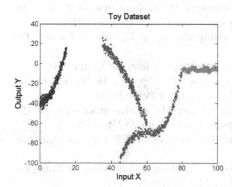
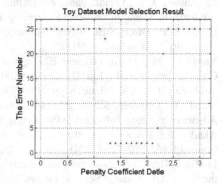

Fig. 9. The toy dataset   **Fig. 10.** Model selection result on toy dataset

4.4   Experimental Conclusion and Penalty Coefficient Choice

It can be summarized from the experimental results above that the performance of our proposed model selection framework heavily relies on the penalty coefficient δ. With a suitable value of δ, our model selection framework works well on both the synthetic datasets and the toy dataset. Besides, the appropriate intervals for δ in these experiments are close to each other and the intersection of these intervals is $(1.3, 1.7)$, which leads to the correct value of C with very high probability. Therefore, $(1.3, 1.7)$ can be an empirical interval of δ for model selection on a new dataset.

5   Conclusion

We have established an effective criterion for model selection of the MGP model, where the log-likelihood and the penalty are much more synchronously balanced in comparison with classical criterions like AIC and BIC. From the experimental results, it can be demonstrated that when the penalty coefficient is within a certain feasible interval, like (1.3,1.7), our proposed SB criterion can obtain the true number of GP components with very high probability, and significantly outperforms AIC and BIC.

Acknowledgement. This work was supported by the National Science Foundation of China under Grant 61171138

References

1. Yuan, C., Neubauer, C.: Variational mixture of Gaussian process experts. In: Advances in Neural Information Processing Systems, pp. 1897–1904 (2008)
2. Tresp, V.: Mixtures of Gaussian processes. In: Advances in Neural Information Processing Systems, vol. 13, pp. 654–660 (2000)
3. Nguyen, T., Bonilla, E.: Fast Allocation of Gaussian Process Experts. In: Proceedings of The 31st International Conference on Machine Learning, pp. 145–153 (2014)
4. Rasmussen, C.E., Ghahramani, Z.: Infinite mixtures of Gaussian process experts. In: Advances in Neural Information Processing Systems, vol. 14, pp. 881–888 (2001)
5. Fergie, M.P.: Discriminative Pose Estimation Using Mixtures of Gaussian Processes. The University of Manchester (2013)
6. Yang, Y., Ma, J.: An efficient EM approach to parameter learning of the mixture of Gaussian processes. In: Liu, D., Zhang, H., Polycarpou, M., Alippi, C., He, H. (eds.) ISNN 2011, Part II. LNCS, vol. 6676, pp. 165–174. Springer, Heidelberg (2011)
7. Meeds, E., Osindero, S.: An alternative infinite mixture of Gaussian process experts. In: Advances in Neural Information Processing Systems, vol. 18, pp. 883–890 (2005)
8. Akaike, H.: A new look at the statistical identification model. IEEE Trans. on Automat. Control 19(6), 716–723 (1974)
9. Liddle, A.R.: Information criterion for astrophysical model selection. Monthly Notices of the Royal Astronomical Society: Letters 377(1), L74–L78 (2007)
10. Dempster, A.P., Laird, N.M., Rubin, D.B.: Maximum likelihood from incomplete data via the EM algorithm. Journal of Royal Statistical Society, Series B (Methodological), 1–38 (1977)
11. Chen, Z., Ma, J., Zhou, Y.: A Precise Hard-Cut EM Algorithm for Mixtures of Gaussian Processes. In: Huang, D.-S., Jo, K.-H., Wang, L. (eds.) ICIC 2014. LNCS, vol. 8589, pp. 68–75. Springer, Heidelberg (2014)
12. Ueda, N., Nakano, R., Ghahramani, Y.Z., Hiton, G.E.: SMEM algorithm for mixture models. Neural Computation 12(9), 2109–2128 (2000)
13. Rasmussen, C.E., Williams, C.K.I.: Gaussian Processes for Machine Learning. The MIT Press, Cambridge (2006)
14. Fergie, M.P.: Discriminative Pose Estimation Using Mixture of Gaussian Processes. The University of Manchester (2013)

Orthogonal Basis Extreme Learning Algorithm and Function Approximation

Ying Li[1], Yan Li[1], and Xiangkui Wan[2]

[1] University of Southern Queensland, Toowoomba, Australia
yingli8qld@yahoo.com.au, yan.li@usq.edu.au
[2] Hubei University of Technology, Wuhan, China
wanxiangkui@163.com

Abstract. A new algorithm for single hidden layer feedforward neural networks (SLFN), Orthogonal Basis Extreme Learning (OBEL) algorithm, is proposed and the algorithm derivation is given in the paper. The algorithm can decide both the NNs parameters and the neuron number of hidden layer(s) during training while providing extreme fast learning speed. It will provide a practical way to develop NNs. The simulation results of function approximation showed that the algorithm is effective and feasible with good accuracy and adaptability.

Keywords: Neural network, orthogonal basis extreme learning, function approximation.

1 Introduction

The neural network (NN) is one of the popular intelligent data analysis methods and a robust way to solve some complex issues with uncertainties and stochastic systems. It is widely used in many areas, such as finance and banking, economics, business, image processing, bionomics, internetworking, engineering, et al. The two biggest challenges of NNs are: how to construct hidden layer(s) or how to decide neuron numbers of hidden layer(s), how to increase the NN learning speed which enables them to meet the challenges of dynamic learning.

For determining the hidden node number some algorithms that construct a NN dynamically have been proposed. The most well known constructive algorithms are dynamic node creation (DNC), the cascade correlation (CC) algorithm, feedforward NN construction (FNNC) algorithm and weight freezing based constructive algorithm.

Huang and Wang discussed a hybrid NN model for rainfall-runoff forecasting based on a Genetic Algorithm (GA). The GA was employed to select the lag period of a time series for NN inputs and the structure of the first NN. Then the first NN is connected to the second NN and the BP algorithm is used to train the second NN [1]. The method optimized the inputs but made a more complex NN in both the structure and the algorithm. Actually one NN is good enough to achieve the required function. Zhang and Li provided a Gegenbauer orthogonal basis NN and its weights direct determination method [2]. It is approved to be a fast way for training a single-input and single-output

X. Hu et al. (Eds.): ISNN 2015, LNCS 9377, pp. 355–364, 2015.
DOI: 10.1007/978-3-319-25393-0_39

(SISO) NN. But it still left the NN structure and the multi-input or multi-output (MIMO) issues as open questions.

Guang-Bin Huang proposed an Extreme Learning Machine (ELM) learning algorithm in 2006 [3]. In theory, this algorithm tends to provide the best generalization performance at extremely fast learning speed. It directly calculates a generalized inverse of the hidden layer output matrix H by using distinct samples. Huang proved that the single hidden layer feedforward neural networks (SLFN) with arbitrarily assigned input weights and hidden layer biases and with almost all nonzero activation functions can universally approximate any continuous functions on any compact input sets [4]. ELM has been successfully applied to a number of real world applications. However the issue regarding the design of ELM network architecture remains open. In some cases it may not be easy to calculate the inverse of H, especially when the samples are very large with some repeat sample values. The distinct samples may not fully represent the real stochastic system.

In this paper, an Orthogonal Basis Extreme Learning (OBEL) algorithm is put forwarded for the solution of above problems. The algorithm derivation is given in the paper. During the training the algorithm can decide both the NNs parameters and the neuron number of hidden layer(s). It has extreme fast learning speed, good accuracy and adaptability. It will provide a practical way to develop NNs. The OBEL algorithm is tested in three functions approximation, single or multiple inputs. The simulation results of function approximation showed that the algorithm is effective and feasible.

2   OBEL Algorithm

2.1   Approximation Problem of SLFN

For a multi-input $x = [x_1, \ldots, x_r]^T \in R^r$ and multi-output $y = [y_1, \ldots, y_m]^T \in R^m$ (MIMO) SLFN with n hidden nodes, its input/output (I/O) relationship can be mathematically modeled as:

$$y = f(x) = \sum_{i=1}^{n} \theta_i \psi_i(x, \omega_i) \qquad (1)$$

Here, $\psi_i(x, \omega_i)$ is the activation function of the ith hidden node. $\omega_i = [\omega_{i1}, \omega_{i2}, \ldots, \omega_{ir}]^T$ is the weight vector connecting the ith hidden node and the input nodes. $\theta_i = [\theta_{i1}, \theta_{i2}, \ldots, \theta_{im}]^T$ is the weight vector connecting the ith hidden node and output nodes.

For N samples $\{(x_j, y_{dj}), j = 1, \ldots, N\}$ with modelling error δ means that $\sum_{i=1}^{N} \|y_j - y_{dj}\| = \delta$, i.e.,

$$y_{dj} = \sum_{i=1}^{n} \theta_i \psi_i(x, \omega_i) + \delta, \quad j = 1, \ldots, N \qquad (2)$$

The above N equations can be written compactly as:

$$Y_d = \Psi(X, \omega)\theta + \delta \qquad (3)$$

Where

$$Y_d = \begin{bmatrix} y_1^T \\ \vdots \\ y_N^T \end{bmatrix}_{N \times m} , \quad \theta = \begin{bmatrix} \theta_1^T \\ \vdots \\ \theta_n^T \end{bmatrix}_{n \times m} , \quad \delta = \begin{bmatrix} \delta_1 \\ \vdots \\ \delta_N \end{bmatrix} , \tag{4a}$$

$$\Psi(X, \omega) = \begin{bmatrix} \psi_{11}(x_1, \omega_1) & \cdots & \psi_{1r}(x_1, \omega_r) \\ \vdots & \cdots & \vdots \\ \psi_{N1}(x_N, \omega_1) & \cdots & \psi_{Nr}(x_N, \omega_r) \end{bmatrix} \tag{4b}$$

If we select the norm cost function

$$J(\omega, \theta) = \|Y_d - \Psi(X, \omega)\theta\| \tag{5}$$

The NN learning algorithm is to find ω^* and θ^* such that

$$\{\omega^*, \theta^*\} = \operatorname{argmin} J(\omega, \theta) \tag{6}$$

2.2   Minimum Norm Least-Squares Solution of SLFN

According to the ELM method, randomly assigns the input weight ω, then (3) can be rewritten as a general linear system

$$Y = \Psi\theta , \quad \delta \to 0 \tag{7}$$

The θ^* is the minimum norm least-squares solution of (7) if

$$\|\Psi\theta^* - Y\| = \min_\theta \|\Psi\theta - Y\| \tag{8}$$

and

$$\theta^* = \Psi^\dagger Y \tag{9}$$

Here $\Psi^\dagger$ is the Moore-Penrose generalized inverse of the matrix Ψ. This can be proved as follows.

Let's split Y into two parts, Y_1 and Y_2. In which Y_1 is the range of Ψ, Y_2 is its orthogonal complement.

$$Y = Y_1 + Y_2 ,$$

$$Y_1 = \Psi\Psi^\dagger Y \in R(\Psi),$$

$$Y_2 = I - \Psi\Psi^\dagger Y \in R(\Psi)^\perp$$

From $\Psi\theta - Y_1 \in R(\Psi)$, there is

$$\|\Psi\theta - Y\|^2 = \|\Psi\theta - Y_1 + (-Y_2)\|^2$$
$$= \|\Psi\theta - Y_1\|^2 + \|Y_2\|^2$$

If and only if θ is the solution of $\Psi\theta = Y_1 = \Psi\Psi^\dagger Y$, then $\theta = \Psi^\dagger Y$ is the least-squares solution of (7).

The complementary solution of (7) is the particular solution (9), plus the kernel or nullspace of Ψ.

$$\mathbb{N}(\Psi) = \mathbb{N}(\Psi^\dagger\Psi)$$
$$= R(I - \Psi^\dagger\Psi)$$
$$= \{(I - \Psi^\dagger\Psi)h : h \in C^N\}$$

The complementary solution of (7) is

$$\theta = \Psi^\dagger Y + (I - \Psi^\dagger\Psi)h : h \in C^N$$

Because

$$\left\|\Psi^\dagger Y\right\|^2 < \left\|\Psi^\dagger Y\right\|^2 + \left\|(I - \Psi^\dagger\Psi)h\right\|^2$$

$$= \left\|\Psi^\dagger Y + (I - \Psi^\dagger\Psi)h\right\|^2,$$

$$(I - \Psi^\dagger\Psi)h \neq 0$$

There for (9) is the minimum norm least-squares solution of (7).

2.3   OBEL Algorithm

However sometimes it may not be easy to calculate out $\Psi^\dagger$, especially when the samples are very large. Let us set the hidden layer activation function $\psi_i(x)$ as an orthogonal polynomial. For N samples $\{(x_j, y_{dj}), j = 1, \ldots, N\}$ and the kth input, there are [5], [6]:

$$X_k = [x_k(1), x_k(2), \ldots, x_k(N)]^T \tag{10}$$

$$F_k = [f_k(x_k(1)), f_k(x_k(2)), \ldots, f_k(x_k(N))]^T \tag{11}$$

$$\Psi_{ik} = [\psi_{ik}(x_k(1)), \psi_{ik}(x_k(2)), \ldots, \psi_{ik}(x_k(N))]^T \tag{12a}$$

$$\Psi_{xik} = [x_k(1)\psi_{ik}(x_k(1)), \ldots, x_k(N)\psi_{ik}(x_k(N))]^T \tag{12b}$$

The following norm and inner product are introduced:

$$\|F_k\| = [\textstyle\sum_{j=1}^N \omega_{kj} f_k^2(x_k(j))]^{1/2} \tag{13a}$$

$$< F_k, \Psi_{ik} > = \textstyle\sum_{j=1}^N \omega_{kj} f_k(x_k(j))\psi_{ik}(x_k(j)) \tag{13b}$$

The cost function is defined as the norm square of learning errors, namely

$$J_k(t) = \|E_k\|^2 = \textstyle\sum_{j=1}^N \omega_{kj}[y_{dj} - f_k(x_k(j))]^2$$

$$= \textstyle\sum_{j=1}^N \omega_{kj}[y_{dj} - \textstyle\sum_{i=1}^n \theta_{ik}\psi_{ik}(x_k(j))]^2 \tag{14}$$

Here t is the learning iteration number. Obviously the NN learning is searching the optimal square approximation of $f(x)$. The linear transfer functions are employed for the input layer and output layer neurons. The orthogonal basis activation function is used for the hidden layer.

Let $\frac{\partial J}{\partial \theta_{ik}} = 0$, i.e.,

$$\sum_{j=1}^{N} \omega_{kj}[y_{dj} - \sum_{i=0}^{n} \theta_{ik}\psi_{ik}(x_k(j))] \cdot \sum_{i=0}^{n} \psi_{ik}(x_k(j))$$

$$\underset{\delta \to 0}{\Longrightarrow} \sum_{j=1}^{N} \omega_{kj}[\sum_{i=0}^{n} \theta_{ik}\psi_{ik}(x_k(j)) - f_k(x_k(j))]\sum_{i=0}^{n} \psi_{ik}(x_k(j)) = 0 \qquad (15)$$

The ith equation in (15) is as follows:

$$\sum_{j=1}^{N} \omega_{kj}[\theta_{ik}\psi_{ik}(x_k(j)) - f_k(x_k(j))] \cdot \psi_{ik}(x_k(j)) = 0$$

The optimal solution of θ_{ik} is

$$\theta_{ik} = \frac{\sum_{j=1}^{N} \omega_{kj}f_k(x_k(j))\psi_{ik}(x_k(j))}{\sum_{j=1}^{N} \omega_{kj}\psi_{ik}(x_k(j))\psi_{ik}(x_k(j))} = \frac{<F_k,\Psi_{ik}>}{<\Psi_{ik},\Psi_{ik}>} \qquad (16)$$

For simplifying the NN structure and the algorithm, an output f_k is built for each input x_k. The total output with weighted coefficients d_k is defined in (17). Here d_k is calculated by the least square method.

$$f = \sum_{k=1}^{r} d_k f_k = \sum_{k=1}^{r} \sum_{i=1}^{n} d_k \theta_{ik}\psi_{ik} \qquad (17)$$

Let $\omega_{kj} = 1$, the recursion formulas are shown in (18), Here $k=1, \ldots, r; i=1, \ldots, n;$ t is the iteration number and $t=0, 1, \ldots.$

$$\psi_{-1k}(x_k) = 0, \quad \psi_{0k}(x_k) = 1, \quad \psi_{1k} = x - \alpha_0, \quad \beta_{0k} = 0, \quad e_0 = y_{d1} \qquad (18a)$$

$$\alpha_{(i+1)k} = \frac{\langle \Psi_{xik},\Psi_{ik}\rangle}{\langle \Psi_{ik},\Psi_{ik}\rangle}, \quad \beta_{ik} = \frac{\langle \Psi_{ik},\Psi_{ik}\rangle}{\langle \Psi_{(i-1)k},\Psi_{(i-1)k}\rangle},$$

$$\psi_{(i+1)k}(x_k) = (x_k - \alpha_{(i+1)k})\psi_{ik}(x_k) - \beta_{ik}\psi_{(i-1)k}(x_k) \qquad (18b)$$

$$\theta_{ik} = \frac{\langle E_{k-1},\Psi_{ik}\rangle}{\langle \Psi_{ik},\Psi_{ik}\rangle} \qquad (18c)$$

$$f_k(x_k) = \sum_{i=1}^{n} \theta_{ik}\psi_{ik}(x_k) \qquad (18d)$$

$$e_k = e_{k-1} - f_k(x_k), \quad J_k(t) = \|E_k\|^2 \qquad (18e)$$

Here $J_k(t)$ in (18e) is decreasing which demonstrated by Zhang [7], i.e. $J_{k+1}(t) \leq J_k(t)$ and $J_k(t+1) \leq J_k(t)$. Therefore $J_k(t)$ can be calculated to a required precision and e_k can converge at the global minimum.

For a dynamic learning algorithm, let's consider a single sample. The cost function defined in (14) is shown as (19).

$$J_k(t) = \|e_k\|^2 = [y_d - \sum_{i=1}^{n} \theta_{ik}\psi_{ik}(x_k)]^2 \qquad (19)$$

The recursion formulas (18b) and (18c) can be modified to the scalar inner product which is shown in (20) and (21).

$$\alpha_{(i+1)k} = \frac{\langle x_k \Psi_{ik}, \Psi_{ik} \rangle}{\langle \Psi_{ik}, \Psi_{ik} \rangle}, \quad \beta_{ik} = \frac{\langle \Psi_{ik}, \Psi_{ik} \rangle}{\langle \Psi_{(i-1)k}, \Psi_{(i-1)k} \rangle} \tag{20}$$

$$\theta_{ik} = \frac{\langle e_{k-1}, \Psi_{ik} \rangle}{\langle \Psi_{ik}, \Psi_{ik} \rangle} \tag{21}$$

The OBEL algorithm can use any orthogonal polynomial as the activation function.

3   OBEL-Based Function Approximation

In this section, the OBEL algorithm is used to approximate the following functions:
Single input function:

$$y(x) = \sin(x), \quad x \in (0, 2\pi) \tag{22}$$

Bessel function:

$$x^2 \ddot{y} + x \dot{y} + (x^2 - a^2)y = 0, \quad x \in [0, 10] \tag{23}$$

Two input function:

$$z = \cos(x) \sin(y), \quad x \in [-2, 2], \quad y \in [-2, 2] \tag{24}$$

The training set and the testing set are created respectively where inputs are uniformly randomly distributed on the interval shown as (22-24).

In the simulations the standard orthogonal polynomials are employed as the hidden activation functions. According to (18) to (21) and OBEL algorithm, the hidden neuron number will automatically increase during learning. To prevent the over large NN structure, a limitation of 20 hidden neuron numbers is set. The training will be stopped when hidden nodes exceeding the limitation.

The Levenberg-Marquardt (LM) algorithm is used as a comparison method. The LM algorithm is often the fastest back-propagation (BP) algorithm in the MATLAB toolbox, and is highly recommended as a first-choice supervised algorithm, although it does require more memory than other algorithms. For comparing the both algorithm at the same accuracy level, the LM training epochs is set to 1000 and the LM training goal is set to 0.001. The activation functions of LM hidden layer are sigmoid functions.

All the simulations for the OBEL and LM algorithms are carried out in the following environment:

a) PC Processor: Intel® Core™ i5 CPU 750 @ 2.67GHz;
b) RAM: 4.00 GB (3.25 GB usable);
c) System: Windows 7 Home Premium SP1 32-bit;
d) Software: MATLAB R2009b (Version 7.9.0.529) 32-bit;

The simulation results of function approximation by using OBEL and LM algorithms are shown as Fig. 1 to Fig. 8. In Fig. 1 to Fig.4 the red line is the target, blue line is the approximation, and green line is approximating error.

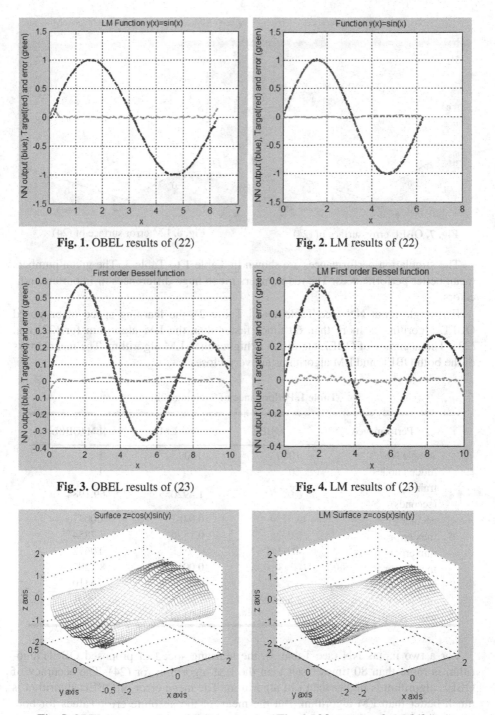

Fig. 1. OBEL results of (22)

Fig. 2. LM results of (22)

Fig. 3. OBEL results of (23)

Fig. 4. LM results of (23)

Fig. 5. OBEL output surface of (24)

Fig. 6. LM output surface of (24)

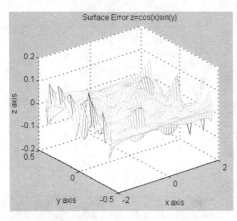
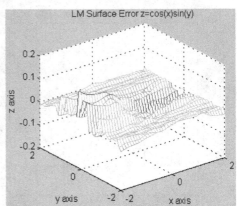

Fig. 7. OBEL error surface of (24)   **Fig. 8.** LM error surface of (24)

The simulation performances are shown in Table 1 to Table 3. The green numbers mean better performances. The red numbers mean huge differences or important indicators.

For single input NN (see Table 1 and Table 2), the learning speed of proposed OBEL algorithm is more than 60 times faster than the LM algorithm for (22) and (23). The accuracy of OBEL algorithm is higher than LM algorithm. The mean errors of the both OBEL and LM algorithms are very close.

Table 1. Performance of Training (22)

Performance	OBEL	LM	|LM/OBEL|
sample No.	1000	1000	1
hidden nodes	20	20	1
training time (seconds)	0.021562	1.492055	69.1984
accuracy	0.0139	0.0287	2.0647
max error	0.0205	0.1596	7.7854
min error	-0.0244	-0.0236	0.9672
mean error	-8.4889e-4	0.0069	8.1283
max |error|	0.0244	0.1596	6.5410
min |error|	2.8889e-5	2.8863e-5	0.9991
mean |error|	0.0126	0.0140	1.1111

For a two inputs NN (see Table 3), the learning speed of proposed OBEL algorithm is more than 80 times faster than the LM algorithm for (24). The accuracy of OBEL algorithm is higher than LM algorithm. The mean error of OBEL algorithm is much better than LM algorithm. But the mean of the absolute error values for both OBEL and LM algorithms are very close.

Table 2. Performance of Training (23)

Performance	OBEL	LM	|LM/OBEL|
sample No.	1000	1000	1
hidden nodes	20	20	1
training time (seconds)	0.021023	1.351731	64.2977
accuracy	0.0178	0.0234	1.3146
max error	0.0257	0.0280	1.0895
min error	-0.0779	-0.1446	1.8562
mean error	0.0031	-0.0025	0.8065
max |error|	0.0779	0.1446	1.8562
min |error|	8.1861e-5	6.4180e-5	0.7840
mean |error|	0.0147	0.0139	0.9456

Table 3. Performance of Training (24)

Performance	OBEL	LM	|LM/OBEL|
sample No.	41×41	41×41	1
hidden nodes	20	20	1
training time (seconds)	0.018131	1.536473	84.7429
accuracy	0.0290	0.0315	1.0862
max error	0.1599	0.1329	0.8311
min error	-0.1599	-0.1024	0.6404
mean error	-8.5467e-18	-0.0015	1.7551e+014
max |error|	0.1599	0.1329	0.8311
min |error|	2.3964e-19	1.6385e-6	6.8373e+012
mean |error|	0.0194	0.0177	0.9124

The simulation results showed that the OBEL algorithm is an extreme fast and feasible method with good accuracy. It can be used as a dynamic learning algorithm for both single input and multiple inputs NNs.

4   Conclusion

In this paper, an OBEL algorithm is developed and the algorithm derivation is given. During training the algorithm can decide both the NNs parameters and the NN structure, i.e. the neuron number of hidden layer. The OBEL algorithm may be used for dynamic learning.

The OBEL algorithm is applied to three functions approximation for both single input and multiple inputs situation. The simulation of LM algorithm is given for comparing the OBEL algorithm. The LM algorithm is often the fastest BP algorithm in the

MATLAB toolbox. The OBEL algorithm is more than 60 times faster than the LM algorithm. The errors of the both OBEL and LM algorithms are very close.

The simulation results showed that the OBEL algorithm has extreme fast learning speed, good accuracy and adaptability. It may provide a practical way to develop NNs. It may exhibit extensive applicability [5], [6]. However the current OBEL algorithm is sensitive to initial input values which will be improved in future study.

Acknowledgement. This work was partially supported by the Science and Technology Program of Guangdong Province, China (No. 2013A022100017) and Natural Science Foundation of Hubei Province, China (No.2015CFB449).

References

1. Huang, G.J., Wang, L.Z.: Hybrid Neural Network Models for Hydrologic Time Series Forecasting Based on Genetic Glgorithm. In: 4th International Conference of Computational Sciences and Optimization, pp. 1347–1350. Kunming (2011)
2. Zhang, Y., Li, W.: Gegenbauer Orthogonal Basis Neural Network and Its Weights-direct-determination Method. Electronics Letters 45, 1184–1185 (2009)
3. Huang, G.B., Zhu, Q.Y., Siew, C.K.: Extreme Learning Machine: A New Learning Scheme of Feedforward Neural Networks. Neurocomputing 70, 489–501 (2006)
4. Huang, G.B., Zhu, Q.Y., Siew, C.K.: Universal Approximation Using Incremental Feedforward Networks with Arbitrary Input Weights. Technical report, Nanyang Technological University, Singapore (2003)
5. Li, Y., Li, Y., Wang, X.: Study on Orthogonal Basis NN-Based Storage Modelling for Lake Hume of Upper Murray River, Australia. In: Wang, X., Pedrycz, W., Chan, P., He, Q., et al. (eds.) ICMLC 2014. CCIS, vol. 481, pp. 431–441. Springer, Heidelberg (2014)
6. Li, Y., Zou, J.X., Zhang, Y.Y.: Adaptive Neural Network Applied to Prediction Modelling of Water Quality. System Engineering 19, 98–93 (2001)
7. Zhang, Y.: NN Based Modelling and Control for Nonlinear System. PhD Thesis, South China University of Technology, Guangzhou (1997)

Large Scale Text Clustering Method Study Based on MapReduce

Zhanquan Sun[1,*], Feng Li[2], Yanling Zhao[1], and Lifeng Song[3]

[1] Shandong Provincial Key Laboratory of Computer Networks, Shandong Computer Science Center(National Supercomputer Center in Jinan), Jinan, 250014, China
[2] Department of History, College of Liberal Arts, Shanghai University, Shanghai, 200436, China
[3] Information communication section, Shandong Provincial Public Security Bureau, Jinan, Shandong, 250010, China
sunzhq@sdas.org

Abstract. Text clustering is an important research topic in data mining. Many text clustering methods have been proposed and obtained satisfactory results. Information Bottleneck algorithm, which is based on information loss, can measure complicated relationship between variables. It is taken as one of the most informative text clustering methods and has been applied widely in practical. With the development of information technology, the scale of text becomes larger and larger. Classical information bottleneck based clustering method will be out of work to process large-scale dataset because of expensive computational cost. For dealing with large scale text clustering problem, a novel clustering method based on MapReduce is proposed. In the method, dataset is divided into sub datasets and deployed to different computational nodes. Each computational node will only process sub dataset. The computational cost can be reduced markedly. The efficiency of the method is illustrated with a practical text clustering problem.

Keywords: Text clustering, Large Scale, MapReduce, Information Bottleneck, Feature selection

1   Introduction

Text Clustering is an unsupervised technique, which groups a collection of documents into a set of clusters, where documents within a cluster share a high level of homogeneity while different clusters exhibit a high level of heterogeneity of information [1]. It has become one of important contents in text mining. In the last decades, many methods have been introduced for text clustering, such as K-means, Clique algorithm, Self Organizing Map (SOM), Information Bottleneck (IB) algorithm and so on[2-4]. Information Bottleneck algorithm, which is based on information loss, can measure complicated relationship between variables. It is taken as one of the most informative

---

* Corresponding author.

© Springer International Publishing Switzerland 2015
X. Hu et al. (Eds.): ISNN 2015, LNCS 9377, pp. 365–372, 2015.
DOI: 10.1007/978-3-319-25393-0_40

text clustering methods and has been applied widely in practice. With the development of information technology, the scale of text dataset becomes larger and larger. Classical information bottleneck based clustering method will be out of work to process large-scale dataset because of expensive computational cost. Efficient parallel algorithms and implementation techniques are the key to meeting the scalability and performance requirements entailed in such large scale data mining analyses. Many parallel algorithms are implemented using different parallelization techniques such as threads, MPI, MapReduce, and mash-up or workflow technologies yielding different performance and usability characteristics [5]. MapReduce is a cloud technology developed from the data analysis model of the information retrieval field. The MapReduce architecture in Hadoop doesn't support iterative Map and Reduce tasks, which is required in many data mining algorithms. Iterative MapReduce architecture software is developed, such as Twister and Spark. Iterative MapReduce supports not only non-iterative MapReduce applications but also an iterative MapReduce programming model [6-7]. Some clustering methods based on MapReduce were proposed, such as k-means, EM, Dirichlet Process Clustering and so on [8]. But the complicated clustering methods are not realized in MapReduce model. In this paper, a novel parallel text clustering method based on parallel IB is proposed to deal with large scale text clustering problems.

The general procedure of text processing is summarized as feature extraction, text representation, clustering process and cluster interpretation. Most textual information is available in the form of natural language. Some text feature selection methods have been developed, such as Document Frequency (DF), Term Contribution (TC), Term variance quality (TVQ) and Term Variance (TV) [9-10]. These unsupervised methods can filter unimportant features efficiently. In this paper, DF and TC are used to select text feature for clustering.

After extracting features, the document collection should be represented using a suitable numerical model in order to be processed by a clustering algorithm. To the best of our knowledge, Vector Space Model (VSM) is the most popular method in representing documents. Tf-idf is taken as the most commonly used feature to represent a text document. In this paper, Tf-idf is used as the feature of a document. It is used to generate document VSM. Parallel IB based on MapReduce is used to cluster the text documents. A paractical text clutering example is analyzed with the proposed method in the end.

2   Feature Selection

Commonly used feature selection methods DF and TC are introduced as follows. They are used to filter unimportant features in clustering.

2.1   Document Frequency (DF)

Document frequency is the number of documents in which a term occurs in a dataset. It is the simplest criterion for term selection and easily scales to a large dataset with linear computation complexity. A basic assumption of this method is that terms appear in minority documents are not important or will not influence the clustering efficiency. It is a simple but effective feature selection method for text categorization.

2.2   Term Contribution (TC)

TF-IDF synthetically considers the frequency of a term in a document and the document frequency of the term. It believes that if a term appears in too many documents, it's too common and not important for clustering. So Inverse Document Frequency is considered. A common form of TF-IDF is:

$$f(t_i, d_j) = \frac{n_{i,j}}{\sum_k n_{k,j}} \log \frac{|D|}{|\{j:t_i \in d_j\}|} \tag{1}$$

The result of text clustering is highly dependent on the documents similarity. So the contribution of a term can be viewed as its contribution to the documents' similarity. The contribution of a term in a dataset is defined as its overall contribution to the documents' similarities.

$$TC(t_i) = \sum_{j,k \cap j \neq k} f(t_i, d_j) \times f(t_i, d_k) \tag{2}$$

2.3   Feature Representation

Tf-idf will be used to generate VSM of document. In the calculation of information loss, VSM should be represented with probability value. So the value of $f(t_i, d_j)$ is normalized as follows.

$$p(t_i, d_j) = \frac{f(t_i, d_j)}{\sum_i (f(t_i, d_j))} \tag{3}$$

3   Iterative MapReduce Model

The computational cost of mutual information of large scale dataset is expensive. Parallel feature selection is the efficient means to solve it. Different MapReduce model will lead to different operation performance. Twister is an iterative MapReduce model that is efficient in dealing with most machine learning methods. Twister's programming model can be described as in figure 1.

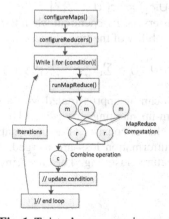

Fig. 1. Twister's programming model

4   Parallel IB Clustering

Classical clustering method based on information bottleneck theory has been applied to many application fields. The clustering method is introduced as follows.

4.1   IB Principle

The IB clustering method states that among all the possible clusters of a given object set when the number of clusters is fixed, the desired clustering is the one that minimizes the loss of mutual information between the objects and the features extracted from them. Let $p(x, y)$ be a joint distribution on the "object" space X and the "feature" space Y. According to the IB principle we seek a clustering $\hat{X}$ such that the information loss $I(X; \hat{X}) = I(X; Y) - I(\hat{X}; Y)$ is minimized. $I(X; \hat{X})$ is the mutual information between X and $\hat{X}$

$$I(X; \hat{X}) = \sum_{x, \hat{x}} p(x) p(\hat{x}|x) \log \frac{p(\hat{x}|x)}{p(\hat{x})} \tag{4}$$

The loss of the mutual information between X and Y caused by the clustering $\hat{X}$ can be calculated as follows.

$$d(x, \hat{x}) = I(X; Y) - I(\hat{X}; Y) =$$
$$\sum_{x, \hat{x}, y} p(x, \hat{x}, y) \log \frac{p(y|x)}{p(y)} - \sum_{x, \hat{x}, y} p(x, \hat{x}, y) \log \frac{p(y|\hat{x})}{p(y)} = ED(p(x, \hat{x})||p(y|\hat{x}) \tag{5}$$

Let c_1 and c_2 be two clusters of symbols. The information loss due to the merging is

$$d(c_1, c_1) = I(c_1; Y) + I(c_2; Y) - I(c_1, c_2; Y) \tag{6}$$

Standard information theory operation reveals

$$d(c_1, c_1) = \sum_{y, j=1,2} p(c_i) p(y|c_i) \log \frac{p(y|c_i)}{p(y|c_1 \cup c_2)} \tag{7}$$

where $p(c_i) = |c_i|/|X|$, $|c_i|$ denotes the cardinality of c_i, $|X|$ denotes the cardinality of object space X, $p(c_1 \cup c_2) = |c_1 \cup c_2|/|X|$.

It assumes that the two clusters are independent when the probability distribution is combined. The combined probability of the two clusters is

$$p(y|c_1 \cup c_2) = \sum_{i=1,2} \frac{|c_i|}{|c_1 \cup c_2|} p(y|c_i) \tag{8}$$

The minimization problem can be approximated with a greedy algorithm. The algorithm is based on a bottom-up merging procedure. The algorithm starts with the trivial clustering where each cluster consists of a single data vector. In each step, the two clusters with minimum information loss are merged. The method is suitable to both sample clustering and feature clustering. The clustering procedure based on IB is shown in figure 2.

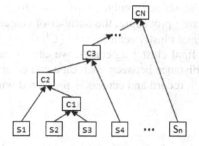

Fig. 2. The clustering procedure of information bottleneck based clustering method

4.2   Parallel IB Based on MapReduce

The computational cost classical IB clustering method is expensive. It is impossible to deal with large scale dataset. Parallel IB clustering method based on MapReduce can improve the computational speed markedly. The method is introduced as follows.

Given data set D with n samples, it is divided into m partitions $D^1, D^2, \cdots, D^m$ with $n_1, n_2, \cdots, n_m$ samples separately. The parallel calculation process based on MapReduce is shown in figure 3. The clustering procedure is summarized as follows.

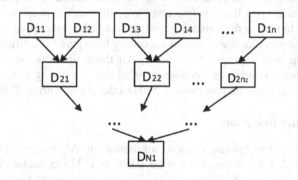

Fig. 3. The calculation process of parallel IB based on MapReduce

(1) In computational node i, operate clustering on data partition D^i based on IB method. The dataset is clustered into N_i ($N_i \in [m, n_i/2]$) clusters. The ratio between the initial size of the sub dataset and the clusters of the sub dataset is denoted by β. It can assure that the dataset scale of each computational node will not larger than previous level. Bigger cluster number N_i means that the information loss will be less. The N_i clustered centers are output to next level.

(2) Arbitrary 2 computational nodes' output are combined in one computational node to generate new clustering dataset. Apply IB clustering method to the new dataset to generate new clusters and new clusters are input to the next level. The clustering result should include the sample size of each sub-centroids which will be used in calculation of the final clustering result. If only one computational node is left, go to next step, or else iterate the step 2.

(3) In the last level, dataset are combined into one group. Operate IB clustering on the dataset. In the clustering procedure, the number of centers is set m. After calculation, we can obtain the final cluster centers $C = \{C^1, C^2, \cdots, C^m\}$.

(4) According to the final clustering center, we can calculate the clustering results through calculating the distance between each data and each center. In the calculation, the distance between each record and centers is measured with information loss.

5   Example

For illustrating the efficiency of the proposed method, a large scale text clustering example is analyzed in this section.

5.1   Data Source

Department of science and technology will collect thousands of project proposals every year. Proposal evaluation is an important task. How to assign a project proposal to a proper expert is a significant problem. Firstly, the project proposals should be clustered. Similar project proposals are put together and assigned to same experts. Traditionally, the projects proposals are assigned according to the keywords. It will render mismatch between proposals and experts. Proposal clustering based on proposal content can improve the reasonability of project evaluation. Clustering the project proposal manually is a burdensome task. We will use the proposed parallel text clustering method to realize the clustering. We collect 958 project proposal samples in this example. These samples are clustered according to experts' classification manually according to the whole proposals' content. All those projects are divided into 5 clusters. The example is analyzed in Shandong cloud computing center. Each node installs Ubuntu Linux OS. The processor is 3GHz Intel Xeon with 8GB RAM.

5.2   Text Feature Selection

We use ICTCLAS tool to operate Chinese segmentation. After segmentation, we get 32535 terms in total. Each document can be taken as a 32535 vector. We apply DF and TC feature selection methods to the high dimension document vector. The threshold value of DF is set 10. After filtering, 3240 terms are left. We apply TC method on the left terms. The threshold value is set 0.04. We obtain 1890 terms finally. We use the 1890 terms to realize text clustering.

5.3   Text Clustering

We apply the parallel text clustering method to the dataset. In the clustering, the ratio value β between the initial size of sub dataset and the clusters is set 0.3. The final cluster number is determined with the introduced method in section 3. In this example, the number of clusters is determined 12. The clustering results are shown as in table 1.

Table 1. Clustering results based on different computational nodes

Number of computational nodes	Computational time	Clustering precision
1	294.04	90.2
2	172.12	89.68
4	126.33	89.45
8	143.42	89.12

For comparison, we use k-means clustering method to realize the clustering problem. The number of centers is set 5 and the initial center is selected randomly from the initial dataset. The clustering is operated on one computational node. The clustering result is shown as in table 2.

Table 2. Clustering results based on different computational nodes

Number of computational nodes	Computational time	Clustering precision
1	74.32	83.21
2	39.36	82.98
4	20.68	82.64

5.4   Result Comparison

From above analysis results we can find that the clustering method based on Information bottleneck is better than classic clustering method. It is more suitable to be used to process text mining problems. But the computational cost of IB is higher than classic clustering method. MapReduce technique can improve the computation speed markedly. But the computational speed is not improved linearly. The increase of computational node will lead to the increase of computational layer, which will increase the computational time. At the same time, the data transfer from previous layer to next layer will cost time. So the number of computational node should be determined properly according to practical problem.

6   Conclusions

Large scale text clustering becomes more and more important in current society. More and more application areas require text clustering. Clustering based on MapReduce program model is the development trend of large scale dataset clustering problems. The parallel clustering methods proposed in this paper can deal with large scale text clustering problems efficiently. The clustering precision based on MapReduce will not affected markedly by the number of computational node. The clustering result is better than other classic clustering method.

Acknowledgement. This work is partially supported by national youth science foundation (No. 61004115), national science foundation (No. 61472230), national youth science foundation (No. 61402271).

References

1. Hotho, A., Nurnberger, A., Paass, G.: A brief survey of text mining. Ldv Forum 20(1), 19–62 (2005)
2. Ranjan, M., Peterson, A.D., Ghosh, P.A.: A systematic evaluation of different methods for initializing the K-means clustering algorithm. IEEE Transactions on Knowledge and Data Engineering, 1–13 (2010)
3. Simon, H.: Self-organizing maps. Neural networks - A comprehensive foundation. Prentice-Hall (1999)
4. Tishby, N., Fernando, C., Bialek, W.: The information bottleneck method. In: The 37th Annual Allerton Conference on Communication, Control and Computing, Monticello, pp. 1–11 (1999)
5. Fox, G.C., Bae, S.H., et al.: Parallel Data Mining from Multicore to Cloudy Grids. High Speed and Large Scale Scientific Computing 18, 311–341 (2010)
6. Zhang, B.J., Ruan, Y., et al.: Applying Twister to Scientific Applications. In: Proceedings of CloudCom, pp. 25–32 (2010)
7. Ekanayake, J., Li, H., et al.: Twister: A Runtime for iterative MapReduce. In: The First International Workshop on MapReduce and its Applications of ACM HPDC, pp. 810–818 (2010)
8. Sun, Z.Q., Fox, G.C.: A parallel clustering method combined information bottleneck theory and centroid-based clustering. Journal of Supercomputing 69(1), 452–467 (2014)
9. Matharage, S., Ganegedara, H., Alahakoon, D.: A scalable and dynamic self-organizing map for clustering large volumes of text data. In: International Joint Conference on Neural Networks, pp. 1–8 (2013)
10. Wu, O., Zuo, H., Zhu, M., et al.: Rank aggregation based text feature selection. In: IEEE International Conference on Web Intelligence, pp. 165–172 (2009)

Representing Data by Sparse Combination of Contextual Data Points for Classification

Jingyan Wang[1,2,3], Yihua Zhou[4], Ming Yin[5], Shaochang Chen[5], and Benjamin Edwards[6]

[1] National Time Service Center, Chinese Academy of Sciences,
Xi' an, Shaanxi 710600, China
[2] Graduate University of Chinese Academy of Sciences, Beijing 100049, China
[3] Provincial Key Laboratory for Computer Information Processing Technology,
Soochow University Suzhou 215006, China
jingbinwang1@outlook.com
[4] Department of Mechanical Engineering and Mechanics,
Lehigh University, Bethlehem, PA 18015, USA
[5] Electronic Engineering College, Naval University of Engineering,
Wuhan 430033, China
[6] Department of Computer Science, Sam Houston State University,
Huntsville, TX 77341, USA
benjamin.edwards1@hotmail.com

Abstract. In this paper, we study the problem of using contextual data points of a data point for its classification problem. We propose to represent a data point as the sparse linear reconstruction of its context, and learn the sparse context to gather with a linear classifier in a supervised way to increase its discriminative ability. We proposed a novel formulation for context learning, by modeling the learning of context reconstruction coefficients and classifier in a unified objective. In this objective, the reconstruction error is minimized and the coefficient sparsity is encouraged. Moreover, the hinge loss of the classifier is minimized and the complexity of the classifier is reduced. This objective is optimized by an alternative strategy in an iterative algorithm. Experiments on three benchmark data set show its advantage over state-of-the-art context-based data representation and classification methods.

Keywords: Pattern classification, Context learning, Nearest neighbors, and Sparse regularization.

1 Introduction

Pattern classification is a major problem in machine learning research [32,5,6,13]. The two most important topics of pattern classification are data representation and classifier learning. Zhang et al. proposed an efficient multi-model classifier for large scale Bio-sequence localization prediction [36]. Zhang et al. developed and optimized association rule mining algorithms and implemented them on paralleled micro-architectural platforms [39,38]. Most data representation and

© Springer International Publishing Switzerland 2015
X. Hu et al. (Eds.): ISNN 2015, LNCS 9377, pp. 373–381, 2015.
DOI: 10.1007/978-3-319-25393-0_41

classification methods are based on single data point. When one data point is considered for representation and classification, all other data points are ignored. However, the other data points other than the data point under consideration, which are called contextual data points, may play important roles in its representation and classification. It is necessary to explore the contexts of data points when they are represented and/or classified. In this paper, we investigate the problem of learning effective representation of a data point from its context guided by its class label, and proposed a novel supervised context learning method using sparse regularization and linear classifier learning formulation.

We propose a novel method to explore the context of a data point, and use it to represent it. We use its k nearest neighbors as its context, and try to reconstruct it by the data points in its context. The reconstruction errors are imposed to be spares. Moreover, the reconstruction result is used as the new representation of this data point. We apply a linear function to predict its class label from the sparse reconstruction of its context. The motivation of this contribution is that for each data point, only a few data points in its context is of the same class as itself. To find the critical contextual data points, we proposed to learn the classifier together with she sparse context. We mode this problem as a minimization problem. In this problem, the context reconstruction error, reconstruction sparsity, classification error, and classifier complexity are minimized simultaneously. We also problem a novel iterative algorithm to solve this minimization problem. We first reformulate it as ist Lagrange formula, and the use an alterative optimization method to solve it.

This paper is organized as follows. In section 2, we introduce the proposed method. In section 3, we evaluate the proposed method experimentally. In section 4, this paper is concluded with future works.

2   Proposed Method

We consider a binary classification problem, and a training set of n data points are given as $\{(\mathbf{x}_i, y_i)\}_{i=1}^n$, where $\mathbf{x}_i \in \mathbb{R}^d$ is a d-dimensional feature vector of the i-th data point, and $y_i \in \{+1, -1\}$ is the class label of the i-th point. To learn from the context of the i-th data point, we find its k nearest neighbors and denote them as $\{\mathbf{x}_{ij}\}_{j=1}^k$, where $\mathbf{x}_{ij}$ is the j-th nearest neighbor of the i-th point. They are further organized as a $d \times k$ matrix $X_i = [\mathbf{x}_{i1}, \cdots, \mathbf{x}_{ik}] \in R^{d \times k}$, where the j-th column is $\mathbf{x}_{ij}$. We represent $\mathbf{x}_i$ by linearly reconstructing it from its contextual points as

$$\mathbf{x}_i \approx \widehat{\mathbf{x}}_i = \sum_{j=1}^k \mathbf{x}_{ij} v_{ij} = X_i \mathbf{v}_i \tag{1}$$

where $\widehat{\mathbf{x}}_i$ is its reconstruction, and v_{ij} is the reconstruction coefficient of the j-th nearest neighbor. $\mathbf{v}_i = [v_{i1}, \cdots, v_{ik}]^\top \in \mathbb{R}^k$ is the reconstruction coefficient vector of the i-th data point. The reconstruction coefficient vectors of all the training points are organized in reconstruction coefficient matrix $V = [\mathbf{v}_1, \cdots, \mathbf{v}_n] \in$

$\mathbb{R}^{k \times n}$, with its i-th column as $\mathbf{v}_i$. To solve the reconstruction coefficient vectors, we propose the following minimization problem,

$$\min_V \left\{ \beta \sum_{i=1}^n \|\mathbf{x}_i - X_i \mathbf{v}_i\|_2^2 + \gamma \sum_{i=1}^n \|\mathbf{v}_i\|_1 \right\}, \tag{2}$$

where β and γ are trade-off parameters. In the objective of this problem, the first term is to minimize the reconstruction error measured by a squared ℓ_2 norm penalty between $\mathbf{x}_i$ and $X_i \mathbf{v}_i$, and the second term is a ℓ_1 norm penalty to the contextual reconstruction coefficient vector $\mathbf{v}_i$.

We design a classifier to classify the i-th data point,

$$f(\widehat{\mathbf{x}}_i) = \mathbf{w}^\top \widehat{\mathbf{x}}_i = \mathbf{w}^\top X_i \mathbf{v}_i \tag{3}$$

where $\mathbf{w} \in \mathbb{R}^d$ is the classifier parameter vector. The following optimization problem is proposed to learn $\mathbf{w}$,

$$\min_{\mathbf{w}, V, \boldsymbol{\xi}} \left\{ \frac{1}{2} \|\mathbf{w}\|_2^2 + \alpha \sum_{i=1}^n \xi_i \right\} \tag{4}$$
$$s.t. \ 1 - y_i \left(\mathbf{w}^\top X_i \mathbf{v} \right) \leq \xi_i, \xi_i \geq 0, i = 1, \cdots, n,$$

where $\frac{1}{2}\|\mathbf{w}\|_2^2$ is the the squared ℓ_2 norm regularization term to reduce the complexity of the classifier, ξ_i is the slack variable for the hinge loss of the i-th training point, $\boldsymbol{\xi} = [\xi_1, \cdots, \xi_n]^\top$ and α is a tradeoff parameter.

The overall optimization problem is obtained by combining the problems in both (2) and (4) as

$$\min_{\mathbf{w}, V, \boldsymbol{\xi}} \left\{ \frac{1}{2} \|\mathbf{w}\|_2^2 + \alpha \sum_{i=1}^n \xi_i + \beta \sum_{i=1}^n \|\mathbf{x}_i - X_i \mathbf{v}_i\|_2^2 + \gamma \sum_{i=1}^n \|\mathbf{v}_i\|_1 \right\} \tag{5}$$
$$s.t. \ 1 - y_i \left(\mathbf{w}^\top X_i \mathbf{v} \right) \leq \xi_i, \xi_i \geq 0, i = 1, \cdots, n.$$

According to the dual theory of optimization, the following dual optimization problem is obtained,

$$\max_{\boldsymbol{\delta}, \boldsymbol{\epsilon}} \min_{\mathbf{w}, V, \boldsymbol{\xi}} \left\{ \frac{1}{2} \|\mathbf{w}\|_2^2 + \alpha \sum_{i=1}^n \xi_i + \beta \sum_{i=1}^n \|\mathbf{x}_i - X_i \mathbf{v}_i\|_2^2 + \gamma \sum_{i=1}^n \|\mathbf{v}_i\|_1 \right.$$
$$\left. + \sum_{i=1}^n \delta_i \left(1 - y_i \left(\mathbf{w}^\top X_i \mathbf{v}_i \right) - \xi_i \right) - \sum_{i=1}^n \epsilon_i \xi_i \right\}, \tag{6}$$
$$s.t. \ \boldsymbol{\delta} \geq 0, \boldsymbol{\epsilon} \geq 0,$$

where $\boldsymbol{\delta} = [\delta_1, \cdots, \delta_n]^\top$, and $\boldsymbol{\epsilon} = [\epsilon_1, \cdots, \epsilon_n]^\top$ are Lagrange multipliers. By setting the partial derivative of $\mathcal{L}$ with regard to $\mathbf{w}$ and ξ_i to zeros, we have

$$\mathbf{w} = \sum_{i=1}^{n} \delta_i y_i X_i \mathbf{v}_i.$$

$$\alpha - \delta_i = \epsilon_i \tag{7}$$

$$\Rightarrow \alpha \geq \delta_i.$$

We substitute (7) to (6) to eliminate $\mathbf{w}$ and $\boldsymbol{\delta}$,

$$\max_{\boldsymbol{\delta}} \min_{V} \left\{ -\frac{1}{2} \sum_{i,j=1}^{n} \delta_i \delta_j y_i y_j \mathbf{v}_i^\top X_i^\top X_j \mathbf{v}_j + \beta \sum_{i=1}^{n} \| \mathbf{x}_i - X_i \mathbf{v}_i \|_2^2 \right.$$

$$\left. + \gamma \sum_{i=1}^{n} \| \mathbf{v}_i \|_1 + \sum_{i=1}^{n} \delta_i \right\} \tag{8}$$

$$s.t. \ \boldsymbol{\alpha} \geq \boldsymbol{\delta} \geq 0.$$

where $\boldsymbol{\alpha} = [\alpha, \cdots, \alpha]^\top$ is a n dimensional vector of all α elements. We solve this problem with the alternate optimization strategy. In each iteration of an iterative algorithm, we fix $\boldsymbol{\delta}$ first to solve V, and then fix V to solve $\boldsymbol{\delta}$.

Solving V When $\boldsymbol{\delta}$ is fixed and only V is considered, we solve $\mathbf{v}_i|_{i=1}^{n}$ one by one, (8) is further reduced to

$$\min_{\mathbf{v}_i} \left\{ -\frac{1}{2} \sum_{i,j=1}^{n} \delta_i \delta_j y_i y_j \mathbf{v}_i^\top X_i^\top X_j \mathbf{v}_j + \beta \| \mathbf{x}_i - X_i \mathbf{v}_i \|_2^2 + \gamma \| \mathbf{v}_i \|_1 \right\}. \tag{9}$$

This problem could be solved efficiently by the modified feature-sign search algorithm proposed by Gao et al. [2].

Solving $\boldsymbol{\delta}$ When V is fixed and only $\boldsymbol{\delta}$ is considered, the problem in (8) is reduced to

$$\max_{\boldsymbol{\delta}} \left\{ -\frac{1}{2} \sum_{i,j=1}^{n} \delta_i \delta_j y_i y_j \mathbf{v}_i^\top X_i^\top X_j \mathbf{v}_j + \sum_{i=1}^{n} \delta_i \right\} \tag{10}$$

$$s.t. \ \boldsymbol{\alpha} \geq \boldsymbol{\delta} \geq 0.$$

This problem is a typical constrained quadratic programming (QP) problem, and it can be solved efficiently by the active set algorithm.

3   Experiments

In this section, we evaluate the proposed supervised sparse context learning (SSCL) algorithm on several benchmark data sets.

3.1   Experiment Setup

In the experiments, we used three date sets, which are introduced as follows:

- **MANET loss data set**: The packet losses of the receiver in mobile Ad hoc networks (MANET) can be classified into three types, which are wireless random errors caused losses, the route change losses induced by node mobility and network congestion. We collect 381 data points for the congestion loss, 458 for the route change loss, and 516 data points for the wireless error loss for this data set. Thus in the data set, there are 1355 data points in total. To extract the feature vector each data point, we calculate 12 features from each data point as in [1], and concatenate them to form a vector.
- **Twitter data set**: The second data set is a Twitter data set. The target of this data set is to predict the gender of the twitter user, male or female, given one of his/her Twitter massage. We collected 53,971 twitter massages in total, and among them there are 28,012 messages sent by male users, and 25,959 messages sent by female users. To extract features from each Twitter message, we extract Term features, linguistic features, and medium diversity features as gender-specific features as in [8].
- **Arrhythmia data set**: The third data set is publicly available at http://arc hive.ics.uci.edu/ml/datasets/Arrhythmia. In this data set, there are 452 data points, and they belongs to 16 different classes. Each data point has a feature vector of 279 features.

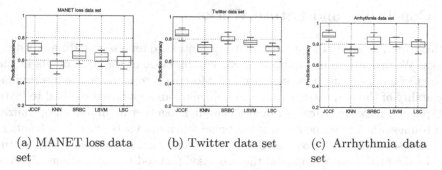

(a) MANET loss data set   (b) Twitter data set   (c) Arrhythmia data set

Fig. 1. Boxplots of prediction accuracy of different context-based algorithms.

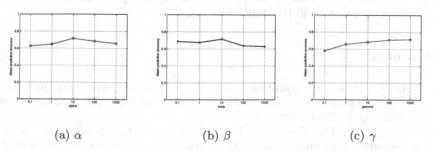

(a) α (b) β (c) γ

Fig. 2. Parameter sensitivity curves.

To conduct the experiments, we used the 10-fold cross validation.

3.2   Experimental Results

Since the proposed algorithm is a context-based classification and sparse representation method, we compared the proposed algorithm to three popular context-based classifiers, and one context-based sparse representation method. The three context-based classifiers are traditional k-nearest neighbor classifier (KNN), sparse representation based classification (SRBC) [26],and Laplacian support vector machine (LSVM) [11]. The context-based sparse representation method is Gao et al.'s Laplacian sparse coding (LSC) [3]. The boxplots of the 10-fold cross validation of the compared algorithms are given in figure 1. From the figures, we can see that the proposed method SSCL outperforms all the other methods on all three data sets. The second best method is SRBC, which also uses sparse context to represent the data point. This is a strong evidence that learning a supervised sparse context is critical for classification problem.

Sensitivity to Parameters. In the proposed formulation, there are three tradeoff parameters, α, β, and γ. We plot the curve of mean prediction accuracies against different values of parameters, and show them in figure 2. From figure 2(a) and 2(b), we can see the accuracy is stable to the parameter α and β. From figure 2(c), we can see a larger γ leads to better classification performances.

4   Conclusion and Future Works

In this paper, we study the problem of using context to represent and classify data points. We propose to use a sparse linear combination of the data points in the context of a data point to represent itself. Moreover, to increase the discriminative ability of the new representation, we develop an supervised method to learn the sparse context by learning it and a classifier together in an unified optimization framework. Experiments on three benchmark data sets show its advantage over state-of-the-art context-based data representation and classification methods. In the future, we will extend the proposed method to applications of information security [33,27,30,29,28,31,34], bioinformatics [25,24,23,12,15,14,7,37,7], computer vision [16,17], and big data analysis using high performance computing [43,18,9,35,4,41,40,39,38,35,10,42,21,20,43,19,22].

References

1. Deng, Q., Cai, A.: Svm-based loss differentiation mechanism in mobile ad hoc networks. In: 2009 Global Mobile Congress, GMC 2009 (2009), doi:10.1109/GMC.2009.5295834
2. Gao, S., Tsang, I.H., Chia, L.T.: Laplacian sparse coding, hypergraph laplacian sparse coding, and applications. IEEE Transactions on Pattern Analysis and Machine Intelligence 35(1), 92–104 (2013)

3. Gao, S., Tsang, I.W., Chia, L.T., Zhao, P.: Local features are not lonely–laplacian sparse coding for image classification. In: 2010 IEEE Conference on Computer Vision and Pattern Recognition (CVPR), pp. 3555–3561. IEEE (2010)
4. Gao, Y., Zhang, F., Bakos, J.D.: Sparse matrix-vector multiply on the keystone ii digital signal processor. In: 2014 IEEE High Performance Extreme Computing Conference (HPEC), pp. 1–6 (2014)
5. Guo, Z., Li, Q., You, J., Zhang, D., Liu, W.: Local directional derivative pattern for rotation invariant texture classification. Neural Computing and Applications 21(8), 1893–1904 (2012)
6. He, Y., Sang, N.: Multi-ring local binary patterns for rotation invariant texture classification. Neural Computing and Applications 22(3-4), 793–802 (2013)
7. Hu, J., Zhang, F.: Improving protein localization prediction using amino acid group based physichemical encoding. In: Rajasekaran, S. (ed.) BICoB 2009. LNCS, vol. 5462, pp. 248–258. Springer, Heidelberg (2009)
8. Huang, F., Li, C., Lin, L.: Identifying gender of microblog users based on message mining. In: Li, F., Li, G., Hwang, S.-W., Yao, B., Zhang, Z. (eds.) WAIM 2014. LNCS, vol. 8485, pp. 488–493. Springer, Heidelberg (2014)
9. Li, T., Zhou, X., Brandstatter, K., Raicu, I.: Distributed key-value store on hpc and cloud systems. In: 2nd Greater Chicago Area System Research Workshop (GCASR). Citeseer (2013)
10. Li, T., Zhou, X., Brandstatter, K., Zhao, D., Wang, K., Rajendran, A., Zhang, Z., Raicu, I.: Zht: A light-weight reliable persistent dynamic scalable zero-hop distributed hash table. In: 2013 IEEE 27th International Symposium on Parallel & Distributed Processing (IPDPS), pp. 775–787 (2013)
11. Melacci, S., Belkin, M.: Laplacian support vector machines trained in the primal. The Journal of Machine Learning Research 12, 1149–1184 (2011)
12. Peng, B., Liu, Y., Zhou, Y., Yang, L., Zhang, G., Liu, Y.: Modeling nanoparticle targeting to a vascular surface in shear flow through diffusive particle dynamics. Nanoscale Research Letters 10(1), 235 (2015)
13. Tian, Y., Zhang, Q., Liu, D.: v-nonparallel support vector machine for pattern classification. Neural Computing and Applications 25(5), 1007–1020 (2014)
14. Wang, J., Li, Y., Wang, Q., You, X., Man, J., Wang, C., Gao, X.: Proclusensem: predicting membrane protein types by fusing different modes of pseudo amino acid composition. Computers in Biology and Medicine 42(5), 564–574 (2012)
15. Wang, J.J.Y., Bensmail, H., Gao, X.: Multiple graph regularized protein domain ranking. BMC Bioinformatics 13(1), 307 (2012)
16. Wang, J.J.Y., Bensmail, H., Gao, X.: Joint learning and weighting of visual vocabulary for bag-of-feature based tissue classification. Pattern Recognition 46(12), 3249–3255 (2013)
17. Wang, J.Y., Almasri, I., Gao, X.: Adaptive graph regularized nonnegative matrix factorization via feature selection. In: 2012 21st International Conference on Pattern Recognition (ICPR), pp. 963–966 (2012)
18. Wang, K., Kulkarni, A., Zhou, X., Lang, M., Raicu, I.: Using simulation to explore distributed key-value stores for exascale system services. In: 2nd Greater Chicago Area System Research Workshop, GCASR (2013)
19. Wang, K., Liu, N., Sadooghi, I., Yang, X., Zhou, X., Lang, M., Sun, X.H., Raicu, I.: Overcoming hadoop scaling limitations through distributed task execution. In: Proc. of the IEEE International Conference on Cluster Computing, Cluster 2015 (2015)

20. Wang, K., Zhou, X., Chen, H., Lang, M., Raicu, I.: Next generation job management systems for extreme-scale ensemble computing. In: Proceedings of the 23rd International Symposium on High-Performance Parallel and Distributed Computing, pp. 111–114 (2014)
21. Wang, K., Zhou, X., Li, T., Zhao, D., Lang, M., Raicu, I.: Optimizing load balancing and data-locality with data-aware scheduling. In: 2014 IEEE International Conference on Big Data (Big Data), pp. 119–128 (2014)
22. Wang, K., Zhou, X., Qiao, K., Lang, M., McClelland, B., Raicu, I.: Towards scalable distributed workload manager with monitoring-based weakly consistent resource stealing. In: Proceedings of the 24rd International Symposium on High-Performance Parallel and Distributed Computing, pp. 219–222. ACM (2015)
23. Wang, S., Zhou, Y., Tan, J., Xu, J., Yang, J., Liu, Y.: Computational modeling of magnetic nanoparticle targeting to stent surface under high gradient field. Computational Mechanics 53(3), 403–412 (2014)
24. Wang, Y., Han, H.C., Yang, J.Y., Lindsey, M.L., Jin, Y.: A conceptual cellular interaction model of left ventricular remodelling post-mi: dynamic network with exit-entry competition strategy. BMC Systems Biology 4(suppl. 1), S5 (2010)
25. Wang, Y., Yang, T., Ma, Y., Halade, G.V., Zhang, J., Lindsey, M.L., Jin, Y.F.: Mathematical modeling and stability analysis of macrophage activation in left ventricular remodeling post-myocardial infarction. BMC Genomics 13(suppl. 6), S21 (2012)
26. Wright, J., Yang, A.Y., Ganesh, A., Sastry, S.S., Ma, Y.: Robust face recognition via sparse representation. IEEE Transactions on Pattern Analysis and Machine Intelligence 31(2), 210–227 (2009)
27. Xu, L., Zhan, Z., Xu, S., Ye, K.: Cross-layer detection of malicious websites. In: Proceedings of the Third ACM Conference on Data and Application Security and Privacy, pp. 141–152. ACM (2013)
28. Xu, L., Zhan, Z., Xu, S., Ye, K.: An evasion and counter-evasion study in malicious websites detection. In: 2014 IEEE Conference on Communications and Network Security (CNS), pp. 265–273. IEEE (2014)
29. Xu, S., Lu, W., Xu, L., Zhan, Z.: Adaptive epidemic dynamics in networks: Thresholds and control. ACM Transactions on Autonomous and Adaptive Systems (TAAS) 8(4), 19 (2014)
30. Xu, S., Lu, W., Zhan, Z.: A stochastic model of multivirus dynamics. IEEE Transactions on Dependable and Secure Computing 9(1), 30–45 (2012)
31. Xu, S., Qian, H., Wang, F., Zhan, Z., Bertino, E., Sandhu, R.: Trustworthy information: concepts and mechanisms. In: Chen, L., Tang, C., Yang, J., Gao, Y. (eds.) WAIM 2010. LNCS, vol. 6184, pp. 398–404. Springer, Heidelberg (2010)
32. Xu, Y., Shen, F., Zhao, J.: An incremental learning vector quantization algorithm for pattern classification. Neural Computing and Applications 21(6), 1205–1215 (2012)
33. Zhan, Z., Xu, M., Xu, S.: Characterizing honeypot-captured cyber attacks: Statistical framework and case study. IEEE Transactions on Information Forensics and Security 8(11), 1775–1789 (2013)
34. Zhan, Z., Xu, M., Xu, S.: A characterization of cybersecurity posture from network telescope data. In: Proceedings of the 6th International Conference on Trustworthy Systems, Intrust, vol. 14 (2014)
35. Zhang, F., Gao, Y., Bakos, J.D.: Lucas-kanade optical flow estimation on the ti c66x digital signal processor. In: 2014 IEEE High Performance Extreme Computing Conference (HPEC), pp. 1–6 (2014)

36. Zhang, F., Hu, J.: Bayesian classifier for anchored protein sorting discovery. In: IEEE International Conference on Bioinformatics and Biomedicine, BIBM 2009, pp. 424–428 (2009)
37. Zhang, F., Hu, J.: Bioinformatics analysis of physicochemical properties of protein sorting signals (2010)
38. Zhang, F., Zhang, Y., Bakos, J.: Gpapriori: Gpu-accelerated frequent itemset mining. In: 2011 IEEE International Conference on Cluster Computing (CLUSTER), pp. 590–594 (2011)
39. Zhang, F., Zhang, Y., Bakos, J.D.: Accelerating frequent itemset mining on graphics processing units. The Journal of Supercomputing 66(1), 94–117 (2013)
40. Zhang, Y., Zhang, F., Bakos, J.: Frequent itemset mining on large-scale shared memory machines. In: 2011 IEEE International Conference on Cluster Computing (CLUSTER), pp. 585–589 (2011)
41. Zhang, Y., Zhang, F., Jin, Z., Bakos, J.D.: An fpga-based accelerator for frequent itemset mining. ACM Transactions on Reconfigurable Technology and Systems (TRETS) 6(1), 2 (2013)
42. Zhao, D., Zhang, Z., Zhou, X., Li, T., Wang, K., Kimpe, D., Carns, P., Ross, R., Raicu, I.: Fusionfs: Toward supporting data-intensive scientific applications on extreme-scale high-performance computing systems. In: 2014 IEEE International Conference on Big Data (Big Data), pp. 61–70 (2014)
43. Zhou, X., Chen, H., Wang, K., Lang, M., Raicu, I.: Exploring distributed resource allocation techniques in the slurm job management system. Illinois Institute of Technology, Department of Computer Science, Technical Report (2013)

A Novel K-Means Evolving Spiking Neural Network Model for Clustering Problems

Haza Nuzly Abdull Hamed[1,*], Abdulrazak Yahya Saleh[2],
and Siti Mariyam Shamsuddin[2]

[1] Soft Computing Research Group, Faculty of Computing, Universiti Teknologi
Malaysia (UTM), Skudai, 81310 Johor, Malaysia
haza@utm.my
[2] UTM Big Data Centre, Universiti Teknologi Malaysia (UTM), Skudai,
81310 Johor, Malaysia
abdulrazakalhababi@gmail.com, mariyam@utm.my

Abstract. In this paper, a novel K-means evolving spiking neural network (K-ESNN) model for clustering problems has been presented. K-means has been utilised to improve the original ESNN model. This model enhances the flexibility of the ESNN algorithm in producing better solutions to overcoming the disadvantages of K-means. Several standard data sets from UCI machine learning are used for evaluating the performance of this model. It has been found that the K-ESNN provides competitive results in clustering accuracy and speed performance measures compared to the standard K-means. More discussion is provided to prove the effectiveness of the new model in clustering problems.

Keywords: Clustering, Evolving Spiking Neural Networks, K-ESNN, K-means, Spiking Neural Network.

1 Introduction

The evolving spiking neural network (ESNN) has been used widely in recent research. The ESNN has several advantages [1], including being a simple, efficient neural model and trained by a fast one-pass learning algorithm. The evolving nature of the model can be updated whenever new data becomes accessible with no requirement to retrain earlier existing samples [2]. However, the ESNN model has not yet been investigated as a clustering method. For this reason, a novel integrated method is proposed to determine the effectiveness of the ESNN in clustering.

Several clustering techniques can be utilised to enhance the ESNN performance output and reduce the consuming time. However, the K-means, proposed by several researchers across different disciplines like [3], has been considered as one of the most popular techniques used in data mining [4]. Many research studies have been proposed due its simplicity and efficiency in various fields [5, 6]. The algorithm begins to initiate the K object to be the centroid object by chance. After that, the Euclidean distance equation is used to test each object compared to the K object. The cluster technique is concerned with grouping objects into classes of similar objects called clusters.

---

* Corresponding author.

© Springer International Publishing Switzerland 2015
X. Hu et al. (Eds.): ISNN 2015, LNCS 9377, pp. 382–389, 2015.
DOI: 10.1007/978-3-319-25393-0_42

There are many challenges in dealing with clustering techniques, such as the choice of the number of clusters. Consequently, any incorrect choice of the number of clusters yields bad clustering results. Additionally, these algorithms suffer from inadequate accuracy when the data set contains clusters with noise and outliers. Based on [7], the greatest challenge with K-means is how to overcome the disadvantages of K-means, such as the poor performance and the sensitivity to outliers by proposing new variants.

Hence, this research aims to find such clusters where the similarity of objects within individual clusters is high, while the similarity of objects from different clusters is low by enhancing ESNN training using K-means. The remaining sections of this paper are formed as follows: sections 2 and 3 explain the related works and methodology used in this paper, respectively, while section 4 elucidates the experimental results; finally, section 5 presents the conclusion and future works.

2   Related Works

The original K-means algorithm has been utilised in many different ways [8, 9]. Some historical issues related to the well-known K-means algorithm in cluster analysis can be found in [10]. However, the K-means has some shortcomings, i.e. the difficulty to predict K-values and different initial partitions can result in different final clusters [11]. The K-means clustering algorithm flowchart is shown in Fig. 1.

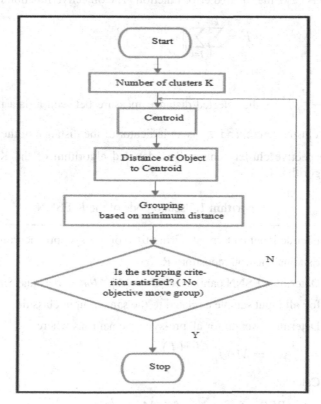

Fig. 1. K-means clustering algorithm flowchart

Many researchers have modified the K-means algorithm in different aspects [12]. A solution for better results could be instead of integrating all the requirements into a single algorithm, trying to build a combination of clustering algorithms[13]. Up to now, no studies have been found that investigate the capabilities of combining K-means with the ESNN for clustering purposes. The ESNN was being tested only for classification purposes [14-17].

3   The Proposed Method K-ESNN

In this hybrid method, K-means enhances ESNN learning by classifying a given data set through a definite number of clusters (K clusters). The key idea is to identify K centroids, one for each cluster. These centroids are supposed to be positioned in a proper manner because different locations produce different outcomes. Consequently, the best alternative is to set them far away from each other. After that, each point belonging to a given data set should be associated with the nearest centroid. This process must be repeated until no point is pending. As a result of this process, it might be observed that the K centroids change their position stepwise until no more changes are done. Finally, this algorithm works toward minimising the objective function, which is, in this case, the squared error function. The objective function is as follows:

$$F = \sum_{j=1}^{k} \sum_{i=1}^{n} \left\| x_i^{(j)} - z_j \right\|^2 \tag{1}$$

where $\left\| x_i^{(j)} - z_j \right\|^2$ is the selected distance measure between a data instance point $x_i^{(j)}$ and the cluster centre, and z_j is an indicator of the distance of the n data points from their respective cluster centres. The detailed algorithm of the K-ESNN is as follows in Algorithm 1.

Algorithm 1. Pseudo-code of the K-ESNN

---

1. Encode input pattern into firing time of pre-synaptic neuron f
2. Establish neuron repository R
3. Determine ESNN parameters $C = [0,1]$, $Mod = [0,1]$ and $Sim= [0,1]$
4. **for all** input sample e related to the same output class **do**
5. Determine weight for all pre-synaptic neurons where:

$$wf = Mod_e^{order(f)}$$

6. Calculate

$$PSP_{max}(e) = \sum wfe * Mod_e^{order(f)}$$

7. Obtain the value of PSP threshold $\theta = PSP_{max}(e) * C$

8. **if** the trained weight vector <= Sim of trained weight in R then

9. Merge weight and threshold value with the most similar neuron

10. $w = \dfrac{w(new) + w * T}{T + 1}$

11. $\theta = \dfrac{\theta(new) + \theta * T}{T + 1}$

12. where T is number of mergers before

13. **else**

14. Add new neuron to output neuron repository R

15. **end if**

16. **end for** (repeat for all instances for other output class)

17. Initiate the K object to be a centroid object by chance.

18. Allocate each object to the nearby group, which is related to the closest centroid.

19. Recalculate the locations of the K centroids.

20. **Repeat** Steps 18 and 19 until the centroids stop moving.

The process starts by initialising the parameters of the ESNN model modulation factor (Mod), similarity parameter (Sim), threshold parameter (C) and fitness function. Moreover, the pre-synaptic neurons of the ESNN model have been discussed in depth in [18, 19]. The selected values for each data set are indicated in Table 1.

Table 1. Selected values of parameters and pre-synaptic neurons in ESNN

Parameter	Normal Range	Hepatitis data set	Wine data set	BTX data set	Appendicitis data set
Mod	0-1	0.9	0.9	0.9	0.9
Sim	0-1	0.1	0.1	0.1	0.1
Threshold	0.5 - 0.9	0.9	0.3	0.45	0.75
Pre-Synaptic Neurons	10-25	15	20	25	20

Besides, it is essential to initialise the number of K points, which belong to the clustering process. In this study, K has been set to be 2 clusters. After that, the ESNN model process is run until achieving the output spikes for each instance.

The ESNN model stands on two principles: possibility of the establishment of new classes and the merging of similarities. The encoding method, which is used for the ESNN, is the population, as explained in [20]. The population distributes a single input value to multiple input neurons, denoted as M. Each input neuron holds a firing time as input spikes. The firing times are calculated, which represent the input neuron e using the intersection of the Gaussian function. The centre is calculated using (2) and the width is computed using (3) with the variable interval of [Emin, Emax]. The parameter β controls the width of each Gaussian receptive field.

$$R = E_{min} + (2*e-3)/2*(Emax - E_{min}) / (M-2). \tag{2}$$

$$\sigma = 1/\beta \ (E_{max} - E_{min)} / (M-2) \ \text{where} \ 1 \leq \beta \leq 2 \tag{3}$$

The model, which was proposed by Thorpe [21] is similar to the Fast Integrate and Fire Model used in this paper. Thorpe's model shows that the earliest spikes received by a neuron will get a better weight depending on the later spikes. When the Post-Synaptic Potential (PSP) exceeds the threshold value, it will fire and become disabled. The computation of PSP of neuron e is presented in (4):

$$U_e = \{ {}^0_{\sum wfe*Mod_e^{order(f)}} {}^{fired}_{else} \tag{4}$$

Where W_{fe} is the weight of the pre-synaptic neuron f; Mod_e is a parameter called modulation factor with an interval of [0, 1] and order (f) represents the rank of the spike emitted by the neuron. The order (f) starts with 0 if it spikes first among all pre-synaptic neurons and increases according to the firing time. In the One-pass Learning algorithm, each training sample creates a new output neuron. Then, all spike point value locations are updated based on associating them to the group that has the closest centroid. Finally, the process of recalculating the positions of K centroids and the positions of spikes is repeated until the stopping criteria is met, which is when the centroids no longer move. The proposed method is evaluated by using several stand-ard data sets which are listed in Table 2 and were downloaded from the machine learning benchmark repository (http://www.ics.uci.edu/~mlearn/MLRepository.html).

Table 2. Summary of data sets which used in this study

Data set	Features	Instances
Hepatitis	19	155
Wine	13	178
BTX	3	63
Appendicitis	7	106

4   Results and Discussion

This section presents the results of the proposed method compared to the standard K-means. The experiments are conducted using hepatitis, wine, BTX and appendicitis data sets. The results of the comparison have been analysed based on their clustering performances and speeds. Table 3 shows the results of the comparison of the K-ESNN and K-means for all chosen data sets in both accuracy and speed performance. As can be seen, the accuracy of the K-ESNN for the hepatitis, wine, BTX and appendecitis data sets is 90.32%, 84.09%, 80.95% and 72.73%, respectively, while it is 83.55%, 56.82%, 66.67% and 45.45% in K-means As illustrated in Fig. 2, the testing clustering accuracy values signify that the K-ESNN resulted in a better performance compared to K-means for all data sets.

Table 3. Results of clustering of chosen data sets

The data set	Algorithm	Clustering performance	
		Accuracy	Time (second)
Hepatitis	K-means	83.55%	0.112
	K-ESNN	**90.32%**	**0.052**
Wine	K-means	56.82%	0.091
	K-ESNN	**84.09%**	**0.073**
BTX	K-means	66.67%	0.064
	K-ESNN	**80.95%**	**0.007**
Appendicitis	K-means	45.45%	0.064
	K-ESNN	**72.73%**	**0.042**

Table 3 shows the results of the clustering of both models corresponding to these data sets. The clustering performance was tested for both models using the hepatitis, wine, BTX and appendicitis data sets. The best performance of accuracy in the K-ESNN is 90.32% for the hepatitis data set, whereas the best performance of accuracy in K-means is 83.55% for the hepatitis data set. The poorest performance of accuracy in the K-ESNN is clearly in the appendicitis data set, as shown below in Fig. 2.

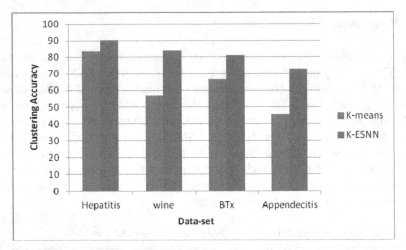

Fig. 2. Comparison between K-ESNN and K-means in terms of clustering accuracy

In addition, the most important advantage found in this comparison is the speed performance. The results of the K-ESNN are 0.052, 0.073, 0.007 and 0.042 seconds for the hepatitis, wine, BTX and appendicitis data sets, respectively, which means it is faster than K-means. There are many factors leading the K-ESNN to be faster than K-means, such as the ESNN nature which makes the clustering faster than the standard K-means.

5   Conclusion and Future Works

In this paper, a novel K-ESNN method was proposed to investigate the capabilities of the ESNN model to be used as a clustering method. A comparative study has been conducted between the K-ESNN and standard K-means to show the performance improvement of the K-ESNN. The accuracy and the speed of K-ESNN gave better results compared to the K-means. More experiments and comparisons are also another approach that can be applied. Moreover, enhancing the proposed K-ESNN method may be a good trend for exploration to obtain more results. Finally, increasing the performance and decreasing the time is an advisable direction to apply the multi-objective optimisation algorithms.

Acknowledgement. This research is supported and funded by Ministry of Education, Malaysia: Fundamental Research Grant Scheme (FRGS/2/2013/ICT07/UTM/02/10).

References

1. Schliebs, S., Defoin-Platel, M., Worner, S., Kasabov, N.: Integrated feature and parameter optimization for an evolving spiking neural network: Exploring heterogeneous probabilistic models. Neural Networks 22, 623–632 (2009)

2. Kasabov, N.K.: NeuCube: A spiking neural network architecture for mapping, learning and understanding of spatio-temporal brain data. Neural Networks 52, 62–76 (2014)
3. MacQueen, J.: Some methods for classification and analysis of multivariate observations. In: Proceedings of the Fifth Berkeley Symposium on Mathematical Statistics and Probability, Oakland, CA, USA, pp. 281–297 (1967)
4. Wu, X., Kumar, V., Quinlan, J.R., Ghosh, J., Yang, Q., Motoda, H., McLachlan, G.J., Ng, A., Liu, B., Philip, S.Y.: Top 10 algorithms in data mining. Knowledge and Information Systems 14, 1–37 (2008)
5. Fayyad, U.M., Piatetsky-Shapiro, G., Smyth, P., Uthurusamy, R.: Advances in knowledge discovery and data mining (1996)
6. Firouzi, B., Niknam, T., Nayeripour, M.: A new evolutionary algorithm for cluster analysis. World Academy of Science, Engineering, and Technology 36, 605–609 (2008)
7. Wu, J.: Advances in K-means clustering: a data mining thinking. Springer Science & Business Media, Heidelberg (2012)
8. Jain, A.K.: Data clustering: 50 years beyond K-means. Pattern Recognition Letters 31, 651–666 (2010)
9. Berkhin, P.: A survey of clustering data mining techniques. In: Grouping Multidimensional Data, pp. 25–71. Springer (2006)
10. Bock, H.-H.: Clustering methods: a history of k-means algorithms. In: Selected Contributions in Data Analysis and Classification, pp. 161-172. Springer (2007)
11. Patel, V.R., Mehta, R.G.: Modified k-means clustering algorithm. In: Das, V.V. (ed.) CIIT 2011. CCIS, vol. 250, pp. 307–312. Springer, Heidelberg (2011)
12. Mandloi, M.: A Survey on Clustering Algorithms and K-Means (July 2014)
13. Kotsiantis, S., Pintelas, P.: Recent advances in clustering: A brief survey. WSEAS Transactions on Information Science and Applications 1, 73–81 (2004)
14. Schliebs, S., Kasabov, N.: Evolving spiking neural network—a survey. Evolving Systems 4, 87–98 (2013)
15. Abdull Hamed, H.N., Kasabov, N., Michlovský, Z., Shamsuddin, S.M.: String Pattern Recognition Using Evolving Spiking Neural Networks and Quantum Inspired Particle Swarm Optimization. In: Leung, C.S., Lee, M., Chan, J.H. (eds.) ICONIP 2009, Part II. LNCS, vol. 5864, pp. 611–619. Springer, Heidelberg (2009)
16. Hamed, H.N., Kasabov, N.: Quantum-inspired particle swarm optimisation for integrated feature and parameter optimisation of evolving spiking neural networks. International Journal of Artificial Intelligence 7, 114–124 (2011)
17. Hamed, H.N., Kasabov, N., Shamsuddin, S.M., Widiputra, H., Dhoble, K.: An extended evolving spiking neural network model for spatio-temporal pattern classification. In: The 2011 International Joint Conference on Neural Networks (IJCNN), pp. 2653–2656. IEEE (2011)
18. Saleh, A.Y., Hameed, H.N.B.A., Najib, M., Salleh, M.: A Novel hybrid algorithm of Differential evolution with Evolving Spiking Neural Network for pre-synaptic neurons Optimization. Int. J. Advance Soft Compu. Appl. 6 (2014)
19. Saleh, A.Y., Shamsuddin, S.M., Hamed, H.N.B.A.: Parameter Tuning of Evolving Spiking Neural Network with Differen-tial Evolution Algorithm. In: International Conference of Recent Trends in Information and Communication Technologies, vol. 13 (2014)
20. Bohte, S.M., Kok, J.N., La Poutre, H.: Error-backpropagation in temporally encoded networks of spiking neurons. Neurocomputing 48, 17–37 (2002)
21. Thorpe, S.: How can the human visual system process a natural scene in under 150ms? experiments and neural network models, pp. 2-9600049. D-Facto public, ISBN (1997)

Prediction of Individual Fish Trajectory from Its Neighbors' Movement by a Recurrent Neural Network

Gang Xiao, Yi Li, Tengfei Shao, and Zhenbo Cheng[*]

Department of Computer Science and Technology,
Zhejiang University of Technology, Hangzhou, China
czb@zjut.edu.cn

Abstract. Individuals in large groups respond to the movements and positions of their neighbors by following a set of interaction rules. These rules are central to understanding the mechanisms of collective motion. However, whether individuals actually use these rules to guide their movements remains untested. Here we show that the real-time movements of individual fish can be directly predicted from their neighbors' motion. We train a recurrent neural network to predict the trajectories of individual fish from input signals. The inputs are projected to the recurrent network as time series representing the movements and positions of neighboring fish. By comparing the data output from the model with the target fish's trajectory, we provide direct evidence that individuals guide their movements via interaction rules. Because the error between the model output and actual trajectory changes when the fish perceive a noxious contaminant, the model is potentially applicable to water quality monitoring.

Keywords: Fish Behavior, Recurrent Neural Network, Force Learning, Water Quality Monitoring.

1 Introduction

Many fish species tend to gather in shoals or schools. A fish shoal can be identified as a spatial aggregation of independently moving fish that are loosely connected to the group, with no mutual attraction between individuals [1]. Shoaling is a complex social behavior that increases the individual fitness of each fish when foraging for food or avoiding prey [2-5]. Synchronous movement of multiple individuals in a group typically leads to collective motion of the fish shoal. Many models attempt to explain how collective motion emerges from interactions among individual fish. In these models, the fish respond to the positions of their neighbors through short-range repulsion and longer-range attraction rules [6-9]. Yet it remains to be seen whether an individual fish's movement can be predicted online using its neighbors' positions and movements.

---

[*] Corresponding author.

© Springer International Publishing Switzerland 2015
X. Hu et al. (Eds.): ISNN 2015, LNCS 9377, pp. 390–397, 2015.
DOI: 10.1007/978-3-319-25393-0_43

In this study, we attempt to predict the motion of the target fish by constructing a movement trajectories reservoir of the target's neighbors using a recurrent neural network. The network receives the trajectories of the target's neighbors as inputs. The neurons within the recurrent network are randomly and sparsely connected. The connections can be either excitatory or inhibitory. The connections within the reservoir are fixed. The movement trajectory of the target fish is predicted by a recursive least squares rule that shapes the readout from the recurrent network [10].

By comparing the prediction errors of the model in untreated water with those in chemically treated water, we might obtain early warning signals of water quality deterioration. We first train the model to generate the trajectory of the target fish, and obtain a threshold (mean training error) between the output and the target fish in untreated water. The model is then applied to treated water, and if the error in the trajectory exceeds the threshold for 10 or more consecutive seconds, an alarm is generated. The test chemical is glyphosate, which is commonly used in agriculture and industry. We find that glyphosate increases the prediction error relative to controls.

2   Materials and Methods

2.1   Test Subjects

The test subjects were 120 adult crucian carp (*Carassius auratus*) obtained from a pet supplier in Hangzhou, China. Crucian carp, belonging to the family Cyprinidae (Telestei), is widely distributed on the Eurasian continent[11]. To easily segment the swimming test fish from their background, we selected only red crucian carp with 3–5-cm body length. The video was captured by a camera positioned 45 cm above the test tank (Fig. 1a). The test fish can swim naturally in the test tank (26 cm wide × 33 cm long × 3.5 cm high; water volume ~3 L). The image frames in the video were converted to gray scale and inverted. A fish was identified by applying an edge-detection algorithm to the thresholded image (Fig. 1d). The position of the identified fish was then tracked at every frame [12]. The trajectory (Fig. 1c) of the fish i tracked through T seconds is given as follows: $S^i = \{(x_1^i, y_1^i), (x_2^i, y_2^i), \dots, (x_T^i, y_T^i)\}$.

In this expression, (x_t^i, y_t^i) represents the coordinates of the centroid of fish i at time t. The average speed in each trial was 121 cm/min. The coordinate origin was set to the lower left corner of the test tank. The x and y coordinates in pixels were converted to cm using a conversion ratio determined by measuring the distance (in pixels) between two corners of the tank. To remove any small spurious changes in position, we also smoothed the coordinates by a Savitzky–Golay smoothing filter with a 6-frame span. The raw videos were processed using custom software written in C++ using the OpenCV library, and all statistical analyses were performed using Matlab2014a.

2.2   The Recurrent Neural Model

The model consists of three interconnected layers: the input layer (IL), the recurrent neural layer (RNL) and the output layer (OL) (Fig. 1b). The IL contains input units that determine the coordinates of each fish. The IL units project to the RNL units through the input synaptic vector $\mathbf{W}^{IL}$. The connective values in the $\mathbf{W}^{IL}$ are randomly drawn from a standard Gaussian distribution $N(0, 0.5)$.

The RNL is a recurrent network with dynamics described by

$$\tau \frac{dx_i}{dt} = -x_i + \sum_{j=1}^{N} W_{ij}^{RNL} r_j + \sum_{j=1}^{J} W_{ij}^{IL} y_j + I_i^{noise}, \tag{1}$$

where r_i ($i = 1, ..., N$) represents the activity of unit i in the RNL, and r_i is a function of the activation variable x_i; specifically, $r_i = \tanh(x_i)$. The RNL size N and the time constant τ are set to 1000 and 10 ms, respectively. The connectivity between the RNL neurons is represented by a sparse $N \times N$ matrix $\mathbf{W}^{RNL}$, with nonzero initial values randomly obtained from Gaussian distribution $N(0, \frac{g^2}{pN})$, where $g = 1.5$ is the synaptic strength scaling coefficient, and $p = 0.1$ is the connection probability among the units. The noise current I^{noise} is an $N \times 1$ random vector drawn from a Gaussian distribution $N(0, 10^{-6})$.

The activities of the RNL units are pooled by the two-unit OL. The activities of the two OL units are determined by

$$z_k = \sum_{i=1}^{N} W_{ki}^{OL} r_i, \quad k = 1, 2. \tag{2}$$

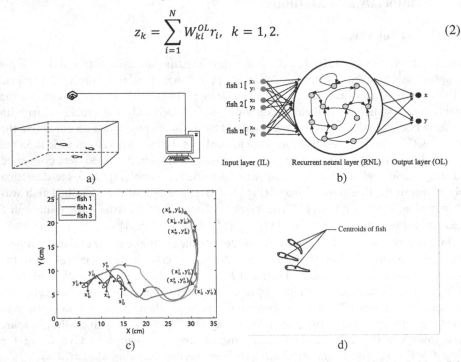

a)

b)

c)

d)

Fig. 1. a) The experimental arena. b) Schematic of the model. c) Tracking of three individual fish in 0.5 min. The three lines represent the trajectories of the three fish, respectively. d) The contour and the centroid of each fish.

2.3   Learning Rule

The connectivity between the RNL and OL units is represented by the vector W_i^{OL}. Initially, the weights are randomly assigned with a uniform distribution. The output weight is then updated during the training process as follows:

$$W_i^{OL}(t) = W_i^{OL}(t-1) - e(t) \sum_{j=1}^{N} P_{ij}(t) r_j(t). \tag{3}$$

In Eq. (3), $e(t)$ is the error signal, defined as

$$e(t) = \sum_{j=1}^{N} W_j^{OL}(t-1) r_j(t) - f(t), \tag{4}$$

where $f(t)$ is the target function of the output unit. P^i is a square matrix that estimates the inverse of the correlation matrix of the presynaptic inputs to unit i, and is updated by

$$P(t) = P(t-1) - \frac{p(t-1)r(t)r^T(t)P(t-1)}{1 + r^T(t)P(t-1)r(t)}. \tag{5}$$

The prediction results were compared by the average root-mean-square (RMS).

3   Results

We first obtained the trajectories of three fish and randomly selected one of them as the target fish. By adjusting the synaptic weights on the output unit, we could train the output unit to generate the desired trajectory of the target fish based on its neighbors' trajectories (Fig.2a, b). The model was trained in 10 trials, each lasting for 60 minutes. According to first-order reduced and controlled error (FORCE) learning [13], the weights of the readout connections changed rapidly, immediately driving the output toward the target trajectory. To evaluate the success of the training, we computed the magnitude of the readout weight vector (Fig. 2c), which reaches a set of static weights that eventually generate the target trajectory without requiring further modification. The learning nearly converges after approximately 15 minutes in each trial.

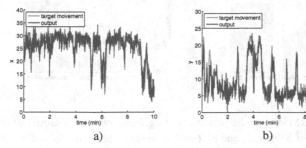

a)   b)

Fig. 2. a) and b) Trajectories of the model's output unit and the target fish throughout the first 10 min after a training trial. c) Length of the readout weight vector |w| after training. d) The RMS error in the network output for training and testing process, respectively.

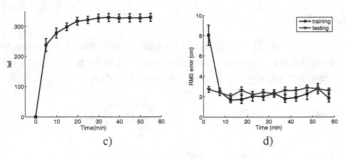

c)   d)

Fig. 2. (*Continued*)

The model was tested in 20 independent testing trials (three test fish per trial) without altering the output weights. The output trajectory was compared with the target fish trajectory. The testing success was evaluated by the RMS in the average of the test trials. As shown in Fig. 2d, the output unit accurately predicted the trajectory of the target fish from the trajectories of its neighbors. We also tested the consistency of the prediction in larger groups of four and five fish. The RMSs were not significantly different ($F(2, 57) = 1.67$, $p = 0.1969$) among 3, 4 and 5 fish groups. These results indicate that an individual fish in the group chooses its movements and positions based on its neighbors' trajectories.

For shoals of four fish, we compared the RMSs of a target fish responding to the trajectories of its two and three nearest neighbors. As shown in Fig. 3a, the average RMSs for the two and three nearest-neighbor inputs were 3.8253 cm and 5.4330 cm, respectively. These differences were statistically significant ($F(1,38) = 27.45$; $p = 6.27e-6$). In addition, for shoals of five fish (Fig. 3b), the average RMS statistically differed between inputting the trajectories of the nearest two and four fish ($F(1,38) = 66.69$; $p = 6.90e-10$). These results indicate that the target fish computes its likely response to each of its nearest neighbors.

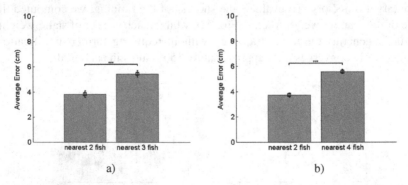

a)   b)

Fig. 3. a) Average RMSs of predictions after testing, based on the trajectories of a) 2 or 3 nearest neighbors in trials of 5 test fish, and b) 2 or 4 nearest neighbors in trials of 6 test fish. All data were acquired from 20 trials, each lasting for 60 minutes. Data are presented as means ± SEM; ***$p < 0.001$.

Having demonstrated that individual fish respond to their neighbors' movements, we then predicted the trajectory of fish exposed to toxic chemicals. As shown in Fig. 4, exposure to glyphosate exerted significant effects on the RMS of the test fish ($F(1,18) = 37.08$; $p = 9.38e-6$). The RMS during glyphosate exposure was almost twice that in the control tests. We also examined the RMS of the trajectories in consecutive 5-min time blocks. Again, the RMS significantly increased when the target fish were exposed to glyphosate, but not significantly affected by the exposure time ($p > 0.5$). Because the error between the model output and the target fish trajectory is sensitive to the environmental conditions, the model is potentially applicable to water quality monitoring.

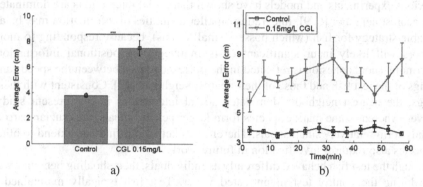

Fig. 4. a) Average RMSs under control conditions and during exposure to 0.15 mg/L CGL. b) Time-course of average RMSs in 5-min time blocks. Glyphosate (0.15 mg/L) was added at 0 min, and continued for 60 min. Prior to exposure, the fish swam for 30 min in control water. The data are those of Fig.4.a, presented as the means ± SEM; ***$p < 0.001$.

4   Discussion

We examined the interactions among shoaling fish by a reverse engineering strategy. Using a recurrent neural network, the strategy predicts the trajectories of individual fish from the movements of their neighbors. The connection within the recurrent neural network is fixed and task-independent. The information is encoded by the network dynamics. The output trajectory depends on an appropriate readout mechanism in the recurrent network dynamics. The model is trained by a learning method based on the recursive least squares rule, which is implemented in the FORCE algorithm [13]. FORCE learning rapidly reduces the magnitude of the difference between the actual and desired outputs. It retains its small value while seeking and eventually finding an accurate set of fixed readout weights requiring no further modification. Thus, we propose that the recurrent neural network is a very useful tool for modeling the trajectories of individual fish.

We also realized that complex collective animal behaviors can emerge from the interactions among individuals, and need not be explicitly coded as a global template. The main interaction rules are the spatial configurations of individuals and the distributions of particular variables such as speed, polarization, tuning rate and effective force [6, 7]. Nonetheless, whether individuals actually employ these interaction rules to guide their movements has not been previously supported by evidence. Here we show that the real-time movements of individual fish are directly predictable from the motions of their neighbors. Our results indicate that the target fish computes its likely response to each of its nearest neighbors. However, our model neglects environmental information and the motion trajectory of the target fish. Previous studies have shown that walls [6] and the target fish's trajectory [14] affect the acceleration and turning behavior of fish. Therefore, the effects of environmental information on fish responses and itself trajectory of the target fish should be identified in future study.

Previous experiments and models have shown that social interactions are dominated by the nearest neighbor [6, 9]. Integrating smaller quantities of information may be an affordable strategy for fish (which possess small brains), because responding to more neighbors will likely incur significant costs in integrating positional information. Furthermore, shoal cohesion is reflected in the correspondence between the speeds and headings of the test fish and those of their nearest neighbors [15]. Consistent with these findings, the nearest neighbors dominated social interactions in our present study. However, whereas some marine species form larger pelagic shoals, the crucian carp in our study form relatively small shoals. Therefore, whether our findings extend to other fish species also requires investigation in future work.

Although the test fish behaved differently as individuals, their shoaling behavior was stable during the control test in untreated water. Test fish typically maintained a distance of nearly one body length from their nearest neighbors. Conversely, exposure to different concentrations of glyphosate induces erratic swimming [16]. Such abnormal behavior contraindicates the use of glyphosate near aquatic environments. The findings indicate that when fish perceive a noxious contaminant, they avoid the area containing the chemical and reduce their shoaling behavior. Because the shoal is diminished under chemical exposure, the error between the model output and the target fish trajectory increases. The increased error signal is a useful indicator of water quality. Indeed, further exposure tests using various test chemicals over longer monitoring periods could enhance the generality of the proposed method in water quality monitoring.

Acknowledgments. This work is supported by the National Natural Science Foundation of China (61272310).

References

1. Viscido, S.V., Parrish, J.K., Grünbaum, D.: Individual Behavior and Emergent Properties of Fish Schools: a Comparison of Observation and Theory. Marine Ecology 273, 239–249 (2004)
2. Landeau, L., Terborgh, J.: Oddity and the 'Confusion Effect' in Predation. Animal Behaviour 34, 1372–1380 (1986)

3. Pitcher, T.J.: Functions of Shoaling Behaviour in Teleosts. In: The Behaviour of Teleost Fishes, pp. 294–337. Springer (1986)
4. Suboski, M.D., Bain, S., Carty, A.E., McQuoid, L.M., Seelen, M.I., Seifert, M.: Alarm Reaction in Acquisition and Social Transmission of Simulated-Predator Recognition by Zebra Danio Fish (Brachydanio Rerio). Journal of Comparative Psychology 104, 12 (1990)
5. Couzin, I.D.: Collective Cognition in Animal Groups. Trends in Cognitive Sciences 13, 36–43 (2009)
6. Herbert-Read, J.E., Perna, A., Mann, R.P., Schaerf, T.M., Sumpter, D.J., Ward, A.J.: Inferring the Rules of Interaction of Shoaling Fish. PNAS 108, 18726–18731 (2011)
7. Katz, Y., Tunstrom, K., Ioannou, C.C., Huepe, C., Couzin, I.D.: Inferring the Structure and Dynamics of Interactions in Schooling Fish. PNAS 108, 18720–18725 (2011)
8. Giardina, I.: Collective Behavior in Animal Groups: Theoretical Models and Empirical Studies. HFSP Journal 2, 205–219 (2008)
9. Bode, N.W.F., Franks, D.W., Wood, A.J.: Limited Interactions in Flocks: Relating Model Simulations to Empirical Data. Journal of the Royal Society 8, 301–304 (2011)
10. Haykin, S.S.: Adaptive filter theory. Pearson Education India (2007)
11. Chen, X.L., Huang, H.J.: Cyprininae. In: Wu, X.W. (ed.) Monographs of Cyprinidae in China. Shanghai People's Press, Shanghai (1982)
12. Xiao, G., Feng, M., Cheng, Z., Zhao, M., Mao, J., Mirowski, L.: Water Quality Monitoring Using Abnormal Tail-Beat Frequency of Crucian Carp. Ecotoxicology and Environmental Safety 111, 185–191 (2015)
13. Sussillo, D., Abbott, L.F.: Generating Coherent Patterns of Activity from Chaotic Neural Networks. Neuron 63, 544–557 (2009)
14. Yoshida, T., Minami, M., Mae, Y.: Fish Catching by Visual Servoing Using Neural Network Prediction. In: 2007 Annual Conference, SICE, Takamatsu, pp. 2372–2378 (2007)
15. Gautrais, J., Ginelli, F., Fournier, R., Blanco, S., Soria, M., Chate, H., Theraulaz, G.: Deciphering Interactions in Moving Animal Groups. PLoS Computational Biology 8, e1002678 (2012)
16. Morgan, J.D., Vigers, G.A., Farrell, A.P., Janz, D.M., Manville, J.F.: Acute Avoidance Reactions and Behavioral Responses of Juvenile Rainbow Trout (Oncorhynchus Mykiss) to Garlon 4®, Garlon 3A® and Vision® herbicides. Environmental Toxicology and Chemistry 10, 73–79 (1991)

Short-Term Wind Speed Forecasting Using a Multi-model Ensemble

Chi Zhang, Haikun Wei, Tianhong Liu, Tingting Zhu, and Kanjian Zhang

Key Laboratory of Measurement and Control of CSE, Ministry of Education,
School of Automation, Southeast University, Nanjing 210096, P.R. China
hkwei@seu.edu.cn

Abstract. Reliable and accurate short-term wind speed forecasting is of great importance for secure power system operations. In this study, a novel two-step method to construct a multi-model ensemble, which consists of linear regression, multi-layer perceptrons and support vector machines, is proposed. The ensemble members first compete with each other in a number of training rounds, and the one with the best forecasting accuracy in each round is recorded. Then, after all the training rounds, the occurrence frequency of each member is calculated and used as the weight to form the final multi-model ensemble. The effectiveness of the proposed multi-model ensemble has been assessed on the real datasets collected from three wind farms in China. The experimental results indicate that the proposed ensemble is capable of providing better performance than the single predictive models composing it.

Keywords: Wind speed forecasting, Model combination, Ensemble, Linear regression, Multi-layer perceptron, Support vector machine.

1 Introduction

Wind energy has been under large-scale development around the world in the last decade. However, a number of challenges in power system operations and planning have been posed, mainly due to the stochastic and intermittent nature of wind speed. One of the possible approaches to address these challenges is to improve the accuracy of wind speed and wind power forecasting [1].

A number of predictive models have been proposed and applied to the forecasting of wind speed. These models can be generally divided into two main groups: physical models and statistical models. Physical models usually employ the geographic and geomorphic conditions as inputs to predict the future wind speed. They are generally good at long-term forecasting and have been successfully applied in large-scale area weather prediction [2]. In contrast, statistical models just use previous historical data to realize the model construction, and they often do well in short-term forecasting. Among the conventional statistical models, the time series models are most widely used in practice [3,4]. In addition, with the development of artificial intelligence techniques, many intelligent

© Springer International Publishing Switzerland 2015
X. Hu et al. (Eds.): ISNN 2015, LNCS 9377, pp. 398–406, 2015.
DOI: 10.1007/978-3-319-25393-0_44

models have also been widely adopted for wind speed prediction, including fuzzy logic (FL) [5], artificial neural networks (ANNs) [6,7], support vector machines (SVM) [8,9], etc.

Recently, a large number of studies can be found in the literature about ensemble (or combined) forecasting models, which combine different predictive models together to attain better performance. In general, these ensemble models tackle the forecasting task in two steps, with the first step being to make predictions using multiple plausible models, and the second step being to combine all the generated forecasts into a single one according to some weighting strategy. However, in most studies, an ensemble for wind prediction usually consists of predictive models taken from one single class, e.g. neural networks [10] or support vector machines [11]. Moreover, the commonly used combination strategy is weighted average or weighted median [12]. In this paper, a novel multi-model ensemble is proposed to combine wind speed forecasts from different kinds of models. The development of the proposed ensemble includes two main steps. The first step consists of testing and recording models from different model classes. For this purpose, both linear (linear regression) and non-linear (multi-layer perceptron and support vector machine) models are utilized. In the second step, the forecasts yielded by the ensemble members are combined according to their prior performance to produce the final estimate of the wind speed. The experimental assessment of the proposed method is carried out on the basis of three real wind speed datasets.

The remainder of the paper is organized as follows. Section 2 briefly describes the ensemble members (single forecasting models) used in the study. Section 3 illustrates the proposed multi-model ensemble. The experimental results are presented and discussed in Section 4. Finally, Section 5 draws the conclusion of this work.

2   Single Forecasting Models

2.1   Linear Regression

Linear regression (LR) is a model for expressing the linear dependence of a response variable on several explanatory variables [13]. It has the form

$$y = w_0 + \sum_{i=1}^{d} w_i x_i = \mathbf{w}^\mathrm{T} \mathbf{x} \tag{1}$$

where w_0 and w_i are unknown model coefficients. In this study, the response variable y in Eq. 1 is y_t, and the explanatory variable x_i is y_{t-i}. In such case, the LR model can also be considered as an autoregressive (AR) model.

2.2   Multi-layer Perceptron

Multi-layer perceptron (MLP) is a feedforward neural network, which consists of an input layer, a hidden layer, and an output layer [14]. With d input neurons, m hidden neurons, the output of one hidden neuron is

$$z_j = g(\sum_{i=0}^{d} w_{i,j} x_i) \qquad (2)$$

where g is a sigmoid function, $w_{i,j}$ is the weight from input neuron i to hidden neuron j, x_i is the ith input and x_0 is always 1. The final output of the MLP is

$$y = \sum_{j=1}^{m} v_j z_j + b \qquad (3)$$

where v_j is the weight from hidden neuron j to the output neuron and b is the bias. Initially, all the weights and biases are assigned with small random values, and then updated according to a learning algorithm. The Levenberg-Marquardt training algorithm [15] is adopted in this study.

2.3   Support Vector Machine

The basic idea of SVM for regression is to map the input data into a high dimensional feature space via a non-linear function and to perform a linear regression in this feature space [16]. The regression is calculated by minimizing the following risk function

$$R[f] = \frac{1}{2}\|\mathbf{w}\|^2 + \frac{1}{2}C \sum_{i=1}^{N} L(y_i, f(\mathbf{x}_i)) \qquad (4)$$

where C is the regularization parameter and $L(y_i, f(\mathbf{x}_i))$ is the ϵ-insensitive loss function, which is defined as

$$L(y_i, f(\mathbf{x}_i)) = \begin{cases} |y_i - f(\mathbf{x}_i)| - \epsilon & |y_i - f(\mathbf{x}_i)| \geq \epsilon \\ 0 & \text{otherwise} \end{cases} \qquad (5)$$

where ϵ controls the width of the error margin allowed. By introducing the Lagrange multipliers α_i, α_i^*, the final form of function $f(\mathbf{x})$ is

$$f(\mathbf{x}) = \sum_{i=1}^{N} (\alpha_i - \alpha_i^*)k(\mathbf{x}_i, \mathbf{x}_j) + b \qquad (6)$$

where k is a kernel function. In this study, the radial basis function (RBF) is selected as the kernel function for SVM model and it is defined as

$$K(\mathbf{x}_i, \mathbf{x}_j) = e^{-\gamma\|\mathbf{x}_i - \mathbf{x}_j\|^2} \qquad (7)$$

where $\|\cdot\|$ denotes the 2-norm and γ is an adjustable parameter to determine the RBF kernel width.

3   The Proposed Ensemble Method

First, in order to obtain the training examples (input-output pairs) from the original wind speed time series, it is necessary to determine the model order (i.e. the number of previous observations used as inputs). Then, all the training examples are randomly permuted K times to obtain K training sets with different orders of examples, where the first 80% of each training set is used as a learning set and the last 20% is used as a validation set. After that, several training rounds are conducted on these different learning sets and validation sets. Specifically, a set of different types of models are built using the learning set and the one that produces the best forecasting performance on the validation set is recorded. Note that the input-output pairs used in each round for model construction and evaluation are different. This procedure can be regarded as a variant of K-fold cross validation. Moreover, in order to introduce diversity to the final ensemble, heterogeneous models are considered, including LR, MLPs and SVMs.

Suppose that the ensemble consists of m candidate members, after repeating the training rounds K times, the occurrence counts of each member, which is the number of being the best forecasting model, $t_i (i = 1, 2, ..., m)$ is recorded. This occurrence counts is used to calculate the weight (occurrence frequency) for each member in the multi-model ensemble. After determining the weights, all the training examples are utilized to retrain the ensemble members. Finally, the output for a given unknown test sample $\mathbf{x}$ is generated by model fusion as

$$f(\mathbf{x}) = w_1 f_1(\mathbf{x}) + w_2 f_2(\mathbf{x}) + ... + w_m f_m(\mathbf{x}) \tag{8}$$

where $w_i (i = 1, 2, ..., m)$ is the weight, which is calculated as the occurrence counts of the corresponding model $f_i (i = 1, 2, ..., m)$ dividing by the total number of the training rounds, i.e.

$$w_i = \frac{t_i}{K} \tag{9}$$

The multi-model ensemble is trained for one-step ahead prediction. For the case of multi-step ahead prediction, the iterative forecasting method is adopted.

4   Experimental Procedure and Results

4.1   Wind Data

The wind speed data used in this study are hourly mean observations collected from three wind farms in China. These measurements were recorded from April 1, 2012 to April 30, 2012. The three wind sites are located in different provinces of China: Jiangsu, Ningxia and Yunnan, respectively. Fig. 1 shows the three hourly wind speed time series, each consisting of 720 data points in total. Each dataset is partitioned into two parts: the first 600 samplings for training and the remaining 120 samplings for testing. In order to assess the performance of the proposed multi-model ensemble quantitatively, three error metrics: root mean square error (RMSE), mean absolute error (MAE), and mean absolute percentage error (MAPE), are employed, which are defined as follows:

$$\text{RMSE} = \sqrt{\frac{\sum\limits_{i=1}^{N}(y_i - \hat{y}_i)^2}{N}} \tag{10}$$

$$\text{MAE} = \frac{1}{N}\sum_{i=1}^{N}|y_i - \hat{y}_i| \tag{11}$$

$$\text{MAPE} = \frac{1}{N}\sum_{i=1}^{N}\left|\frac{y_i - \hat{y}_i)}{y_i}\right| \times 100\% \tag{12}$$

where N is the testing sample size, y_i and $\hat{y}_i$ are the observed and predicted values at time period i, respectively.

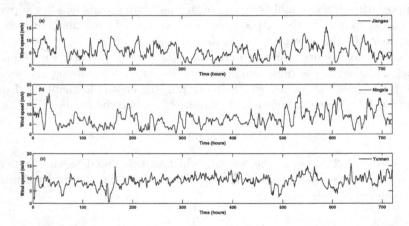

Fig. 1. The hourly mean wind speed data in: (a) Jiangsu, (b) Ningxia, and (c) Yunnan

4.2 Single Model Training

The hyperparameters in MLP and SVM is determined by 5-fold cross validation. Specifically, all the training examples are first divided into 5 roughly equal parts. For each $k = 1, 2, ..., 5$, the model is built using the other 4 parts, and the error is computed by the kth part. The cross validation error for a specific hyperparameter (or hyperparameter combination) is obtained by averaging over 5. The above procedure is repeated for different values of hyperparameters (or hyperparameter combinations) and the one that generates the smallest error is selected. After choosing the hyperparameters, all the training data are utilized to estimate the model coefficients.

4.3   Multi-model Ensemble Training

In this paper, the candidate models which constitute the multi-model ensemble are: (1) LR, (2) MLP, with one hidden layer and the number of neurons in the hidden layer varies from 1 to 10 with an increment of one (totally 10 MLPs), (3) SVM, with $\epsilon \in \{0, 0.01, 0.1\}$ and $\gamma \in \{2^{-2}, 2^{-1}, 2^0, 2^1, 2^2\}$ (totally 15 SVMs). Note that the regularization parameter C is fixed according to [17]. As a result, 26 candidate models are built and assessed in each training round and the one that produces the minimum error on the validation set is recorded. The number of training rounds is set to 50.

4.4   Comparison of Prediction Results

For purpose of comparison, Bayesian information criterion (BIC) is used to determine the model order, such that all the models compared have the same training examples for model construction. In this paper, the values of model order for all the three wind speed time series are 2 according to BIC. The prediction results of test set in Jiangsu produced by the best single models (LR, MLP and SVM) and the proposed multi-model ensemble are provided in Fig. 2. It can be seen that no single model could keep the best forecasting performance for all the prediction horizons. Meanwhile, there is also no optimal single model at a given forecasting horizon in terms of all the error measures. However, when comparing the performance of the component models with that of the proposed multi-model ensemble, it can be found that the multi-model ensemble has better or the closest accuracy to the optimal component model in terms of any metrics and any prediction horizons. Specifically, in terms of MAE, the multi-model ensemble has the best performance from 1-step to 6-step ahead prediction. And in

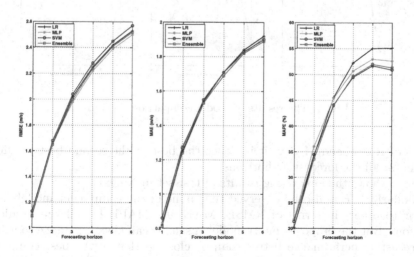

Fig. 2. The results of model comparison in Jiangsu.

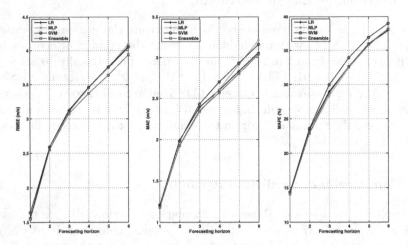

Fig. 3. The results of model comparison in Ningxia.

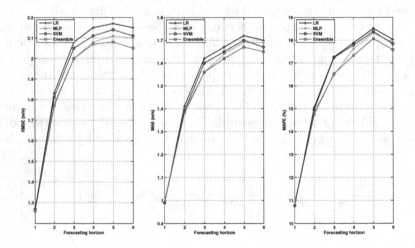

Fig. 4. The results of model comparison in Yunnan.

terms of RMSE and MAPE, it is either the best or the second best predictive model for all the forecasting horizons.

Fig. 3 shows the forecasting results of test set in Ningxia.

Similarly, it can also be observed that none of the component models is superior to others in terms of RMSE, MAE and MAPE for all the prediction horizons. But the proposed multi-model ensemble can always manage to achieve a forecasting performance better than or close to that of the best component model. Specifically, in this forecasting case, in terms of RMSE, the multi-model ensemble has the best accuracy from 1-step to 6-step ahead prediction. And in terms of MAE and MAPE, it is either the best or the second best predictive model for all the forecasting horizons.

Fig. 4 presents the prediction results of test set in Yunnan. Again, it can be found that the performance of the proposed multi-model ensemble is always better than or close to that of the best component model. And in this forecasting case, the superiority of the proposed ensemble is more significant as in terms of RMSE and MAE, it generates the smallest values for all the forecasting horizons.

5 Conclusion

In this paper, a novel multi-model ensemble for short-term wind speed prediction is presented. The ensemble consists of three different types of models, including linear regression, multi-layer perceptrons and support vector machines. These models with different hyperparameters are tested in a number of training rounds to determine their weights in the final ensemble. Based on the wind speed datasets from three wind farms in China, it is found that by applying the proposed method, the forecasting performance is better than or close to that of the optimal component model in terms of RMSE, MAE and MAPE for both one-step and multi-step ahead predictions. Thus, the proposed multi-model ensemble provides an effective way to solve the model selection problem in short-term wind speed prediction.

Acknowledgement. This research is supported by the National Natural Science Foundation of China (Grant No. 61374006), and by the Major Program of National Natural Science Foundation of China (Grant No. 11190015).

References

1. Jung, J., Broadwater, R.P.: Current status and future advances for wind speed and power forecasting. Renewable and Sustainable Energy Reviews 31, 762–777 (2014)
2. Lazić, L., Pejanović, G., Zivković, M.: Wind forecasts for wind power generation using the Eta model. Renewable Energy 35(6), 1236–1243 (2010)
3. Huang, Z., Chalabi, Z.S.: Use of time-series analysis to model and forecast wind speed. Journal of Wind Engineering and Industrial Aerodynamics 56(2), 311–322 (1995)
4. Erdem, E., Shi, J.: ARMA based approaches for forecasting the tuple of wind speed and direction. Applied Energy 88(4), 1405–1414 (2011)
5. Mohandes, M., Rehman, S., Rahman, S.M.: Estimation of wind speed profile using adaptive neuro-fuzzy inference system (ANFIS). Applied Energy 88(11), 4024–4032 (2011)
6. Welch, R.L., Ruffing, S.M., Venayagamoorthy, G.K.: Comparison of feedforward and feedback neural network architectures for short term wind speed prediction. In: International Joint Conference on Neural Networks, pp. 3335–3340. IEEE, Atlanta (2009)
7. Alanis, A.Y., Ricalde, L.J., Sanchez, E.N.: High Order Neural Networks for wind speed time series prediction. In: International Joint Conference on Neural Networks, pp. 76–80. IEEE, Atlanta (2009)

8. Mohandes, M.A., Halawani, T.O., Rehman, S., Hussain, A.A.: Support vector machines for wind speed prediction. Renewable Energy 29(6), 939–947 (2004)
9. Zhao, P., Xia, J., Dai, Y., He, J.: Wind speed prediction using support vector regression. In: 5th IEEE Conference on Industrial Electronics and Applications, pp. 882–886. IEEE, Taichung (2010)
10. Li, G., Shi, J., Zhou, J.: Bayesian adaptive combination of short-term wind speed forecasts from neural network models. Renewable Energy 36(1), 352–359 (2011)
11. Heinermann, J., Kramer, O.: Precise wind power prediction with SVM ensemble regression. In: Wermter, S., Weber, C., Duch, W., Honkela, T., Koprinkova-Hristova, P., Magg, S., Palm, G., Villa, A.E.P. (eds.) ICANN 2014. LNCS, vol. 8681, pp. 797–804. Springer, Heidelberg (2014)
12. Barrow, D.K., Crone, S.F., Kourentzes, N.: An evaluation of neural network ensembles and model selection for time series prediction. In: International Joint Conference on Neural Networks, pp. 1–8. IEEE, Barcelona (2010)
13. Hastie, T., Tibshirani, R., Friedman, J.: The elements of statistical learning. Springer, New York (2009)
14. Wei, H.K.: Theory and Methods for Neural Networks Architecture Design. National Defence Industry Press, Beijing (2005)
15. Haykin, S.S.: Neural networks and learning machines. Pearson Education, Upper Saddle River (2009)
16. Smola, A.J., Schölkopf, B.: A tutorial on support vector regression. Statistics and Computing 14(3), 199–222 (2004)
17. Cherkassky, V., Ma, Y.: Practical selection of SVM parameters and noise estimation for SVM regression. Neural Networks 17(1), 113–126 (2004)

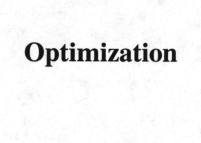

Optimization

Power Control Optimization Method for Transmitted Signals in OFDM Communication Systems

Jing Gao, Xiaochen Ding, and Xin Song

School of Information Science and Engineering, Northeastern University,
110004 Shenyang, China
{summergj,xiaochend,xsong}@126.com

Abstract. Orthogonal frequency division multiplexing (OFDM) introduces large peak power of transmitted signals in time, which can result in significant signal distortion in the presence of nonlinear amplifiers. Partial transmit sequence (PTS) are well-known techniques for peak-power reduction in OFDM. However, the exhaustive search of phase factors in conventional PTS causes high computational complexity. In this paper, we present a suboptimal strategy for combining partial transmitted sequences that achieve good balance between computational complexity and power control performance. The simulation results show that the proposed algorithm can not only reduces the PAPR significantly, but also decreases the computational complexity

Keywords: OFDM, Peak Power, PAPR, PTS.

1   Introduction

Orthogonal frequency-division multiplexing (OFDM) has been standardized in many wireless applications with high-speed data transmission due to its various advantages such as high spectral efficiency and robustness to channel fading [1].Hence, this transmission technique has been proposed in various wireless communication standards such as digital audio and video broadcasting, asymmetric-digital-subscriber-line modems, and wireless local-area-networks systems, such as the IEEE 802.11 and IEEE 802.16. Additionally, OFDM is being considered for future broadband applications [2].

Due to the large number of subcarriers, OFDM systems have a large dynamic signal range with a very high peak-to-average power ratio (PAPR). As a result, several proposals have been suggested and studied in the literature. For instance, we find clipping with filtering, block coding, optimization with tone reservation (TR), selected mapping, partial transmit sequence (PTS) and others [3-5]. Among these methods, PTS scheme is the most efficient approach and a distortionless scheme for PAPR reduction by optimally combining signal subblocks [6-8]. In PTS, the transmitted signal is made to have low PAPR by optimally combining signal subblocks, which introduce additional complexity but provide improved PAPR statistics with little cost in efficiency. To address the computational complexity, many suboptimal PTS techniques have been developed. The iterative flipping PTS (IPTS) in [9] has computational complexity linearly proportional to the number of subblocks.

© Springer International Publishing Switzerland 2015
X. Hu et al. (Eds.): ISNN 2015, LNCS 9377, pp. 409–417, 2015.
DOI: 10.1007/978-3-319-25393-0_45

In this paper, a suboptimal approach based on heuristic optimization is proposed to obtain better PAPR reduction performance with less complexity compared with the conventional PTS scheme. The rest of this paper is organized as follows. Section 2 defines PAPR of OFDM signals, and PTS algorithm is showed in this study. Section 3 shows how the problem of PAPR reduction be formulated as an optimization problem for OFDM systems, and suboptimal algorithm is proposed. The performance of the proposed algorithm is evaluated through computer simulations in section 4. Finally, Section 5 summarizes and concludes the paper.

2   OFDM Systems and PTS Technique

With OFDM modulation, a block of N data symbols (one OFDM symbol) $\{X_n, n = 0,1,\cdots,N-1\}$ will be transmitted in parallel such that each modulates a different subcarrier from a set $\{f_n, n = 0,1,\cdots,N-1\}$. The N subcarriers are orthogonal, i.e. $f_n = n\Delta f$, where $\Delta f = 1/NT$ and T is the symbol period. The complex envelope of the transmitted OFDM signal is given as

$$x_n(t) = \frac{1}{\sqrt{N}} \sum_{n=0}^{N-1} X_n e^{j2\pi f_n t} \qquad 0 \le t \le NT \tag{1}$$

The PAPR of the transmitted OFDM signal of (1) is defined as

$$\text{PAPR} = \frac{\max\limits_{0 \le t \le NT} |x_n(t)|^2}{E[|x_n(t)|^2]} = \frac{\max\limits_{0 \le t \le NT} |x_n(t)|^2}{\frac{1}{NT} \int\limits_0^{NT} |x_n(t)|^2 \, dt} \tag{2}$$

where E[·] denotes the expected value. Then, the complementary cumulative distribution function (CCDF), is the probability that the PAPR of an OFDM symbol exceeds the given threshold PAPR_0, which can be expressed as

$$\text{CCDF} = \Pr(\text{PAPR} > \text{PAPR}_0) \tag{3}$$

In a typical OFDM system with PTS technique to reduce the PAPR, the input data block X is partitioned into M disjoint subblocks, which are represented by the vectors $X^{(m)} = \{X_0^{(m)} X_1^{(m)} \cdots X_{N-1}^{(m)}\}$, therefore

$$X = \sum_{m=0}^{M-1} X^{(m)} \tag{4}$$

then the subblocks $X^{(m)}$ are transformed into time-domain partial transmit sequences by IFFTs. These partial sequences are independently rotated by phase factors $b_m = e^{j\theta_m}$, $\theta_m \in \{\frac{2\pi k}{W} |_{k=0,1,\cdots W-1}\}$. The object is to optimally combine the M subblocks to obtain the OFDM signals with the lowest PAPR

$$x = \sum_{m=0}^{M-1} b_m x^{(m)}$$ (5)

Assuming that there are W phase angles to be allowed, thus there are $D = W^M$ alternative representations for an OFDM symbol. The block diagram of the PTS technique is shown in Fig.1.

Cimini and Sollenberger's iterative PTS (IPTS) technique is developed as a sub-optimum technique for PTS. The IPTS technique only use binary phase factors, which reduces the search complexity significantly, but there is some gap between its PAPR performance and that of the ordinary PTS technique.

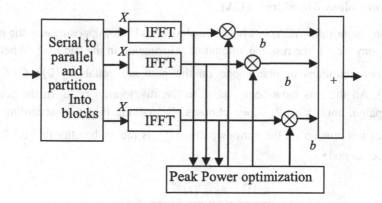

Fig. 1. Block diagram of the PTS technique

3   Suboptimal PTS Algorithm

The searching sequences of PTS can be formulated as a combinatorial optimization problem. Then, an Ant colony-annealing search based PTS algorithm, is proposed to achieve better PAPR reduction with low complexity.

3.1   Mathematics Model of PTS Algorithm

The optimization problem of PTS, which is trying to find the aggregate of phase factors vector b_m to yield the OFDM signals with the minimum PAPR, can be considered as the combinatorial optimization problem. In other words, the objective function (6) is to minimize the PAPR of the transmitted OFDM signals. Constraint (7) ensures the phase factors to be a finite set of values $0 \le \theta_m < 2\pi \, (0 \le m \le M-1)$

To minimize

$$f(b) = \|x\|_\infty^2 = \left\| \sum_{m=0}^{M-1} b_m x^{(m)} \right\|_\infty^2 \tag{6}$$

Subject to (7)

$$b = \{e^{j\theta_m}\}^{M-1} \tag{7}$$

where $\theta_m \in \{\frac{2\pi k}{W}|_{k=0,1,\cdots W-1}\}$.

3.2   Ant Colony Algorithm (ACA)

Ant colony algorithm (ACA) is a heuristic algorithm [10]. Supposing m is the number of ant colony. τ_{ij} is the residual amount of information in the sub road. When $t = 0$, various concentrations of pheromone on the path are equal, $\tau_{ij}(0) = C$ (C is the constant). All the ants have been placed to the distribution node. In the course of the campaign, ant k ($k = 1, 2, \cdots, m$) chooses the moving direction according to the amount of information on the various paths. P_{ij}^k is the probability that ant k moves from node i to node j.

$$P_{ij}^k = \begin{cases} \dfrac{\left[\tau_{ij}(t)\right]^\alpha * \left[\eta_{ij}(t)\right]^\beta}{\sum\limits_{s \in allow_k} \left[\tau_{is}(t)\right]^\alpha * \left[\eta_{is}(t)\right]^\beta}, & s \in allow_k \\ 0 & , \quad s \notin allow_k \end{cases} \tag{8}$$

The natural ant colony is different from the artificial ant that has a memory function. Set $tabu_k$ ($k = 1, 2, \cdots, m$) was utilized to record the current traversed nodes and dynamically adjusted by the process of evolution. η_{ij} is the heuristic function, which denotes the expectation of ant transferred from node i to node j. $allow_k$ ($k = 1, 2, \cdots, m$) denotes the nodes allow to be accessed of ant k. As time progresses, the element declining until empty, it means that all nodes have access to finished. α is the pheromone importance factor, the larger of the value, pheromone concentration play a greater role in metastasis. β is the heuristic function importance factor, the larger the value, heuristic function play a   greater role in metastasis.

As time goes on, the previous pheromone left gradually decays. After the completion of a cycle, the pheromone was adjusted according to equation 9.

$$\begin{cases} \tau_{ij}(t+1) = (1-\rho)\tau_{ij}(t) + \Delta\tau_{ij} \\ \Delta\tau_{ij} = \sum\limits_{k=1}^{n} \Delta\tau_{ij}^k \end{cases}, \quad 0 < \rho < 1 \tag{9}$$

ρ denotes pheromone evaporation degree. Where, $\tau_{ij}(t+1)$ is the amount of pheromone on the path, $\Delta\tau_{ij}$ is the pheromone increment of this cycle path.

Ant cycle system model through the use of the whole path information (the total length of path) calculate the degree of pheromone release. In ant cycle system model, the formula is show as equation 10.

$$\Delta\tau_{ij}^{k} = \begin{cases} Q/L_{k}, & \text{if ant } k \text{ travel}(i, j)\text{in the circle} \\ 0, & \text{others} \end{cases} \tag{10}$$

3.3   Simulated Annealing (SA)

Simulated annealing algorithm (SA) is a kind of heuristic random optimum algorithm [11]. It attempts to avoid being trapped in a local optimum by sometimes allowing the temporal acceptance of inferior solutions. The acceptance or rejection of an inferior solution is probabilistically determined by Metropolis and repeats sampling process with the temperature declining and finally gains the problem's global optimal solution.

The steps of the SA algorithm are illustrated as follow.

a) Initial solution: Select a initial state, the main control parameters need to be set are cooling rate k, the initial temperature T_0, the end temperature T_{end} and each iteration times M.
Generation a new solution: Perturb the current state ω randomly to generate a new solution ω'.
Metropolis criterion: Calculate the increment $\Delta f = f(\omega') - f(\omega)$, $f(\omega)$ is the evaluation function.

$$P = \begin{cases} \exp(-\Delta f / T), & \Delta f < 0 \\ 1, & \Delta f \geq 0 \end{cases} \tag{11}$$

If $\Delta f < 0$, accept the new path with probability 1, else accept the new path with probability P.

Cooling: Gradually reduce the temperature T by cooling rate i.e. $T = k*T$, make sure let the $T \to 0$, if $T < T_{end}$ then output the optimal solution, the program would be terminated.

3.4   Hybrid Algorithm ACA-SA

ACA-SA algorithm using ant colony algorithm to achieve precise searching local optima, and the global optimal judgment based on simulated annealing. The flowchart of hybrid algorithm is show in Figure 2.

As above mentioned, the steps, in which the ACA-SA is presented for searching the optimal combination of phase factors in PTS algorithm. Initially, the phase factors and ACA-SA algorithm parameters are specified. Using randomly generated initial phase factors, the efficiency of the PAPR reduction is determined by means of the ACA-SA algorithm.

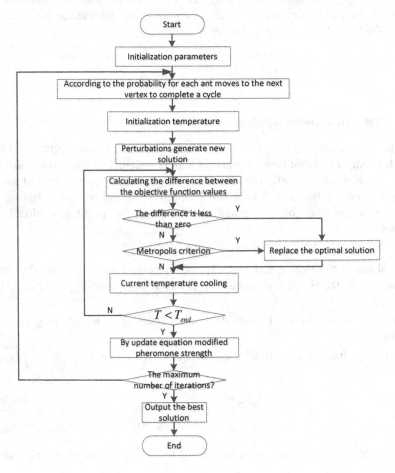

Fig. 2. The flowchart of hybrid algorithm

4   Simulation Results

In this section, we present some simulations to demonstrate the performance of the PTS technique based on ant colony-simulated annealing (ACA-SA) search algorithm. We assume that random QAM modulated OFDM symbols were generated with $N = 256$ subcarriers. The transmitted signal is oversampled by a factor of $L = 4$. Some results of the CCDF are simulated for the OFDM system, in which $M=16$ subblocks employing random partition and the phase factors vector b in $\{+1, -1\}^M$ uniformly distributed random variable are used for PTS. Ant colony size (the number of ants) is set to $m = 50$, pheromone importance factor $\alpha = 6$, heuristic function importance factor $\beta = 7$. pheromone evaporation factor $\rho = 0.1$, pheromone total release $Q = 50$, cooling rate $k = 0.95$, the initial temperature $T_0 = 20$, end temperature $T_{end} = 0.1 \wedge 5$, iteration times $iter\_num = 20$.

The curves labeled by "PTS" are obtained by Monte Carlo searching with full enumeration of W^M phase factors, the curves labeled by "IPTS" are obtained by iterative searching of M phase factors with the Iterative PTS technique, and the curves labeled by "ACA-SA" are obtained by the proposed technique.

In Fig.3, It is shown that when $Pr(PAPR > PAPR_0) = 10^{-4}$, the $PAPR_0$ of the original OFDM is 11.5dB, PTS is 9.4dB with exhaustive searching number $2^{16} = 65536$, while ACA-SA is 10.5dB with iteration number $iter\_num = 20$. Therefore, ACA-SA technique can offer good PAPR reduction with low complexity.

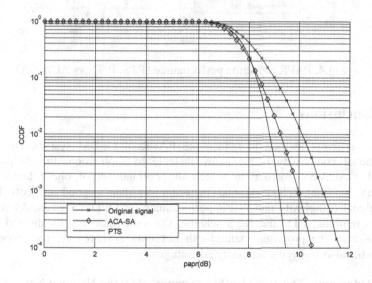

Fig. 3. PAPR reduction performance (PTS vs ACA-SA)

The performance of Iterative PTS (IPTS) algorithm is compared with ACA-SA in Fig. 4. When $Pr(PAPR > PAPR_0) = 10^{-4}$, the $PAPR_0$ of the original OFDM is 11.8dB, IPTS is 10.6dB with iterative searching number 16, PTS is 8.9dB with exhaustive searching number $2^{16} = 65536$, while ACA-SA is 9.6dB with search number $iter\_num = 20$. It is evident that the ACA-SA can provide the better performance of PAPR reduction than that of IPTS while keeping similar low complexity.

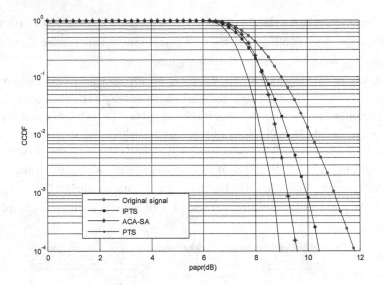

Fig.4. PAPR reduction performance (PTS, IPTS vs ACA-SA)

5   Conclusions

PTS is an efficient method to reduce the PAPR of OFDM system, but its high computational complexity is the main obstacle to application. In this paper, we presented a ACA-SA method to search the optimal set of phase factors of all subcarriers for the PTS technique in order to obtain good tradeoff between computational complexity and PAPR performance for OFDM signals. As compared to the conventional PTS and IPTS scheme, the simulation results showed that the performance of the proposed method can not only achieved relative good PAPR reduction but also enjoyed complexity advantages.

Acknowledgments. This work has been supported by the National Natural Science Foundation of China under Grant No.61403069 and No.61473066, Natural Science Foundation of Hebei Province under Grant No. F2014501055, the Program of Science and Technology Research of Hebei University No. ZD20132003, the Program for New Century Excellent Talents in University No. NCET-12-0103.

References

1. Tao, J., Wu, Y.Y.: An Overview: Peak-to-Average Power Ratio Reduction Techniques for OFDM signals. IEEE Transactions on Broadcasting, 257–268 (2008)
2. Ou, J.L., Zeng, X.P., Tian, F.C., Wu, H.W.: Simplified Repeated Clipping and Filtering with Spectrum Mask for PAPR Reduction. Journal of Convergence Information Technology, 251–259 (2011)
3. Yang, L., Hu, W.J., Soo, K.K., Siu, Y.M.: Swapped SLM scheme for reducing PAPR of OFDM systems. Electronics Letters, 1608–1609 (2014)

4. Islam, M.A., Ahmed, N., Ahamed, N.U., Rahman, M., Aljunid, S.A.: PAPR Reduction in an OFDM system using Recursive Clipping and Filtering Technique. WSEAS Transactions on Communications, pp. 291–297 (2014)
5. Wang, C.L., Kum, S.J.: Novel Conversion Matrices for Simplifying the IFFT Computation of an SLM-Based Reduction Scheme for OFDM Systems. IEEE Transactions on Communications, 1903–1907 (2009)
6. Duanmu, C.J., Chen, H.T.: Reduction of the PAPR in OFDM Systems by Intelligently Applying Both PTS and SLM Algorithms. Wireless Personal Communications, 849–863 (2014)
7. Hong, E., Park, Y.G., Lim, S.C., Har, D.S.: Adaptive phase rotation of OFDM signals for PAPR reduction. IEEE Transactions on Consumer Electronics, 1491–1495 (2011)
8. Ku, S.J.: Low-Complexity PTS-Based Schemes for PAPR Reduction in SFBC MIMO-OFDM Systems. IEEE Transactions on Broadcasting, 650–658 (2014)
9. Cimini Jr., L.J., Sollenberger, N.R.: Peak-to-average power ratio reduction of an OFDM signal using partial transmit sequences. IEEE Communications Letters, 86–88 (2000)
10. Liao, T.J., Socha, K., et al.: Ant colony optimization for mixed-variable optimization problems. IEEE Transactions on Evolutionary Computation, 503–518 (2014)
11. Tran, T.D., Zelinka, I., Vo, H.D.: Use of Simulated Annealing for Adaptive Control System. International Journal of Energy Optimization and Engineering, 42–54 (2013)

A Neurodynamic Optimization Approach to Bilevel Linear Programming

Sitian Qin[1,3], Xinyi Le[2], and Jun Wang[2,3]

[1] Department of Mathematics, Harbin Institute of Technology at Weihai, China
[2] Department of Mechanical and Automation Engineering,
The Chinese University of Hong Kong, Hong Kong
[3] School of Control Science and Engineering,
Dalian University of Technology, Dalian, China
qinsitian@163.com,
{xyle,jwang}@mae.cuhk.hk

Abstract. This paper presents new results on neurodynamic optimization approach to solve bilevel linear programming problems (BLPPs) with linear inequality constraints. A sub-gradient recurrent neural network is proposed for solving the BLPPs. It is proved that the state convergence time period is finite and can be quantitatively estimated. Compared with existing recurrent neural networks for BLPPs, the proposed neural network does not have any design parameter and can solve the BLPPs in finite time. Some numerical examples are introduced to show the effectiveness of the proposed neural network.

Keywords: Bilevel linear programming problem, sub-gradient recurrent neural network, convergence in finite time.

1   Introduction

The well-known bilevel programming problems (BPPs) have received increasing attention since their important applications in various aspects (see [5]). However, the BPPs have been proved to be NP-hard, and their formulation has inherent difficulties even with respect to the notion of a solution ([2]). Even the simplest model in bilevel programming, i.e., the linear bilevel program, is strongly NP-hard. Hence, it is a challenge for us to find an effective approach to solve such NP-hard problem.

Recently, many researchers derived serval numerical algorithms to solve the BPPs (see [8,13,18]). In [8], an augmented Lagrangian multiplier method was applied to solve a smoothed nonlinear program to obtain an approximate optimal solution of the nonlinear bilevel programming problems. In [13], using the merit function technique, Muu et al. proposed a branch-and-bound algorithm for finding a global optimal solution to the bilevel convex quadratic problem. Wan et al. in [18] proposed a novel evolutionary distribution algorithm for solving a special class of nonlinear bilevel programming problems.

© Springer International Publishing Switzerland 2015
X. Hu et al. (Eds.): ISNN 2015, LNCS 9377, pp. 418–425, 2015.
DOI: 10.1007/978-3-319-25393-0_46

However, the conventional numerical algorithm is usually less effective for BPPs, especially for large-scale BPPs. That is because the computing time of the conventional numerical algorithm greatly depends on the dimension and the structure of BPPs. Then, one possible and promising approach to solve BPPs in real time is to employ recurrent neural networks based on circuit implementation [4]. Since the seminal work in [9], recurrent neural networks for the BPPs and their engineering applications have been widely investigated (see [6,7,11,17]). For example, in [17], Shih etc proposed a recurrent neural network to solve the BPPs based on the approach proposed by Hopfield and Tank [9]. Meanwhile, by introducing a changed perturbed Fischer-Burmeister function, the authors in [11] proposed neural network to get the approximate optimal solution of the nonlinear BPPs. He et al. in [6] proposed a neural network to solve convex quadratic bilevel programming problems (CQBPPs), which is modeled by a nonautonomous differential inclusion. In [6,7,11], it was all shown that the equilibrium points sequence of the proposed neural networks can approximately converge to an optimal solution of CQBPP under certain conditions. However, they didn't prove the stability of the equilibrium point. It was only proved that the cluster point set of the state of the proposed neural network belongs to its equilibrium point set. It is well known that convergence in finite time for neural networks plays an important role in optimization. That is because in most of case we would like to obtain the accurate optimal solutions only in finite time. And convergence in finite time is of special interest for designing real-time neural optimization solvers. Neural networks with finite-time convergence has been investigated extensively in the literature ([14]). However, as far as we know, there are few literatures concerning neural networks with finite-time convergence for solving bilevel programming problems.

Motivated by above works, in this paper, we will propose a sub-gradient recurrent neural network to solve bilevel linear programming problems. The structure of this paper is outlined as follows. In section II, we present some related preliminaries. In section III and IV, we introduce bilevel linear programming problems and the related sub-gradient recurrent neural network. In section V, we present several numerical simulations to show the effectiveness of our results. Finally, the main conclusions drawn in the paper are summarized.

2 Preliminaries

Here, we present some definitions and properties, which are needed in the remainder of this paper. We refer readers to [1] for more thorough discussions.

A function $f : \mathbb{R}^n \to \mathbb{R}$ is said to be *regular* at x, if for all $v \in \mathbb{R}^n$, the usual one-sided directional derivative $f'(x; v)$ exists and

$$f'(x; v) = \limsup_{y \to x_0, t \downarrow 0} \frac{f(y + tv) - f(y)}{t}.$$

Let $f : \mathbb{R}^n \to \mathbb{R}$ be a convex function. For any $x \in \mathbb{R}^n$, the *subdifferential* of f at x is defined as $\partial f(x) = \{\xi \in \mathbb{R}^n : f(x) - f(y) \leq \langle \xi, x - y \rangle, \forall y \in \mathbb{R}^n\}$.

We next introduce two important set-valued maps k and h, which are the subdifferential of functions $|t|$ and $t^+ = \max\{t, 0\}$,

$$k(t) = \partial|t| = \begin{cases} 1 & t > 0; \\ -1 & t < 0; \\ [-1, 1] & t = 0. \end{cases} \qquad h(t) = \partial t^+ = \begin{cases} 1 & t > 0; \\ 0 & t < 0; \\ [0, 1] & t = 0. \end{cases} \tag{1}$$

Regular function has a very important property (i.e. chain rule), which has been used in many papers (see [15]).

Lemma 1. *(Chain rule) If $V(x) : \mathbb{R}^n \to \mathbb{R}$ is regular and $x(t) : [0, +\infty) \to \mathbb{R}^n$ is absolutely continuous on any compact interval of $[0, +\infty)$, then $x(t)$ and $V(x(t)) : [0, +\infty) \to \mathbb{R}$ are differentiable, and*

$$\dot{V}(x(t)) = \langle \xi, \dot{x}(t) \rangle, \quad \forall \xi \in \partial V(x(t)),$$

for a.e. $t \in [0, +\infty)$.

3 Bilevel Linear Programming Problem

In this paper, we will study the following bilevel linear programming problems (BLPP),

$$
\begin{aligned}
(UP) \min_{x,y} \ & F(x, y) = c_1^T x + d_1^T y \\
\text{s.t.} \ & A_1 x + B_1 y \le b_1, \\
(LP) \ & \begin{cases} y \in \arg\min_{y} f(x, y) = c_2^T x + d_2^T y \\ \text{s.t.} \ A_2 x + B_2 y \le b_2 \end{cases}
\end{aligned}
\tag{2}
$$

where $c_1, c_2 \in \mathbb{R}^n, d_1, d_2 \in \mathbb{R}^m, A_1 \in \mathbb{R}^{p \times n}, B_1 \in \mathbb{R}^{p \times m}, A_2 \in \mathbb{R}^{q \times n}, B_2 \in \mathbb{R}^{q \times m}, b_1 \in \mathbb{R}^p, b_2 \in \mathbb{R}^q$. The term (UP) above is called upper level problem and (LP) above is called lower level problem, and correspondingly the terms x, y in (2) are the upper level variable and the lower level variable respectively. Throughout the rest of the paper, we make the following assumptions:

Assumption 1. *The constraint region of the BLPP (2)*

$$S = \{(x^T, y^T) : A_1 x + B_1 y \le b_1, A_2 x + B_2 y \le b_2\}$$

is nonempty and compact.

Based on the assumption 1, the BLPP (2) has at least an optimal solution. According to the Karush-Kuhn-Tucker (KKT) Theorem, BLPP (2) can be degenerated to the following form,

$$
\begin{aligned}
\min_{x,y} \ & F(x, y) = c_1^T x + d_1^T y \\
\text{s.t.} \ & A_1 x + B_1 y \le b_1, \quad A_2 x + B_2 y \le b_2, \quad d_2 + B_2^T \mu = 0, \\
& -\mu^T (A_2 x + B_2 y - b_2) = 0, \quad -\mu \le 0,
\end{aligned}
\tag{3}
$$

where $\mu \in \mathbb{R}^q$ is a parameter. In general, the problem (3) is one element of special classes of mathematical programming problems called mathematical programs with equilibrium constraints (MPECs). MPECs are special classes of non-linear programming problems with equilibrium conditions in the constraints, and are inherently ill-posed (see [12,16]). We next introduce the following assumption,

Assumption 2. *The matrix B_2 is invertible and $(B_2^T)^{-1}d_2 \leq 0$.*

Based on the above assumption 2 and by (3), we have that $\mu = -(B_2^T)^{-1}d_2$. For convenience, we suppose

$$z = (x^T, y^T)^T, \quad C = (c_1^T, d_1^T)^T, \quad A = \begin{pmatrix} A_1 & B_1 \\ A_2 & B_2 \end{pmatrix},$$
$$b = (b_1^T, b_2^T)^T, \quad B = (d_2^T B_2^{-1} A_2, d_2^T), d = d_2^T B_2^{-1} b_2. \tag{4}$$

Hence, the problem (3) can be expressed as the following simple form,

$$\begin{aligned} \min & \; C^T z \\ \text{s.t. } & Az \leq b, \quad Bz = d. \end{aligned} \tag{5}$$

Then, according to Theorem 1 in [3], the following lemma holds.

Lemma 2. *z is an optimal solution of the linear programming problem (5) if and only if there exist $\lambda \in \mathbb{R}^{p+q}$ and $\nu \in \mathbb{R}^1$ such that*

$$C - A^T\lambda - B^T\nu = 0, Az \leq b, \lambda \leq 0, \lambda^T(Az - b) = 0, Bz = d. \tag{6}$$

From (6), one has $A^T\lambda = C - B^T\nu$ and $\lambda^T Az = \lambda^T b$, which implies that $(C - B^T\nu)^T z = \lambda^T b$. Combing with (6), we have $C^T z - \nu^T d = \lambda^T b$. Hence,

Lemma 3. *z is an optimal solution of the linear programming problem (5) if and only if there exist $\lambda \in \mathbb{R}^{p+q}$ and $\nu \in \mathbb{R}^1$ such that*

$$\begin{aligned} C - A^T\lambda - B^T\nu &= 0, \; Az \leq b, \\ C^T z - \nu^T d &= \lambda^T b, \quad \lambda \leq 0, \quad Bz = d. \end{aligned} \tag{7}$$

4   Recurrent Neural Network for BLPP (2)

In this section, based on Lemma 3, we will propose one recurrent neural network for solving BLPP (2) in finite time. For convenience, we let $X := (z^T, \lambda^T, \nu^T)^T$, and construct an energy function as follows,

$$\begin{aligned} E(X) := & \; \|C - A^T\lambda - B^T\nu\|_1 + \|(Az - b)^+\|_1 \\ & + \|\lambda^+\|_1 + |C^T z - \nu^T d - \lambda^T b| + \|Bz - d\|_1, \end{aligned} \tag{8}$$

where $\lambda^+ = (\lambda_1^+, ..., \lambda_{p+q}^+)^T$ and $t^+ = \max\{t, 0\}$.

Lemma 4. *$E(X) \geq 0$, and $E(X)$ is a convex function. $E(X) = 0$ if and only if $X := (z^T, \lambda^T, \nu^T)^T$ is a KKT point of the linear programming problem (5).*

After simple calculation, we have

$$\partial E(X) = \begin{pmatrix} A^T H(Az - b) + CK(C^T z - \nu^T d - \lambda^T b) + B^T K(Bz - d) \\ -AK(C - A^T\lambda - B^T\nu) + H(\lambda) - bK(C^T z - \nu^T d - \lambda^T b) \\ -BK(C - A^T\lambda - B^T\nu) - dK(C^T z - \nu^T d - \lambda^T b) \end{pmatrix} \quad (9)$$

where $K(u) = (k(u_1), ..., k(u_N))^T$, $H(v) = (h(v_1), ..., h(v_M))^T$ with appropriate dimension N and M. Here, k, h are defined in (1).

Based on the above results, we propose a sub-gradient recurrent neural network to solve the BLPP (2), with the following dynamical equations

$$\frac{d}{dt}X(t) \in -\varepsilon\partial E(X(t)), \quad (10)$$

where ε is a positive constant.

Theorem 1. $(x^{*T}, y^{*T})^T$ *is an optimal solution of the BLPP (2) if and only if there exist* $\lambda^* \in \mathbb{R}^{p+q}$ *and* $\nu^* \in \mathbb{R}^1$ *such that* $X^* := (x^{*T}, y^{*T}, \lambda^{*T}, \nu^{*T})^T$ *is an equilibrium point of recurrent neural network (10).*

Proof. Suppose that $z^* = (x^{*T}, y^{*T})^T$ is an optimal solution of the BLPP (2). From Lemma 3 and Lemma 4, there exist $\lambda^* \in \mathbb{R}^{p+q}$ and $\nu^* \in \mathbb{R}^1$ such that $E(z^*, \lambda^*, \nu^*) = 0$. Hence, by the non-negativity of E, $X^* := (z^{*T}, \lambda^{*T}, \nu^{*T})$ is a global minimal point of E. Then, $0 \in \partial E(X^*)$. That is, X^* is an equilibrium point of recurrent neural network (10).

Conversely, let $X^* := (x^{*T}, y^{*T}, \lambda^{*T}, \nu^{*T})^T$ be an equilibrium point of recurrent neural network (10). From the definition of the subdifferential in Preliminary, we have $E(X^*) - E(X) \le 0$, for any X. By the assumption 1 and lemma 4, it is clear that $z^* = (x^{*T}, y^{*T})^T$ is an optimal solution of the BLPP (2).

In this paper, we let $\mathbb{M}$ be the equilibrium set of neural network (10), i.e., $\mathbb{M} = \{X : 0 \in \partial E(X)\}$. According to the assumption 1, the set $\mathbb{M}$ is nonempty. By $m(\partial E(X))$, we denote the element of $\partial E(X)$ with smallest norm, i.e., $m(\partial E(X)) \in \partial E(X)$ and $\|m(\partial E(X))\| = \min\{\|\eta\| : \eta \in \partial E(X)\}$.

Lemma 5.

$$M_0 := \inf_{X \notin \mathbb{M}} \|m(\partial E(X))\| > 0 \quad (11)$$

Proof. Firstly, K and H in (9) have finite values, since k, h defined in (1) only take a finite number of different values. Hence, $\partial E(X)$ in (9) takes a finite number of different values. Similarly, $m(\partial E(X))$ also takes a finite number of different values. Without loss of generality, we assume that the values of $\|m(\partial E(X))\|$ are,

$$0, M_1, M_2, ..., M_N.$$

On the other hand, if $X \notin \mathbb{M}$, we have $\|m(\partial E(X))\| \ne 0$. Therefore,

$$\inf_{X \notin \mathbb{M}} \|m(\partial E(X))\| = \min\{M_1, M_2, ..., M_N\} > 0.$$

Theorem 2. *For any initial point X_0, the state $X(t)$ of neural network (10) is convergent to an equilibrium point of neural network (10) in finite time.*

Proof. Let

$$T = \frac{E(X(0))}{\varepsilon M_0^2}. \tag{12}$$

If $X(t) \notin \mathbb{M}$ for $t \in [0, T)$, by lemma 5, $\|m(\partial E(X(t)))\| \geq M_0$, for a.e. $t \in [0, T)$. Hence, by chain rule (i.e., Lemma 1), we have

$$\tfrac{d}{dt} E(X(t)) = -\varepsilon \|m(\partial E(X(t)))\|^2 \leq -\varepsilon M_0^2, \tag{13}$$

for a.e. $t \in [0, T)$. Integrating above inequality between 0 and T, we have $E(X(T)) \leq E(X(0)) - \varepsilon M_0^2 T = 0$, which implies that $E(X(T)) = 0$ by the positivity of E. Then, from Lemma 4 and Theorem 1, $X(T)$ is an equilibrium point of neural network (10). In this case, we suppose

$$Y(t) = \begin{cases} X(t), & \text{if } t \in [0, T) \\ X(T), & \text{if } t \geq T \end{cases} \tag{14}$$

It is obvious that $Y(t)$ is a state of neural network (10) with initial point $Y(0) = X_0$. Then, by the uniqueness of the solution of neural network (10),

$$X(t) = Y(t), \text{ for all } t \geq 0.$$

Hence, the state $X(t)$ is convergent to the equilibrium point $X(T)$ at time T.

On the other hand, if there exists $t_0 \in [0, T)$ such that $X(t_0) \in \mathbb{M}$. It is clear that $X(t_0)$ is an equilibrium point of neural network (10). Similarly, we obtain that $X(t)$ is convergent to the equilibrium point $X(t_0)$ at time t_0.

5   Comparisons and Numerical Examples

Example 1: Consider a bi-level optimization problem as follows [19]

$$\begin{aligned} (UP) \min_{x,y} F(x,y) &= 2x - 11y \\ \text{s.t.} \quad x - 2y &\leq 4, \quad 2x - y \leq 24, \\ 3x + 4y &\leq 96, \quad x + 7y \leq 126, \\ -4x + 5y &\leq 65, \quad x + 4y \leq 8, \\ (LP) \begin{cases} y \in \arg\min_y f(x,y) = x + 3y \\ \text{s.t. } 2x - y \leq 24 \end{cases} \end{aligned} \tag{15}$$

Fig. 1 depicts the transient behaviors of the bi-level optimization problem from any initial states. The optimal solution $[x, y]^T$ of (15) is $[17.455, 10.909]^T$. The constraints are active. Simulation results have shown the globally finite time convergent property of the proposed model.

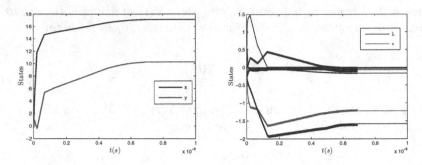

Fig. 1. Transient behaviors of state variables of linear bi-level optimization problem in Example 1.

Example 2: Consider a supply chain model discussed in [10] with two distribution centers and one assembly factory. This study assumes that distribution centers belong to the upper level, while assembly factories are the lower level. the corresponding model can be formulated as the following bi-level optimization problem:

$$(UP) \min_{x,y} F(x, y) = 135y_1 + 195y_2$$
$$\text{s.t.} \quad x_1 + x_2 \geq 50, 0 \leq x_1 \leq 30$$
$$0 \leq x_2 \leq 20, y_1, y_2 \geq 0 \tag{16}$$
$$(LP) \begin{cases} y \in \arg\max_y f = -40x_1 - 50x_2 + 100y_1 + 150y_2 \\ \text{s.t.} \quad y_1 \leq x_1, y_2 \leq x_2 \end{cases}$$

Fig. 2 depicts the transient behaviors of the bi-level optimization problem from any initial states. The optimal solution is $x = [20, 30]^T$, $y = [20, 30]^T$. This application has shown great potential of the proposed method in economic and management optimization problems.

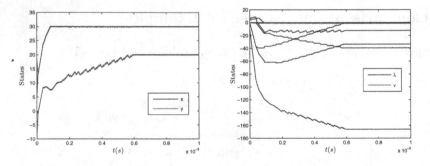

Fig. 2. Transient behaviors of state variables of linear bi-level optimization problem in Example 2.

Acknowledgment. The work described in the paper was supported by the Research Grants Council of the Hong Kong Special Administrative Region, China (Project no. CUHK416812E) and the national science fund of grant 61403101, China.

References

1. Aubin, J.P., Cellina, A.: Differential inclusions. Springer, Berlin (1984)
2. Calvete, H.I., Gale, C.: On linear bilevel problems with multiple objectives at the lower level. Omega 39(1), 33–40 (2011)
3. Cheng, L., Hou, Z.G., Lin, Y., Tan, M., Zhang, W.C., Wu, F.: Recurrent neural network for non-smooth convex optimization problems with application to the identification of genetic regulatory networks. IEEE Trans. Neural Networks 22(5), 714–726 (2011)
4. Cichocki, A., Unbehauen, R.: Neural networks for optimization and signal processing. Wiley (1993)
5. Colson, B., Marcotte, P., Savard, G.: An overview of bilevel optimization. Annals of Operations Research 153(1), 235–256 (2007)
6. He, X., Li, C., Huang, T., Li, C.: Neural network for solving convex quadratic bilevel programming problems. Neural Networks 51, 17–25 (2014)
7. He, X., Li, C., Huang, T., Li, C., Huang, J.: A recurrent neural network for solving bilevel linear programming problem. IEEE Transactions on Neural Networks and Learning Systems 25(4), 824–830 (2014)
8. Jiang, Y., Li, X., Huang, C., Wu, X.: An augmented lagrangian multiplier method based on a chks smoothing function for solving nonlinear bilevel programming problems. Knowledge-Based Systems 55, 9–14 (2014)
9. Kennedy, M.P., Chua, L.O.: Neural networks for nonlinear programming. IEEE Trans. Circuits and Systems I: Regular Papers 35(5), 554–562 (1988)
10. Kuo, R., Huang, C.: Application of particle swarm optimization algorithm for solving bi-level linear programming problem. Computers & Mathematics with Applications 58(4), 678–685 (2009)
11. Lv, Y., Chen, Z., Wan, Z.: A neural network for solving a convex quadratic bilevel programming problem. Journal of Computational and Applied Mathematics 234(2), 505–511 (2010)
12. Mersha, A.G., Dempe, S.: Feasible direction method for bilevel programming problem. Optimization 61(5), 597–616 (2012)
13. Muu, L.D., Quy, N.V.: A global optimization method for solving convex quadratic bilevel programming problems. J. Global Optimization 26(2), 199–219 (2003)
14. Qin, S., Xue, X.: Global exponential stability and global convergence in finite time of neural networks with discontinuous activations. Neural Processing Letters 29(3), 189–204 (2009)
15. Qin, S., Xue, X.: Dynamical analysis of neural networks of subgradient system. IEEE Trans. Automat. Contr. 55(10), 2347–2352 (2010)
16. Scholtes, S.: Convergence properties of a regularization scheme for mathematical programs with complementarity constraints. SIAM Journal on Optimization 11(4), 918–936 (2001)
17. Shih, H.S., Wen, U.P., Lee, S., Lan, K.M., Hsiao, H.C.: A neural network approach to multiobjective and multilevel programming problems. Computers & Mathematics with Applications 48(1-2), 95–108 (2004)
18. Wan, Z., Mao, L., Wang, G.: Estimation of distribution algorithm for a class of nonlinear bilevel programming problems. Information Sciences 256, 184–196 (2014)
19. Wen, U.P., Hsu, S.T.: Linear bi-level programming problems–a review. Journal of the Operational Research Society, 125–133 (1991)

A Nonlinear Neural Network's Stability Analysis and Its kWTA Application

Yinhui Yan[1,2]

[1] Shenzhen Airlines Co., Ltd., Shenzhen Bao'an International Airport, Shenzhen, China
[2] College of Civil Aviation, Nanjing University of Aeronautics and Astronautics, Nanjing, China
A15601@shenzhenair.com

Abstract. In this paper, the stability of a novel nonlinear neural network solving linear programming problems is studied. We prove that this nonlinear neural network is stable in the sense of Lyapunov under certain conditions. Inspired by the study of this neural network, we propose a novel neural system to solving the k-winners-take-all (kWTA) problem. Numerical simulations demonstrate that the effectiveness and good performance of our new kWTA neural network.

Keywords: Nonlinear Neural Network, Lyapunov Stability, Linear Programming, kWTA.

1 Introduction

Linear programming (LP) studies the optimization of the linear objective function satisfying linear equality or linear inequality constraints. It has been successfully applied to various fields such as transportation [1], energy [2], telecommunication [3], and manufacturing [4]. The linear programming method was first developed by Leonid Kantorovich in 1939. Traditional methods to solve LP problems include George B. Dantzig's the simplex method, John Von Neumann's theory of duality, Narendra Karmarkar's new interior-point method etc. In 1986, Hopfield and Tank in their paper [5] proposed a new approach to solve LP problems by using recurrent neural networks. The main advantage of this method is that it can be implemented by using analog electronic circuits, possibly on a VLSI (very large-scale integration) circuit, which can operate in parallel. With comparison to traditional approaches which may involve an iterative process and require long computational time, this model can potentially provide an optimal solution in real time. After their pioneer work [5] [6], numerous neural network models have been proposed to tackle LP problems. Kennedy and Chua [7] developed a neural network for solving nonlinear programming problems based on Karush-Kuhn-Tucker (KKT) optimal conditions. By using a penalty parameter it's solution usually can approximates the optimal solution. Unfortunately it is same as the exact solution only when the penalty parameter is very large. Later Maa and Shanblatt [8] extend this penalty based method by using a two-phase model and ensure that the model converges to the optimal solution. However, their model is more complex and still requires careful parameter selection. To overcome these drawbacks, Xia [9] proposed a primal and

© Springer International Publishing Switzerland 2015
X. Hu et al. (Eds.): ISNN 2015, LNCS 9377, pp. 426–435, 2015.
DOI: 10.1007/978-3-319-25393-0_47

dual model to solving this problem. For other approaches, Zhang [10] invented a lagrangian neural network based on the theory of lagrangian multipliers. In this model slack variables are introduced as new variables to cope with inequality constraints, this may result in high dimensional problem thus require more computation. Unlike previous approaches using a fixed parameter, Wang etc. [11] used a time-variant temperature to design a deterministic annealing neural network to solve the linear programs. In International Symposium on Mathematical Programming, 2000, Nguyen [12] presents a novel recurrent neural network model to solve linear optimization problem. Compared with Xia's model, Nguyen's model not only retains the advantages of Xia's model but also has a more intuitive economic interpretation and much faster convergence. More details on this neural network is given in Section II. The most attractive thing for this model is its high convergence speed. However, the author did not give the stability analysis of his proposed neural network system. More recently Nguyen's approach has been adopted and extended in [13]–[15]. This paper will study the stability of Nguyen's model and prove that under certain conditions this neural network is stable in the sense of Lyapunov. For the background and details of neural networks, we refer to [16]–[25]. The rest of this paper is organized as follows: Section II gives a detailed description of Nguyen's model and some comments on this model. Section III presents the main result on the stability of Nguyen's dynamical neural network. Section IV applies the idea of this new model to solve the kWTA problem. In the end, Section V gives a summary of this paper and points out some future research directions.

2 Model Description

Consider a linear programming problem with the following standard form:

$$\text{Find } \mathbf{x} \text{ which maximizes} : \mathbf{b}^T \mathbf{x} \tag{1}$$
$$\text{with the constraints} : \mathbf{A}\mathbf{x} \le \mathbf{c}, \mathbf{x} \ge 0$$

where $\mathbf{x}$ and $\mathbf{b} \in \mathbb{R}^n$, $\mathbf{A} \in \mathbb{R}^{m \times n}$, and $\mathbf{c} \in \mathbb{R}^m$.

The dual problem of (1) is

$$\text{Find } \mathbf{y} \text{ which minimizes} : \mathbf{c}^T \mathbf{y} \tag{2}$$
$$\text{with the constraints} : \mathbf{A}^T \mathbf{y} \ge \mathbf{b}, \mathbf{y} \ge 0.$$

In the paper [12], the author proposed a new recurrent network for solving the above linear programming problem based on a nonlinear dynamic system which is described by

$$\begin{cases} \dot{\mathbf{x}} = \mathbf{b} - \mathbf{A}^T(\mathbf{y} + k\dot{\mathbf{y}}), \mathbf{x} \ge 0, & (3a) \\ \dot{\mathbf{y}} = -\mathbf{c} + \mathbf{A}(\mathbf{x} + k\dot{\mathbf{x}}), \mathbf{y} \ge 0, & (3b) \end{cases}$$

where $\dot{\mathbf{x}}$ and $\dot{\mathbf{y}}$ are used to denote the derivatives of x and y with respect to the time variable t, i.e., $\frac{dx}{dt}$, $\frac{dy}{dt}$ respectively, k is a positive real number. Nguyen's neural network consists of two layers of neurons, i.e., primal neurons and dual neurons. The outputs from one layer are the inputs to the other layer. The inputs of the primal neurons are composed of the dual neuron's outputs and their derivatives, while the inputs of the dual neurons are composed of the primal neuron's outputs and their derivatives. Due to the involvement of these derivatives, Nguyen's neural network model is a nonlinear dynamic system. This feature enables the system to handle large discrete time steps without becoming unstable and thus lead to high convergence rate. As for the complexity, Nguyen's network entails about $n + m$ adders and $2n \cdot m$ multipliers which is only about half of Xia's neural network [9].

The main property of the neural system described by (3a) and (3b) is stated in the following theorem [12]:

Theorem 1: If the neural network whose dynamics is described by the differential equations (3a) and (3b) converges to a stable state, then the convergence will be the optimal solutions for the LP problem (1) and its dual problem (2).

In [12] the author uses the Euler method to solving the dynamic systems (3a) and (3b). Some experiments show that system has high convergence rate. For an example with 5 variables and 4 constraints, Nguyen's model converges to the exact solution only after about 1000 iterations, while Xia's model requires about 200,000 iterations to get the same solution.

3   Stability Analysis

Theorem 2: If the matrix $\mathbf{A}\mathbf{A}^T$ or $\mathbf{A}^T\mathbf{A}$ is non-singular and $k > 0$, then the neural network described by (3a) and (3b) is asymptotic stable.

Proof. The dynamic system equations (3a) and (3b) can be rewritten as

$$\begin{pmatrix} \mathbf{I} & k\mathbf{A}^T \\ -k\mathbf{A} & \mathbf{I} \end{pmatrix} \begin{pmatrix} \dot{\mathbf{x}} \\ \dot{\mathbf{y}} \end{pmatrix} = \begin{pmatrix} 0 & -\mathbf{A}^T \\ \mathbf{A} & 0 \end{pmatrix} \begin{pmatrix} \mathbf{x} \\ \mathbf{y} \end{pmatrix} + \begin{pmatrix} \mathbf{b} \\ -\mathbf{c} \end{pmatrix},$$

where $\mathbf{I}$ is the identity matrix. Multiplying $\begin{pmatrix} \mathbf{I} & -k\mathbf{A}^T \\ k\mathbf{A} & \mathbf{I} \end{pmatrix}$ to both sides of the above equation, we have

$$\begin{pmatrix} \mathbf{I} + k^2\mathbf{A}^T\mathbf{A} & 0 \\ 0 & \mathbf{I} + k^2\mathbf{A}\mathbf{A}^T \end{pmatrix} \begin{pmatrix} \dot{\mathbf{x}} \\ \dot{\mathbf{y}} \end{pmatrix} = \begin{pmatrix} \mathbf{I} & -k\mathbf{A}^T \\ k\mathbf{A} & \mathbf{I} \end{pmatrix} \begin{pmatrix} 0 & -\mathbf{A}^T \\ \mathbf{A} & 0 \end{pmatrix} \begin{pmatrix} \mathbf{x} \\ \mathbf{y} \end{pmatrix} + \begin{pmatrix} \mathbf{b} - k\mathbf{A}^T\mathbf{c} \\ k\mathbf{A}\mathbf{b} - \mathbf{c} \end{pmatrix}.$$

It's obvious that the matrix $\begin{pmatrix} \mathbf{I} + k^2\mathbf{A}^T\mathbf{A} & 0 \\ 0 & \mathbf{I} + k^2\mathbf{A}\mathbf{A}^T \end{pmatrix}$ is positive definite, therefore we only need to consider the eigenvalues of the matrix $\begin{pmatrix} \mathbf{I} & -k\mathbf{A}^T \\ k\mathbf{A} & \mathbf{I} \end{pmatrix} \begin{pmatrix} 0 & -\mathbf{A}^T \\ \mathbf{A} & 0 \end{pmatrix}$ in order to study the stability of the dynamic system defined by the Equations (3a) and (3b). Let

$$\mathbf{M} = \begin{pmatrix} 0 & -\mathbf{A}^T \\ \mathbf{A} & 0 \end{pmatrix},$$

then we have

$$\begin{pmatrix} \mathbf{I} & -k\mathbf{A}^T \\ k\mathbf{A} & \mathbf{I} \end{pmatrix} \begin{pmatrix} 0 & -\mathbf{A}^T \\ \mathbf{A} & 0 \end{pmatrix} = (\mathbf{I} + k\mathbf{M})\mathbf{M}.$$

Now we would like to study the properties of the eigenvalues of the matrix $(\mathbf{I}+k\mathbf{M})\mathbf{M}$. Suppose λ be an eigenvalue of $(\mathbf{I}+k\mathbf{M})\mathbf{M}$ with the corresponding non-zero eigenvector V, by the definition of eigenvector, we get

$$(\mathbf{I} + k\mathbf{M})\mathbf{V} = \lambda \mathbf{V}.$$

Multiplying both sides of the above equation by $\overline{\mathbf{V}}^T$, we obtain

$$\bar{\mathbf{V}}^T (\mathbf{I} + k\mathbf{M})V = \lambda \bar{\mathbf{V}}^T V.$$

We use an overline $\bar{z}$ to denote the complex conjugate of z here and hence after. Without loss of generality we may assume that $\overline{\mathbf{V}}^T V = 1$, thus we have

$$\lambda = \overline{\mathbf{V}}^T (\mathbf{I} + k\mathbf{M})\mathbf{M}V = \overline{\mathbf{V}}^T \mathbf{M}V + k\overline{\mathbf{V}}^T M^2 V.$$

For the first term in of the last equation, since $\overline{\mathbf{M}} = \mathbf{M}$ and $\mathbf{M}^T = -\mathbf{M}$, we have $\overline{\overline{\mathbf{V}}^T \mathbf{M}V}^T = (V^T \mathbf{M}\overline{V})^T = \overline{\mathbf{V}}^T \mathbf{M}^T V = -\overline{\mathbf{V}}^T \mathbf{M}V$. Thus we conclude that the real part of the entity $\overline{\mathbf{V}}^T \mathbf{M}V$ is equal to 0. For the second term in the last equation, since $\overline{\overline{\mathbf{V}}^T \mathbf{M}^2 V}^T = (V^T \mathbf{M}^2 \overline{V})^T = \overline{\mathbf{V}}^T \mathbf{M}^2 V$, we claim that the item $k\overline{\mathbf{V}}^T \mathbf{M}^2 V$ is a real number.

Note that

$$\mathbf{M}^2 = \begin{pmatrix} 0 & -\mathbf{A}^T \\ \mathbf{A} & 0 \end{pmatrix} \begin{pmatrix} 0 & -\mathbf{A}^T \\ \mathbf{A} & 0 \end{pmatrix} = \begin{pmatrix} -\mathbf{A}^T \mathbf{A} & 0 \\ 0 & -\mathbf{A}\mathbf{A}^T \end{pmatrix}.$$

If $\mathbf{A}^T \mathbf{A}$ or $\mathbf{A}\mathbf{A}^T$ is non-singular (note that rank($\mathbf{A}^T \mathbf{A}$) = rank($\mathbf{A}\mathbf{A}^T$)), we conclude that the matrix $\mathbf{M}^2$ is negative positive and the real part of $\overline{\mathbf{V}}^T (\mathbf{I}+k\mathbf{M})\mathbf{V}$ is negative. Thus we have proved all the eigenvalues of the matrix $\begin{pmatrix} \mathbf{I} & -k\mathbf{A}^T \\ k\mathbf{A} & \mathbf{I} \end{pmatrix} \begin{pmatrix} 0 & -\mathbf{A}^T \\ \mathbf{A} & 0 \end{pmatrix}$ have negative real parts and the dynamic system defined by (3a) and (3b) is asymptotic stable. Moreover the convergence rate of this neural dynamic system is proportional to the constant coefficient k.

4   KWTA Application

Winner-take-all (WTA) is an operation that seeks the maximum item of multiple input signals. Such operation has been successfully applied to various fields such as associative memories [26], cooperative models of binocular stereos [27] and feature selection [28]. K-winners-take-all (KWTA) is an extension of Winner-take-all operation in the sense that it selects the k ($1 \le k < n$) largest input signals from the n total inputs instead just the maximum one. In the recent decades, many kinds of neural networks have been designed to solve the kWTA problem such as the primal and dual network [29], the dual network [30] and the linear programming based neural network [31]. The main research direction is how to improve the convergence speed and simplify the the architecture complexity.

Mathematically, kWTA can be modelled as a function as follows:

$$x_i = f(u_i) = \begin{cases} 1 & \text{if } u_i \in \{k \text{ largest elements of } \mathbf{u}\}, \\ 0 & \text{otherwise.} \end{cases}$$

for $i = 1, 2, \ldots n$, where $\mathbf{u} \in \mathbb{R}^n$ and $k \in \{1, 2, \ldots, n-1\}$. The function can be further written by the following binary integer program:

$$\begin{aligned} \text{Minimize} \quad & -\mathbf{u}^T\mathbf{x}, \\ \text{subject to} : \quad & \mathbf{e}^T\mathbf{x} = k, \\ & x_i \in \{0, 1\}, i = 1, 2, \ldots, n, \end{aligned} \tag{4}$$

where $\mathbf{u} = [u_1, u_2, \ldots, u_n]^T$, $\mathbf{e} = [1, 1, \ldots, 1]^T \in \mathbb{R}^n$, $\mathbf{x} = [x_1, x_2, \ldots, x_n]^T \in \mathbb{R}^n$ and k is a positive integer less than n.

In order to solve this program using the dynamic system approach, we extend it to the continuous case under some conditions. Towards such objective, we give the following theorem:

Theorem 3: The integer programming problem (4) is equivalent to the following linear programming problem if the kth largest element of $\mathbf{u}$ is strictly larger than the $(k+1)^{th}$ largest element of $\mathbf{u}$

$$\begin{aligned} \text{Minimize} \quad & -\mathbf{u}^T\mathbf{x}, \\ \text{subject to} : \quad & \mathbf{e}^T\mathbf{x} = k, \\ & x_i \in [0, 1], i = 1, 2, \ldots, n. \end{aligned} \tag{5}$$

Proof. Without loss of generality, we assume that $u_1, u_2, \ldots, u_k$ are the k largest elements and $u_1 \geq u_2 \geq \ldots \geq u_k$, then the optimal solution of the problem (4) is $x_1 = x_2 = \ldots = x_k = 1$ and $x_{k+1} = x_{k+2} = \ldots = x_n = 0$. To prove this solution is also the optimal solution to problem (5). First, we use the constraint $\sum_{i=1}^n x_i = k$ to solve x_{k+1} in terms of $x_1, \ldots, x_k, x_{k+2}, \ldots, x_n$, giving $x_k = k - \sum_{i \neq k+1} u_i x_i$. Then we use this relationship to substitute for x_{k+1} in the objective function, saying

$$-\mathbf{u}^T\mathbf{x} = -ku_{k+1} + \sum_{i=1}^k (-u_i + u_{k+1})x_i + \sum_{i=k+2}^n (-u_i + u_{k+1})x_i.$$

Since the coefficients of $x_1, x_2, \ldots, x_k$ are negative and the coefficients of $x_{k+2}, x_{k+3}, \ldots, x_n$ are nonnegative in the objective function, we conclude that the solution of the minimizing problem (5) is $u_1 = u_2 = \ldots = u_k = 1$.

We remark that the *Theorem 1* on the equivalence of (4) and (5) presented in [31] is not correct in general.

The lagrangian function of this minimization problem (5) can be written as

$$\mathcal{L}(\mathbf{x}, y, \mathbf{z}) = -\mathbf{u}^T\mathbf{x} - y(\mathbf{e}^T\mathbf{x} - k) - \mathbf{z}^T(-\mathbf{Ix} + \mathbf{e}), \tag{6}$$

where $\mathbf{z} \in \mathbb{R}_+^n = \{\mathbf{z} \in \mathbb{R}^n | \mathbf{z} \geq 0\}$ and $y \in \mathbb{R}$ are Lagrangian multipliers. According to the Karush-Kuhn-Tucker (KKT) conditions [32] [33], $\mathbf{x}^\star$ is a solution of (5) if and only if there exist $y^\star \in \mathbb{R}^m$, $\mathbf{z}^\star \in \mathbb{R}_+^p$ so that $(\mathbf{x}^\star, y^\star, \mathbf{z}^\star)$ satisfies the following conditions:

$$-\mathbf{u} - y^\star\mathbf{e} + \mathbf{z}^\star \geq 0,$$
$$\mathbf{x}^{\star T}\left(-\mathbf{u}^T - y^\star\mathbf{e} + \mathbf{z}^\star\right) = 0,$$
$$k - \mathbf{e}^T\mathbf{x}^\star = 0$$
$$-\mathbf{e} + \mathbf{I}\mathbf{x}^\star \leq 0,$$
$$\mathbf{z}^{\star T}\left(-\mathbf{e} + \mathbf{I}\mathbf{x}^\star\right) = 0. \tag{7}$$

We propose a recurrent neural network for solving the primal and dual problem as follows:

$$\begin{cases} \dot{\mathbf{x}} = \mathbf{u} + (\mathbf{y} + \lambda\dot{\mathbf{y}})\mathbf{e} - (\mathbf{z} + \lambda\dot{\mathbf{z}}), \mathbf{x} \geq 0, & (8a) \\ \dot{y} = k - \mathbf{e}^T(\mathbf{x} + \lambda\dot{\mathbf{x}}), & (8b) \\ \dot{\mathbf{z}} = (\mathbf{x} + \lambda\dot{\mathbf{x}}) + \mathbf{e}, \mathbf{z} \geq 0, & (8c) \end{cases}$$

where λ is a positive constant. The architecture of the proposed neural network model is shown in Fig. 1.

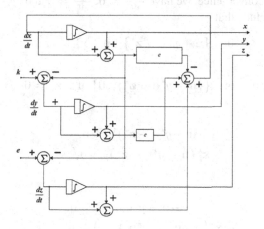

Fig. 1. Block diagram of the neural network (8a, 8b, and 8c)

The proposed neural network consists of two layers of neurons, i.e., primal neurons and dual neurons. The outputs from one layer are the inputs to the other layer. The inputs of the primal neurons are composed of the dual neuron's outputs and their derivatives, while the inputs of the dual neurons are composed of the primal neuron's outputs and their derivatives. Due to the involvement of these derivatives, this neural network model is a nonlinear dynamic system. The convergence property of the system is stated by the following theorem.

Theorem 4: If the neural network whose dynamics guided by the differential equations (8a, 8b, and 8c) converges to a steady state $(\mathbf{x}^*, y^*, \mathbf{z}^*)$, then $\mathbf{x}^*$ will be the optimal solution of the primal LP problem (5) and the Lagrangian multipliers y^* and $\mathbf{z}^*$ the optimal solution of the dual of the LP problem.

Proof. Let $\mathbf{x}_i$ be the ith component of $\mathbf{x}$, then the equation (8a) can be written as

$$\frac{d\mathbf{x}_i}{dt} = \{\mathbf{u} + (y + \lambda\dot{y})\mathbf{e} - (\mathbf{z} + \lambda\dot{\mathbf{z}})\}_i \text{ if } \mathbf{x}_i > 0, \tag{9}$$

$$\frac{d\mathbf{x}_i}{dt} = \max\{\{\mathbf{u} + (y + \lambda\dot{y})\mathbf{e} - (\mathbf{z} + \lambda\dot{\mathbf{z}})\}_i, 0\} \text{ if } \mathbf{x}_i = 0. \tag{10}$$

Note that (10) is to ensure that $\mathbf{x}$ will bounded from below by 0. Let $\mathbf{x}^*$, $\mathbf{y}^*$ and $\mathbf{z}^*$ be the limit of $\mathbf{x}(t)$, $y(t)$ and $\mathbf{z}(t)$ respectively. In other words

$$\lim_{t\to\infty} \mathbf{x}(t) = \mathbf{x}^*, \tag{11}$$

$$\lim_{t\to\infty} y(t) = y^*, \tag{12}$$

$$\lim_{t\to\infty} \mathbf{z}(t) = \mathbf{z}^*. \tag{13}$$

By the definition of convergence, we have $\frac{d\mathbf{x}^*}{dt} = 0$, $\frac{dy^*}{dt} = 0$ and $\frac{d\mathbf{z}^*}{dt} = 0$. From Eqns. (9) and (10) we conclude that

$$0 = \{\mathbf{u} + y^*\mathbf{e} - \mathbf{z}^*\}_i \text{ if } \mathbf{x}_i^* > 0, \tag{14}$$

$$0 = \max\{\{\mathbf{u} + y^*\mathbf{e} - \mathbf{z}^*\}_i, 0\} \text{ if } \mathbf{x}_i^* = 0. \tag{15}$$

In other words:

$$(\mathbf{u} + y^*\mathbf{e} - \mathbf{z}^*))_i \leq 0, \tag{16}$$

$$\mathbf{x}_i^* (\mathbf{u} + y^*\mathbf{e} - \mathbf{z}^*)_i = 0, \tag{17}$$

or

$$-\mathbf{u} - y^*\mathbf{e} + \mathbf{z}^* \geq 0, \tag{18}$$

$$\mathbf{x}^{*T} \left(-\mathbf{u}^T - y^*\mathbf{e} + \mathbf{z}^*\right) = 0. \tag{19}$$

Similarly, from Eqns. (8b) and (8c), we have:

$$k - \mathbf{e}^T\mathbf{x}^* = 0, \tag{20}$$

$$-\mathbf{e} + \mathbf{I}\mathbf{x}^* \leq 0, \tag{21}$$

$$\mathbf{z}^{*T} \left(-\mathbf{e} + \mathbf{I}\mathbf{x}^*\right) = 0. \tag{22}$$

By KKT conditions in (7) and conditions provided in (19-22) we have shown that x^* and (y^*, z^*) are the optimal solutions for the problem (5) and its dual problem respectively. This completes the proof of the theorem.

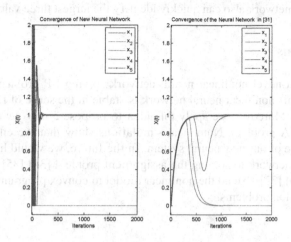

Fig. 2. Convergence of the kWTA network in Example 1

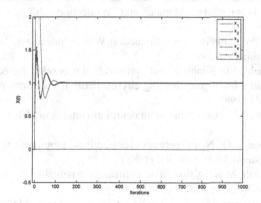

Fig. 3. Convergence of the kWTA network in Example 2

To demonstrate the behavior and properties of the proposed nonlinear neural network model, we consider two examples with $n = 5$ and $k = 3$. All the simulations are conducted with MATLAB 12c. We use the Euler method to solve the neural system of ordinary differential equations (8a, 8b, and 8c). In the first example, the inputs are set to be $u_i = i$ ($i = 1, 2, 3, 4, 5$). The transient trajectories of $\mathbf{x}$ is shown on the left side in Fig. 2. It is obvious that the steady state is $[0, 0, 1, 1, 1]^T$. This means that our neural network successfully identifies three largest elements, i.e., u_3, u_4 and u_5. Furthermore this experiment demonstrates that our neural system converges very fast and goes to the equilibrium point after about 200 iterations. In comparison, it takes more than 1000 iterations to reach the steady output using the neural network ([31]) as shown on the right side in Fig. 2. For the second example with inputs $u_i = 10 * sin(2\pi * (0.3 + 0.2 * (i - 1)))$ $(i = 1, 2, 3, 4, 5)$, the simulation result depicted in Fig. 3 shows that the proposed kWTA network also can quickly identify the largest three values.

5   Conclusions

The stability of a novel nonlinear neural network solving LP problems is analyzed. We proved that this nonlinear neural network is stable in the sense of Lyapunov under some conditions. Moreover, we apply this idea to propose a new neural network for solving the kWTA problem. Numerical simulations show that the effectiveness and good performance of our new neural system. In the future, we would like to apply the proposed neural network to solving the assignment problem [34] [35] and maximum flow problem [36] [37], extend the nonlinear model to convex programming and more general optimization problems.

References

1. Dorfman, R., Samuelson, P.A. and Solow, R. M.: Linear Programming and Economic Analysis. Dover Publications (1987)
2. Matousek, J., Gärtner, B.: Understanding and Using Linear Programming. Springer (2006)
3. Gass, S.I.: Linear Programming: Methods and Applications, 5th edn. Dover Publications (2010)
4. Sultan, A.: Linear Programming: An Introduction With Applications, 2nd edn. CreateSpace Independent Publishing Platform (2011)
5. Tank, D.W., Hopfield, J.J.: Simple neural optimization networks: An A/D converter, signal decision circuit, and a linear programming circuit. IEEE Transactions on Circuits and Systems 33(5), 533–541 (1986)
6. Hopfield, J.J., Tank, D.W.: Computing with neural circuits: A model. Science 233, 625–633 (1986)
7. Kennedy, M.P., Chua, L.O.: Neural networks for nonlinear programming. IEEE Transactions on Circuits and Systems 35(5), 554–562 (1988)
8. Maa, C.Y., Schanblatt, M.A.: A two-phase optimization neural network. IEEE Transactions on Neural Network 3(6), 1003–1009 (1992)
9. Xia, Y.: A new neural network for solving linear programming problems and its application. IEEE Transactions on Neural Networks 7(2), 525–529 (1996)
10. Zhang, S., Constantinides, A.G.: Lagrange programming neural networks. IEEE Transactions on Circuits and Systems II 39(7), 441–452 (1992)
11. Wang, J.: A deterministic annealing neural network for convex programming. Neural Networks 5(4), 962–971 (1994)
12. Nguyen, K.V.: A Nonlinear Neural Network for Solving Linear Programming Problems. In: International Symposium on Mathematical Programming, ISMP 2000, Atlanta, GA, USA (2000)
13. Suresh, S., Mani, V., Omkar, S.N., Kim, H.J.: Parallel Video Processing Using Divisible Load Scheduling Paradigm. Journal of Broadcast Engineering 10(1), 83–102 (2005)
14. Senthilnath, J., Omkar, S.N., Mani, V., Katti, A.R.: Cooperative communication of UAV to perform multi-task using nature inspired techniques. In: IEEE Symposium on Computational Intelligence for Security and Defense Applications (CISDA), pp. 45–50 (2013)

15. Yan, Y.: A New Nonlinear Neural Network for Solving QP Problems. In: Zeng, Z., Li, Y., King, I. (eds.) ISNN 2014. LNCS, vol. 8866, pp. 347–357. Springer, Heidelberg (2014)
16. Taylor, J.G.: Mathematical Approaches to Neural Networks. North-Holland (1993)
17. Harvey, R.L.: Neural Network Principles. Prentice Hall (1994)
18. Veelenturf, L.: Analysis and Applications of Artifical Neural Networks. Prentice Hall (1995)
19. Rojas, R., Feldman, J.: Neural Networks A Systematic Introduction. Springer (1996)
20. Mehrotra, K., Mohan, C.K., Ranka, S.: Elements of Artificial Neural Networks. MIT Press (1997)
21. Haykin, S.: Neural Networks A Comprehensive Foundation, 2nd edn. Prentice Hall (1998)
22. Michel, A., Liu, D.: Qualitative Analysis and Synthesis of Recurrent Neural Networks. CRC Press (2001)
23. Hagan, M.T., Demuth, H.B., Beale, M.H.: Neural Network Design. Martin Hagan (2002)
24. Gurney, K.: An Introduction to Neural Networks. CRC Press (2003)
25. Graupe, D.: Principles of Artificial Neural Networks, 2nd edn. World Scientific Pub. Co. Inc. (2007)
26. Krogh, A., Hertz, J., Palmer, R.G.: Introduction to the Theory of Neural Computation. Addison-Wesley, Redwook (1991)
27. Marr, D., Poggio, T.: Cooperative computation of stereo disparity. Science 195, 283–328 (1977)
28. Yuille, A.L., Geiger, D.: The Handbook of Brain Theory and Neural Networks. MIT Press (2002)
29. Xia, Y., Feng, G., Wang, J.: A primal-dual neural network for online resolving constrained kinematic redundancy in robot motion control. IEEE Transactions on Systems, Man and Cybernetics 35(1), 54–64 (2005)
30. Xia, Y., Wang, J.: A general projection neural network for solving monotone variational inequalities and related optimization problems. IEEE Transactions on Neural Networks 15(2), 318–328 (2004)
31. Gu, S., Wang, J.: A K-Winners-Take-All Neural Network Based on Linear Programming Formulation. In: Proceedings of International Joint Conference on Neural Networks, Orlando, Florida, USA (2007)
32. Boyd, S., Vandenbeghe, L.: Convex Optimization. Cambridge University Press (2004)
33. Bertsekas, D.P., Tsitsiklis, J.N.: Parallel and Distributed Computation: Numerical Methods. Prentice-Hall (1989)
34. Wang, J.: Analogue neural network for solving the assignment problem. Electronics Letters 28(11), 1047–1050 (1992)
35. Hu, X., Wang, J.: Solving the assignment problem with the improved dual neural network. In: Liu, D., Zhang, H., Polycarpou, M., Alippi, C., He, H. (eds.) ISNN 2011, Part I. LNCS, vol. 6675, pp. 547–556. Springer, Heidelberg (2011)
36. Effati, S., Ranjbar, M.: Neural network models for solving the maximum flow problem. Applications and Applied Mathematics 3(3), 149–162 (2008)
37. Nazemi, A., Omidi, F.: A capable neural network model for solving the maximum flow problem. Journal of Computational and Applied Mathematics 236(14), 3498–3513 (2012)

Continuous-Time Multi-agent Network for Distributed Least Absolute Deviation*

Qingshan Liu[1], Yan Zhao[2], and Long Cheng[3]

[1] School of Automation, Huazhong University of Science and Technology,
Wuhan 430074, Hubei, China
qsliu@hust.edu.cn
[2] Department of Basic Courses, Wannan Medical College,
Wuhu 241000, Anhui, China
[3] State Key Lab. of Management and Control for Complex Systems,
Institute of Automation, Chinese Academy of Sciences,
Beijing 100190, China

Abstract. This paper presents a continuous-time multi-agent network for distributed least absolute deviation (DLAD). The objective function of the DLAD problem is a sum of many least absolute deviation functions. In the multi-agent network, each agent connects with its neighbors locally and they cooperate to obtain the optimal solutions with consensus. The proposed multi-agent network is in fact a collective system with each agent being considered as a recurrent neural network. Simulation results on a numerical example are presented to illustrate the effectiveness and characteristics of the proposed distributed optimization method.

Keywords: Distributed least absolute deviation, multi-agent network, nonsmooth optimization, consensus.

1 Introduction

Least absolute deviation (LAD) or L_1-norm criterion provides in many applications in science and engineering, including signal and image processing, system identification, parameter estimation, source localization and regression [1,2]. Recently, the increasing interest in performing distributed optimization based on multi-agent systems has raised wide investigations in the literature (e.g., see [3,4,5] and references therein). The goal of distributed optimization is to minimize a sum of local objective functions, which can be used to perform large-scale optimization problems in the distributed manner. For distributed optimization, most of the works build on consensus algorithms which are described by discrete-time dynamics to search for optimal solutions [3,6]. Moreover, a few recent works on continuous-time dynamics for distributed optimization are presented [5,7].

* This work was supported in part by the National Natural Science Foundation of China under Grant 61473333, by the Program for New Century Excellent Talents in University of China under Grant NCET-12-0114, and by the Fundamental Research Funds for the Central Universities of China under Grant 2015QN035.

© Springer International Publishing Switzerland 2015
X. Hu et al. (Eds.): ISNN 2015, LNCS 9377, pp. 436–443, 2015.
DOI: 10.1007/978-3-319-25393-0_48

More recently, researchers are especially interested in the dynamical behavior analysis of multi-agent networks, including consensus, cooperation, and competition (e.g., see [8,9,10] and references therein).

Since 1980s, neurodynamic optimization based on continuous-time recurrent neural networks are widely investigated [11,12,13,14,15,16,17,18]. Among them, Tank and Hopfield [11] proposed a neurodynamic optimization approach for linear programming. Kennedy and Chua [12] presented a recurrent neural network for nonlinear optimization. From then on, the research on this topic has been well developed and numerous neural network models have been designed for optimization. The recurrent neural networks seek the optimal solutions in parallel. In general, the collective recurrent neural networks, which each neural network connects with its neighbors, can be considered as a multi-agent network.

In this paper, we are concerned with design of continuous-time multi-agent network for solving the LAD problems based on distributed optimization. In the literature, for LAD problems, either convergence speed of the algorithms is slow, or the network parameters have to be adjusted. If the LAD problem is solved with distributed optimization, we can design some parallel algorithms to solve the problems with efficiency. Moreover, the projection method for neurodynamic optimization [13,18,19] is used to perform the L_1-norm, which corresponds to a continuous piecewise-linear activation function. Compared with the subgradient method for nonsmooth optimization [6,20], the continuous activation function can take faster convergence rate for distributed optimization.

2   Problem Formulation and Multi-agent Network Modeling

We consider the distributed least absolute deviation as the following optimization problem:

$$\begin{aligned}
\text{minimize} \quad & \sum_{i=1}^{m} \|C_i x - d_i\|_1, \\
\text{subject to} \quad & x \in \bigcap_{i=1}^{m} \Omega_i,
\end{aligned} \tag{1}$$

where $x \in \mathbb{R}^n$, $C_i \in \mathbb{R}^{p_i \times n}$ and $d_i \in \mathbb{R}^{p_i}$, and $\Omega_i \subseteq \mathbb{R}^n$ is a closed convex set, which is assumed to be a hyper-box or hyper-sphere.

Assume that the network is consisted of m agents with connected undirected graph, denoted by $\mathcal{G}$. Each agent is assigned a local least absolute deviation function $\|C_i x - d_i\|_1$ and constraint set Ω_i ($i = 1, 2, \ldots, m$). The objective for the network is to cooperatively solve the LAD problem with consensus.

Let $x_i \in \mathbb{R}^n$ ($i = 1, 2, \ldots, m$) denotes the estimate of agent i about the value of the solution to problem (1), $X = (x_1, x_2, \cdots, x_m) \in \mathbb{R}^{n \times m}$ to be a matrix with column vector x_i ($i = 1, 2, \ldots, m$), and $\tilde{x} = \text{vec}(X) \in \mathbb{R}^{mn}$ to be the vectorization of matrix X which converts X into a column vector obtained by stacking the columns of X on top of one another. Then the following lemma is derived directly.

Lemma 1. *[5] Let $L_m \in \mathbb{R}^{m \times m}$ be the Laplacian matrix of connected graph $\mathcal{G}$ and $L = L_m \otimes I \in \mathbb{R}^{mn \times mn}$, where $\otimes$ is the Kronecker product and I is n-dimensional identity matrix. The problem (1) is equivalent to the following optimization problem:*

$$\begin{aligned} \text{minimize } & \|C\tilde{x} - d\|_1, \\ \text{subject to } & L\tilde{x} = 0, \ \tilde{x} \in \Omega, \end{aligned} \tag{2}$$

where C is the block diagonal matrix of C_1, C_2 to C_m (i.e., $C = diag\{C_1, C_2, \ldots, C_m\}$), $d = ((d_1)^T, (d_2)^T, \ldots, (d_m)^T)^T$ and $\Omega = \prod_{i=1}^{m} \Omega_i$ defined by the Cartesian product.

In (1), let $p = p_1 + p_2 + \cdots + p_m$. Then $C \in \mathbb{R}^{p \times mn}$ and $d \in \mathbb{R}^p$ in (2). Next, a necessary and sufficient condition for the optimal solutions to problem (2) is described as follows.

Theorem 1. *$\tilde{x}^* \in \mathbb{R}^{mn}$ is an optimal solution to problem (2) if and only if there exist $\tilde{y}^* \in \mathbb{R}^p$ and $\tilde{z}^* \in \mathbb{R}^{mn}$ such that $(\tilde{x}^*, \tilde{y}^*, \tilde{z}^*)$ satisfies*

$$\begin{cases} \tilde{x}^* = \phi(\tilde{x}^* - C^T \tilde{y}^* - L\tilde{z}^*), \\ \tilde{y}^* = g(\tilde{y}^* + C\tilde{x}^* - d), \\ L\tilde{x}^* = 0, \end{cases} \tag{3}$$

where ϕ and g are projection operators from $\mathbb{R}^n$ to Ω and $\mathbb{R}^p$ to $[-1, 1]^p$ respectively.

Proof: The proof is similar to that of Theorem 1 in [7].   □

According to (3), the continuous-time multi-agent network is proposed as the following differential equations ($i = 1, 2, \ldots, m$):

$$\begin{cases} \dfrac{dx_i}{dt} = 2\{-x_i + \phi_i[x_i - C_i^T \tilde{g}_i - \displaystyle\sum_{j=1, j \neq i}^{m} a_{ij}(x_i + z_i - x_j - z_j)]\}, \\ \dfrac{dy_i}{dt} = -y_i + g_i(y_i + C_i x_i - d_i), \\ \dfrac{dz_i}{dt} = x_i, \end{cases} \tag{4}$$

where $\tilde{g}_i = g_i(y_i + C_i x_i - d_i)$ and a_{ij} is the connection weight between agents i and j in the network.

The multi-agent network (4) can be equivalently written as the following compact form:

$$\begin{cases} \dfrac{d\tilde{x}}{dt} = 2\{-\tilde{x} + \phi[\tilde{x} - C^T g(\tilde{y} + C\tilde{x} - d) - L(\tilde{x} + \tilde{z})]\}, \\ \dfrac{d\tilde{y}}{dt} = -\tilde{y} + g(\tilde{y} + C\tilde{x} - d), \\ \dfrac{d\tilde{z}}{dt} = x. \end{cases} \tag{5}$$

3   Consensus Analysis

In this section, the consensus of the multi-agent network in (5) (or (4)) is analyzed using the Lyapunov method.

First, the projection operators in (3) have the following property.

Lemma 2. *[21] The following inequality holds:*

$$[u - \phi(u)]^T [\phi(u) - v] \geq 0, \quad \forall u \in \mathbb{R}^n, v \in \Omega.$$

Next, the convergence of the system in (5) (or (4)) is presented as follows.

Theorem 2. *The state vector $\tilde{x}$ of the multi-agent network (5) (or (4)) is globally convergent to an optimal solution to problem (2).*

Proof: Let $\tilde{x}^* \in \mathbb{R}^{mn}$ be an optimal solution to problem (2). According to Theorem 1, there exist $\tilde{y}^* \in \mathbb{R}^p$ and $\tilde{z}^* \in \mathbb{R}^{mn}$ such that the equations in (3) hold.

Consider the following Lyapunov function

$$\begin{aligned}
V(\tilde{x}, \tilde{y}, \tilde{z}) = {} & \varphi(\tilde{x}, \tilde{y}, \tilde{z}) - \varphi(\tilde{x}^*, \tilde{y}^*, \tilde{z}^*) - [\tilde{x} - \tilde{x}^*]^T [C^T g(\tilde{y}^* + C\tilde{x}^* - d) \\
& + L\tilde{z}^*] - (\tilde{y} - \tilde{y}^*)^T g(\tilde{y}^* + C\tilde{x}^* - d) - (\tilde{z} - \tilde{z}^*)^T L\tilde{z}^* \\
& + \frac{1}{2} \left[\|\tilde{x} - \tilde{x}^*\|^2 + \|\tilde{y} - \tilde{y}^*\|^2 + (\tilde{z} - \tilde{z}^*)^T L(\tilde{z} - \tilde{z}^*) \right],
\end{aligned} \quad (6)$$

where $\varphi(\tilde{x}, \tilde{y}, \tilde{z}) = \psi(\tilde{y} + C\tilde{x} - d) + (\tilde{x} + \tilde{z})^T L(\tilde{x} + \tilde{z})/2$. Here, ψ is defined as

$$\psi(\tilde{y} + C\tilde{x} - d) = \sum_{j=1}^{p} \psi(\tilde{y}_j + C_j\tilde{x} - d_j),$$

with $\psi(\tilde{y}_j + C_j\tilde{x} - d_j)$ as follows $(j = 1, 2, \ldots, p)$

$$\psi(\tilde{y}_j + C_j\tilde{x} - d_j) = \begin{cases} \tilde{y}_j + C_j\tilde{x} - d_j, & \tilde{y}_j + C_j\tilde{x} - d_j > 1, \\ (\tilde{y}_j + C_j\tilde{x} - d_j)^2/2, & \tilde{y}_j + C_j\tilde{x} - d_j \in [-1, 1], \\ -(\tilde{y}_j + C_j\tilde{x} - d_j), & \tilde{y}_j + C_j\tilde{x} - d_j < -1, \end{cases}$$

where C_j, $\tilde{y}_j$ and d_j are the jth rows of C, $\tilde{y}$ and d respectively. From the formula of $\psi(\cdot)$, we get that $\psi(\cdot)$ is convex and differentiable with $\nabla\psi(\cdot) = g(\cdot)$.

Then we have

$$\nabla\varphi(\tilde{x}, \tilde{y}, \tilde{z}) = \begin{pmatrix} C^T g(\tilde{y} + C\tilde{x} - d) + L(\tilde{x} + \tilde{z}) \\ g(\tilde{y} + C\tilde{x} - d) \\ L(\tilde{x} + \tilde{z}) \end{pmatrix}.$$

The gradient of $V(\tilde{x}, \tilde{y}, \tilde{z})$ is

$$\nabla V(\tilde{x}, \tilde{y}, \tilde{z}) = \begin{pmatrix} C^T \tilde{g} + L(\tilde{x} + \tilde{z}) - C^T \tilde{g}^* - L\tilde{z}^* + \tilde{x} - \tilde{x}^* \\ \tilde{g} - \tilde{g}^* + \tilde{y} - \tilde{y}^* \\ 2L(\tilde{z} - \tilde{z}^*) + L\tilde{x} \end{pmatrix},$$

where $\tilde{g} = g(\tilde{y} + C\tilde{x} - d)$ and $\tilde{g}^* = g(\tilde{y}^* + C\tilde{x}^* - d)$.

According to the chain rule, the derivative of $V(\tilde{x}, \tilde{y}, \tilde{z})$ along the solution of system (5) is

$$\dot{V}(\tilde{x}(t), \tilde{y}(t), \tilde{z}(t))$$
$$= \nabla_{\tilde{x}} V(\tilde{x}, \tilde{y}, \tilde{z})^T \left(\frac{d\tilde{x}}{dt}\right) + \nabla_{\tilde{y}} V(\tilde{x}, \tilde{y}, \tilde{z})^T \left(\frac{d\tilde{y}}{dt}\right) + \nabla_{\tilde{z}} V(\tilde{x}, \tilde{y}, \tilde{z})^T \left(\frac{d\tilde{z}}{dt}\right).$$

Then

$$\dot{V}(\tilde{x}(t), \tilde{y}(t), \tilde{z}(t)) = 2[C^T \tilde{g} + L(\tilde{x} + \tilde{z}) - C^T \tilde{g}^* - L\tilde{z}^* + \tilde{x} - \tilde{x}^*]^T [-\tilde{x} + \tilde{\phi}]$$
$$+ (\tilde{g} - \tilde{g}^* + \tilde{y} - \tilde{y}^*)^T (-\tilde{y} + \tilde{g}) + [2L(\tilde{z} - \tilde{z}^*) + L\tilde{x}]^T \tilde{x},$$

where $\tilde{\phi} = \phi[\tilde{x} - C^T g(\tilde{y} + C\tilde{x} - d) - L(\tilde{x} + \tilde{z})]$.

We next prove the following equality regarding the argument of the right hand side of the previous equation

$$2[C^T \tilde{g} + L(\tilde{x} + \tilde{z}) - C^T \tilde{g}^* - L\tilde{z}^* + \tilde{x} - \tilde{x}^*]^T [-\tilde{x} + \tilde{\phi}]$$
$$+ (\tilde{g} - \tilde{g}^* + \tilde{y} - \tilde{y}^*)^T (-\tilde{y} + \tilde{g}) + [2L(\tilde{z} - \tilde{z}^*) + L\tilde{x}]^T \tilde{x}$$
$$= 2[C^T \tilde{g} + L(\tilde{x} + \tilde{z}) - C^T \tilde{g}^* - L\tilde{z}^* + \tilde{x} - \tilde{x}^*]^T [\tilde{\phi} - \tilde{x}^*]$$
$$+ 2[C^T \tilde{g} + L(\tilde{x} + \tilde{z}) - C^T \tilde{g}^* - L\tilde{z}^* + \tilde{x} - \tilde{x}^*]^T [\tilde{x}^* - \tilde{x}]$$
$$+ (\tilde{g} - \tilde{g}^* + \tilde{y} - \tilde{y}^*)^T (-\tilde{y} + \tilde{g}) + 2[L(\tilde{z} - \tilde{z}^*)]^T \tilde{x} + \tilde{x}^T L\tilde{x}.$$

Let $J_1 = 2[C^T \tilde{g} + L(\tilde{x} + \tilde{z}) - C^T \tilde{g}^* - L\tilde{z}^* + \tilde{x} - \tilde{x}^*]^T [\tilde{\phi} - \tilde{x}^*]$, $J_2 = 2[L(\tilde{x} + \tilde{z}) - L\tilde{z}^* + \tilde{x} - \tilde{x}^*]^T [\tilde{x}^* - \tilde{x}] + 2[L(\tilde{z} - \tilde{z}^*)]^T \tilde{x} + \tilde{x}^T L\tilde{x}$, and $J_3 = 2(C^T \tilde{g} - C^T \tilde{g}^*)^T (\tilde{x}^* - \tilde{x}) + (\tilde{g} - \tilde{g}^* + \tilde{y} - \tilde{y}^*)^T (-\tilde{y} + \tilde{g})$.

Then

$$\dot{V}(\tilde{x}(t), \tilde{y}(t), \tilde{z}(t)) = J_1 + J_2 + J_3. \tag{7}$$

For J_1, we have

$$J_1 = -2[\tilde{x} - C^T \tilde{g} - L(\tilde{x} + \tilde{z}) - \tilde{\phi}]^T [\tilde{\phi} - \tilde{x}^*] - 2(C^T \tilde{g}^* + L\tilde{z}^*)^T (\tilde{\phi} - \tilde{x}^*)$$
$$+ 2(\tilde{x} - \tilde{\phi} + \tilde{x} - \tilde{x}^*)^T (\tilde{\phi} - \tilde{x} + \tilde{x} - \tilde{x}^*)$$
$$= -2[\tilde{x} - C^T \tilde{g} - L(\tilde{x} + \tilde{z}) - \tilde{\phi}]^T [\tilde{\phi} - \tilde{x}^*] - 2(C^T \tilde{g}^* + L\tilde{z}^*)^T (\tilde{\phi} - \tilde{x}^*)$$
$$- 2\|\tilde{x} - \tilde{\phi}\|^2 + 2\|\tilde{x} - \tilde{x}^*\|^2.$$

In Lemma 2, let $u = \tilde{x} - C^T \tilde{g} - L(\tilde{x} + \tilde{z})$ and $v = \tilde{x}^*$, then $[\tilde{x} - C^T \tilde{g} - L(\tilde{x} + \tilde{z}) - \tilde{\phi}]^T [\tilde{\phi} - \tilde{x}^*] \geq 0$. Since $\tilde{x}^*$ is an optimal solution to problem (2), we have $(C^T \tilde{g}^* + L\tilde{z}^*)^T (\tilde{\phi} - \tilde{x}^*) \geq 0$ due to $\tilde{g}^* = \tilde{y}^*$ and $\tilde{\phi} \in \Omega$. Then $J_1 \leq -2\|\tilde{x} - \tilde{\phi}\|^2 + 2\|\tilde{x} - \tilde{x}^*\|^2$.

For J_2, we have

$$J_2 = 2[L(\tilde{x} + \tilde{z}) - L\tilde{z}^* + \tilde{x} - \tilde{x}^*]^T [\tilde{x}^* - \tilde{x}] + 2[L(\tilde{z} - \tilde{z}^*)]^T \tilde{x} + \tilde{x}^T L\tilde{x}$$
$$= 2[L(\tilde{x} + \tilde{z}) - L\tilde{z}^*]^T [\tilde{x}^* - \tilde{x}] + 2(\tilde{x} - \tilde{x}^*)^T (\tilde{x}^* - \tilde{x})$$
$$+ 2[L(\tilde{z} - \tilde{z}^*)]^T \tilde{x} + \tilde{x}^T L\tilde{x}$$
$$= 2(\tilde{x} - \tilde{x}^*)^T (\tilde{x}^* - \tilde{x}) - \tilde{x}^T L\tilde{x},$$

where the last equality holds since $L\tilde{x}^* = 0$.

For J_3, from (3) we have $\tilde{g}^* = \tilde{y}^*$, then

$$
\begin{aligned}
J_3 &= 2(C^T\tilde{g} - C^T\tilde{g}^*)^T(\tilde{x}^* - \tilde{x}) + (\tilde{g} - \tilde{g}^* + \tilde{y} - \tilde{y}^*)^T(-\tilde{y} + \tilde{g}) \\
&= 2(\tilde{g} - \tilde{y}^*)^T(C\tilde{x}^* - C\tilde{x}) + (\tilde{g} + \tilde{y} - 2\tilde{y}^*)^T(-\tilde{y} + \tilde{g}) \\
&= 2(\tilde{g} - \tilde{y}^*)^T(C\tilde{x}^* - C\tilde{x}) - (\tilde{y} - \tilde{g})^T(\tilde{y} - \tilde{g}) + 2(\tilde{g} - \tilde{y}^*)^T(-\tilde{y} + \tilde{g}) \\
&= -\|\tilde{y} - \tilde{g}\|^2 - 2(\tilde{g} - \tilde{y}^*)^T(\tilde{y} - \tilde{g} + C\tilde{x} - C\tilde{x}^*) \\
&= -\|\tilde{y} - \tilde{g}\|^2 - 2(\tilde{g} - \tilde{y}^*)^T(\tilde{y} + C\tilde{x} - d - \tilde{g}) - 2(\tilde{g} - \tilde{y}^*)^T(d - C\tilde{x}^*).
\end{aligned}
$$

In Lemma 2, let $u = \tilde{y} + C\tilde{x} - d$ and $v = \tilde{y}^*$, then $(\tilde{y} + C\tilde{x} - d - \tilde{g})^T(\tilde{g} - \tilde{y}^*) \geq 0$. According to (3), since $\tilde{y}^* = g(\tilde{y}^* + C\tilde{x}^* - d)$, then $\tilde{y}^*$ satisfies the following variational inequality

$$(\tilde{y} - \tilde{y}^*)^T(-C\tilde{x}^* + d) \geq 0, \ \forall \tilde{y} \in [-1, 1]^p,$$

which follows $(\tilde{g} - \tilde{y}^*)^T(d - C\tilde{x}^*) \geq 0$ due to $\tilde{g} \in [-1, 1]^p$. Then $J_3 \leq -\|\tilde{y} - \tilde{g}\|^2$.

Consequently, we have

$$J_1 + J_2 + J_3 \leq -2\|\tilde{x} - \tilde{\phi}\|^2 - \|\tilde{y} - \tilde{g}\|^2 - \tilde{x}^T L\tilde{x}.$$

Combining with (7), it derives that

$$\dot{V}(\tilde{x}(t), \tilde{y}(t), \tilde{z}(t)) \leq -2\|\tilde{x} - \tilde{\phi}\|^2 - \|\tilde{y} - \tilde{g}\|^2 - \tilde{x}^T L\tilde{x}. \tag{8}$$

Since the remainder of the proof is similar to that of Theorem 2 in [7], it is omitted here. □

On one hand, according to Lemma 1, if the undirected graph $\mathcal{G}$ is connected, the optimization problems in (1) and (2) are equivalent. On the other hand, the state vectors $x_i(t)$ $(i = 1, 2, \ldots, m)$ of the multi-agent network are globally convergent to an optimal solution to problem (2). Consequently, the state vectors $x_i(t)$ $(i = 1, 2, \ldots, m)$ can reach consensus at an optimal solution to problem (1) if the undirected graph of the multi-agent network is connected.

4   An Illustrative Example

In the following, the proposed multi-agent network is utilized to solve an LAD problem based on distributed optimization to show its performance and characteristics.

Example. Consider the following L_1-minimization problem:

$$
\begin{aligned}
\text{minimize } & \|Ax - b\|_1, \\
\text{subject to } & -1 \leq x_l \leq 1, \ l = 1, 2, 3,
\end{aligned}
$$

where

$$
A = \begin{pmatrix} 1 & 3 & 1 & 1 & 0 \\ -1 & 1 & -2 & 3 & 4 \\ 1 & -1 & 4 & 3 & 6 \end{pmatrix}^T, \ b = \begin{pmatrix} 1 & 2 & -1 & 1 & -1 \end{pmatrix}^T.
$$

Since the objective function $\|Ax - b\|_1$ is a sum of $A_i x - b_i$ $(i = 1, 2, \ldots, 5)$, where A_i and b_i are the ith rows of A and b respectively, we use a five-agent network with the circular connectivity to solve this problem. The agent i is allocated the objective function $|A_i x - b_i|$. The connection weight between agents i and j is set as 1 if they are connected and 0 otherwise. The transient behaviors of the five agents on state vector x are shown in Fig. 1, which shows that the multi-agent network reaches a consensus with respect to x at the optimal solution $x^* = (0.5106, 0.1809, -0.2872)^T$.

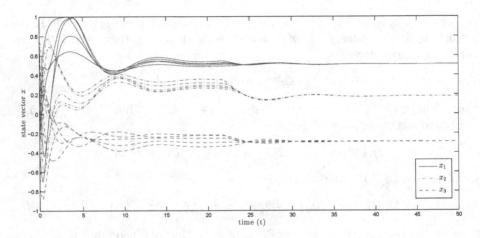

Fig. 1. Consensus of the state vector x of the multi-agent network for solving the problem in the example.

5   Concluding Remarks

This paper presents a continuous-time multi-agent network described by a collective dynamic system for distributed least absolute deviation. It is proven that the multi-agent network reaches consensus at the optimal solutions if the graph of the network is undirected and connected. Numerical example is elaborated to show the effectiveness and performance of the proposed method.

References

1. Cochocki, A., Unbehauen, R.: Neural Networks for Optimization and Signal Processing. John Wiley & Sons, New York (1993)
2. Boyd, S., Parikh, N., Chu, E., Peleato, B., Eckstein, J.: Distributed optimization and statistical learning via the alternating direction method of multipliers. Foundations and Trends in Machine Learning 3, 1–122 (2011)
3. Nedic, A., Ozdaglar, A., Parrilo, P.A.: Constrained consensus and optimization in multi-agent networks. IEEE Transactions on Automatic Control 55, 922–938 (2010)

4. Zhu, M., Martínez, S.: On distributed convex optimization under inequality and equality constraints. IEEE Transactions on Automatic Control 57, 151–164 (2012)

5. Gharesifard, B., Cortés, J.: Distributed continuous-time convex optimization on weight-balanced digraphs. IEEE Transactions on Automatic Control 59, 781–786 (2014)

6. Lobel, I., Ozdaglar, A.: Distributed subgradient methods for convex optimization over random networks. IEEE Transactions on Automatic Control 56, 1291–1306 (2011)

7. Liu, Q., Wang, J.: A second-order multi-agent network for bound-constrained distributed optimization. IEEE Transactions on Automatic Control (2015), doi:10.1109/TAC.2015.2416927

8. Lin, P., Jia, Y.: Consensus of a class of second-order multi-agent systems with time-delay and jointly-connected topologies. IEEE Transactions on Automatic Control 55, 778–784 (2010)

9. Mei, J., Ren, W., Ma, G.: Distributed coordination for second-order multi-agent systems with nonlinear dynamics using only relative position measurements. Automatica 49, 1419–1427 (2013)

10. Yu, W., Ren, W., Zheng, W.X., Chen, G., Lü, J.: Distributed control gains design for consensus in multi-agent systems with second-order nonlinear dynamics. Automatica 49, 2107–2115 (2013)

11. Tank, D., Hopfield, J.: Simple neural optimization networks: An a/d converter, signal decision circuit, and a linear programming circuit. IEEE Transactions on Circuits and Systems 33, 533–541 (1986)

12. Kennedy, M., Chua, L.: Neural networks for nonlinear programming. IEEE Transactions on Circuits and Systems 35, 554–562 (1988)

13. Xia, Y., Wang, J.: A general projection neural network for solving monotone variational inequalities and related optimization problems. IEEE Transactions on Neural Networks 15, 318–328 (2004)

14. Hu, X., Wang, J.: Solving the assignment problem using continuous-time and discrete-time improved dual networks. IEEE Transactions on Neural Networks and Learning Systems 23, 821–827 (2012)

15. Liu, Q., Wang, J.: A one-layer projection neural network for nonsmooth optimization subject to linear equalities and bound constraints. IEEE Transactions on Neural Networks and Learning Systems 24, 812–824 (2013)

16. Liu, Q., Dang, C., Huang, T.: A one-layer recurrent neural network for real-time portfolio optimization with probability criterion. IEEE Transactions on Cybernetics 43, 14–23 (2013)

17. Liu, Q., Huang, T.: A neural network with a single recurrent unit for associative memories based on linear optimization. Neurocomputing 118, 263–267 (2013)

18. Liu, Q., Huang, T., Wang, J.: One-layer continuous- and discrete-time projection neural networks for solving variational inequalities and related optimization problems. IEEE Transactions on Neural Networks and Learning Systems 25, 1308–1318 (2014)

19. Liu, Q., Wang, J.: A projection neural network for constrained quadratic minimax optimization (2015), doi:10.1109/TNNLS.2015.2425301

20. Nedic, A., Ozdaglar, A.: Distributed subgradient methods for multi-agent optimization. IEEE Transactions on Automatic Control 54, 48–61 (2009)

21. Kinderlehrer, D., Stampacchia, G.: An Introduction to Variational Inequalities and Their Applications. Academic, New York (1982)

A Fully Complex-Valued Neural Network for Rapid Solution of Complex-Valued Systems of Linear Equations

Lin Xiao[1], Weiwei Meng[2], Rongbo Lu[1], Xi Yang[1], Bolin Liao[1], and Lei Ding[1]

[1] College of Information Science and Engineering,
Jishou University, Jishou 416000, China
[2] Department of Computer and Information Sciences,
Delaware State University, USA
xiaolin860728@163.com

Abstract. In this paper, online solution of complex-valued systems of linear equations is investigated in the complex domain. Different from the conventional real-valued neural network, which is only designed for real-valued linear equations solving, a fully complex-valued gradient neural network (GNN) is developed for online complex-valued systems of linear equations. The advantages of the proposed complex-valued GNN model decrease the unnecessary complexities in theoretical analysis, real-time computation and related applications. In addition, the theoretical analysis of the fully complex-valued GNN model is presented. Finally, simulative results substantiate the effectiveness of the fully complex-valued GNN model for online solution of the complex-valued systems of linear equations in the complex domain.

Keywords: complex domain, simulation verification, complex-valued linear system, neural network.

1 Introduction

Complex-valued systems of linear equations arise in many important science and engineering applications [1,2,3], such as in neuro-fuzzy inference system (e.g., [4]), in human action recognition (e.g., [5]), blind signal extraction (e.g., [6]), and in the numerical solution of (stiff) systems of ordinary differential equations using implicit Runge-Kutta methods.

In mathematics, the problem of complex-valued systems of linear equations can be generally formulated as

$$Az(t) = b \in \mathbb{C}^n, \tag{1}$$

where $z(t) \in \mathbb{C}^n$ is an unknown complex-valued vector to be obtained, complex matrix $A \in \mathbb{C}^{n \times n}$ and complex vector $b \in \mathbb{C}^n$ are complex-valued coefficients of (1). For convenience, let $z^* \in \mathbb{C}^n$ denote the theoretical solution of (1).

Note that, in most of past literatures on online solution of complex-valued systems of linear equations [7,8,9,10], complex matrix coefficient $A \in \mathbb{C}^{n \times n}$ and

© Springer International Publishing Switzerland 2015
X. Hu et al. (Eds.): ISNN 2015, LNCS 9377, pp. 444–451, 2015.
DOI: 10.1007/978-3-319-25393-0_49

complex vector coefficient $b \in \mathbb{C}^n$ are often divided into their real and imaginary parts, and then processed respectively. Specifically, complex-valued matrix $A \in \mathbb{C}^{n \times n}$ is treated as the combination of its real and imaginary parts, and complex-valued vector $b \in \mathbb{C}^n$ is also treated as the combination of its real and imaginary parts, i.e., $A = A_{\mathrm{re}} + jA_{\mathrm{im}}$ and $b = b_{\mathrm{re}} + jb_{\mathrm{im}}$ with $j = \sqrt{-1}$ denoting an imaginary unit [also $z(t) = z_{\mathrm{re}}(t) + jz_{\mathrm{im}}(t)$]. Therefore, complex-valued linear equation system (1) is equivalent to the following one:

$$[A_{\mathrm{re}} + jA_{\mathrm{im}}][z_{re}(t) + jz_{\mathrm{im}}(t)] = b_{\mathrm{re}} + jb_{\mathrm{im}}, \tag{2}$$

where $A_{\mathrm{re}} \in \mathbb{R}^{n \times n}$, $A_{\mathrm{im}} \in \mathbb{R}^{n \times n}$, $z_{\mathrm{re}} \in \mathbb{R}^n$, $z_{\mathrm{im}} \in \mathbb{R}^n$, $b_{\mathrm{re}} \in \mathbb{R}^n$ and $b_{\mathrm{im}} \in \mathbb{R}^n$.

In addition, since the real and imaginary parts of the left-side and right-side of equation (2) always equal, according to the conventional processing method [11], the following real-valued systems of linear equation can be derived from equation (2) as

$$\begin{cases} A_{\mathrm{re}}z_{\mathrm{re}}(t) - A_{\mathrm{im}}z_{\mathrm{im}}(t) = b_{\mathrm{re}} \in \mathbb{R}^n, \\ A_{\mathrm{re}}z_{\mathrm{im}}(t) + A_{\mathrm{im}}z_{\mathrm{re}}(t) = b_{\mathrm{im}} \in \mathbb{R}^n, \end{cases}$$

which is equivalently expressed in a compact matrix-vector form as

$$\begin{bmatrix} A_{\mathrm{re}} & -A_{\mathrm{im}} \\ A_{\mathrm{im}} & A_{\mathrm{re}} \end{bmatrix} \begin{bmatrix} z_{\mathrm{re}}(t) \\ z_{\mathrm{im}}(t) \end{bmatrix} = \begin{bmatrix} b_{\mathrm{re}} \\ b_{\mathrm{im}} \end{bmatrix} \in \mathbb{R}^{2n}. \tag{3}$$

For presentation simplification, matrix $B \in \mathbb{R}^{2n \times 2n}$, vectors $x(t) \in \mathbb{R}^{2n}$ and $d \in \mathbb{R}^{2n}$ are used to denote the above variables; i.e.,

$$B = \begin{bmatrix} A_{\mathrm{re}} & -A_{\mathrm{im}} \\ A_{\mathrm{im}} & A_{\mathrm{re}} \end{bmatrix}, \quad x(t) = \begin{bmatrix} z_{\mathrm{re}}(t) \\ z_{\mathrm{im}}(t) \end{bmatrix}, \quad d = \begin{bmatrix} b_{\mathrm{re}} \\ b_{\mathrm{im}} \end{bmatrix}.$$

Therefore, solving the complex-valued linear system problem is equivalently converted to the following real-valued linear system problem solving:

$$Bx(t) = d \in \mathbb{R}^{2n}. \tag{4}$$

Then, most of the reported methods [12,13,14,15] for solving real-valued system of linear equation can be applied to online solution of the complex-valued linear equation system (1). However, the corresponding vector space dimension and computation complexity have increased one times for the equivalent real-valued linear system (4) solving, as compared with the complex-valued linear equation system (1) solving. In this paper, a fully complex-valued gradient neural network (GNN) is developed for online solution of complex-valued linear equation directly. This method does not require a equivalent conversion and directly solves the complex-valued linear system problem. More importantly, the fully complex-valued gradient neural network decreases the unnecessary complexities in theoretical analysis, real-time computation, and related applications.

2   Fully Complex-Valued GNN Model

In conventional design processing, when applied to online solution of complex-valued systems of linear equations, complex matrix coefficient $A \in \mathbb{C}^{n \times n}$ and complex vector coefficient $b \in \mathbb{C}^n$ are often divided into their real and imaginary parts, and treated separately in the real domain. However, the fully complex-valued gradient neural network (GNN) is based on the original complex-valued linear equation system (1) instead of a equivalent real-valued linear equation system (4). In addition, the fully complex-valued GNN model needs not such a conversion process. Specifically, the design procedure of fully complex-valued GNN model can be presented as follows.

First, the following the scalar-valued nonnegative energy function is defined (which is based on the original complex-valued linear equation system):

$$\varepsilon(t) = \|Az(t) - b\|_2^2/2. \tag{5}$$

Then, according to the gradient-based design method used in the real domain [12,13,14], a complex-valued gradient algorithm is designed to evolve along a negative gradient descent direction of this energy function until the minimum point can be reached. Obviously, the negative gradient of this energy function can be derived as follows:

$$-\frac{\partial\|Ax(t) - b\|_2^2/2}{\partial z(t)} = -A^{\mathrm{H}}(Az(t) - b). \tag{6}$$

Third, by using the above negative gradient, we can construct the following fully complex-valued GNN model for online solution of the complex-valued linear equation system (1):

$$\dot{z}(t) = -\gamma A^{\mathrm{H}}(Az(t) - b), \tag{7}$$

where design parameter $\gamma > 0$ is used to scale the convergence rate of the fully complex-valued GNN model, and complex state vector $z(t) \in \mathbb{C}^n$, starting from initial state $z(0) \in \mathbb{C}^n$, corresponds to the theoretical solution $z^*] \in \mathbb{C}^n$ of (1).

2.1   Convergence Analysis

It is worth pointing out that convergence performance is of primary importance for a neural-network model to be successfully used [16,17,18,19,20]. Thus, in this subsection, we investigate the convergence performance of the fully complex-valued GNN model (7), which is presented through the following theorem.

Theorem 1. Consider the complex-valued system of linear equation (1). Starting from any initial state $z(0) \in \mathbb{C}^n$, complex state solution $z(t)$ of the fully complex-valued GNN model (7) converges exponentially to the theoretical solution z^* of (1). Besides, the exponential convergence rate is the product of the minimum eigenvalue α of $A^{\mathrm{H}}A$ and the value of γ.

Proof: Let $\tilde{z}(t) = z(t) - z^*$ denote the difference between the complex state solution $z(t)$ generated by the fully complex-valued GNN model (7) and the theoretical solution z^* of linear equation system (1). Then, we have

$$z(t) = \tilde{z}(t) + z^* \in \mathbb{C}^{n \times n} \text{ and } \dot{z}(t) = \dot{\tilde{z}}(t) \in \mathbb{C}^{n \times n}.$$

Substituting the above two equations to (7); and in view of equation $Az^* - b = 0$, we further obtain:

$$\dot{\tilde{z}}(t) = -\gamma A^{\mathrm{H}} A \tilde{z}(t). \tag{8}$$

Thus, we can define a Lyapunov function candidate $v(t)$ as below:

$$v(t) = \|A\tilde{z}(t)\|_2^2 / 2 = (A\tilde{z}(t))^{\mathrm{H}} (A\tilde{z}(t)) / 2 \geqslant 0.$$

Obviously, such a Lyapunov function is positive definite, because $v(t) > 0$ for any $\tilde{z}(t) \neq 0$, and $v(t) = 0$ only for $\tilde{z}(t) = 0$.

In addition, its time derivative can be derived as below:

$$\dot{v}(t) = \frac{\mathrm{d}v}{\mathrm{d}t} = (A\tilde{z}(t))^{\mathrm{H}} (A\dot{\tilde{z}}(t)) = -\gamma \|A^{\mathrm{H}} A \tilde{z}(t)\|_2^2 \leqslant 0, \tag{9}$$

which guarantees the final negative-definiteness of $\dot{v}(t)$. In other words, $\dot{v}(t) < 0$ for any $\tilde{z}(t) \neq 0$, and $\dot{v}(t) = 0$ only for $\tilde{z}(t) = 0$. Besides, we also have $v(t) \to +\infty$, when $\|\tilde{z}\| \to +\infty$. According to Lyapunov theory [12,13,14], complex vector $\tilde{z}$ globally asymptotically converges to zero. On the other hand, due to $\tilde{z}(t) = z(t) - z^*$, we have the following equivalent result: complex state vector $z(t)$ is globally asymptotically convergent to the theoretical solution z^*.

Furthermore, given $\alpha > 0$ as the minimum eigenvalue of $A^{\mathrm{H}} A$, from the above equation, we can obtain:

$$\begin{aligned}
\dot{v}(t) &= -\gamma \|A^{\mathrm{H}} A \tilde{z}(t)\|_2^2 \\
&= -\gamma (A\tilde{z}(t))^{\mathrm{H}} A A^{\mathrm{H}} A \tilde{z}(t) \\
&\leqslant -\alpha\gamma (A\tilde{z}(t))^{\mathrm{H}} A \tilde{z}(t) \\
&= -\alpha\gamma \|A\tilde{z}(t)\|_2^2 \\
&= -2\alpha\gamma v(t).
\end{aligned} \tag{10}$$

In view of equation (10), we can further obtain its analytic solution:

$$v(t) \leqslant v(0) \exp(-2\alpha\gamma t).$$

Moreover, $v(t) = (A\tilde{z}(t))^{\mathrm{H}} A\tilde{z}(t)/2 \geqslant \alpha \tilde{z}(t)^{\mathrm{H}} \tilde{z}(t)/2 = \alpha \|\tilde{z}(t)\|_2^2 / 2$, and $v(0) = \|A\tilde{z}(0)\|_2^2 / 2 \leqslant \|A\|_2^2 \|\tilde{z}(0)\|_2^2 / 2$. Therefore,

$$\alpha \|\tilde{z}(t)\|_2^2 / 2 \leqslant v(t) \leqslant v(0) \exp(-2\alpha\gamma t),$$

which can be further simplified as

$$\|\tilde{z}(t)\|_2 = \|z(t) - z^*\|_2 \leqslant \frac{\|A\|_2 \|\tilde{z}(0)\|_2}{\sqrt{\alpha/2}} \exp(-\alpha\gamma t).$$

Obviously, as $t \to \infty$, $z(t) \to z^*$ exponentially with rate $\alpha\gamma$, which implies that, starting from any randomly-generated initial complex state matrix $z(0)$, complex state matrix $z(t)$ of the fully complex-valued GNN model (7) converges globally and exponentially to the theoretical solution z^* of (1) with rate $\alpha\gamma > 0$. The proof is thus complete. $\qquad\square$

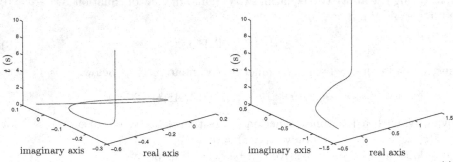

(a) First element of complex vector $z(t)$ (b) Second element of complex vector $z(t)$

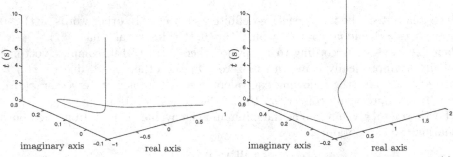

(c) Third element of complex vector $z(t)$ (d) Fourth element of complex vector $z(t)$

Fig. 1. Transient behavior of $z(t)$ synthesized by the fully complex-valued GNN model (7) starting with a randomly-generated initial state.

3   Illustrative Verification

In the above section, the fully complex-valued GNN model (7) has been presented for computing online complex-valued linear system (1). In this section, one illustrative example is presented and the corresponding computer-simulation results are provided for substantiating the efficacy of the fully complex-valued GNN model (7).

Without loss of generality, let us consider the following complex-valued Vandermonde matrix:

$$A = \begin{bmatrix} -0.7597 + 0.6503j & -0.8391 - 0.5440j & 0.2837 - 0.9589j & 1.0000 \\ 0.7597 + 0.6503j & -0.8391 + 0.5440j & -0.2837 - 0.9589j & 1.0000 \\ 0.7597 - 0.6503j & -0.8391 - 0.5440j & -0.2837 + 0.9589j & 1.0000 \\ 0 - 1.0000j & -1.0000 & 0 + 1.0000j & 1.0000 \end{bmatrix}.$$

and a randomly-generated vector:

$$b = \begin{bmatrix} 1.0000, & 0.2837 + 0.9589j, & 0.2837 - 0.9589j, & 0 \end{bmatrix}^{\mathrm{T}}.$$

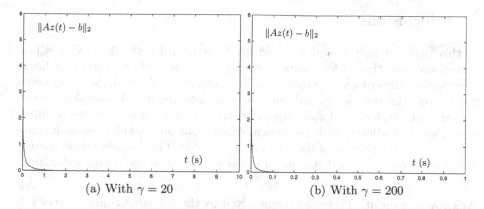

Fig. 2. Transient behavior of the residual error $\|Az(t) - b\|_2$ synthesized by the fully complex-valued GNN model (7) with $\gamma = 20$ and $\gamma = 200$.

To verify the effectiveness of the fully complex-valued GNN model (7), the theoretical solution z^* of the above complex-valued linear equation system is directly given out below:

$$z^* = \begin{bmatrix} -0.4683 - 0.2545j, 1.2425 + 0.3239j, -0.6126 + 0.0112j, 1.5082 + 0.4683j \end{bmatrix}$$

First, starting from a randomly-generated initial complex vector $z(0) \in \mathbb{C}^4$, the fully complex-valued GNN model (7) is applied to computing the above complex-valued linear system under the conditions of design parameter $\gamma = 20$. The computing results in MATLAB are shown in Figs. 1 and 2. From Fig. 1, we can see that each element of complex vector $z(t) \in \mathbb{C}^4$ synthesized by the fully complex-valued GNN model (7) converges to a certain numerical value after a short time. As compared with the theoretical solution given in the above, these numerical values are exactly the theoretical solutions of complex-valued linear system (1). Simulative results demonstrate that the fully complex-valued GNN model (7) is effective on solving the complex-valued systems of linear equations.

Besides, Fig. 2(a) gives the corresponding transient behavior of the residual error $\|Az(t) - b\|_2$ under the same conditions (i.e., the same initial value and the same design parameter γ). As observed from Fig. 2(a), the residual error $\|Az(t) - b\|_2$ of the above complex-valued linear system synthesized by the fully complex-valued GNN model (7) converges to zero after about 1.2 s. This result also means that the solution of the fully complex-valued GNN model (7) can fit with the theoretical solution of the above complex-valued linear system very well. It is worth pointing out that, as shown in Fig. 2(b), the convergence speed of the fully complex-valued GNN model (7) can be accelerated with time. Specifically, the convergence time of the fully complex-valued GNN model (7) is shortened from 1.2 s to 0.12 s when the value of design parameter γ increases from 20 to 200. In summary, we can draw a conclusion from the above simulation results that the fully complex-valued GNN model (7) is effective on computing online complex-valued systems of linear equations.

4   Conclusions

In this paper, a fully complex-valued gradient neural network (GNN) is presented and investigated for computing online complex-valued systems of linear equations in the complex domain. The fully complex-valued GNN model needs not convert the complex-valued linear system into the double-complexity real-valued linear system, and can solves directly complex-valued systems of linear equations. In addition, both theoretical discussions and simulative results substantiate the effectiveness of the fully complex-valued GNN model for computing online the complex-valued systems of linear equations in the complex domain.

Acknowledgment. This work is supported by the Research Foundation of Education Bureau of Hunan Province, China (Grant No. 15B192), the National Natural Science Foundation of China(Grants No. 61503152, 61563017, 61363073 and 61363033), and the Experiment Teaching Reform Foundation of Jishou University, China (Grant No. 2015SYJG034). In addition, the authors would like to thank the editors and anonymous reviewers for their valuable suggestions and constructive comments which have really helped the authors improve very much the presentation and quality of this paper.

References

1. Day, D., Heroux, M.A.: Solving Complex-Valued Linear Systems via Equivalent Real Formulations. SIAM J. Sci. Comput. 23, 480–498 (2000)
2. Owe, A., Andrey, K.: Real Valued Iterative Methods for Solving Complex Symmetric Linear Systems. Numer. Linear Algebra Appl. 7, 197–218 (2000)
3. Chen, X., Song, Q.: Global Stability of Complex-Valued Neural Networks with Both Leakage Time Delay and Discrete Time Delay on Time Scales. Neurocomputing 121, 254–264 (2013)
4. Subramanian, K., Savitha, R., Suresh, S.: A Complex-Valued Neuro-Fuzzy Inference System and Its Learning Mechanism. Neurocomputing 123, 110–120 (2014)
5. Venkatesh Babu, R., Suresh, S., Savitha, R.: Human Action Recognition Using a Fast Learning Fully Complex-Valued Classifier. Neurocomputing 89, 202–212 (2012)
6. Durán-Díaz, I., Cruces, S., Sarmiento-Vega, M.A., Aguilera-Bonet, P.: Cyclic maximization of Non-Gaussianity for Blind Signal Extraction of Complex-Valued Sources. Neurocomputing 74, 2867–2873 (2011)
7. Zhang, Y., Chen, Z., Chen, K.: Convergence Properties Analysis of Gradient Neural Network for Solving Online Linear Equations. Acta Automatica Sinica 35, 1136–1139 (2009)
8. Yi, C., Zhang, Y.: Analogue Recurrent Neural Network for Linear Algebraic Equation Solving. Electron. Lett. 44, 1078–1079 (2008)
9. Zhang, Y., Chen, K.: Global Exponential Convergence and Stability of Wang Neural Network for Solving Online Linear Equations. Electron. Lett. 44, 145–146 (2008)
10. Liao, B., Zhang, Y.: Different Complex ZFs Leading to Different Complex ZNN Models for Time-Varying Complex Generalized Inverse Matrices. IEEE Trans. Neural Netw. Learning Syst. 25, 1621–1631 (2014)

11. Liao, W., Wang, J., Wang, J.: A Recurrent Neural Network for Solving Complex-Valued Quadratic Programming Problems With Equality Constraints. In: Tan, Y., Shi, Y., Tan, K.C. (eds.) ICSI 2010, Part II. LNCS, vol. 6146, pp. 321–326. Springer, Heidelberg (2010)
12. Zhang, Y.: Revisit the Analog Computer and Gradient-Based Neural System for Matrix Inversion. In: Proceedings of IEEE International Symposium on Intelligent Control, pp. 1411–1416 (2005)
13. Guo, D., Yi, C., Zhang, Y.: Zhang Neural Network Versus Gradient-based Neural Network for Time-Varying Linear Matrix Equation Solving. Neurocomputing 74, 3708–3712 (2011)
14. Zhang, Y., Shi, Y., Chen, K., Wang, C.: Global Exponential Convergence and Stability of Gradient-based Neural Network for Online Matrix Inversion. Appl. Math. Comput. 215, 1301–1306 (2009)
15. Zhang, Y., Ge, S.S.: Design and Analysis of a General Recurrent Neural Network Model for Time-Varying Matrix Inversion. IEEE Trans. Neural Netw. 16, 1477–1490 (2005)
16. Zhang, Y., Ke, Z., Xu, P., Yi, C.: Time-varying Square Roots Finding via Zhang Dynamics Versus Gradient Dynamics and the Former's Link and New Explanation to Newton-Raphson Iteration. Inform. Process. Lett. 110, 1103–1109 (2010)
17. Xiao, L., Zhang, Y.: Two New Types of Zhang Neural Networks Solving Systems of Time-Varying Nonlinear Inequalities. IEEE Trans. Circuits Syst. I 59, 2363–2373 (2012)
18. Xiao, L., Zhang, Y.: From Different Zhang Functions to Various ZNN Models Accelerated To Finite-Time Convergence for Time-Varying Linear Matrix Equation. Neural Process. Lett. 39, 309–326 (2014)
19. Xiao, L., Lu, R.: Finite-time Solution to Nonlinear Equation Using Recurrent Neural Dynamics with a Specially-Constructed Activation Function. Neurocomputing 151, 246–251 (2015)
20. Xiao, L.: A Finite-Time Convergent Neural Dynamics for Online Solution of Time-Varying Linear Complex Matrix Equation. Neurocomputing 167, 254–259 (2015)

Novel Approaches and Applications

Sparse Representation
via Intracellular and Extracellular Mechanisms[*]

Jiqian Liu[**] and Chengbin Zeng

School of Information Engineering, Guizhou Institute of Technology,
Guiyang, Guizhou Province, 550003, P.R. China
{Liujiqian,zengchengbin}@git.edu.cn

Abstract. Sparse representation in sensory cortex has been well verified and its capability of yielding response properties of single neurons is also demonstrated. In order to improve sparse representation to be more neurally plausible, we reconsider several response properties of single neurons, especially the cross orientation suppression and surround suppression. A new sparse representation model using intracellular and extracellular neural mechanisms is presented. Simulation results of the presented model explain physiological observations very well.

Keywords: Sparse representation, cross orientation suppression, surround suppression, membrane current.

1 Introduction

Sparse representation by neuronal populations in primary visual cortex has been heavily investigated during the last decades. Recently, many response properties of single neurons are reported to be well simulated by sparse representation [1-3], giving insight into the unified understanding of the response properties of both neuronal populations and single neurons. For further study in this area, some response properties of single neurons need to be reconsidered, especially those related to lateral inhibition. The reason is that the functional importance of lateral inhibition in neural processing has been well documented in previous studies [4, 5]. In this paper, two well-known phenomena which are closely related to lateral inhibition, the cross orientation suppression and surround suppression, are taken into account.

In cross orientation suppression, simple cell responses to the preferred orientation are suppressed by a superimposed orthogonal stimulus. It was thought that lateral inhibition from neurons preferring different orientations is responsible for the occurrence of this phenomenon. But measurements with intracellular recordings show that synaptic inhibition is primarily tuned to similar orientations [6]. In fact, the

[*] This work was supported by the Science Research Foundation for High-level Talents of Guizhou Institute of Technology XJGC20130902, and in part by the Science and Technology Foundation of Guizhou Province J[2014]2081 and by the Innovation Team of Guizhou Provincial Eduction Department under Grant No [2014] 34.
[**] Corresponding author.

X. Hu et al. (Eds.): ISNN 2015, LNCS 9377, pp. 455–462, 2015.
DOI: 10.1007/978-3-319-25393-0_50

orthogonal stimulus decreases instead of increasing the cortical inhibition [7]. Cross orientation suppression occurs very fast with a latency almost identical to that of the feedforward excitation [8], indicating that cross orientation suppression may originate in the excitatory input. A purely feedforward model with the contrast saturation and rectification in LGN neuron responses has revealed to be able to account for a lot of properties of cross orientation suppression [9, 10]. But this model predicts the loss of cross orientation suppression at low contrasts, which is in contrast to the case of neuronal populations [11].

Another phenomenon relating to cortical inhibition is surround suppression where the presence of stimuli in the non-classical receptive field of cortical neurons can suppress their spiking responses. Different from cross orientation suppression, the latency of surround suppression is much longer than the onset of the center response [8]. This cannot be explained by subcortical mechanisms. Cortical inhibition increased by the surround stimulus provides a possible explanation for this effect. But intracellular recordings observe a similar decrease in both synaptic excitation and inhibition in surround suppression [12] as that in cross orientation suppression. But in another research, increasing the stimulus width did increase the membrane potential [13].

In summary, we notice that cortical inhibition is found to be reduced with cortical excitation in both cross orientation and surround suppression by intracellular studies. Moreover, the conflicting observations between intracellular and extracellular recordings have been reported on orientation selectivity besides surround suppression. Intracellular recordings found that excitation and inhibition have similar tuning for orientation [6], but extracellular recordings suggest that intracortical inhibition is more broadly tuned than excitation [14]. One possible explanation for these paradoxes could be that intracellular and extracellular mechanisms perform different roles in neural information processing. Based on these analyses, this paper presents our new sparse representation model constructed with intracellular and extracellular mechanisms.

2   The Sparse Representation Model

As pointed out by Olshausen and Field [15], there were only two global objectives need to be optimized for sparse representation, that the representation is sparse and that the representation error is small. Sparse representation using l_0-norm constraint can be expressed as

$$E = \min\left\{(x - Wy)^2 + \alpha\|y\|_0\right\},\tag{1}$$

where x is the input signal, W is a matrix whose ith column is the basis vector w_i, y is a vector of coefficients, and α is the trade-off. There are plenty of methods have been proposed for solving this problem. In iterative thresholding algorithms (ITA), the input vector is approximated iteratively while the representation is made sparser by a thresholding mechanism at each iteration step [16-18]. The main calculation step in such methods is typical, which is given as follows:

$$y(n+1) = y(n) + \mu\left(W^T x - W^T W y(n)\right).\tag{2}$$

where μ is the step length. The main purpose of this equation is to minimize the representation error. This step together with a thresholding process, which cuts off the small outputs and makes the representation sparser, can yield sparse representation of the input signal.

The implementation of the Eq. (2) in neural systems could be realized by extracellular mechanisms of cortical excitation and inhibition. That is, feedforward excitation $W^T x$ is suppressed by the cortical inhibition $W^T W y(n)$, where $W^T W$ is the lateral inhibitory coefficient matrix. These extra excitation and inhibition are integrated to establish the extra environment of cortical neurons. Let v denote the integral of feedforward excitation and lateral inhibition (without the self-inhibition) of a cortical neuron. Thus

$$v = W^T x - (W^T W - I) y(n). \tag{3}$$

As pointed out above, this external integration is responsible for the minimization of the representation error.

We suppose that the external integration v will elicit excitatory membrane current I^+, and inhibitory membrane current I^-. Both of these membrane currents, I^+ and I^-, are defined to be positive. Then the internal state $u = I^+ - I^-$ is a function of the external integration: $u = f(v)$. As for the relationship between u and v, we introduce the generalized tanh-function as follows: when the external integration v is zero, the excitatory and inhibitory membrane currents are balanced, that is, $I^+ = I^- = I_0$ where I_0 is a constant satisfying $0 < I_0 < 1$; while when $v \neq 0$, the membrane currents are calculated by

$$\begin{cases} I^+ = \dfrac{e^v}{e^v + \alpha e^{-v}} \\[2mm] I^- = \dfrac{e^{-v}}{\alpha e^v + e^{-v}} \end{cases} \tag{4}$$

where I_0 is named the zero state parameter and $\alpha = (1 - I_0)/I_0$. Especially, when $I_0 = 1/2$, we have $\alpha = 1$ and $u = I^+ - I^- = \tanh(v)$. Thus, the generalized tanh-function $u = f(v)$ is given by

$$f(v) = \mathrm{gtanh}(v) = \frac{e^v}{e^v + \alpha e^{-v}} - \frac{e^{-v}}{\alpha e^v + e^{-v}}, \qquad \alpha > 0. \tag{5}$$

This function is a sigmoid function (see Fig. 1a). The relationship between v and u, I^+, I^- is shown in Fig. 1b. The excitatory membrane current I^+ is positively related to the external integration v while the inhibitory membrane current I^- is just the opposite.

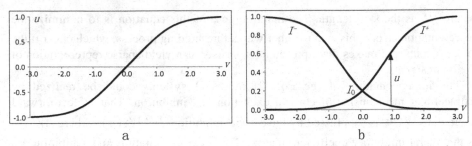

a   b

Fig 1. Sketch showing the relationship between the internal state u, the excitatory membrane current I^+, the inhibitory membrane current I^- and the external integration v.

We assume that the spike response y of cortical neurons is modulated to match the internal state. The acceleration of this modulation is proportional to the difference between the internal state and the spike response. In summarize, the presented sparse representation model using intracellular and extracellular mechanisms, named Intracellular and Extracellular Mechanisms based Sparse Representation (IEM-SR), is given as follows

$$\begin{cases} v = W^T x - \left(W^T W - I \right) y \\ u = I^+ - I^- = \text{gtanh}(v) \\ \Delta y = \mu(u - y), u > 0 \end{cases} \quad (6)$$

When this process converges, $y=u=\text{gtan}(v)$. In the special case where $\alpha = 1$ and the magnitude of v is small enough, we have $u = \text{gtanh}(v) \approx v$. The schematic diagram of the presented model is shown in Fig. 2.

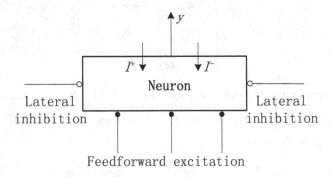

Fig. 2. Schematic diagram of the neuron response model. External integration of feedforward excitationand lateral inhibition elicits membrane currents. The excitatory membrane current I^+ and the inhibitory membrane current I^- set up the internal state of the neuron. The spike response y is modulated to match the internal state.

3   Experiments

We validate the IEM-SR model by simulating several well-known response properties of single neurons, including cross orientation suppression within CRF, and surround suppression from nCRF, and also the sharp and contrast invariant orientation tuning. The visual stimuli and Gabor-like connection weights are generated using the same code published by Spratling [19]. A family of 32 Gabor functions covering 8 orientations and 4 phases is used as the connection weights $\{w_i\}$, which are normalized to unit length. The DoG filtered visual stimulus x is subject to a saturating nonlinearity:

$$x' = \tanh(2\pi x). \tag{7}$$

Both the visual stimuli $\{x'\}$ and the Gabor functions are then processed to be non-negative to generate x and W.

During the simulation, we found the zero state parameter I_0 has a close relationship with the neuronal activity level. For orientation-tuning experiments, we assume I_0 is proportional to the feedforward excitation. That is, I_0 is given by $\lambda w_i x$ where λ is a constant.

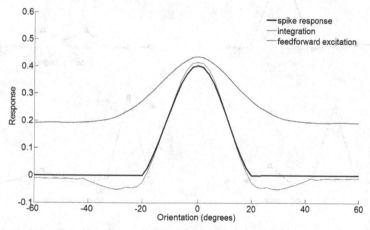

Fig. 3. Orientation tunings of feedforward excitation, integration and spike response. The spike response matches the integration very well when the latter is positive. Feedforward excitation is sensitive to a wide range of stimuli and exhibits less orientation selectivity than the spike response.

A comparison between orientation tunings of feedforward excitation, integration and spike response of an activated output neuron is shown in the Fig. 3. As we see, the neuron will respond to the full range of orientations and have a much broader tuning width without lateral inhibition. This is consistent with that the tuning width of cortical excitation is broader than that of the spike response [20]. The sharp orientation selectivity of the spike responses can be reduced by blocking cortical inhibition [21].

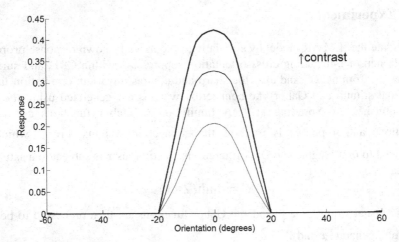

Fig. 4. Illustration of the contrast invariance of orientation selectivity. The tuning with remains unchanged with increasing stimulus contrast.

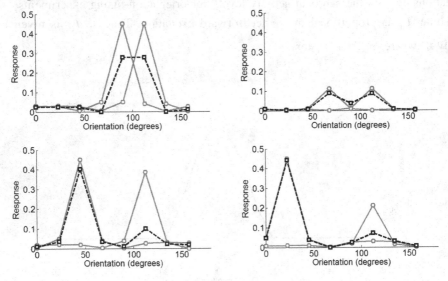

Fig. 5. Illustration of cross orientation suppression using stimuli with different orientation and contrast. From top to bottom and left to right, the contrasts of the two components of the superimposed stimulus are: 1, 1; 0.1, 0.1; 1, 0.5; 1, 0.2. The suppression effect is still observable at low contrasts.

It has been suggested that the orientation selectivity of simple cells originates purely from the excitatory convergence of LGN afferents. According to this theory, the tuning width of orientation selectivity should be widened with increasing contrast of stimuli, which on the contrary was not observed in experiments [22]. This phenomenon is known as the contrast invariance of orientation selectivity. The simulation result given by the presented model is shown in Fig. 4.

Cross orientation suppression was thought to be responsible for contrast invariance of orientation selectivity [14, 23]. But intracellular recording observations rule out this hypothesis. Although a purely feedforward model can account for many response properties of cross orientation suppression, it failed to predict the suppression at low contrasts [11]. In fact, non-negative data processing of LGN input is enough to explain this disagreement. The simulating results are shown in Fig. 5.

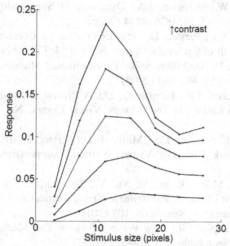

Fig. 6. Contrast dependent size tuning of surround suppression. The optimal size of the tuning curve is the one that elicits the peak response.The optimal sizes at low contrasts are larger than those at high contrasts.

To simulate the contrast dependent size tuning of surround suppression [24], only the Gabor function with the same orientation as the center stimulus is used for the simplicity of computation. This is reasonable because surround suppression is narrowly tuned to the preferred orientation. As shown in Fig. 6, the optimal stimulus sizes at high contrasts are smaller than those at low contrasts.

References

1. Hunt, J.J., Dayan, P., Goodhill, G.J.: Sparse Coding Can Predict Primary Visual Cortex Receptive Field Changes Induced by Abnormal Visual Input. PLoS Computational Biology 9, e1003005 (2013)
2. Giorno, A.D., Zhu, M., Rozell, C.J.: A Sparse Coding Model of V1 Produces Surround Suppression Effects in Response to Natural Scenes. BMC Neuroscience 14, P335 (2013)
3. Zhu, M., Rozell, C.J.: Visual Nonclassical Receptive Field Effects Emerge From Sparse Coding in A Dynamical System. PLoS Computational Biology 9, e1003191 (2013)
4. Olshausen, B.A.: Principles of Image Representation in Visual Cortex. In: Chalupa, L.M., Werner, J.S. (eds.) The Visual Neurosciences, pp. 1603–1615 (2003)

5. Liu, J., Jia, Y.: A Lateral Inhibitory Spiking Neural Network for Sparse Representation in Visual Cortex. In: Zhang, H., Hussain, A., Liu, D., Wang, Z. (eds.) BICS 2012. LNCS, vol. 7366, pp. 259–267. Springer, Heidelberg (2012)
6. Anderson, J.S., Carandini, M., Ferster, D.: Orientation Tuning of Input Conductance, Excitation, and Inhibition in Cat Primary Visual Cortex. J. Neurophysiol. 84, 909–926 (2000)
7. Priebe, N.J., Ferster, D.: Mechanisms Underlying Cross-orientation Suppression in Cat Visual Cortex. Nat. Neurosci. 9, 552–561 (2006)
8. Smith, M.A., Bair, W., Movshon, J.A.: Dynamics of Suppression in Macaque Primary Visual Cortex. J. Neurosci. 26, 4826–4834 (2006)
9. Finn, I.M., Priebe, N.J., Ferster, D.: The Emergence of Contrast-Invariant Orientation Tuning in Simple Cells of Cat Visual Cortex. Neuron 54, 137–152 (2007)
10. Priebe, N.J., Ferster, D.: Inhibition, Spike Threshold, and Stimulus Selectivity in Primary Visual Cortex. Neuron 57, 482–497 (2008)
11. MacEvoy, S.P., Tucker, T.R., Fitzpatrick, D.: A Precise Form of Divisive Suppression Supports Population Coding in The Primary Visual Cortex. Nat. Neurosci. 12, 637–645 (2009)
12. Ozeki, H., Finn, I.M., Schaffer, E.S., Miller, K.D., Ferster, D.: Inhibitory Stabilization of The Cortical Network Underlies Visual Surround Suppression. Neuron 62, 578–592 (2009)
13. Haider, B., Krause, M.R., Duque, A., Yu, Y., Touryan, J., Mazer, J.A.: Synaptic and Network Mechanisms of Sparse and Reliable Visual Cortical Activity During Nonclassical Receptive Field Stimulation. Neuron 65, 107 (2010)
14. Sompolinsky, H., Shapleyt, R.: New Perspectives on The Mechanisms for Orientation. Current Opinion in Neurobiology 7, 514–522 (1997)

Load Balancing Algorithm Based on Neural Network in Heterogeneous Wireless Networks

Xin Song[1], Liangming Wu[2], Xin Ren[1], and Jing Gao[1]

[1] Engineering Optimization and Smart Antenna Institute,
Northeastern University at Qinhuangdao, 066004, China
[2] North Automation Control Technology Institute, Taiyuan 030006, China
Sxin78916@mail.neuq.edu.cn

Abstract. Some load balancing algorithms in heterogeneous wireless networks can not consider the problems arising from the admission control of new service and service transfer of heavy load networks. To solve these problems, we propose a load balancing algorithm based on neural networks. This algorithm is used to conduct prediction through network load rate and achieve the network admission of new service by combining an admission control optimization algorithm. Moreover, by analyzing network performance, some services of heavy load network are transferred to overlay light load network. The simulation results indicate that our algorithm can well realize the load balancing of heterogeneous wireless network and provide high resource utilization.

Keywords: Heterogeneous wireless network, Load balancing, Admission control, Blocking probability.

1   Introduction

With the rapid development of mobile communication technology, a large amount of wireless networks are available. Although these networks provide varying types of communication ways and network access manners, many information isolated islands are likely to be produced if they fail to effectively realize interconnection and intercommunication. Therefore they cannot offer the communication services of ensuring end-to-end QoS, which greatly reduce integrate utility of networks and customers' service experiences. In this context, heterogeneous integrated wireless networks have been the developing trend of communication network [1-2]. For the integrated wireless networks have to fully use the complementarity between networks, and efficiently utilize limited wireless network resources, wireless resource management (RSM) has been a key research. As a key technology of RSM, load balancing algorithms are proposed, which can effectively promote resources utilization, increase network volume, and reduce network block probability [3-5].

In this paper, an IASA load balancing mechanism to maximize the joint utility of integrated heterogeneous networks is proposed, which performs load status monitoring and evaluation for access gateways and heterogeneous networks [6]. According to the characteristics of heterogeneous networks and user demand for

© Springer International Publishing Switzerland 2015
X. Hu et al. (Eds.): ISNN 2015, LNCS 9377, pp. 463–472, 2015.
DOI: 10.1007/978-3-319-25393-0_51

seamlessly connection, an adaptive vertical handoff algorithm based on compensation time is proposed [7] Jiao et al. propose a quality of service (QoS) aware load-balancing algorithm for efficient network resources dispatching in heterogeneous wireless networks. The proposed algorithm based on the characteristics of different wireless services defines a utility function for each terminal to represent its QoS experience and a utility function for each radio access network to represent its load level [8]. An optimized algorithm based on vertical handoff (VHO) prediction approach in heterogeneous wireless networks is proposed, which defines objective function considering received signal strength, user equipment velocity, load and cost per user bandwidth [9]. An adaptive threshold load balancing scheme provides a two-sage load balancing strategy. It makes call and assign to UMTS and WLAN in a proper probability based on assigning services and dynamic network changes [10].

In this paper, the load balancing algorithm is presented, which can rationally arrange service admission into candidate network by predicating the load rate of each network and use optimization control strategy of service admission. To decrease vertical handoff times and improve the load balancing level of the system, our algorithm uses an assisted dynamic load transfer (ADLT) algorithm to set reasonable load transfer amount. The simulation results indicate the algorithm can significantly improve the system performances.

2   Load Balancing Algorithm

2.1   Load Predication

We use the radial basis function neural network (RBFNN) to obtain load predication. Data preprocessing is conducted to make up the drawback of RBFNN in convergence speed. Then, RBFNN is used to predicate the load rate of each network.

The RBFNN is a special three-layered feed-forward network, which consists of the input layer, the output layer, and the hidden layer as shown in Fig. 1. In the hidden layer, the nonlinear functions that perform this transformation are usually taken to be Gaussian functions of appropriately chosen means and variances.

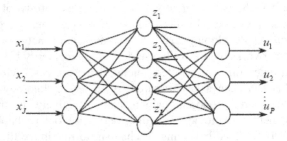

Fig. 1. Architecture of a RBFNN

The weights from the hidden layer to the output layer are identified by following a supervised learning procedure. Assume that the input layer, the hidden layer, and the output layer have N, L, M nodes respectively. The network output vector is given by

$$u_m = \sum_{l=1}^{L} \omega_{l,m} \exp^{\frac{\|x-c(l)\|^2}{\sigma_l^2}} \quad (m = 1, \cdots, M) \tag{1}$$

where $x = [x_1, x_2, \cdots, x_N]$ is the input vector to the network, $c(l)$ and σ_l^2 are mean and standard deviation of the l Gaussian function, $\{\omega_{l,m}, l = 1, \cdots, L; m = 1, \cdots, M\}$ is the weight from the m node in the output layer to the l node in the hidden layer. Training samples are divided into L classes by K learning algorithm, $c(l)$ is the l clustering center vector, and σ_l^2 is the average distance to the first few nearest neighbors of the means of the other Gaussian functions.

In mutual information based data selection algorithm (MIDS), a minimum redundancy- maximum relevance (MRMR) principle is used in the selection of variables. By using the principle, each variable selected can carry more high value information which is not contained in other variables as less as possibly. Moreover, the condition of output maximum relevance can be satisfied. RBFNN is trained by taking the variables selected via MIDS as input data. Then, trained data can be further used to predict load rate of wireless networks.

2.2 The Optimization Model of Admission Control

Assuming that there are m candidate admission networks, the residual bandwidth resource for each candidate network denotes C_j. $j = \{1, 2, ..., m\}$. If there are n new services to be accessed, the bandwidth resource for each new service is b_i, $i = \{1, 2, ..., n\}$. While the objective of the admission control of new service is to reasonably allocate n services into m candidate admission network. The optimization of admission control is [11]:

$$\text{Min} f(X) = \frac{1}{m} \sum_{j=1}^{m} \left(R_{load}(j) - \frac{1}{m} \sum_{j=1}^{m} R_{load}(j) \right)^2$$

$$s.t. \begin{cases} \text{Max} \sum_{j=1}^{m} \sum_{i=1}^{n} x_{ij}, & i = 1, 2, \cdots, n \ j = 1, 2, \cdots, m \\ \sum_{i=1}^{n} x_{ij} b_i \leq C_j & j = 1, 2, \cdots, m \\ \sum_{j=1}^{m} x_{ij} \leq 1, & i = 1, 2, \cdots, \text{n} \end{cases} \tag{2}$$

where $R_{load}(j)$ is the load rate of each candidate admission network and obtained by using load predication, while x_{ij} is the case when ith service is accessed into jth wireless network r_j. When service s_i is accessed into r_j, we have $x_{ij} = 1$, otherwise, $x_{ij} = 0$. The solution to equation (2) is $n \times m$ 0/1 matrix.

2.3   The Solution Based on Genetic Algorithm

(1) Genetic coding

In this research, the chromosome coding used in the optimization model of admission control is binary encoding. The solution of the model is encoded into a binary sequence. By spatially converting the solution space X in equation (2), we obtain

$$X' = [x_{11}, \cdots, x_{n1}, x_{12}, \cdots, x_{n2}, \cdots, x_{1m}, \cdots x_{nm}] \tag{3}$$

(2) Fitness function

When encoding based on above steps, the adaptability function of the optimization model for admission control is defined

$$F(X) = G(f(X)) + \sum_{i=1}^{n} S_i(X) + \sum_{j=1}^{m} R_j(X) + \sum_{i=1}^{n} T_i(X) \tag{4}$$

where $G(f(X))$ is objective function of the model

$$G(f(X)) = \frac{1}{1 + \xi f(X)} \tag{5}$$

$S_i(X)$ is a penalty function when ith service S_i is simultaneously accessed into multiple candidate networks and presented as

$$S_i(X) = \begin{cases} \alpha\left(1 - \sum_{j=1}^{m} x_{ij}\right), & 1 - \sum_{j=1}^{m} x_{ij} < 0 \\ 0, & 1 - \sum_{j=1}^{m} x_{ij} \geq 0 \end{cases} \tag{6}$$

$R_j(X)$ is a penalty function when the resources which are occupied by the new service loaded are not satisfied the certain constraint

$$R_j(X) = \begin{cases} \beta\left(C_i - \sum_{i=1}^{n} x_{ij}b_i\right), & C_i - \sum_{i=1}^{n} x_{ij}b_i < 0 \\ 0, & C_i - \sum_{i=1}^{n} x_{ij}b_i \geq 0 \end{cases} \tag{7}$$

$T_i(X)$ is a penalty function when new service is blocked.

$$T_i(X) = \delta\zeta\left(\sum_{j=1}^{m} x_{ij} - 1\right)b_i \tag{8}$$

(3) Genetic operation

Basic genetic algorithm consists of three operators: selection, crossover and mutation. The strategies for the operators are illustrated as follows.

1) Selection

The selection strategy is based on the applicability proportion. The probability that individual i is selected

$$p_i = \frac{F_i}{\sum\limits_{i=1}^{N} F_i} \tag{9}$$

where F_i applicability value of individual i, and N is individual number of population.

2) Crossover

A consistent crossover method with probability of P_{cro} is used. By setting screen series, it is determined that chromosome of substring individual needs to inherit the chromosomes corresponding to the individuals in two main strings. The screen sequence is a 0/1 sequence which has an equivalent length to individual coding series and can be generated randomly.

3) Mutation

In this research, site mutation is utilized. The crossover probability P_{cro} is in the range from 0.65 to 0.85. The mutation probability P_{mut} is 0.01~0.2. The two probabilities usually accelerate the convergence speed by using ways such as dynamic value and stepwise decrease.

Through steps above-mentioned, the optimal matrix obtained is the solution X of the optimization model for admission networks in this research. By solving x_{ij} in the matrix, the relationship of new service and network admission can be acquired.

2.4   Assisted Dynamic Load Transfer (ADLT) Based on the Utility Function

According to load characteristics, B_i is total amount of network r_i. For the load rate of r_i, it is assumed that r_i is loaded Kth RT services and L NRT services, and the bandwidth resources occupied by the kth RT services and lth NRT services are $B_{RT}(k)$ and $B_{NRT}(l)$ respectively. The load rate of r_i is therefore expressed as

$$R_{load}(i) = \frac{\sum\limits_{k=1}^{K} B_{RT}(k) + \sum\limits_{l=1}^{L} B_{NRT}(l)}{B_i} \tag{10}$$

In addition to ensuring that dynamic blocking probability of service is in an allowable range, the service quality degree of RT service endowed by r_i: $G_{RT}(i)$ is

$$G_{RT}(i) = \begin{cases} \dfrac{\lg P_i}{\lg P_{Max}}, & P_i > P_{Max} \\ 1, & P_i \le P_{Max} \end{cases} \tag{11}$$

This research merges the requirements of network load rate and ensuring degree of QoS in RT service by using a geometric mean method. Moreover, the utility function of r_i in RT service is defined

$$U_{RT}(i) = \sqrt{(1 - R_{load}(i))G_{RT}(i)} \tag{12}$$

Similarly, the utility function of r_i in NRT service is defined

$$U_{NRT}(i) = \sqrt{(1 - R_{load}(i))G_{NRT}(i)} \tag{13}$$

where $G_{NRT}(i)$ is ensuring degree of QoS in NRT service.

In order to fulfill consistence analysis, utility functions of different types of services are converged. Firstly, the functions of $U_{RT}(i)$ and $U_{NRT}(i)$ are standardized to obtain $U'_{RT}(i)$ and $U'_{NRT}(i)$.

The triangle module operator is applied to enhance the characteristics of similar information and harmonicity of contradiction information. Through converging varying utility functions, comprehensive utility function of network $U(i)$ is obtained [12]

$$U(i) = \frac{U'_{RT}(i) * U'_{NRT}(i)}{1 - (1 - U'_{RT}(i))(1 - U'_{NRT}(i))} \tag{14}$$

By analysis of comprehensive utility level of each network in heterogeneous integrated networks, the highest comprehensive performance level r_i and lowest comprehensive performance level r_j are obtained.

The load amount ∇_{ji} which is transferred from heavy load network to light load network is presented as:

$$\nabla_{ji} = \min\left((\eta_H - R_{load}(i) + \beta)C_i, (R_{load}(j) - \eta_H + \beta)C_j\right) \tag{15}$$

where η_H is transfer trigger threshold, and η_M is reference threshold.

In load transfer process, the proposed algorithm uses aided load transfer factor β to decrease the load rate of heavy load network to a reasonable range. Thus, the probability of service handoff can be reduced through rationally control of load transfer amount. In this paper, the operated load transfer service is marked. In following process, the services without marks are given first priority in selection.

3   Simulation Results

To validate the performance of the proposed algorithm, the model is constructed and overlaid by four networks: WCDMA, TD-LTE, WiMAX and McWiLL. The load rate $R_{load}(j)$ of each network at initial status is predicted by using MIDS algorithm. New services arrive randomly based on Poisson distribution. The other parameters are shown in Table 1.

Table 1. Setting of simulating parameters

Items	Parameters
Network number M	10
Network coverage way	Overlay
Network capacity (Mbps)	50-100
Predicted network load rate	0.01-1.00
Bandwidth required by RT service (Kbps)	64
Bandwidth required by NRT service (Kbps)	5
Arrival rate of RT service (call/s)	0.1-1.0
Arrival rate of NRT service	1-10
Average service time	90-110

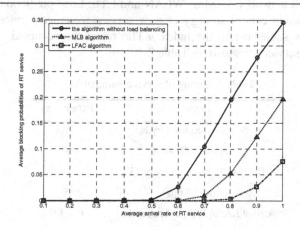

Fig. 2. Average blocking probability of RT service

The average blocking probabilities using three algorithms with the increasing average arrival rate of RT service are shown in Fig 2. With the increase of average arrival rate of RT service, the algorithm without load balancing techniques, leads to the increase of the blocking probability. By using the proposed algorithm, the blocking probability is lowest. With the increase of arrival rate, the blocking probability decreases more apparently.

The average transmission time of NRT service is demonstrated in Fig 3. As the average arrival rate of NRT service increases, the average transmission time of NRT service increases. In contrast, the average transmission time for NRT service using the proposed algorithm is obviously less than that of MLB. This results show that the admission control by using load prediction can acquire better performance.

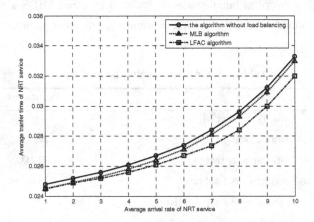

Fig. 3. Average transfer time of NRT service

The proposed algorithm is compared with existing MLB [13] to verify its performance. The network model simulated consists of $N = 20$ overlaid networks from three types of networks: WiMax, WLAN and LTE. Load prediction algorithm is used as the admission control strategy of new service. Average service time of the service distributes subjecting to the index μ. The resources occupied by single RT service and NRT service are shown in Table 2.

Table 2. Setting of service parameters

Items	Parameters
Bandwidth required by RT service (Kbps)	64
Bandwidth required by NRT service (Kbps)	5
Arrival rate of RT service (call/s)	0.1-1.0
Arrival rate of NRT service (call/s)	1-10
Average service time of service(s)	90-110

The resources occupied by single RT service and NRT service are shown in Table 2. The average blocking probabilities of three algorithms are shown in Fig 4. In the case of small amount of services, each network is endowed with sufficient network resources, which showing little block. In this context, the average blocking probability of service approximates to 0. With increasing services, available network resources decrease gradually, which leading to rising blocking probability of new service. As the service amount of MLB algorithm is more than 80, the average blocking probability increases sharply. In our algorithm, to avoid the occurrence of frequent handoff of service, this research uses load prediction method to effectively alleviate the status of heavy load network through accurate analysis of the comprehensive performance of target networks and aided dynamic load transfer. The services transferred are marked. This can well decrease the block resulting from frequent handoff of service, ensure QoS of handoff, and greatly reduce the average blocking probability.

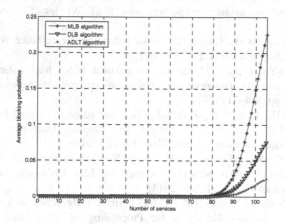

Fig. 4. Average blocking probability

4   Conclusions

PTS is an efficient method to reduce the PAPR of OFDM system, but its high computational complexity is the main obstacle to application. In this paper, we presented a ACA-SA method to search the optimal set of phase factors of all subcarriers for the PTS technique in order to obtain good tradeoff between computational complexity and PAPR performance for OFDM signals. As compared to the conventional PTS and IPTS scheme, the simulation results showed that the performance of the proposed method can not only achieved relative good PAPR reduction but also enjoyed complexity advantages.

Acknowledgments. This work is supported by Program for New Century Excellent Talents in University no. NCET-12-0103, the National Nature Science Foundation of China under Grant no. 61473066 and No.61403069, the Fundamental Research Funds for the Central Universities under Grant No. N130423005, Natural Science Foundation of Hebei Province under Grant No. F2014501055, the Program of Science and Technology Research of Hebei University No. ZD20132003.

References

1. Cao, J., Zhang, C.: Heterogeneous Wireless Networks. In: Seamless and Secure Communications over Heterogeneous Wireless Networks, pp. 9–26. Springer, New York (2014)
2. Steele, R., Nofal, M.: Teletraffic performance of microcellular personal communication networks, Communications, Speech and Vision. Communications, Speech and Vision, IEE Proceedings I. IET 139(4), 448–461 (2008)
3. Xiao, Z., Zhang, Y., Lv, Z.: Study of Self-optimizing Load Balancing in LTE-Advanced Networks. In: Tan, H. (ed.) Knowledge Discovery and Data Mining. AISC, vol. 135, pp. 217–222. Springer, Heidelberg (2012)
4. Skehill, R., Barry, M., Kent, W., et al.: The common RRM approach to admission control for converged heterogeneous wireless networks. IEEE Wireless Communications 14(2), 48–56 (2007)
5. Song, W., Zhuang, W.H., Cheng, Y.: Load balancing for cellular/ WLAN integrated networks. IEEE Network 21(1), 27–33 (2007)
6. Rong, C., Xiao-Yu, D., Jie, M., et al.: An optimal IASA load balancing scheme in heterogeneous wireless networks. In: Communications and Networking in China (CHINACOM), pp. 714–719 (2011)
7. Jin, L., Zhang, H., Yang, L.X., Zhu, H.B.: A novel adaptive vertical handoff algorithm based on UMTS and WLAN. Journal of Nanjing University of Posts and Telecommunications (Natural Science) 33(4), 13–18 (2013)
8. Jiao, Y., Yi, K.C., Ma, M.D., Ma, Y.H., Dong, X.: QoS-aware load-balancing algorithm for heterogeneous wireless networks. Journal of Jilin University (Engineering and Technology Edition) 43(3), 794–800 (2013)
9. Johnson, S.B., Nath, S., Velmurugan, T.: An Optimized Algorithm for Vertical Handoff in Heterogenenous Wireless Networks. In: Proceeding of 2013 IEEE Conference on Information and Communication Technologies (ICT 2013), pp. 1206–1210 (2013)
10. Yongjing, Z., Kui, Z., Cheng, C., et al.: An Adaptive Threshold Load Balancing Scheme for the End-to-End Reconfigurable System. Wireless Personal Communication 46, 47–65 (2008)
11. Sheng, J., Tang, L.R., Hao, J.H.: Hybrid Load Balancing Algorithm Based on Service Transformation and Admission Control in Heterogeneous Wireless Networks. Acta Electronica Sinica 41(2), 321–328 (2013)
12. Sheng, J., Tang, L.R.: A triangle module operator and fuzzy logic based handoff algorithm for heterogeneous wireless networks. In: ICCT 2010, Nanjing, China, pp. 488–491 (2010)
13. Nasri, R., Altman, Z.: Handover adaptation for dynamic load balancing in 3GPP long term evolution systems. In: Proceeding of MoMM 2007, pp. 145–153 (2007)
14. Badia, L., Zorzi, M., Gazzini, A.: A model for threshold comparison call admission control in third generation cellular systems. In: International Conference on Communications, ICC 2003, pp. 1664–1668. IEEE (2003)

Real-Time Multi-Application Network Traffic Identification Based on Machine Learning

Meihua Qiao[1,2], Yanqing Ma[2,*], Yijie Bian[1], and Ju Liu[2]

[1] Business School, Hohai University, Nanjing ,210098, China
qiaomh1@chinaunicom.cn
[2] Suzhou Research Institute, Shandong University, Suzhou, 215021, China
xiaohu2000777@163.com

Abstract. In this paper, kinds of network applications are first analyzed, and some simple and effective features from the package headers of network flows are then generated by using the method of time window. What is more, three kinds of machine learning algorithms, which are support vector machine (SVM), back propagation (BP) neural network and BP neural network optimized by particle swarm optimization (PSO), are developed respectively for training and identification of network traffic. The experimental results show that traffic identification based on SVM can not only quickly generate classifier model, but also reach the accuracy of more than 98% under the condition of small sample. Moreover, the method proposed by this paper can measure and identify Internet traffic at any time and meet the needs of identifying real-time multi-application.

Keywords: network traffic identification, machine learning, SVM, BP neural network, PSO.

1 Introduction

In recent years, various peer to peer network (P2P) applications are emerging along with the developing and maturing of P2P technology such as BT download, thunderbolt and Emule etc. These applications result in insufficient bandwidth, and network congestion, which degraded seriously Quality of Experience (QoE) and Quality of Service (QoS) of the networks. The identification of network traffic plays an important role for both network operators and service providers to improve QoS and QoE. So, the research of the network traffic identification has an attract attention in both the academic and application fields. The key problem in Network Traffic Identification is how to rapidly process large amounts of data and how to correctly identify a variety of network application.

There exist four categories of methods for network traffic identification, i.e., port mapping based, deep packet inspection based, flow recognition behavior

* This work was supported by the Science and Technology Plan of Suzhou (SYG201443) and the Research Fund for the Doctoral Program of Higher Education (20130131110029).

© Springer International Publishing Switzerland 2015
X. Hu et al. (Eds.): ISNN 2015, LNCS 9377, pp. 473–480, 2015.
DOI: 10.1007/978-3-319-25393-0_52

based and machine learning based method [1]. (1) The port mapping based traffic identification methods are simple and efficient, which are capable of real-time identification of network applications. However, with a large number of random ports, the network address translation (NAT) and widely using of agent technology [2], this method is rapid failed [3]. (2) The deep packet inspection based traffic identification methods are easy to implement [4,5], which the recognition accuracy rate is much higher than the port mapping based method. (3) The behavior characteristics based traffic identification is applied by observing the traffic how to connect and interact in the network protocol layer [6,7]. (4) Traffic identification methods by machine learning apply data mine ability of machine learning algorithms in [8], which extract potential, effective and implied features from huge and the complex data of network flows. As in ref. [9], 248 traffic features were put forward by Andrew Moore cooperated with others.

In this paper, the time window method proposed is applied to acquire concise and effective features from package head of the network flow data for various application types in the network. Three machine learning algorithms, supporting vector machine (SVM), back propagation (BP) neural network and the particle swarm optimization (PSO) combined with BP neural network are provided to train and recognize network traffic respectively. The comparison of experimental results show that, traffic identification method based on SVM can not only quickly model to construct classifiers, but also can achieve the recognition rate above 98% in the case of small scale sample.

2   Types of Network Traffic Feature Analysis and Identification Procedure Design

Without loss of generality for the traffic identification, we focus on 6 kinds of network applications which have larger bandwidth requirements. Table 1 shows the 6 kinds of network applications needed to be identified and the corresponding test cases.

Table 1. Network application types

No.	Types of network applications	Test cases
1	P2P Multimedia or Download	Storm player, Xunlei Thunder
2	Non-P2P Multimedia or Download	YouTubeVideo, Local Download
3	WWW(Web Browsing)	Sogou, IE Browsing
4	Online Games(Client)	Demi-Gods and Semi-Devils (online clientgame)
5	Video Calling/Conference	QQ Video
6	File Sharing(LAN)	QQ file transmitting and sharing

2.1   The Features of Network Traffic

(1) Time Window Method
Define the time slot as 1 second. The real time network flow captured in 1 second is simply counted. We obtain the variation of the network flow within a time

window $\tau = n$ and up to $n = 15$ seconds. The flow is divided into stability region and the peak region traffic according to flow mean in the time window. The detailed description of the time window is in [10]. We can get the basic data such as number and length of packages in every second of this window. With that, we can analyze the stability and explosive in the time window τ.

(2) Network Flow Characteristics
With the six kinds of network applications in Table 1, eleven network flow characteristics[10] are generated and selected from the time window. Those characteristics can be defined as uplink or downlink which is determined by terms of the source address of the data packet. Assume an intranet as the local, and the external Internet as the remote. If the source address is the local IP, the data flow is considered to be uplink, i.e. data is uploaded. On the other side, if the source address is the remote IP, the data flow is regarded as downlink, and the data is downloaded.

2.2 Design of Network Traffic Identification Process

According to the application types and features selected from the identification, the network flow identification scheme based on machine learning can be divided into two processes, i.e., off-line training and on-line real-time classification. The designed overall architecture of network traffic identification is shown in Figure 1.

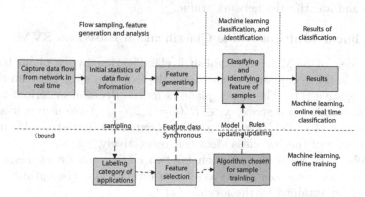

Fig. 1. The designed overall architecture of network traffic identification

(1) The Off-Line Training
Owing to the heavy computing burden and time consuming of machine learning, the training process is offline. The detailed steps of offline training show as following. Firstly, data packet are captured from the network route. Then, data packets are initially analyzed (e.g., counting numbers of packets, computing packet, etc.). Thirdly, extracting reasonable and stable sample from a large number of data obtained. Fourthly, label categories of the application are labeled. Fifthly, a variety of features of the sample data are generated, and effective and reasonable features are selected. Finally, the appropriate machine learning algorithm is used to train the eigenvalues of the sample, to generate classification rules and build the classifier model.

(2) Real-Time and Online Classifying

Traffic identification is performed online, and data packets are captured through the time window in real-time. When network application type of flow has been rapidly identified, the results will be feedbacked. The specific steps of real-time classification of online are as following: Firstly, capture data packet from the network route in real time by making use of capture software (such as Libpcap function library under Linux system). Secondly, analyze the data packet simply (e.g., numbers and lengths of the packets). Thirdly, synchronize the effective feature types from training synchronization module selection, and generate the associated eigenvalues. Fifthly, update the classification rule from the training module and classification model, classify and identify the eigenvalues of the sample. Finally, feedback recognition results.

3   Algorithms for Network Traffic Identification

Network traffic identification technologies are developing the classification and feature extraction algorithms along their development. On the basis of a variety of network applications, the recognition accuracy, complexity and real-time requirements, we select the three kinds of machine learning algorithm, which are SVM, BP neural network, BP neural network combined with PSO optimization, to analyze and identify the network traffic.

3.1   Nonlinear Multiple Value Classification Based on SVM

The main idea of SVM is to establish a classification hyper plane based on the samples in the region and to maximize the isolation edge between the positive and negative [11]. The hyper plane is as the decision criterion. Suppose there are n samples in a space; $\mathbf{x_i} \in R^d (i = 1, 2, \cdots, n)$ are d dimensional input vectors; $y_i \in \{+1, -1\}(i = 1, 2, \cdots, n)$ is expected output. "+1" and "-1" represent results of the two class identifier, respectively.

The problem of SVM classification for two classes can be expressed in an objective function with constraints as a linear formula (1). The optimal decision function can be obtained by the formula (2).

$$
\begin{aligned}
\min_{\alpha} \quad & \frac{1}{2}\sum_{i=1}^{n}\sum_{j=1}^{n} y_i y_j \alpha_i \alpha_j (\mathbf{x_i} \cdot \mathbf{x_j}) - \sum_{i=1}^{n}\alpha_i \\
s.t. \quad & \sum_{i=1}^{n} y_i \alpha_i = 0, \\
& \alpha_i \geq 0, \ i = 1, \cdots, n.
\end{aligned}
\tag{1}
$$

$$
f(\mathbf{x}) = \mathrm{sgn}\{(\omega \cdot \mathbf{x}) + b\} = \mathrm{sgn}\{\sum_{i=1}^{n}\alpha_i y_i (\mathbf{x_i} \cdot \mathbf{x}) + b\}
\tag{2}
$$

Where, the weight vector ω is adjustable, b is offset of the hyperplane, and α_i are Lagrange multipliers

For the case of network traffic linear non separable, on one hand, slack variables may be introduced in the constraint conditions of classification hyper plane;

on the other hand the nonlinear problem in the original space can be transformed into a linear problem in high dimensional space through the nonlinear kernel function, then solve the optimal hyperplane in a higher dimensional space. With the idea above introducedthe SVM objective function, constraint condition and the optimal decision functions of nonlinear classification problems for two classes can be summarized as follows:

$$\min_{\alpha} \quad \frac{1}{2} \sum_{i=1}^{n} \sum_{j=1}^{n} y_i y_j \alpha_i \alpha_j K(\mathbf{x_i}, \mathbf{x_j}) - \sum_{i=1}^{n} \alpha_i$$
$$s.t. \quad \sum_{i=1}^{n} y_i \alpha_i = 0, \tag{3}$$
$$0 \le \alpha_i \le C, \ i = 1, \cdots, n.$$

$$f(\mathbf{x}) = \text{sgn}\{\sum_{i=1}^{n} \alpha_i y_i K(\mathbf{x_i}, \mathbf{x}) + b\} \tag{4}$$

Where, C is the penalty parameter, $K(\mathbf{x_i}, \mathbf{x_j})$ is a kernel function. In generally, radial basis function is chosen as the kernel function. Duality and convex programming are used for the solution of the optimization problem, and the detail description can be found in [10].

The classic SVM is a simple two value classifier, which can only identify two kinds of network application type. However, network traffic identification is a typical multi-classification problem, so the structure of multiple value SVM nonlinear classifiers can be established through optimal hyperplane method of between any of two categories (One-Against-One).

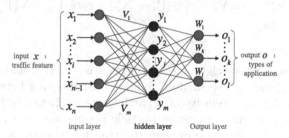

Fig. 2. Traffic identification by 3 layers BP neural networks

3.2   Nonlinear Multiple Flows Classification Based on BP Neural Network

It has been proved theoretically by Hornik that the BP neural network of three layers can be adjusted by the number of the hidden layer nodes and connection weights to approximate any nonlinear function with any precision [12]. Therefore, BP neural network of three layers has satisfied the network traffic identification problem of nonlinear multivariate classification.

In Figure 2, the input $\mathbf{x} = [x_1, x_2, \cdots, x_i, \cdots, x_{n-1}, x_n]^T$ is feature of network flow, $\mathbf{x} = [x_1, x_2, \cdots, x_i, \cdots, x_{n-1}, x_n]^T$ is numerical results for the hidden layer. And the output $\mathbf{o} = [o_1, o_2, \cdots, o_k, \cdots, o_l]$ is the number of network application types. The suitable weight matrix $\mathbf{V} = [\mathbf{V}_1, \mathbf{V}_2, \cdots, \mathbf{V}_j, \cdots, \mathbf{V}_m]$ and $\mathbf{W} = [\mathbf{W}_1, \mathbf{W}_2, \cdots, \mathbf{W}_k, \cdots, \mathbf{W}_l]$ are obtained by constantly training samples with gradient descent and error back propagation algorithm in the BP neural networks.

The number of hidden layer nodes is obtained by training samples under the condition of the recognition accuracy. With the number of nodes, a reliable BP neural network model can be determined.

3.3 Traffic Identification Based on BP Neural Network with PSO Optimization (BP-PSO)

The adjustment of connecting weights in BP neural network is based on error gradient. Local optimum is taken for every adjustment. Therefore the decision that the training process is going to get into the local minima or entering the global minimum mainly depends on the initial weights of BP neural network. If the initialization of weights are determined randomly, then it is difficult to predict the error convergence results of BP network. In this regard, the PSO algorithm is a stochastic global optimization technique [13,14] which searches BP neural network weight appropriate initialization, the training process of BP network can more easily get into the global minimum.The iterative process of each time is shown in update formula (5) and formula (6). Where, the velocity V_i and position X_i of itself are updated through individual extremum P_i and population global extremum P_g.

$$\mathbf{V}_i^{k+1} = \omega \mathbf{V}_i^k + c_1 r_1 (\mathbf{P}_i^k - \mathbf{X}_i^k) + c_2 r_2 (\mathbf{P}_g^k - \mathbf{X}_i^k) \tag{5}$$

$$\mathbf{X}_i^{k+1} = \mathbf{X}_i^k + \mathbf{V}_i^{k+1} \tag{6}$$

Where, ω is the inertia weight, k is the particle number; c_1 and c_2 are training constant; and r_1 and r_2 are random number distributions between $[0, 1]$. In general, the position and velocity of a particle are set limits to $[-X_{\max}, X_{\max}]$ and $[-V_{\max}, V_{\max}]$ in order to prevent particle population searching disordered. At the same time, this can also prevent the transfer function of the BP neural network into the saturated zone.

Through the optimizing of the initial weights by PSO algorithm, the BP neural network in training process can get into the global minimum. So, the BP neural network based on PSO optimization can accurately identify all kinds of network applications in the traffic identification.

4   Results and Analysis of Network Traffic Identification

Using the window time method, network flows are sampled in different time segment when 6 kinds of typical network applications are running stably. 4 groups of samples are obtained. Each group of samples has 6 kinds of network application types and each network application types have 100 samples, so 4 sample sets consist of 2400 samples. Respectively 60, 120, 240, 480, 960 and 1920 samples were randomly selected as the training set, the remain rest corresponding samples are as a test set.

As usual, the True Positives (TP)[15] as the accurate rate is adopted as the effectiveness evaluation of network traffic identification methods.

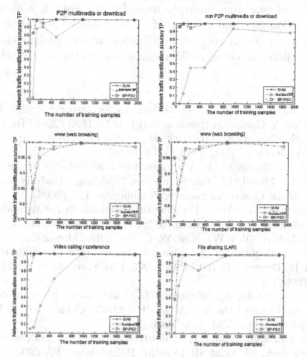

Fig. 3. Identification accuracy TP comparison of SVM, standard BP and BP-PSO

In Figure 3, according to the 6 network applications, three kinds of algorithms, SVM, BP and BP-PSO, are used respectively for traffic identification. Through the comparative analysis, the accuracy of TP recognition of three algorithms will grow with the increase in the number of training samples; both SVM and BP-PSO are better than that of standard BP, contributed to the high and stable accuracy, especially the identification accuracy of SVM algorithm in the case of small sample remains above 98%.

5   Conclusions

The limitations and defects of existing traffic identification methods based on port mapping, deep packet inspection and behavioral characteristics are analyzed. Real time Multi-Application Network Traffic Identification method Based on Machine Learning is proposed. To achieve the requirements of real-time network traffic identification, the method uses time window only from the network flow data of packet head to get the simple and effective features. The training and recognition are performed respectively through SVM, BP neural network and BP neural network optimized by PSO and machine learning algorithms. The implementation process is simple for network flow. The characteristics of network flows can be extracted while the traffic is being identified. Comparative experimental results show that the identify process and results can meet the requirements of real-time multiple uses. Future work is aimed to improve constantly the rationality and validity of "time window" feature extraction for the unique characteristics of network application, and to make the network flow identification methods based on SVM more stable and more robust.

References

1. Zhao, G., Ji, Z., Xu, C.: Survey of Techniques for Internet Traffic Identification. Journal of Chinese Computer Systems 31(8), 1514–1520 (2010) (in Chinese)
2. Schulzrinne, H., Casner, S., Frederick, R., et al.: RTP: A Transport Protocol for Real-Time Applications. RFC 1889, IETF (1996)
3. Madhukar, A., Williamson, C.: A longitudinal study of P2P traffic classification. In: Proc. of the 14th IEEE Int. Symp. on Modeling, Analysis and Simulation, pp. 179–188. IEEE Computer Society, Washington, DC (2006)
4. Sen, S., Spatscheck, O., Wang, D.: Accurate, scalable in network identification of P2P traffic using application signatures. In: Proc. of 13th International Conference on World Wide Web (WWW), New York, NY, USA (May 2004)
5. Moore, A.W., Papagiannaki, K.: Toward the accurate identification of network applications. In: Dovrolis, C. (ed.) PAM 2005. LNCS, vol. 3431, pp. 41–54. Springer, Heidelberg (2005)
6. Karagiannis, T., Broido, A., Faloutsos, M., et al.: Transport layer identification of P2P traffic. In: Proc. of the 4th ACM SIGCOMM Conference on Internet Measurement, pp. 121–134. ACM, New York (2004)
7. Karagiannis, T., Papagiannaki, K., Faloutsos, M.: BLINC: multilevel traffic classification in the dark. In: ACM SIGCOMM, Philadelphia, PA (2005)
8. Auld, T., Moore, A.W., Gull, S.F.: Bayesian neural networks for internet traffic classification. IEEE Transactions on Neural Network 18(1), 223–239 (2007)
9. Moore, A.W., Zuev, D.: Discriminators for use in Flow-based classification. Technical Report IRC-TR-04-028, Intel Research, Cambridge (2004)
10. Ma, Y.: Methods and Implementations of Network Traffic Identification Based on Machine Learning. Master thesis, Shandong University (2014) (in Chinese)
11. Cristianini, N., Shawe-Taylor, J.: An Introduction to Support Vector Machine An Introduction to Support Vector Machines and Other Kernel-based Learning Methods. Publishing House of Electronics Industry, Beijing (2004) (Chinese Version, Translated by G. Li, M. Wang, H. Ceng)
12. Hornik, K.M., Stinchcombe, M., White, H.: Multilayer feed forward networks are universal approximators. Neural Networks 2(2), 359–366 (1989)
13. Kennedy, J., Eberhart, R.C.: Particle Swarm Optimization. In: Proceedings of IEEE International Conference on Neural Networks, pp. 1942–1948 (1995)
14. Zhang, G., Li, Y.M.: Cooperative particle swarm optimizer with elimination mechanism for global optimization of multimodal problems. In: Proceedingds of IEEE Congress on Evolutionary Computation (CEC), Beijinag, China, pp. 210–217 (2014)
15. Nguyen, T., Armitage, G.: A Survey of Techniques for Internet Traffic Classification using Machine Learning. IEEE Communications Surveys & Tutorials 10(4), 56–76 (2008)

A New Virus-Antivirus Spreading Model

Bei Liu and Chuandong Li*

College of Electronic and Information Engineering,
Southwest University, 400715, Chongqing, China
cdli@swu.edu.cn

Abstract. Indeed, countermeasures, as well as computer viruses, could spread in the network. This paper aims to investigate the effect of propagation of countermeasures on viral spread. For the purpose, a new virus-antivirus spreading model is proposed. The global asymptotic stability of the virus-free equilibrium is proved when the threshold is below the unity, and the existence of the viral equilibrium is shown when the threshold exceeds the unity. The influences of different model parameters on the threshold are also analyzed. Numerical simulations imply that the propagation of countermeasures contributes to the suppress of viruses, which is consistent with the fact.

Keywords: computer virus, countermeasures, virus-antivirus spreading model, equilibrium, global asymptotic stability

1 Introduction

As is known to all, computer virus has brought huge damage to human society. Thus, it is crucially important to find effective strategies to contain the spread of viruses. Based on the intriguing analogy between computer viruses and their biological counterparts, multifarious epidemic models of computer viruses, ranging from conventional models such as SIS models [1,2], SIR models[3,4], $SIRS$ models [5,6], $SLBS$ models [7,8], SIC models [9,10], to unconventional models such as delayed models [11,12,13], impulsive models [7,14], stochastic models [7,15] and the network-based models [16] have been proposed.

To examine the CMC strategy put forward by Chen and Carley [17], Zhu et al [9] proposed a mixing propagation model of computer viruses and countermeasures. The SIC model, however, ignores the marked difference between latent computers and breaking-out computers. To overcome the defect, this paper proposed an $SLBC$ model (see Figure 1). This model is proved to have a virus-free equilibrium when the threshold is below the unity and a viral equilibrium when the threshold exceeds the unity. Numerical simulations imply that the propagation of countermeasures is conducive to inhibiting the prevalence of computer viruses.

The rest of the paper is organized as follows. Section 2 elaborates the new model. Section 3 proves the global stability of the virus-free equilibrium and the

---

* Corresponding author.

© Springer International Publishing Switzerland 2015
X. Hu et al. (Eds.): ISNN 2015, LNCS 9377, pp. 481–488, 2015.
DOI: 10.1007/978-3-319-25393-0_53

existence of the viral equilibrium. Numerical simulations are displayed in Section 4. Section 5 summarizes the work.

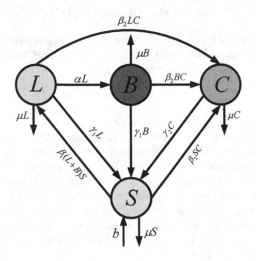

Fig. 1. The state transition diagram of the new model.

2   Model Formulation

For the purpose of modeling, the following hypotheses are imposed.

(*H*1) All external computers are susceptible.

(*H*2) External computers enter the Internet at constant rate b.

(*H*3) Internal computers leave the Internet at constant rate μ.

(*H*4) Every latent computer breaks out with constant probability α.

(*H*5) Due to possible contacts with latent (resp. breaking-out) computers, at time t every susceptible computer gets infected with probability $\beta_1(L+B)$.

(*H*6) Every susceptible, latent or breaking-out computer acquires countermeasures with constant probability β_2C.

(*H*7) Due to system reinstallation, every infected computer becomes susceptible with constant probability γ_1.

(*H*8) Every computer equipped with countermeasures loses immunity with constant probability γ_2.

From this collection of hypotheses, one can obtain the following differential equations

$$\begin{cases} \dot{S} = b - \mu S + \gamma_1(L+B) + \gamma_2 C - \beta_1(L+B)S - \beta_2 SC \ , \\ \dot{L} = \beta_1(L+B)S - \gamma_1 L - \mu L - \alpha L - \beta_2 LC \ , \\ \dot{B} = \alpha L - \gamma_1 B - \mu B - \beta_2 BC \ , \\ \dot{C} = \beta_2(S+L+B)C - \mu C - \gamma_2 C \ . \end{cases} \tag{1}$$

where S, L, B represent, at time t, the average numbers of susceptible, latent, breaking-out computers, respectively, and C represents uninfected computers that have temporary immunity.

The basic reproduction number, R_0, is defined as the average number of susceptible computers that are infected by a single infected computer during its life span. From the above model, one can derive the basic reproduction number R_0 as

$$R_0 = \frac{\beta_1(N^* - C^*)}{\gamma_1 + \mu + \beta_2 C^*} .$$

Let $N = S + L + B + C, N^* = \frac{b}{\mu}, C^* = N^* - \frac{\mu + \gamma_2}{\beta_2}$. Then, system (1) can be written as the following limiting system [18]:

$$\begin{cases} \dot{L} = -\beta_1(L + B)^2 + [\beta_1(N^* - C^*) - (\gamma_1 + \alpha + \mu + \beta_2 C^*)]L \\ \qquad + \beta_1(N^* - C^*)B , \\ \dot{B} = \alpha L - (\gamma_1 + \mu)B - \beta_2 BC^* . \end{cases} \qquad (2)$$

Bellow we mainly consider the existence and global stability of the equilibrium points in regard to the positively invariant region: $\Omega = \{(L, B) \in R_+^2 : L + B \leq N^*\}$.

3   Theoretical Analysis

In this section, the dynamical properties of system (2) would be studied.

3.1   The Virus-Free Equilibrium

It is obvious that system (2) always has a virus-free equilibrium $E^0(0,0)$, i.e. $L = B = 0$. Next, we would show its global stability.

Theorem 1. $E^0(0,0)$ *is locally asymptotically stable with respect to Ω when $R_0 < 1$.*

Proof 1. *For the linearized system of system (2) at E^0, the corresponding Jacobian matrix is*

$$J_{E^0} = \begin{pmatrix} k_1 - \alpha & k_1 - k_2 \\ \alpha & k_2 \end{pmatrix} ,$$

where

$$k_1 = \beta_1(N^* - C^*) - (\gamma_1 + \mu + \beta_2 C^*), k_2 = -(\gamma_1 + \mu + \beta_2 C^*) .$$

The characteristic equation of J_{E^0} is

$$(\lambda - k_1)[\lambda - (k_2 - \alpha)] = 0 ,$$

whose roots are

$$\lambda_1 = k_1 , \quad \lambda_2 = k_2 - \alpha .$$

Clearly, all of the roots are negative. The claimed result follows from the Lyapunov theorem[19].

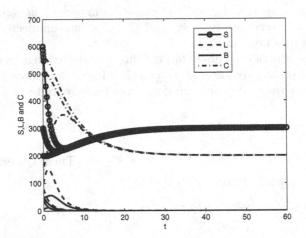

Fig. 2. Time plots of S, L, B, C for a common system with three different initial conditions when $R_0 < 1$.

We are ready to study the global stability of E^0.

Theorem 2. $E^0(0,0)$ *is globally asymptotically stable with respect to Ω when $R_0 < 1$.*

Proof 2. *Define $D(L, B) = 1/L$. Let*

$$f_1(L, B) = -\beta_1(L+B)^2 + [\beta_1(N^* - C^*) - (\gamma_1 + \alpha + \mu + \beta_2 C^*)]L + \beta_1(N^* - C^*)B ,$$

$$f_2(L, B) = \alpha L - (\gamma_1 + \mu)B - \beta_2 BC^* .$$

Then

$$\frac{\partial(Df_1)}{\partial L} + \frac{\partial(Df_2)}{\partial B} = -\beta_1 - \frac{\beta_1 B(N^* - C^* - B)}{L^2} - \frac{\gamma_1 + \mu + \beta_2 C^*}{L} < 0 .$$

By the Bendixson-Dulac criterion [19], system (2) admits no periodic orbit that lies in the interior of Ω.

Now, consider an arbitrary point $(\overline{L}, \overline{B})$ on $\partial\Omega$. There are three possibilities which are displayed as follows.

(1) $0 < \overline{L} < N^*$, $\overline{B} = 0$. Then $dB/dt|_{(\overline{L},\overline{B})} = \alpha\overline{L} > 0$,

(2) $0 < \overline{B} < N^*$, $\overline{L} = 0$. Then $dL/dt|_{(\overline{L},\overline{B})} = \beta_1\overline{B}(N^* - C^* - \overline{B}) > 0$,

(3) $\overline{L} + \overline{B} = N^*$, $\overline{L} \neq 0$, $\overline{B} \neq 0$. Then $d(L + B)/dt|_{(\overline{L},\overline{B})} = -(\beta_1 C^* + \beta_2 C^* + \gamma_1 + \mu)N^* < 0$. Taking into account the smoothness of all orbits, system (2) has no periodic orbit that passes through a point on $\partial\Omega$.

Hence, the claimed result follows from the generalized Poincaré − Bendixson theorem [19].

Remark 1. *Theorem 2 implies that computer viruses would disappear eventually. Figure 2 verify the result.*

3.2   The Viral Equilibrium

When $R_0 > 1$, it's easy to verify that system (2) has a unique viral equilibrium $E^*(L^*, B^*)$, i.e. $L^* + B^* > 0$, where

$$L^* = \frac{(R_0 - 1)(\gamma_1 + \mu + \beta_2 C^*)^2}{\beta_1(\gamma_1 + \alpha + \mu + \beta_2 C^*)} \;,\; B^* = \frac{\alpha L^*}{\gamma_1 + \mu + \beta_2 C^*} \;.$$

Next, let us consider the local stability of E^*.

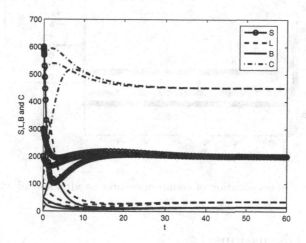

Fig. 3. Time plots of S, L, B, C for a common system with three different initial conditions when $R_0 > 1$.

Theorem 3. *E^* is locally asymptotically stable with respect to Ω when $R_0 > 1$.*

Proof 3. *For the linearized system of system (2) at E^*, the corresponding Jacobian matrix is*

$$J_{E^*} = \begin{pmatrix} -k_1 - \alpha & -k_1 - k_2 \\ \alpha & k_2 \end{pmatrix} \;,$$

where

$$k_1 = \beta_1(N^* - C^*) - (\gamma_1 + \mu + \beta_2 C^*) \;,\; k_2 = -(\gamma_1 + \mu + \beta_2 C^*) \;.$$

The characteristic equation of J_{E^*} is

$$(\lambda - (-k_1))[\lambda - (k_2 - \alpha)] = 0 \;,$$

whose roots are

$$\lambda_1 = -k_1 \ , \ \lambda_2 = k_2 - \alpha \ .$$

Obviously, all of the roots are negative. The claimed result follows from the Lyapunov theorem[19].

Remark 2. *Figure 3 implies that E^* is probably globally asymptotically stable with respect to Ω when $R_0 > 1$.*

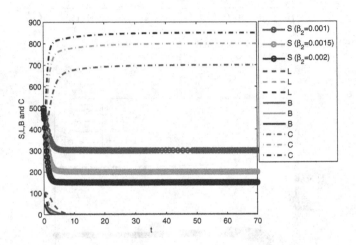

Fig. 4. The effect of propagation of countermeasures on viral spread when $R_0 < 1$.

4   Further Discussions

Some numerical simulations are conducted in this section. From Figure 4-5, it is concluded that the CMC strategy is effective in eradicating viruses. What's more, The effects of model parameters on R_0 are as follows.

$$\frac{\partial(R_0)}{\partial(\beta_1)} > 0 \ , \ \frac{\partial(R_0)}{\partial(\gamma_1)} < 0 \ , \ \frac{\partial(R_0)}{\partial(\beta_2)} < 0 \ , \ \frac{\partial(R_0)}{\partial(b)} < 0 \ ,$$

$$\frac{\partial(R_0)}{\partial(\gamma_2)} = \frac{\mu\beta_1(\gamma_1 + \mu + \beta_2 C^*) + \beta_1(N^* - C^*)}{(\gamma_1 + \mu + \beta_2 C^*)^2} > 0 \ ,$$

$$\frac{\partial(R_0)}{\partial(\mu)} = \frac{\beta_1\mu^2(\gamma_1 + \mu + \beta_2 C^*) + b\beta_1\beta_2^2(N^* - C^*)}{\beta_2\mu^2(\gamma_1 + \mu + \beta_2 C^*)^2} > 0 \ .$$

Clearly, R_0 is decreasing with γ_1, and β_2, respectively, and is increasing with β_1. Thus, some effective measures are presented below. (1) Update the antivirus software timely, (2) Reinstall the operating system when necessary, (3) Do not click the unknown links.

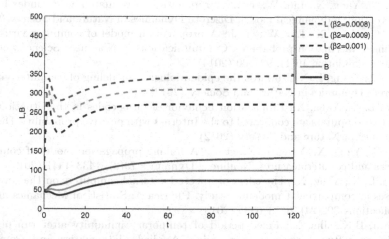

Fig. 5. The effect of propagation of countermeasures on viral spread when $R_0 > 1$.

5 Conclusions

A new virus-antivirus spreading model has been proposed in this paper. This model has a virus-free equilibrium which is globally asymptotically stable when the threshold is below the unity and a viral equilibrium when the threshold exceeds the unity. Numerical simulations imply that the propagation of countermeasures conduces to the containment of viruses. On this basis, some useful advice has been posed.

Acknowledgments. This publication was made possible by NPRP grant ♯ NPRP 4-1162-1-181 from the Qatar National Research Fund (a member of Qatar Foundation). The statements made herein are solely the responsibility of the authors. This work was also supported by Natural Science Foundation of China (grant no: 61374078) and Natural Science Foundation Project of Chongqing CSTC (Grant No. cstc2014jcyjA40014)

References

1. Kephart, J.O., White, S.R.: Directed-graph epidemiological models of computer viruses. In: 1991 IEEE Computer Society Symposium on Security Privacy, pp. 343–359. IEEE Computer Society, Oakland (1991)
2. Billings, L., Spears, W.M., Schwartz, I.B.: A unified prediction of computer virus spread in connected networks. Physics Letters A 297, 261–266 (2002)
3. Ren, J., Yang, X., Zhu, Q., et al.: A novel computer virus model and its dynamics. Nonlinear Analysis: Real World Applications 13, 376–384 (2012)
4. Zhu, Q., Yang, X., Ren, J.: Modeling and analysis of the spread of computer virus. Communications in Nonlinear Science and Numerical Simulation 17, 5117–5124 (2012)

5. Gan, C., Yang, X., Liu, W., et al.: Propagation of computer virus under human intervention: a dynamical model. Discrete Dynamics in Nature and Society (2012)
6. Gan, C., Yang, X., Liu, W., et al.: A propagation model of computer virus with nonlinear vaccination probability. Communications in Nonlinear Science and Numerical Simulation 19(1), 92–100 (2014)
7. Yang, X., Yang, L.-X.: Towards the epidemiological modeling of computer viruses. Discrete Dynamics in Nature and Society (2012)
8. Yang, L.-X., Yang, X.: Propagation behavior of virus codes in the situation that infected computers are connected to the Internet with positive probability. Discrete Dynamics in Nature and Society (2012)
9. Zhu, Q., Yang, X., Yang, L.-X., et al.: A mixing propagation model of computer viruses and countermeasures. Nonlinear Dynamics 73(3), 1433–1441 (2013)
10. Yang, L.-X., Yang, X.: The effect of infected external computers on the spread of viruses: a compartment modeling study. Physica A: Statistical Mechanics and its Applications 392(24), 6523–6535 (2013)
11. Mishra, B.K., Jha, N.: Fixed period of temporary immunity after run of antimalicious software on computer nodes. Applied Mathematics and Computation 190(2), 1207–1212 (2007)
12. Mishra, B.K., Saini, D.K.: SEIRS epidemic model with delay for transmission of malicious objects in computer network. Applied Mathematics and Computation 188(2), 1476–1482 (2007)
13. Han, X., Tan, Q.: Dynamical behavior of computer virus on Internet. Applied Mathematics and Computation 217(6), 2520–2526 (2010)
14. Zhang, C., Zhao, Y., Wu, Y.: An impulse model for computer viruses. Discrete Dyn. Nat. Soc. (2012)
15. Zhang, C., Zhao, Y., Wu, Y., Deng, S.: A stochastic dynamic model of computer viruses. Discrete Dyn. Nat. Soc. (2012)
16. Pastor-Satorras, R., Vespignani, A.: Epidemic spreading in scale-free networks. Physics Review Letters 86(14), 3200–3203 (2001)
17. Chen, L.C., Carley, K.M.: The impact of countermeasure propagation on the prevalence of computer viruses. IEEE Transactions on Systems, Man, and Cybernetics, Part B: Cybernetics 34(2), 823–833 (2004)
18. Thieme, H.R.: Asymptotically autonomous differential equations in the plane. Rocky Mt. J. Math. (1994)
19. Robinson, R.C.: An Introduction to Dynamical System: Continuous and Discrete. American Mathematical Soc. (2004)

Exploring Feature Extraction and ELM in Malware Detection for Android Devices

Wei Zhang[1], Huan Ren[1,2], Qingshan Jiang[1], and Kai Zhang[1,2]

[1] Shenzhen Institutes of Advanced Technology, Chinese Academy of Sciences, Shenzhen, 518055, China
[2] University of Science and Technology of China, Heifei, 230051, China

Abstract. A huge increase in the number of mobile malware brings a serious threat to Internet security, as the adoption rate of mobile device is soaring, especially Android device. A variety of researches have been developed to defense malware, but the mobile device users continuously suffer private information leak or economic losses from malware. Recently, a large number of methods have been proposed based on static or dynamic features analysis combining with machine learning methods, which are considered effective to detect malware on mobile device. In this paper, we propose an effective framework to detect malware on Android device based on feature extraction and neural network calssifier. In this framework, we take use of static features to represent malware and utilize extreme learning machine (ELM) algorithm to learn the neural network. We first extract features from the malware, and then utilize three different feature extraction methods including principal component analysis (PCA), Karhunen-Loève transform (KLT) and independent component analysis (ICA) to transform the feature matrix into new feature spaces and generate three new feature matrixes. For each feature matrix, we construct En base classifiers by using ELM. Finally, we utilize Stacking method to combine the results. Experimental results suggest that the proposed framework is effective in detecting malware on Android device.

Keywords: Feature extraction, Android malware detection, ELM, Stacking method.

1 Introduction

Malware torments Internet users persistently, which is one of the major threats on the Internet. The popularity of smart phones prompt explosive growth of malware, especially on the Android platform. Malware is short for malicious software, which is secretly inserted into a system by using the system vulnerabilities for sensitive information or financial gain. Reported by Lookout [1] malware grew substantially in the U.S, and malware on android platform have increased in 2014 by 75% compared with the year 2013.

As the popularity of Android malware has led to enormous security problems, many researchers and security organizations have dedicated to detecting malware on Android platform. Machine learning is considered as an effective tool in

© Springer International Publishing Switzerland 2015
X. Hu et al. (Eds.): ISNN 2015, LNCS 9377, pp. 489–498, 2015.
DOI: 10.1007/978-3-319-25393-0_54

malware detection. Machine learning methods make use of the features extracted from the malware to find patterns as well as relations among them. The features used in machine learning methods are usually extracted by static analysis [2,3] or dynamic analysis [4,5]. Static analysis takes use of the the source code of the malware to generate the features, which need not perform an executable file. On the contrary, dynamic analysis generates the features in the process of performing the executable file. However, because of the resource constraints, it is not straight forward to analyze and capture suspicious information on mobile device [6].

In order improve the detection accuracy and investigate classification properties of the static features in different feature spaces, we propose a new framework to detect malware on Android platform. In this framework, we first extract Dalvik instructions from the source code, and then three different feature extraction methods are utilized to transform the feature to three new feature matrix. Then, we construct EN base classifier for each feature matrix by ELM. Finally, ELM is also utilized to combine the results. In summary, our main contributions in this paper are as follows:

- This research carries out an effective framework to detect malware for Android device.
- This research utilizes three different feature extracion methods to transform the original features into new feature spaces, and employs all the transformed features to construct ensemble learning model.
- This research introduces ELM to malware detection on Android platform, which is an effective and high accuracy algorithm to learn the neural network classifier.

The remainder of this paper is organized as follows. Section 2 gives the related works of malware detection on Android platform. Section 3 describes the proposed framework and reviews theoretical backgrounds about PCA, KLT, ICA for feature extraction, ELM algorithm for learning the neural networks and Stacking for base classifiers combination. Section 4 tests the performance of the proposed framework for malware detection on Android platform. Finally, section 5 concludes the paper.

2   Relative Works

Currently, a great quantity of machine learning methods have been used in malware detection for Android device. Zhao et al. [5] proposed a framework based on software behavior signature by utilizing Support Vector Machine(SVM) classifier. They applied different famous malware to evaluate the proposed detection framework and obtained the best results of 93.33% true positive rate and 3.7% false positive rate. Sahs et al. [7] extracted a list of requested permissions from the APK file by using the open source project 'Androguard', and constracuted the detection model by using SVM. They obtained low false negative rate, but suffered high positive rate.

Bayesian classifier is also often used to construct the malware detection model. Yerima et al. [8] analyzed static code and extracted features from the APK files by implementing a Java-based Android package profiling tool. They constructed API call detector, Linux system commands detector and permissions detector. They extracted 58 properties to train the detection model by using Bayesian classifier. Their experiments obtained the best results of 90.6% true positive rate and 6.3% false positive rate. Sharma et al. [9] detected malware combining API calls with permissions, and constructed the detection model by using Naive Bayes classifier and k-Nearest Neighbour(kNN) classifier. They first extracted APT calls and permissions from the malware and then made use of correlation based feature selection and information gain method to select features. They obtained the result that kNN classifier achieved higher accuracy than Naive Bayesian classifier combining with both of above mentioned feature selection methods. However, Naive Bayes performed better in terms of true positive rate.

Aafer et al. [10] extracted a combination of API, package and parameter level information features, and compared four different classifiers including ID3, C4.5, kNN and SVM. They concluded that kNN was the best performing model, which generated the best accuracy above 99% and obtained false positive rate 2.2%. Wu et al. [11] utilized requested permissions and Intent messages passing to be the features, and regarded components (Activity, Service, Receiver) as entry points drilling down for tracing API Calls related to permissions. They employed k-means algorithm to enhance the malware modeling capability. Then, they implemented kNN algorithm to classify the application as benign or malicious. They generated the accuracy of 97.87%, but encountered 87.39% recall.

Barrera [12] extracted 119 permission requests from a real-world dataset of 1,100 applications, and then employed Self-Organizing Map (SOM) algorithm to support component planes analysis, which can provide interesting usage patterns. Yu et al. [13] developed malware detection method based on neural networks. They systematically compared the permission requests from application requests trained by feedforward neural networks and system calls trained by recurrent neural networks to capture the behavior of applications. Mas'ud et al. [14] evaluated five sets of feature selection combining with five different machine learning classifiers including NB, kNN, Decision Tree (J48), Multi-Layer Perceptron (MLP) and Random Forest (RF). Their research took system calls as the features, and the experiment results showed that MLP combining with feature selection generated the best performance.

Ozdemir et al. [15] implemented an ensemble learning approach for Android malware detection. They first separately extracted static features including native API calls and Dalvik Byte API calls, and dynamic features. Then, they took use of feature selection to select the most informative features. They took different classifier as the base learner to train the static and dynamic data sets, results of these base learners were combined in the scope of ensemble learning, and they obtained the best accuracy of 97.33%. Sheen et al. [16] implemented a detecting malware system on Android platform by using ensemble learning model based static features. They first extracted API calls and permissions requested from APK file to create the feature sets, and then utilized a filter for feature selection. They trained on an ensemble of classifiers using a collaborative approach for arriving at the final decision, and achieved the true positive rate of 98.9% and recall of 98.8%. Kang et al. [17] proposed a method to classify malware family by using Random Forest, which was based on Dalvik Bytecode.

The above mentioned researches took static features or dynamic features to detect malware based on machine learning. Feature selection is employed in some literatures to select best subset for high performance, and machine learning methods are implemented to construct detection model. In this paper, we follow mainstream researches and propose a novel framework to detect malware on Android platform, which is based on feature extraction methods and ELM algorithm.

3   Framework and Theoretical Backgrounds

In this section, we first give the framework of the proposed malware detection system for Android device. Then, we introduce theoretical backgrounds about feature extraction, and describe the ELM algorithm for learning neural network. Finally, we briefly describe the Stacking method for ensemble learning.

3.1   Framework Description

The framework of our malware detection system for Android device is summarized in Fig. 1. Basically, the system first utilizes Dalvik instructions extractor to extract Dalvik instructions from the collected samples including malware .apk and normal .apk files. Then, the original features are transformed into new feature spaces by using three different feature extraction methods, and three feature matrixes are obtained. For each feature matrix, we randomly select sub-set samples En time to train base classifiers using ELM algorithm. Finally, Stacking method is utilized to combine the results of base classifiers. The main components of the framework are described below.

- Dalvik instructions extraction module: An APK is the Android package, which is a compressed 'ZIP' bundle of files typically consisting of Android-Manifest.xml and classes.dex. The classes.dex file holds the complete bytecode to be interpreted by Dalvik VM [8]. This module unpackes the APK file and decompiles the classes.dex file to extract Dalvik instructions.
- Feature extraction module: This module utilizes PCA, KLT and ICA to transform the original feature matrix into new feature spaces. Three feature matrixes are generated from this module.
- Ensemble learning module: This module select sub-set En times to generate En classifiers for each feature matrix. ELM is taken as the base classifier.
- Decision module: This module employs Stacking method to combine the result of each classifier, and ELM is also taken as the classifier.

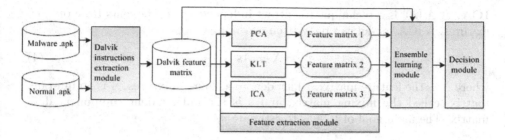

Fig. 1. Framework of malware detection system for Android device

3.2 Feature Selection Methods

PCA. PCA is defined as a linear projection that transforms the original high-dimensional features into lower-dimensional features. In the processing of projection, the variance of the projected data is maximized [18]. Denote a set of N samples $X = \{X_1, X_2, \ldots, X_N\}$, where $X_i = [X_i(1), X_i(2), \ldots, X_i(d)]^T \in \mathbb{R}^d$ and d is the number of features. PCA linearly transforms each sample X_i into a new one $y_i \in \mathbb{R}^d$ [19],

$$y_i = U^T X_i \tag{1}$$

where U is a $d \times d$ orthogonal matrix. The i-th column u_i of U is the i-th eigenvector of the sample covariance matrix,

$$C = \frac{1}{d} \sum_{i=1}^{d} X_i X_i^T \tag{2}$$

In summery, PCA first calculates the covariance matrix of X, and denotes it as S. Then, calculate eigenvalues and eigenvectors of S, and denote them as $\lambda_1, \lambda_2, \ldots, \lambda_d$ and $u_1, u_2, \ldots, u_d$. Finally, we can calculate the new feature by using equation(1).

KLT. KLT can be considered as the generalized PCA. In PCA, the generation matrix is the covariance matrix of original feature matrix X. KLT can take other matrixes as the generation matrix. This paper takes the malware detection as a classification problem. Therefore, we take the inter-class scatter matrix as the generation matrix, which could make the samples more easier to be distinguished in the new feature space,

$$S_w = \sum_{i=1}^{L} P_i E[(X - \bar{m}_i)(X - \bar{m}_i)^T] \tag{3}$$

where L is number of classes, which is equal to 2 in this research, P_i is the probability of i-th class and $\bar{m}_i$ is the mean of the i-th class. Then, we calculate eigenvalues and eigenvectors of S_w, and denote them as $\lambda_1, \lambda_2, \ldots, \lambda_d$ and $u_1, u_2, \ldots, u_d$. Finally, we can calculate the new feature by using equation(1).

ICA. ICA [20] is a technique to extract independent components from original features. A ICA model can be written as,

$$X = As \tag{4}$$

where X is the feature matrix, which denote as linear mixtures, A is an full-rank matrix, called the mixing matrix, and s is the independent component data matrix. The main goal of ICA is to estimate s,

$$\hat{s} = UX \tag{5}$$

where U is un-mixing matrix. In this paper, we utilize FastICA [21] algorithm to estimate s. FastICA [21] algorithm takes maximizing the negentropy of s as the criterion to estimate s, which is approximated via the bellow contrast function [19],

$$J_G(u_i) = [E\{G(u_i^T X_i)\} - E\{G(\nu)\}]^2 \tag{6}$$

where u_i is a d dimensional vector, comprising one of the rows of matrix U. ν is a standardized Gaussian variable. G is non-quadratic function. The commonly used functions for G are $G_1(s) = \frac{1}{\gamma_1} logcosh(\gamma_1 s)$, $G_2(s) = -\frac{1}{\gamma_2} exp(-\gamma_2 s^2/2)$ and $G_3(s) = \frac{1}{4}s^4$, where γ_1 ($1 <= \gamma_1 <= 2$) and γ_2 ($\gamma_2 \approx 1$) are parameters. Then, u_i can be estimated by maximizing $J_G(u_i)$,

$$u_i^+ = E\{X_i g(u_i^T X_i)\} - E\{g'(u_i X_i)\} u_i \tag{7}$$

$$u_i^* = \frac{u_i^+}{\|u_i^+\|} \tag{8}$$

where u_i^* is a new estimated value of u_i, g and g' are respectively the first and second derivatives of G. Based on the maximal negentropy principal, the matrix U can be computed by maximizing the sum of one-unit contrast function and taking into account the constraint of decorrelation [21].

3.3   Neural Networks for Classification

This research employs neural networks as the classifier, and takes the idea that some parameters are considered as random values to train the single feed-forward neural networks. This idea for training neural networks is original from Schmidt et al.[22]. Then, Huang et al.[23] proposed ELM and elaborated this idea to handle classification problems[24]. ELM is based on least square to train the networks [23]. ELM first randomly assigns for input weights and biases, and analytically determines the output weights of the SLFNs [23]. The steps of ELM to learn neural networks for malware are as follows.

- In the learning step, randomly choose the hidden node parameters and computes the output of the jth hidden neuron,

$$\mathbf{h_{ij}} = g(\mathbf{w_j x_i} + \mathbf{b_j}) \qquad i = 1, 2, 3, \dots, N; j = 1, 2, 3, \dots, K \tag{9}$$

where $\mathbf{w_j} = [w_{j1}, w_{j2}, \ldots, w_{jn}]^T$ is the weight vector connecting the jth hidden neuron and the input data $\mathbf{x_i}$, $\mathbf{x_i} = [x_{i1}, x_{i2}, \ldots, x_{in}]^T$ is the ith sample with n features, $\mathbf{b_j}$ denotes the bias of the jth hidden neuron, N denotes total number of input samples, K is the number of hidden neuron and g is the activation function.

The hidden layer output matrix is denoted as $\mathbf{H}$, and $\mathbf{H} = \{h_{ij}\}$. The weight vector $\hat{\beta}$ connects the hidden and output neurons,

$$\hat{\beta} = \mathbf{H}^\dagger T \tag{10}$$

where $\mathbf{H}^\dagger$ denotes the Moore-Penrose generalized inverse operation of $\mathbf{H}$, T is the class label, and $T = [t_1, t_2, , \ldots, t_N]^T$.

- In the detection steps: the class label $T_{Unlabelled}$ of given unlabeled software is generated by,

$$T_{Unlabelled} = \mathbf{H}_{Unlabelled}\hat{\beta} \tag{11}$$

where $\mathbf{H}_{Unlabelled}$ can be calculated by formula (9).

3.4 Stacking for Combination

Stacking is a technique to combining classifiers by learning [25]. Stacking consists of two levels which are base learner as level-0 and stacking model learner as level-1. Base learner utilizes a set of base classifiers to learn from data set. The outputs of each of the classifier are collected to create a new data set, which is taken as the input of the stacking model learner(level-1). The labels of the original data set are still regarded as the labels of new data set.

4 Experiments

4.1 Experimental Setup

This research constructs experiments to evaluate the proposed framework. In the experiment, malware samples are collected from Android Malware GenomeProject [26], and normal samples are downloaded from Google Play and Androidonline [27]. To evaluate the performance of proposed approach, we employ the evaluation metrics of accuracy and F-measure. Accuracy measures the percentage of correctly classified malwares and normal ones. F-measure is the harmonic mean of precision and recall. Formulas are as follows,

$$Accuracy = \frac{TP + TN}{TP + TN + FP + FN} \tag{12}$$

$$F - measure = \frac{2 * PR}{P + R} \tag{13}$$

where TP is the number of correctly classified malwares, FN is the number of malwares that are misclassified as normal ones, FP is the number of normal ones that are misclassified as malwares, and TN is the number of correctly classified normal ones. $P = TP/(TP + FP)$ is precision and $R = TP/(TP + FN)$ is recall.

The Dalvik instructions module is carried out in JAVA. Feature extraction module, Ensemble classification module and Decision module are carried out in MATLAB 2012b environment running in Intel Core2 CPU clocked at 2.89 GHz with 1.8GB RAM. ELM code is download from Huang 's homepage [28].

4.2 Experiment Results

We construct this experiment to evaluate our proposed framework. We perform ELM, ELM with PCA, ELM with KLT, ELM with ICA, Stacking without feature extraction as comparisions. We randomly select 80% samples as the traning date, and the other 20% samples are taken as testing data. For PCA and KLT, we set the number of feature from 5 to 230 to select the highest accuracy. When the number is 50 and 45, PCA and KLT separately reach maximum accuracy value. We set the iteration times to be 1000 in ICA. From the 78th independent component, the iteration will not converge. Therefore, we take 77 independent components to be the features in ICA. As the input weights and biases are randomly assigned, we perform 100 simulations for each combinations. The average values of accuracy and F-measure are given in table 1.

Table 1. Experiment results

Classifier	Accuracy	F-measure
ELM	93.56 ±1.07	93.43
ELM+PCA	92.10 ±1.21	93.81
ELM+KLT	93.79 ±0.93	90.73
ELM+ICA	93.55 ±1.02	92.94
Stacking(without feature extraction)	95.11 ±0.99	93.58
Stacking(with feature extraction)	97.52 ±0.87	97.52

It is shown that the F-measure of ELM with PCA is slight higher than single ELM, but the accuracy of former is lower than latter. The accuracy of ELM with KLT is lower than ELM, but the f-measure is higher. ELM with ICA generates slight lower results whether in terms of accuracy or f-measure. Stacking without feature extraction generates higher accuracy and f-measure than single ELM, which indicates ensemble learning performs better than single ELM. Stacking with feature extraction outperforms other combinations, which indicates our proposed framework is effective to improve the malware detection results for Android device.

5   Conclusions

The proliferation of malware on Android devices critically influences network security. In this paper, we propose a new framework of malware detection by employing feature extraction and machine learning. We implement PCA, KLT and ICA to transform the feature matrix of Dalvik instructions into different feature spaces and generate three new feature matrixes, and construct detection on original feature matrix and new feature matrixes by employing ELM-based ensemble learning. The experiments show that the proposed framework performs well and improve the accuracy and f-measure of malware detection for Android devices.

Acknowledgements. This research work is supported by Special Funds on Guangdong Province Chinese Academy of Sciences Comprehensive Strategic Cooperation, Key Technologies and Industrialization NO. 2013B091300019. We would like to thank Android Malware GenomeProject for supplying us with the malware samples.

References

1. 2014 Mobile Threat Report (2014), https://www.lookout.com/static/ee_images/Consumer_Threat_Report_Final_ENGLISH_1.14.pdf
2. Bartel, A., Klein, J., Le Traon, Y., et al.: Dexpler: converting android dalvik byte-code to jimple for static analysis with soot. In: Proceedings of the ACM SIGPLAN International Workshop on State of the Art in Java Program Analysis, pp. 27–38. ACM (2012)
3. Nath, H.V., Mehtre, B.M.: Static Malware Analysis Using Machine Learning Methods. In: Martínez Pérez, G., Thampi, S.M., Ko, R., Shu, L. (eds.) SNDS 2014. CCIS, vol. 420, pp. 440–450. Springer, Heidelberg (2014)
4. Amos, B., Turner, H., White, J.: Applying machine learning classifiers to dynamic android malware detection at scale. In: Proceedings of 2013 9th International Wireless Communications and Mobile Computing Conference (IWCMC), pp. 1666–1671. IEEE (2013)
5. Zhao, M., Ge, F., Zhang, T., Yuan, Z.: AntiMalDroid: An efficient SVM-based malware detection framework for android. In: Liu, C., Chang, J., Yang, A., et al. (eds.) ICICA 2011, Part I. CCIS, vol. 243, pp. 158–166. Springer, Heidelberg (2011)
6. Shabtai, A., Kanonov, U., Elovici, Y., et al.: "Andromaly": a behavioral malware detection framework for android devices. Journal of Intelligent Information Systems 38(1), 161–190 (2012)
7. Sahs, J., Khan, L.: A machine learning approach to android malware detection. In: 2012 European Proceedings of Intelligence and Security Informatics Conference (EISIC), pp. 141–147. IEEE (2012)
8. Yerima, S.Y., Sezer, S., McWilliams, G., et al.: A new android malware detection approach using bayesian classification. In: Proceedings of 2013 IEEE 27th International Conference on Advanced Information Networking and Applications (AINA), pp. 121–128. IEEE (2013)
9. Sharma, A., Dash, S.K.: Mining API Calls and Permissions for Android Malware Detection. In: Gritzalis, D., Kiayias, A., Askoxylakis, I. (eds.) CANS 2014. LNCS, vol. 8813, pp. 191–205. Springer, Heidelberg (2014)
10. Aafer, Y., Du, W., Yin, H.: DroidAPIMiner: Mining API-level features for robust malware detection in android. In: Zia, T., Zomaya, A., Varadharajan, V., Mao, M. (eds.) SecureComm 2013. LNICST, vol. 127, pp. 86–103. Springer, Heidelberg (2013)

11. Wu, D.J., Mao, C.H., Wei, T.E., et al.: Droidmat: Android malware detection through manifest and api calls tracing. In: Proceedings of 2012 Seventh Asia Joint Conference on Information Security (Asia JCIS), pp. 62–69. IEEE (2012)

12. Barrera, D., Kayacik, H.G., van Oorschot, P.C., et al.: A methodology for empirical analysis of permission-based security models and its application to android. In: Proceedings of the 17th ACM Conference on Computer and Communications Security, pp. 73–84. ACM (2010)

13. Yu, W., Ge, L., Xu, G., et al.: Towards Neural Network Based Malware Detection on Android Mobile Devices. In: Cybersecurity Systems for Human Cognition Augmentation. Advances in Information Security, pp. 99–117. Springer International Publishing (2014)

14. Mas'ud, M.Z., Sahib, S., Abdollah, M.F., et al.: Analysis of features selection and machine learning classifier in android malware detection. In: proceedings of 2014 International Conference on Information Science and Applications (ICISA), pp. 1–5. IEEE (2014)

15. Ozdemir, M., Sogukpinar, I.: An Android Malware Detection Architecture based on Ensemble Learning. Transactions on Machine Learning and Artificial Intelligence 2(3), 90–106 (2014)

16. Sheen, S., Anitha, R., Natarajan, V.: Android based malware detection using a multifeature collaborative decision fusion approach. Neurocomputing 151, 905–912 (2015)

17. Kang, B., Kang, B.J., Kim, J., et al.: Android malware classification method: Dalvik bytecode frequency analysis. In: Proceedings of the 2013 Research in Adaptive and Convergent Systems, pp. 349–350. ACM (2013)

18. Bishop, C.M.: Pattern recognition and machine learning. Springer, Heidelberg (2006)

19. Cao, L.J., Chong, W.K.: Feature extraction in support vector machine: a comparison of PCA, XPCA and ICA. In: Proceedings of the 9th International Conference on Neural Information Processing, ICONIP 2002, pp. 1001–1005. IEEE (2002)

20. Du, K.L., Swamy, M.N.S.: Independent component analysis. In: Neural Networks and Statistical Learning, pp. 419–450. Springer, London (2014)

21. Hyvarinen, A.: Fast and robust fixed-point algorithms for independent component analysis. IEEE Transactions on Neural Networks 10(3), 626–634 (1999)

22. Schmidt, W.F., Kraaijveld, M., Duin, R.P.W.: Feedforward neural networks with random weights. In: Proceedings of Conference on 11th IAPR International, pp. 1–4. IEEE (1992)

23. Huang, G.B., Zhu, Q.Y., Siew, C.K.: Extreme learning machine: a new learning scheme of feedforward Neural Networks. In: Proceedings of IEEE International Joint Confrence on Neural Networks, pp. 985–990. IEEE (2004)

24. Huang, G.B., Zhou, H., Ding, X., et al.: Extreme learning machine for regression and multiclass classification. IEEE Transactions on Systems, Man, and Cybernetics, Part B: Cybernetics 42(2), 513–529 (2012)

25. Zhou, Z.H.: Ensemble methods: foundations and algorithms. CRC Press (2012)

26. Android Malware GenomeProject, http://www.malgenomeproject.org/

27. Androidonline, http://www.androidonline.net

28. http://www.ntu.edu.sg/home/egbhuang/elm_codes.html

Data-Driven Optimization of SIRMs Connected Neural-Fuzzy System with Application to Cooling and Heating Loads Prediction

Chengdong Li[1], Weina Ren[1], Jianqiang Yi[2], Guiqing Zhang[1], and Fang Shang[1]

[1] School of Information and Electrical Engineering, Shandong Jianzhu University, Jinan, China
chengdong.li@foxmail.com
[2] Institute of Automation, Chinese Academy of Sciences, Beijing, China

Abstract. In modeling, prediction and control applications, the single-input-rule-modules (SIRMs) connected fuzzy inference method can efficiently tackle the rule explosion problem that conventional fuzzy systems always face. In this paper, to improve the learning performance of the SIRMs method, a neural structure is presented. Then, based on the least square method, a novel parameter learning algorithm is proposed for the optimization of the SIRMs connected neural-fuzzy system. Further, the proposed neural-fuzzy system is applied to the cooling and heating loads prediction which is a popular multi-variable problem in the research domain of intelligent buildings. Simulation and comparison results are also given to demonstrate the effectiveness and superiority of the proposed method.

Keywords: data-driven optimization, single input rule module, least square method, fuzzy system, cooling and heating loads.

1 Introduction

For some complex systems or processes, it is quite difficult to obtain their mechanism models. However, such systems or processes can generate large amount of data every day. Nowadays, data driven methods which study how to use such data to realize the modeling, optimization and control, are attracting more and more attentions [1-4]. As the fuzzy system has universal approximation ability and can easily combine our experience knowledge into the constructed model, it is a powerful tool for the data driven modeling, optimization and control [5].

On the other aspect, when we construct the data driven fuzzy systems, we usually face the rule explosion problem as the complex systems or processes often have large number of variables. There are two popular ways to overcome the rule explosion problem in conventional fuzzy systems. The first way is to use the hierarchical structure, which is named the hierarchical fuzzy system [6-8], while the other one is to utilize the single input rule modules (SIRMs) connected structure, which is called SIRMs connected fuzzy system [9-14].

The SIRMs connected fuzzy system   (SIRM-FS) is firstly proposed by J. Yi, et al. [9-14] to tackle the rule explosion problem and to simplify the design process of

© Springer International Publishing Switzerland 2015
X. Hu et al. (Eds.): ISNN 2015, LNCS 9377, pp. 499–507, 2015.
DOI: 10.1007/978-3-319-25393-0_55

fuzzy controllers. In the SIRM-FS, we first construct a SIRM for each input variable, and then we obtain the final output through weighting the outputs of different SIRMs. In the SIRM-FS, the number of fuzzy rules is linearly increased with the increase of input variables. So, the number of fuzzy rules can be greatly reduced. Due to its simple structure, the SIRM-FS has been wildly applied to various domains, such as stabilization control of different kinds of inverted pendulums [9-13], the anti-swing and positioning control of the overhead traveling crane [14]. In recent years, H. Seki, et al. [15] have extended this method to the functional one, and we have extended this method to the type-2 fuzzy case [16]. Also, some fundamental properties of the SIRM-FSs have been explored, for example, the relationships between the SIRM-FSs and other types of fuzzy systems (e.g. Takagi-Sugeno fuzzy systems) [16-18], the continuity, monotonicity and stability of the SIRM-FSs [19-22]. In order to utilize the SIRM-FSs as the modeling or prediction tools, in [23-25], some parameter learning algorithms have been proposed and applied to the nonlinear function identification, medical diagnosis and the thermal comfort prediction. However, all these parameter learning algorithms are based on the steepest descent methods which have some limitations, e.g. low convergence speed, local optimization, etc.

In order to improve the learning performance of the SIRMs method, in this study, we present a neural structure for the SIRM-FSs. Then, to accomplish the parameter learning of the neural-fuzzy system, we propose a novel parameter training algorithm based on the least square method which has fast convergence speed and can find global optimal solutions. Further, we apply the proposed neural-fuzzy system to the cooling and heating loads prediction problem. The cooling and heating loads are influenced by eight factors including relative compactness, surface area, wall area, roof area, overall height, orientation, glazing area, glazing area distribution etc. [26]. So, it is quite difficult to construct conventional fuzzy models for them. Simulation results show that the proposed method is effective. And, comparisons with other methods demonstrate that the proposed method has its superiorities.

2   SIRMs Connected Neural-Fuzzy System

In this Section, we will first introduce the SIRM-FSs, and then present the SIRMs connected neural-fuzzy system.

2.1   SIRMs Connected Fuzzy System

The fuzzy system with n inputs and one output is considered here. It can be easily extended to the multi-input-multi-output case.

A SIRM-FS with n input variables $x_1, x_2, ..., x_n$ is composed of n SIRMs. The SIRMs for input variables $x_1, x_2, ..., x_n$ can be expressed as [9-14]

$$SIRM-i: \quad \left\{ R_i^{j_i} : if \quad x_i = \tilde{A}_i^{j_i} \quad then \quad f_i = c_i^{j_i} \right\}_{j_i=1}^{m_i} \tag{1}$$

where $\tilde{A}_k^{j_k}$ is the antecedent fuzzy set of rule j_k in SIRM-k. And, $c_k^{j_k}$ is a crisp value of the consequent part of rule j_k in SIRM-k and can be seen as the center of the consequent fuzzy set of rule j_k . In this study, the triangular membership function (MF) is adopted for the fuzzy set $\tilde{A}_k^{j_k}$.

A SIRM can be seen as a single-input-single-output fuzzy system [9-14]. Suppose that, in each SIRM, the singleton fuzzifier, the product inference engine and the weighted average defuzzifier are used. Then the fuzzy inference output of SIRM-k can be computed as [9-14]

$$f_k(x_k) = \frac{\sum_{j_k=1}^{m_k} \mu_{\tilde{A}_k^{j_k}}(x_k) c_k^{j_k}}{\sum_{j_k=1}^{m_k} \mu_{\tilde{A}_k^{j_k}}(x_k)} \qquad k = 1, 2, ..., n. \qquad (2)$$

Finally, the output of the SIRM-FS can be calculated as [9-14]

$$y(X) = \sum_{k=1}^{n} \omega_k f_k(x_k), \qquad (3)$$

where ω_k is the importance degree of SIRM-k, and $X = (x_1, x_2, \cdots, x_n)$. The importance degrees can express the different roles of the input variables on system performance .

2.2 SIRMs Connected Neural-Fuzzy System

Corresponding to the SIRM-FS, the structure of the SIRMs connected neural-fuzzy system is depicted in Fig. 1. The SIRMs connected neural-fuzzy system consists of four layers. The first layer is the input layer. The second layer is the membership function layer which is used to compute the firing strength of each rule in all SIRMs. The third layer is SIRMs layer. This layer is utilized to compute the output of every SIRM through combing the firing strength and the weights between layer 2 and layer 3. The fourth layer is the output layer. This layer is used to calculate the final output through combing the outputs of layer 3 and the weights between layer 3 and layer 4.

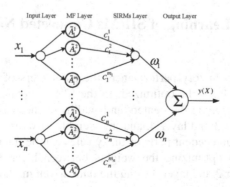

Fig. 1. Structure of the SIRMs connected neural-fuzzy system

The final output of this SIRMs connected neural-fuzzy system can be calculate as

$$y(X) = \sum_{k=1}^{n} \omega_k \frac{\sum_{j_k=1}^{m_k} \mu_{\tilde{A}_k^{j_k}}(x_k) c_k^{j_k}}{\sum_{j_k=1}^{m_k} \mu_{\tilde{A}_k^{j_k}}(x_k)} \tag{4}$$

From (3) and (4), the final output result $y(X)$ can be given in a vector form as:

$$y(X) = \sum_{k=1}^{n} \omega_k f_k(x_k) = F(X)^{\mathrm{T}} W \tag{5}$$

where $F(X) = [f_1(x_1), f_2(x_2), \cdots, f_n(x_n)]^{\mathrm{T}}$ is the inference output vector of all SIRMs,

and $W = [\omega_1, \omega_2, \cdots, \omega_n]^{\mathrm{T}}$ is a n-by-1 vector of the weights between layer 3 and layer 4.

From (4), the final output result $y(X)$ can also be given in another vector form as

$$y(X) = \sum_{k=1}^{n} \omega_k \frac{\sum_{j_k=1}^{m_k} \mu_{\tilde{A}_k^{j_k}}(x_k) c_k^{j_k}}{\sum_{j_k=1}^{m_k} \mu_{\tilde{A}_k^{j_k}}(x_k)} = \varphi(X)^{\mathrm{T}} C \tag{6}$$

where $C = [c_1^1, \ldots, c_1^{m_1}, \cdots, c_k^1, \cdots, c_k^{m_k}, \cdots, c_n^1, \cdots, c_n^{m_n}]^{\mathrm{T}}$ is a vector of the weights between

layer 2 and layer 3, and $\varphi(X)$ can be computed as

$$\varphi(X) = \left[\frac{\omega_1 \mu_{\tilde{A}_1^1}(x_1)}{\sum_{j_1=1}^{m_1} \mu_{\tilde{A}_1^{j_1}}(x_1)} \cdots \frac{\omega_1 \mu_{\tilde{A}_1^{m_1}}(x_1)}{\sum_{j_1=1}^{m_1} \mu_{\tilde{A}_1^{j_1}}(x_1)} \cdots \frac{\omega_n \mu_{\tilde{A}_n^1}(x_n)}{\sum_{j_n=1}^{m_n} \mu_{\tilde{A}_n^{j_n}}(x_n)} \cdots \frac{\omega_n \mu_{\tilde{A}_n^{m_n}}(x_n)}{\sum_{j_n=1}^{m_n} \mu_{\tilde{A}_n^{j_n}}(x_n)} \right]^{\mathrm{T}} \tag{7}$$

From (5) and (6), we can observe that the output of the SIRMs connected neural-fuzzy system is linear with respect to the weights between layer 3 and layer 4 (importance degrees) and the weights between layer 2 and layer 3 (the consequent parts of the rules in all SIRMs).

3   Parameter Learning of SIRMs Connected Neural-Fuzzy System

In order to obtain satisfactory performance, the parameters of the SIRMs connected neural-fuzzy system should be optimized. In the SIRMs connected neural-fuzzy system, the parameters include the centers and widths of the fuzzy sets in layer 2, the weights between layer 3 and layer 4, and the weights between layer 2 and layer 3. As the fuzzy sets in the antecedent parts of fuzzy rules can be manually obtained, in this section, we focus on optimizing the weights between layer 3 and layer 4 and the weights between layer 2 and layer 3 using the data driven method.

Suppose that there are N input–output data points $(X^t, y^t) = (x_1^t, x_2^t, \cdots, x_n^t, y^t)$ $t = 1, \cdots, N$. And, the training criteria is chosen to minimize the following squared error function:

$$E = \sum_{t=1}^{N} \left(y(X^t, W, C) - y^t \right)^2 \tag{8}$$

where $y(X^t)$ is the inference output value by the SIRMs connected neural-fuzzy system.

From (5) and (8), we can derive that

$$E = \sum_{t=1}^{N} \left(F(X^t)^T W - y^t \right)^2 = \left(\Gamma W - Y^0 \right)^T \left(\Gamma W - Y^0 \right) = \left\| \Gamma W - Y^0 \right\|_2^2 \tag{9}$$

where $Y^0 = \left[y^1, y^2, \cdots, y^N \right]^T$, and $\Gamma = \left[F(X^1), F(X^2), \cdots, F(X^N) \right]^T$.

Similarly, from (6) and (8), we can obtain that

$$E = \sum_{t=1}^{N} \left(\varphi(X)^T C - y^t \right)^2 = \left(\Phi C - Y^0 \right)^T \left(\Phi C - Y^0 \right) = \left\| \Phi C - Y^0 \right\|_2^2 \tag{10}$$

where $\Phi = \left[\varphi(X^1), \varphi(X^2), \cdots, \varphi(X^N) \right]^T$.

The data-driven optimization of the SIRMs connected neural-fuzzy system can be realized by solving the following problem:

$$\min_{W,C} \sum_{t=1}^{N} \left(y(X^t, W, C) - y^t \right)^2. \tag{11}$$

To solve this optimization problem, we propose the following training algorithm which is also shown in Fig. 2.

- **Step 1:** Set the maximum iterative epoch and the training accuracy. And, randomly initialize the parameter vector C between layer 2 and layer 3.
- **Step 2:** Compute the matrix Γ. And then, solve the optimization problem $\min_{W} \left\| \Gamma W - Y^0 \right\|_2^2$. This is a least square problem [27]. The best value of W can be obtained as $W = \Gamma^+ Y^0$, where Γ^+ is a generalized MP inverse matrix of Γ.
- **Step 3:** Compute the matrix Φ. And then, solve the optimization problem $\min_{C} \left\| \Phi C - Y^0 \right\|_2^2$. This is also a least square problem [27]. The best value of C can be obtained as $C = \Phi^+ Y^0$, where Φ^+ is a generalized MP inverse matrix of Φ.
- **Step 4:** If the iterative epoch number is arrived or the training accuracy is satisfied, then stop; otherwise, go to Step 2.

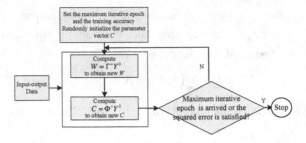

Fig. 2. The least square method based learning algorithm

4   Cooling and Heating Loads Prediction

In this section, we will apply the proposed SIRMs connected neural-fuzzy system and its learning algorithm to the cooling and heating loads prediction. Also, comparisons with linear regression method and BP neural networks are made to show the superiority of the proposed method.

4.1   Problem Description and Prediction Model Design

Accurate cooling load (CL) and heating load (HL) predictions are necessary to optimize building designs and set appropriate heating and cooling equipments' specifications. However, it is not an easy task to construct the accurate CL and HL prediction models as they are affected by many factors, such as the relative compactness, the surface area, the wall area, the roof area, the overall height, the orientation, the glazing area, and the glazing area distribution [26]. In [26], Xifara and Tsanas have collected the CL and HL data. We can use such data to construct the data driven models for the CL and HL prediction. Totally, 768 data pairs are generated in [26]. Each data pair has eight input variables including the relative compactness, the surface area, the wall area, the roof area, the overall height, the orientation, the glazing area, and the glazing area distribution, and two output variables including the CL and HL. The number of the input variables is large. So, it is suitable to choose the SIRMs connected fuzzy method.

In this study, we firstly project the input parts of the 768 data pairs into [-1, 1] through the following formula:

$$\hat{x}_i^j = 2 \frac{x_i^j - \min x_i^j}{\max x_i^j - \min x_i^j} - 1 \tag{12}$$

where $i = 1, \cdots, 8,$ and $j = 1, \cdots, 768$.

We randomly choose 700 data pairs to train the SIRMs connected neural-fuzzy system, while the left 68 data pairs are used to test the constructed model. In the SIRMs connected neural-fuzzy system, each SIRM has five rules, i.e. five fuzzy sets are used to partition each input domain. Hence, there are totally 40 (5*8) fuzzy rules, and 48 parameters in C and W are needed to be trained. The initial values of the weights between layer 2 and layer 3 are randomly generated.

For comparison, the BP neural network (BPNN) and the multi-variable linear regression (LR) method are also adopted to predict the CL and HL. The BPNN has three layers and totally 27 nodes with 161 parameters to be trained. In this study, the maximum learning iteration number of the BPNN is set to be 1000.

To compare the performances of the three models, the root-mean-square error (RMSE) as follows is chosen as the comparison index

$$RMSE = \sqrt{\sum_{t=1}^{T}\left(y(X^{t})-y^{t}\right)^{2}\Big/T} \qquad (13)$$

where T is the number of the training data or testing data, $y(X^{t})$ **is the** predicted value from the models – the SIRMs connected neural-fuzzy system, the BP neural network and the linear regressor.

4.2   Simulation Results and Comparisons

The above simulation process was run ten rounds. The training and testing RMSE values and their means and standard deviations (Std.) of the three methods for CL and HL are listed in Table 1.

From this table, we can observe that the SIRMs connected neural-fuzzy system with the proposed training method can give satisfactory performance. Compared with the LR method and the BP neural network method, the performance of the SIRMs connected neural-fuzzy system is much better, although the BP neural network have more parameters than the SIRMs connected neural-fuzzy system (161 vs. 48).

Table 1. Comparisons of the three models for CL and HL

		Methods	1	2	3	4	5	mean	Std.
Cooling load	Training	LR	3.2189	3.2338	3.2490	3.1942	3.2295	3.2251	0.0204
		BP	2.4133	2.0656	2.4706	2.5653	2.0880	2.3206	0.2292
		SIRM	1.7038	1.7146	1.7275	1.6948	1.7113	1.7104	0.0122
	Testing	LR	3.3534	3.2065	3.0409	3.5890	3.2416	3.2863	0.2029
		BP	2.7486	2.3653	2.4100	2.6980	2.2693	2.4982	0.2124
		SIRM	1.8054	1.6982	1.5459	1.9011	1.7270	1.7355	0.1320
Heating load	Training	LR	2.9766	2.9150	2.9860	2.9775	2.9456	2.9601	0.0295
		BP	1.6719	1.5338	2.0399	1.5468	1.4396	1.6464	0.2350
		SIRM	1.0063	1.0125	1.0118	1.0087	1.0184	1.0115	0.0045
	Testing	LR	2.7603	3.3736	2.6504	2.7349	3.0652	2.9169	0.2997
		BP	1.4290	2.0568	1.8092	1.5045	1.5791	1.6757	0.2562
		SIRM	1.0414	0.9766	0.9872	1.0170	0.9119	0.9868	0.0490

In this application, the conventional fuzzy system is not suitable to be selected to predict the CL and HL. The reason for this is that this application has 8 input variables which will make the number of fuzzy rules of the conventional fuzzy system be so huge that it is difficult to generate the fuzzy rules and train their parameters.

5   Conclusion

In this study, to enhance the learning ability of the SIRMs connected fuzzy system, a neural structure has been presented. From its input-output mapping, it can be observed that the output of the SIRMs connected neural-fuzzy system is linear with respect to its parameters including the consequent parameters of fuzzy rules in all SIRMs and the importance degrees. To optimize such parameters, a least square method based learning algorithm has been proposed. This algorithm can efficiently overcome the drawbacks of the steepest descent method based learning algorithms. The proposed method has also been verified by the cooling and heating loads prediction problem. In the future, we will theoretically study the approximation ability of the SIRMs method and explore how to improve its approximation and generalization abilities.

Acknowledgments. This work is supported by National Natural Science Foundation of China (61473176, and 61273149), the Natural Science Fund of Shandong Province for Outstanding Young Talents in Provincial Universities (ZR2015JL021), and the Open Program from the State Key Laboratory of Management and Control for Complex Systems (20140102).

References

1. Wang, D.: Robust data-driven modeling approach for real-time final product quality prediction in batch process operation. IEEE Trans. on Industrial Informatics 7(2), 371–377 (2011)
2. Wang, Z., Liu, D.: A data-based state feedback control method for a class of nonlinear systems. IEEE Trans. on Industrial Informatics 9(4), 2284–2292 (2013)
3. Hou, Z.S., Wang, Z.: From model-based control to data-driven control: survey, classification and perspective. Information Sciences 235, 3–35 (2013)
4. Song, R., Xiao, W., Zhang, H.: Multi-objective optimal control for a class of unknown nonlinear systems based on finite-approximation-error ADP algorithm. Neurocomputing 119, 212–221 (2013)
5. Mendel, J.M.: Uncertain rule-based fuzzy logic systems: introduction and new directions. Prentice-Hall, Upper Saddle River (2001)
6. Wang, L.-X.: Analysis and design of hierarchical fuzzy systems. IEEE Trans. on Fuzzy Systems 7(9), 617–624 (1999)
7. Joo, M.G., Lee, J.S.: A class of hierarchical fuzzy systems with constraints on the fuzzy rules. IEEE Trans. on Fuzzy Systems 13(2), 194–203 (2005)
8. Hagras, H.: A hierarchical type-2 fuzzy logic control architecture for autonomous mobile robots. IEEE Trans. on Fuzzy Systems 12(4), 524–539 (2004)

9. Yi, J., Yubazaki, N.: Stabilization fuzzy control of inverted pendulum systems. Artificial Intelligence in Engineering 14, 153–163 (2000)
10. Yi, J., Yubazaki, N., Hirota, K.: Upswing and stabilization control of inverted pendulum system based on the sirms dynamically connected fuzzy inference model. Fuzzy Sets and Systems 122, 139–152 (2001)
11. Yi, J., Yubazaki, N., Hirota, K.: Stabilization control of series-type double inverted pendulum systems using the sirms dynamically connected fuzzy inference model. Artificial Intelligence in Engineering 15, 297–308 (2001)
12. Yi, J., Yubazaki, N., Hirota, K.: A proposal of sirms dynamically connected fuzzy inference model for plural input fuzzy control. Fuzzy Sets and Systems 125, 79–92 (2002)
13. Yi, J., Yubazaki, N., Hirota, K.: A new fuzzy controller for stabilization of parallel-type double inverted pendulum system. Fuzzy Sets and Systems 126, 105–119 (2002)
14. Yi, J., Yubazaki, N., Hirota, K.: Anti-swing and positioning control of overhead traveling crane. Information Sciences 155, 19–42 (2003)
15. Seki, H., Ishii, H., Mizumoto, M.: On the generalization of single input rule modules connected type fuzzy reasoning method. IEEE Trans. on Fuzzy Systems 16(5), 1180–1187 (2008)
16. Li, C., Yi, J.: SIRMs based interval type-2 fuzzy inference systems: properties and application. International Journal of Innovative Computing, Information and Control 6(9), 4019–4028 (2010)
17. Seki, H., Mizumoto, M., Yubazaki, N.: On the property of single input rule modules connected type fuzzy reasoning method. IEICE Trans. on Fundamentals J89-A, 557–565 (2006)
18. Seki, H., Mizumoto, M.: On the equivalence conditions of fuzzy inference methods—Part 1: basic concept and definition. IEEE Trans. on Fuzzy Systems 19(6), 1097–1106 (2011)
19. Li, C., Zhang, G., Yi, J., Wang, T.: On the properties of SIRMs connected type-1 and type-2 fuzzy inference systems. In: 2011 IEEE International Conference on Fuzzy Systems, FUZZ-IEEE 2011, pp. 1982–1988 (2011)
20. Seki, H., Ishii, H., Mizumoto, M.: On the monotonicity of single input type fuzzy reasoning methods. IEICE Trans. on Fundamentals E90-A, 1462–1468 (2007)
21. Seki, H., Ishii, H., Mizumoto, M.: On the monotonicity of fuzzy-inference methods related to T–S inference method. IEEE Trans. on Fuzzy Systems 18(3), 629–634 (2010)
22. Li, C., Yi, J., Wang, T.: Stability analysis of SIRMs based type-2 fuzzy logic control systems. In: The 2010 IEEE World Congress on Computational Intelligence (WCCI 2010), pp. 2913–2918 (2010)
23. Seki, H.: Nonlinear identification using single input connected fuzzy inference model. Procedia Computer Science 22, 1121–1125 (2013)
24. Seki, H., Mizumoto, M.: SIRMs connected fuzzy inference method adopting emphasis and suppression. Fuzzy Sets and Systems 215, 112–126 (2013)
25. Li, C., Wang, M., Zhang, G.: Prediction of thermal comfort using SIRMs connected type-2 fuzzy reasoning method. ICIC Express Letters 7(4), 1401–1406 (2013)
26. Tsanas, A., Xifara, A.: Accurate quantitative estimation of energy performance of residential buildings using statistical machine learning tools. Energy and Buildings 49, 560–567 (2012)
27. Nelles, O.: Nonlinear system identification, pp. 35–67. Springer, Berlin (2001)

Erratum to: Interlinked Convolutional Neural Networks for Face Parsing

Yisu Zhou, Xiaolin Hu, and Bo Zhang

Erratum to:
Chapter "Interlinked Convolutional Neural Networks
for Face Parsing" in: X. Hu et al. (Eds.):
Advances in Neural Networks – ISNN 2015, **LNCS 9377,**
https://doi.org/10.1007/978-3-319-25393-0_25

The original version of this paper unfortunately contains a mistake in Table 1. The correction information must be read as follows:

Table 1. Comparison with other models (F-Measure)

Model	Eye	Eyebrow	Nose	In mouth	Upper lip	Lower lip	Mouth (all)	Face Skin	Overall
[13]	0.533	n/a	n/a	0.425	0.472	0.455	0.687	n/a	n/a
[14]	0.679	0.598	0.890	0.600	0.579	0.579	0.769	n/a	0.733
[15]	0.770	0.640	0.843	0.601	0.650	0.618	0.742	0.886	0.738
[16]	0.743	0.681	0.889	0.545	0.568	0.599	0.789	n/a	0.746
[5]	**0.785**	0.722	**0.922**	0.713	0.651	0.700	0.857	0.882	0.804
iCNNs	0.778	**0.863**	0.920	**0.777**	**0.824**	**0.808**	**0.889**	n/a	**0.845**

The updated online online version of this chapter can be found at
https://doi.org/10.1007/978-3-319-25393-0_25

Author Index

Printed in the United States
by Baker & Taylor Publisher Services